Operation and Maintenance of Marine Diesel Engines, Auxiliary Equipment, and Related Systems U.S. Army

Department of Defense

Pentagon
Publishing

***TC 55-509**

Training Circular
No. 55-509

Headquarters
Department of the Army
Washington, D.C., 4 June 2009

Contents

DISTRIBUTION RESTRICTION: Approved for public release; distribution is unlimited.

***This publication supersedes FM 55-509, 3 October 1986.**

Figures

Page

Tables

Preface

This training circular (TC) provides information on the principles of operation and maintenance of marine diesel engines, auxiliary equipment, and related systems. General instructions and precautions requiring special attention are included for guidance for those responsible for training personnel. No attempt has been made to cover all unit models. Specific technical manuals, lubrication orders, or manufacturer's instructions issued with equipment will fully cover required operational and maintenance procedures.

This TC is designed for all Soldiers in the marine engineering field. It also provides information for military occupational specialties (MOSs) 88L and 881A.

This TC reinforces good marine engineman practices. A good knowledge of marine electricity helps maintain the health and welfare of the crew by promoting the safe operation of the many electrical systems onboard a vessel.

This TC applies to the Active Army, the Army National Guard/Army National Guard of the United States, and the United States Army Reserve unless otherwise stated.

The proponent of this publication is Headquarters (HQ), United States Army Training and Doctrine Command (TRADOC). Submit comments and recommendations for improving this publication on DA Form 2028 (Recommended Changes to Publications and Blank Forms) directly to Commander, Training Directorate, Transportation Training Division, ATTN: ATCL-TDM, 2221 Adams Avenue, Fort Lee, VA 23801-2102.

Chapter 1

PERSONNEL, DUTIES, AND RECORDS

INTRODUCTION

1-1. Close coordination and cooperation between the engine room and deck departments are required to ensure the safe operation of a vessel regardless of its size. The master of the vessel will notify the chief engineer of estimated times of arrival or departure as much in advance as possible. If necessary to slow down or stop any engine department machinery affecting the navigation of a vessel, the chief engineer or watch engineer will first notify the bridge. The master or deck watch officer will approve or disapprove, depending on the safety of the change in operation at the time. These basic rules also apply aboard small craft manned by an operator (coxswain) and a senior marine engineman. This chapter discusses the duties and requirements of the chief engineer, assistant engineer, and engineer as well as certain records that must be maintained.

CHIEF ENGINEER

1-2. The engine department comes under the supervision of the chief engineer who is the senior member. This position may be held by a warrant officer (along with assistant engineers) aboard a large vessel or an enlisted man aboard a small craft. In the chain of command, the chief engineer is responsible to the master of the vessel for efficient and economical operation of the engine room machinery, auxiliary machinery, and specific deck equipment. The chief engineer, with the aid of his/her assistants, supervises the administrative and technical activities of the engine room to ensure crew discipline and operational efficiency. When three assistant engineers are assigned, the chief engineer normally does not stand a watch when the vessel is underway.

ADMINISTRATIVE DUTIES

1-3. The chief engineer's administrative duties consist of the following:

- The chief engineer supervises preparation of the engineer department logbook. Rough log entries are made by the watch stander, oiler, or engineman. These entries may be transferred daily to a smooth log at the discretion of the chief engineer. The rough log is the official log which is the document subpoenaed in the event of a marine casualty. The smooth log is regarded as hearsay evidence. Actions required to be logged are listed in AR 56-9.
- Under the supervision of the chief engineer, all records and reports are prepared and maintained as required by the provisions of DA Pamphlet 750-8. Administrative record files of the engine room include vessel manufacturer's instructions, technical manuals (TMs) covering specific equipment, and various drawings pertaining to the vessel.
- The chief engineer establishes and maintains watch schedules for the engine room.

- The chief engineer directs and supervises maintenance and repair of engine room machinery, electrical machinery, ship navigation equipment, and any deck machinery under his/her jurisdiction. The care of refrigeration equipment; batteries; hand and machine tools; supplies of fuel; lubricants, and freshwater; and engine department running supplies and repair parts are the chief engineer's responsibility. The chief engineer must report to the master of the vessel any defects which are beyond his/her ability to repair. He/She requests repairs for items requiring overhaul or repair beyond the capability of the crew (AR 750-59).
- Interference with the official technical performance of engineering duties of the engineering department will not be tolerated, particularly when the vessel is underway. This is in line with the safety of the vessel and of the individuals concerned. Personnel not on official duty or not connected with maintenance and operation of engine room equipment must get the permission of the chief engineer or watch engineer to enter the engine room.

Technical Duties

1-4. The chief engineer and assistant engineers should have expert knowledge of technical subjects connected with operation and maintenance of engines and related equipment. They should know all operating details pertaining to main and auxiliary equipment. This includes the following:

- Fuel capacity and type.
- Freshwater and lubricating oil storage tank capacities.
- Transfer rates.
- Speed.
- Fuel consumption factors.
- Engine revolutions per minute (RPM).

They should be thoroughly familiar with drawings pertaining to machinery and equipment installed on the vessel.

ASSISTANT ENGINEER

1-5. The assistant engineer assists the chief engineer and ensures that the chief engineer's instructions are fully and carefully carried out. When acting in the absence of the chief engineer, this person observes all regulations and performs all duties normally assigned to the chief engineer.

In Port

1-6. The assistant engineer works days if sea watches are broken while the vessel is in port. He/She works under the supervision of the chief engineer on repair and maintenance of machinery pertaining to the engine department. The chief engineer assigns routine duties to assistant engineers. This usually consists of preventive maintenance and repair of the main and auxiliary machinery in the room and other spaces, including deck machinery.

Underway

1-7. The assistant engineer stands watch in the engine room to supervise operation of the machinery and make appropriate entries in the rough log. As watch engineer, he/she must carry out all orders given to him/her by the chief engineer. The assistant engineer needs to report any discrepancies and deficiencies that may hamper navigation and movement of the vessel directly to the bridge and then immediately inform the chief engineer. All other crewmembers will report directly to the chief engineer.

ENGINEMEN

IN PORT

1-8. The engineman works under the supervision of the engineers on all maintenance and repair of machinery. Cleaning stations are usually set up in the engine room, with an engineman assigned to each station to keep it clean. The engineman also makes appropriate entries in the engine room rough log to include gauge readings of the auxiliaries in operation. While the vessel is in port, the engineman's watches are normally unbroken, as no machinery is in operation except the auxiliary generator. In port, the engineman normally stands his/her assigned watch alone. He/She may also stand gangway watch while in port. He/She immediately reports to the watch or standby engineer any major changes in gauge pressures or temperatures or any other unusual occurrences.

UNDERWAY

1-9. When the vessel is underway, the engineman will hand-oil all machinery requiring this care. All instruments and gauges of the various systems are carefully checked to ensure that proper temperatures and pressures are being maintained. Other duties are assigned as required, governed by the type of duties in the engine room.

MARINE CERTIFICATION

1-10. Personnel assigned to an operating Army watercraft must possess a valid US Army Marine License (USAML) for their skill level and pay grade. Watercraft operators will operate only those vessels for which they are licensed in accordance with AR 56-9 and DA Pamphlet 611-21. AR 56-9 prescribes the responsibilities, policies, and procedures for authorization, assignment, operation, maintenance, and safety of Army watercraft. It also defines the procedures for verifying the qualifications of Army marine personnel and for issuing the USAML.

MARINE QUALIFICATION BOARD (MQB)

1-11. The MQB is responsible for evaluating the sea service of watercraft personnel and preparing, administering, and grading the appropriate examinations. The evaluation of sea service is based on the information listed in the Marine Service Book.

DECK AND ENGINE LOGBOOKS

1-12. Deck engine department logbooks are maintained to provide a permanent legal record of operations and conditions of the vessel and the status of its cargo, crew, and/or passengers. All occurrences of importance, interest, or historical value concerning the crew, passengers, operation, and safety of Army watercraft will be recorded daily. This will be done by watches in three types of deck logbooks.

DA Form 4640 (Harbor Boat Deck Department Log for Class A&B Vessels) is required for use on class A and class C-1 vessels. DA Form 5273 (Harbor Boat Deck and Engine Log for Class B Vessels) is required for use on all class B vessels. DA Form 5273 may also be used on the deck or liquid barge design BG 231B and the refrigerator barges BR 7010 and BR 7016. Logbooks will be prepared in accordance with instructions contained in AR 56-9 and DA Pamphlet 750-8. AR 27-series contain requirements for preserving the ship's deck and engine logs and other pertinent records for use in claims against the United States. Such claims may be for damage caused by an Army watercraft and for affirmative claim by the United States for damage to Army property caused by other vessels or floating objects. When a log (or any portion of the log) is to be used in litigation or is to be withheld for any other legal proceedings, HQDA (DAJAZA) will be notified. When the log is no longer required for the legal proceedings, it will be returned to the installation having command over the Army watercraft that was involved. Commanders having assigned watercraft will periodically review log requirements to ensure that logs are maintained in accordance with the provisions of AR 56-9. Amphibians and watercraft under 30 feet do not require maintenance of logs, provided adequate records as prescribed in DA Pamphlet 750-8 are maintained by the unit.

MAINTENANCE AND RETENTION OF LOGBOOKS

1-13. The ship's engineering log will be presented each day to the chief engineer. Should any inaccuracies or omissions be noticed, the chief engineer will have corrections made. After corrections have been made, the chief engineer will approve the log by placing his/her signature on the page. After the log has been approved by the master of the vessel, no changes or additions will be made without his/her permission or direction. Any such changes or additions must be made by the engineer on whose watch the matter under consideration occurred.

- When a correction is necessary, a single line will be drawn through the original entry (in red ink) so that the entry remains legible. The correct entry will then be made legibly and clearly. Corrections will be initialed by the person making the original entry and by the master or coxswain.
- Entries will also be made of all drills and inspections prescribed in CFR 46, paragraph 97.35.
- Reserve component (RC) nondrill dates will be noted in the log, along with the vessel location, and annotated "nonduty days". Logs will be made available to the area maintenance support activity personnel for entries when applicable.
- An engine department smooth log may be maintained at the discretion of the chief engineer. All signatures appearing in the rough log shall be entered in the smooth log in the same hand.
- The rough log will be retained for 5 years as the onboard record concerning the deck and engine departments. At the end of this period, this log will be disposed of in accordance with AR 25-400-2.
- All trash departing from the vessel is logged in a trash logbook. Log in what kind of trash, how bagged, and where disposed of in accordance with AR 56-9.
- Changes to individual log sheets in logbooks can be made by ruling lines in with ink and then making appropriate entries on them. Changes are required for floating cranes to show the number and weight of heavy lifts made and any other entries appropriate to the type of service in which employed.

OIL RECORD BOOK

1-14. In addition to required deck and engine logs, a record will be maintained on all class A-1 and class A-2 vessels and fuel barges of ballasting or cleaning of bunker fuel tanks and disposal of oily residues from those tanks. Other exceptional discharges of oil will also be recorded. These activities will be recorded on the blank form located in CG-4602A (Oil Record Book for Ships). Instructions for use of the form are contained therein, along with applicable laws relating to oil pollution regulations. Supplies of US Coast Guard publications will be furnished by the Coast Guard on request. Class B and class C-1 vessels will have this information recorded in their logbooks. Details will be recorded and underlined in red ink by the person in charge. Oil record forms will be retained for 5 years.

WATCHSTANDING

1-15. Aboard ship, marine enginemen spend much of their working time as watchstander. How you stand your watch is very important to the reliability of the engineering plant and the entire ship. To properly stand an engine room watch, you must be skilled in detecting unusual noises, vibrations, or odors which may indicate faulty machinery operation. You must also be skilled in taking appropriate and prompt corrective measures. You must be ready, in emergencies, to act quickly and independently; you must know the ship's piping systems and how, where, and why they are controlled. You must also know the following about each piece of machinery:

- How it is constructed.
- How it operates.

- How it fits into the engineering plant.
- Where related equipment is controlled.

1-16. You must be skilled in reading and interpreting measuring instruments. You must understand how and why protective devices function (for example, relief valves, speed-limiting governors, overspeed trips, and cut-in and cutout devices). You must be able to recognize and remove fire hazards, stow gear that is adrift, and keep deck plates clean and dry. You must never attempt to operate a defective piece of equipment. You must report all unsafe conditions to the space and/or plant supervisor. Whatever your watch, you are responsible for the following:

- Knowing the status of every piece of machinery at your station.
- Promptly handling any necessary change in speed or setup.
- Recording correctly all data concerning the operation and maintenance of the machinery.

1-17. You must be sure that the log is up to date, that the status boards are correct, that you know what machinery is operating, and that you know what the night orders and standing orders are before you relieve the watch. Above all, if you do not know -- ask. If a noise, odor, or condition seems abnormal to you but you are not certain whether it could be a problem, call your immediate watch supervisor. He/She is responsible for backing you up. One the best ways to gain the respect and confidence of your supervisors and shipmates is to stand a good watch. Relieve the watch on time or even a little early if possible, but do not try to relieve the watch first and figure out the score later. The same applies when you are being relieved; do not be in a big hurry to take off. Be sure your relief understands the situation completely. Before being relieved, make sure your station is clean and squared away. These little considerations will not only get you a good reputation, but will also improve the overall quality of watchstanding within the department.

SOUNDINGS

1-18. Soundings are taken for the purpose of measuring the amount of fuel in a tank or the amount of water in that fuel. A sounding tube is usually made of 1 1/2-inch pipe. The lower end is fitted at a low point in the compartment, tank, or void which the tube serves. The upper end terminates in a flush deck plate which is usually located on the main or second deck; it is closed with a threaded plug or cap. Tubes are made as straight as possible, but some are necessarily curved. Some sounding tubes, particularly those serving spaces under the engine rooms and firerooms, cannot be extended to the upper decks because of the construction of the ship. These tubes terminate in risers which extend approximately 3 feet above the compartment served; they have gate valve closures. Soundings are taken with a sound rod or sounding tape, whichever is provided for that purpose aboard ship. Normally, the sounding rod is approximately 6 feet long and is made of 12-inch lengths of 1/2-inch brass or bronze rods. It is lowered with a chain or line. The sounding tape is a steel tape, coiled on a reel suitable for holding while the tape is lowered. The tape is weighted at the end to facilitate lowering into the sounding tube. To prevent cross contamination, separate clearly marked sounding devices are used for fuel, potable water, and ballast tanks. Water is relatively hard to see on a brass or bronze sounding rod. Before you take a sounding, dry the rod or tape thoroughly and coat it with water detection paste. When the paste becomes wet, it turns to a distinctly visible light brown color. For example, if there are 6 inches of water in a tank when you take a sounding, the light brown color of the paste will be distinctly visible up to the 6-inch mark, while the remainder of the sounding rod will still be covered with the white paste. The paste method is used only where water may be present. When you have coated the sounding rod with paste, lower it down the tube until it touches the bottom; immediately draw it back. Do not drop the rod or tape to the bottom, but lower it slowly so that it hits bottom without injuring the rod or tape. This also prevents premature wear of the striker plate. If the ship is rolling, try to lower and hoist the rod or tape while the ship is on an even keel. After you have taken the sounding, enter the reading in the sounding log.

Chapter 2

MARINE ENGINEERING SAFETY

INTRODUCTION

2-1. This chapter describes the importance of safety, safety rules, and safe operating procedures. These three things are essential to engineering operations.

ACCIDENT PREVENTION

2-2. An accident is any unplanned or unintended event that results in personal injury and/or damage to equipment. It is caused by an individual's not doing something that should have been done or by doing something that should not have been done. The causes of the action or inaction may be due to the lack of technical skills, the lack of self-discipline, or both. Doing a job correctly is also doing the job safely. A near-accident is an accident that almost happened. It is an occurrence which, except for location or timely action, would have resulted in damage to equipment or material or personal injury. It does serve notice that a hazardous condition exists which could result in a future accident. The near-accident is significant because it serves as a warning. Ignoring the conditions that cause near-accidents is a sure invitation to a real accident. Accident prevention (safety) is the process of eliminating accident-producing causes before an accident occurs. Accidents are prevented by personnel performing tasks correctly and therefore in the most desirable manner. This results from successful technical and disciplinary training. Accidents can be prevented if all personnel cooperate in eliminating unsafe acts and conditions. Each person must become an accident-prevention specialist, trained in recognizing and correcting dangerous conditions and in avoiding unsafe acts. These actions must become habitual, for personal habits in any environment will determine the chances for accident involvement. It is for these reasons that high standards of cleanliness and orderliness be maintained. Insistence on good housekeeping and frequent inspections, both formal and informal, to ensure adherence to these standards will serve both to reveal unsafe conditions and to correct them.

BASIC TYPES OF UNSAFE ACTS

Removing or Making Safety Devices Inoperative

2-3. Equipment guards and other safety devices are not installed unless a serious hazard requires their installation. It is a serious unsafe practice to render such devices inoperative by removing or tampering with them. Examples include disconnecting speed-limiting governors to gain speed, gagging pressure relief valves to stop leaks, making limit switches inoperative to gain more room for movement, and permanently removing machinery guards to make adjustments or to make lubrication easier.

Using Defective Tools or Equipment

2-4. Most tools and equipment become defective or unsafe in time through normal wear and sometimes because of misuse or abuse. When they do, they should be repaired or replaced. Continued use of defective equipment invites accidents. Examples of this unsafe practice include using tools with loose handles, chisels with mushroomed heads, and portable electric tools with frayed cable or an unsuitable ground.

Using Tools or Equipment Unsafely

2-5. Personnel may often use sound tools and equipment unsafely. Frequently, they may use equipment or a tool for a purpose other than that for which it was designed, such as using a screwdriver as a chisel or punch. Just as frequently, they may use equipment or a tool for the right purpose but in a wrong way, such as hammering with the side of a hammer or exerting abrupt force on a wrench.

Taking an Unsafe Position or Posture

2-6. Many accidents occur because people put themselves in hazardous positions relative to the things around them. Examples include working directly underneath of overhaul work being done, walking under crane lifts, and standing too close to a tool-swinging shipmate. An unsafe posture concerns how people position their bodies, not where they position them. A typical example of unsafe posture is lifting a heavy object with legs straight and back arched down to the load.

Servicing of Moving, Energized, or Otherwise Hazardous Equipment

2-7. Equipment should never be repaired or serviced when it is moving, energized, or pressurized. Failure to shut down, de-energize, or depressurize equipment before repairing, cleaning, lubricating, adjusting, or inspecting can be fatal. Whenever such equipment is worked on, appropriate securing and tag-out procedures should be followed.

Engaging in Horseplay

2-8. Many accidents resulting in serious injuries are caused by horseplay of one kind or another. Such actions must be recognized as a basic type of unsafe act. Roughhousing, throwing objects, splashing water, and similar hazardous antics should not be tolerated.

Failing to Wear Personal Protective Equipment

2-9. When personal protective equipment is prescribed and issued, the decision has been made because of past accident experience. Failure to wear hardhats, safety goggles, face shields, and protective clothing is an open invitation to serious injury.

WARNING

Wear your ear plugs or ear protection when working in the engine room or around operating machinery. It will save your hearing. Wear your safety goggles or your face shield when chipping paint or working with the lathe or grindstone. It will save your eyes.

REPORTING INJURIES

2-10. Any accident, no matter how minor, will be reported to the chief engineer, if available, or to the vessel master. This action is required so that the incident can be recorded in the engineering log.

SAFETY PRECAUTIONS

GENERAL HOUSEKEEPING STANDARDS

2-11. Good housekeeping goes hand in hand with safety and efficiency. It is a combination of orderliness and cleanliness. To promote orderliness, ensure that all items have assigned places for storage are kept in their assigned places when not in use and that they are stowed in their assigned places in a ready-for-sea condition.

HANDLING MATERIAL

2-12. When working on shipboard machinery, the following things can be done to prevent accidents.

- Materials should never be thrown from elevated places to the deck. Suitable lowering equipment should always be used.
- Lifting or lowering operations being performed by several persons should be done on signal from one person, and then only when all personnel are in the clear.
- Never overload a chain hoist, block and tackle, crane, or any other lifting device. Defective or broken straps should never be used.
- Before any material is handled, it should be examined and personnel should be protected against sharp edges, protruding points, or other factors likely to cause accidents.
- Before opening a fitting or piece of equipment, be sure that all valves which permit entrance of such items as steam, water, air, and oil into the fittings of equipment are closed, tagged, and secured by locking or wiring to prevent accidental opening.
- Keep work areas clear of stray gear such as tools, fire hazards, and oil.
- Warn others of any known unsafe conditions.

Note: Chain hoists are usually provided to lift heavy objects. However, if you must remove a heavy object by hand, do it properly. Stand close to the load with your feet solidly placed and slightly apart. With knees bent, grasp the object firmly and then lift by straightening your legs, keeping your back as vertical as possible. Never stand on a slippery deck to lift a heavy object or to reach it.

WORKING WITH HANDTOOLS

2-13. Certain precautions should be taken when working with handtools. Normally, there should be no problems when working with these tools, but there are certain conditions under which they may constitute a danger. One source of danger, often neglected or ignored, is the use of handtools that are no longer considered serviceable. Tools having plastic or wooden handles that are cracked, chipped, splintered, or broken may result in injuries to personnel from cuts, bruises, particles striking the eye, and so on. Such tools should be condemned, replaced, or if at all possible, repaired, before they cause accidents.

USING POWER TOOLS

2-14. To protect yourself and others next to you, take special precautions when using all power tools. Never attempt to use these tools until you have had adequate instruction in their use and until you have been authorized to use them. Experimentation with power tools can be dangerous to you and can result in damage to the tools. If a power tool slips, you can easily injure or lose a finger or a hand. Always use goggles or a face shield when buffing or grinding, or when there is danger of flying particles. There is no exception to this rule.

- Do not use electrical equipment or machines with frayed or otherwise deteriorated insulation. Electrically driven portable machinery and fixed electrical equipment must have the frame grounded.
- Aboard ship, electrically driven portable handtools must not be operated unless they are equipped with ground wire connections between their metal housings and the steel structure of the ship. There is no exception to this rule. These tools must be equipped with an approved type of grounding plug. You must inspect the attached cable and plug of a tool before you use it to make certain that the insulation and contacts are in good condition. Tears, chafing, exposed insulated conductors, and damaged plugs are sufficient reasons for replacement.
- NEVER operate a power tool in wet conditions.

PREVENTING OIL FIRES

2-15. An oil fire can be caused by the ignition of oil or oil vapor in any place where oil is allowed to spray when under pressure or where it is allowed to collect by leakage or spillage from the system. While specifications for diesel fuel (flashpoint 140^{0} Fahrenheit) are designed to make shipboard storage and handling as safe as possible, if the flashpoint is exceeded, an explosive mixture with air may occur. The blended fuels, depending on their viscosities, may require little or no preheating for either pumping or burning. Distillate fuels should be pumped and burned without preheating. All fuels and lubricating oils are extremely dangerous if atomized (sprayed) through a leak in a flange or gasket. If the spray strikes a hot surface, fire and possibly an explosion will result. Be sure you know about the fuel and oil piping systems in your space and ensure that all flange safety shields are installed. The following precautions must be taken to prevent oil fires:

- Do not allow oil to accumulate any place. Guard against accumulation of oil in drip pans, under pumps, in bilges, or on deck plates.
- If leakage from lubricating oil systems occurs at any time, immediately repair the leak or, in case of a large leak, stop the oil pumps.
- Any oil spilled must be wiped up at once.
- Electrical apparatus shall be inspected frequently and any condition likely to cause sparking shall be corrected before the apparatus is used again.
- Have the Oil Pollution Act posted by the fire and bilge pump.
- Do not pump bilges at night.
- In port, pump bilges only to a shore sludge processing plant, a barge or a tanker truck, and have crewmembers stationed topside to watch discharge. The topside observer must be able to communicate with the person controlling the pumping. Use of intrinsically safe hand held radios is the most common form of device to use for most communication.
- Have hull valves locked and precautions posted by the valves.
- Get permission from the chief engineer before pumping bilges.
- Flush bilge pumping hoses before disconnecting.

WORKING IN CLOSED SPACES

2-16. Since you normally think of breathing as an automatic body function, you seldom worry about whether the air is safe (or unsafe) to breathe. However, aboard ship this is a real problem which has cost many lives. This is especially true in spaces that are not well ventilated or that have been closed for appreciable lengths of time. Unventilated storerooms, double bottoms, tanks, cofferdams, pontoons, voids, and cold boilers are typical "iron coffins." The air you breathe is composed of several gases. Under normal conditions, approximately 20 percent of the air content is oxygen. If for any reason this percentage is reduced much below 16 percent, you cannot survive. Fire, rusting, the drying of paint, and the decomposition of organic material can all contribute to oxygen deficiency. If you step into a closed compartment and there is insufficient oxygen in the air to support life, the results will be painfully swift. You will be immediately weakened. Your body may not respond even to your most desperate efforts to escape. If the oxygen content is particularly low, you may have time for only a few futile gasps before losing consciousness. Death may be only a breath away in an oxygen-deficient atmosphere. However, be aware that even if the air does contain enough oxygen, it may also contain concentrations of other gases which are flammable or toxic, or both. Flammable or toxic gases and vapors may be just as deadly as the lack of oxygen.

WARNING

Always have permission from your supervisor before entering a compartment that has been sealed for a long time. The compartment must be checked for oxygen and/or gas content prior to entry.

ELECTRICAL SAFETY

2-17. Any person working around electrical circuits and equipment must always observe safety precautions to avoid injury caused by contact with electrically charged objects. Electrical shock due to contact with an energized circuit can cause serious injury. Even low voltage circuits (115 volts and below) can cause death upon contact, especially if current passes through the chest. Shipboard conditions are particularly conducive to severe shocks because the body is likely to be in contact with the ship's metal structure and the body's resistance to electricity may be low because of perspiration or damp clothing. Always be alert and extremely careful when working near electrical circuits aboard ship. Be extremely careful to prevent circuits. Short circuits may be caused by accidentally placing or dropping a metal tool, flashlight case, or some other conducting object on an energized line. The arc or the fire that may result can cause extensive damage to equipment and injury to personnel. When working around electrical equipment, keep in mind that electricity strikes without warning, that hurrying reduces caution and invites accidents, and that every electrical circuit is a potential source of danger. Working on energized and de-energized equipment requires training, lock-out, tag-out procedures, and standing operating procedures.

RESCUE OF VICTIMS

2-18. Even when safety precautions are observed, accidents may occur. Always be familiar with the procedures to follow when rescue from electrical contact is necessary and injury from electricity has been received. Rescuing a person who is in contact with an electrically charged object is likely to be a difficult and dangerous job. The first thing to do is to turn off the source of the power supply. NEVER touch the victim's body, the charged object, or any other object that may be conducting electricity. You may be electrocuted yourself!

> **WARNING**
>
> **Do not attempt to administer first aid or come in physical contact with an electric shock victim before the victim has been removed from the live conductor.**

2-19. When attempting to administer first aid to an electric shock victim, first shut off the high voltage. If you cannot shut off the high voltage, remove the victim immediately without touching them, observing the following precautions:

- Protect yourself with dry insulating material.
- Use a dry board, belt, dry clothing, or other available nonconductive material to free the victim from the live wire. Do not touch the victim.
- After you have removed the victim from live conductor, administer artificial respiration as described below.

RESUSCITATION AND ARTIFICIAL RESPIRATION

2-20. The following instructions on cardiopulmonary resuscitation (CPR) for electric shock were furnished by the Bureau of Medicine and Surgery. See also FM 4-25.11 for complete coverage of CPR and treatment of burn and shock victims. Artificial resuscitation, after electric shock, includes artificial respiration to reestablish breathing and external heart massage to reestablish heart beat and blood circulation. To aid a victim of electric shock after removal from contact with the electricity, immediately apply mouth-to-mouth artificial respiration. If there is no pulse, immediately apply heart massage. Do not waste precious seconds by carrying the victim from a cramped, wet, or isolated location to a roomier, drier, or more frequented location. If desired, breathe into the victim's mouth through a cloth or a handkerchief. If assistance is available, take turns breathing into the victim's mouth and massaging the heart.

CARDIAC ARREST (LOSS OF HEARTBEAT)

2-21. If the subject has suffered an electric shock and has no heartbeat, cardiac arrest has occurred. Absence of any pulse at the wrist or in the neck will verify cardiac arrest. Associated with this, the pupils of the eyes will be dilated and respiration will be weak or stopped. The subject may appear to be dead. Under these circumstances, severe brain damage will occur in four minutes unless circulation is reestablished by cardiac massage.

CLOSED CHEST CARDIAC MASSAGE

2-22. This method has been adopted as practical and can be administered by anyone who is properly instructed. The object in closed chest cardiac massage is to squeeze the heart through the chest wall, thereby emptying it to create a peripheral pulse. This must be done about 60 times each minute. Perform the following to administer closed chest cardiac massage:

- Lay the subject face up. A firm surface such as the deck is preferred.
- Expose subject's chest.
- Kneel beside the victim; feel for lower end of his/her sternum (breastbone).
- Place one of your hands across breastbone so the heel of your hand covers the lower part.
- Place your other hand on top of the first so that the fingers point toward the victim's neck.
- With your arms nearly straight, rock forward so that a controlled amount of your body weight is transmitted through your arms and hands to the victim's breastbone. The amount of pressure to apply will vary with the subject. Apply pressure as smoothly as possible. With an adult victim, depress the chest wall 2 to 3 inches with each pressure application.

- Repeat each application of pressure about 60 to 80 times per minute. An assistant should be ventilating the subject's lungs preferably with pure oxygen under intermittent positive pressure; otherwise, with mouth-to-mouth resuscitation. However, closed chest massage will cause some ventilation of the lungs. Therefore, if you are alone, you must concentrate on the massage until help can arrive.
- Direct other assistants, when available, to keep checking the patient's pulse. Use the least amount of pressure that will produce an effective pulse beat. The pupils will become smaller when effective cardiac massage is being performed.
- Pause occasionally to determine whether a spontaneous heartbeat has returned.

Note: Since the heart cannot recover unless it is supplied with oxygenated blood, cardiac massage must be accompanied by mouth-to-mouth artificial respiration. When there is only one operator, the cardiac massage must be interrupted every half-minute or so to institute rapid mouth-to-mouth breathing for three or four respirations.

CAUTION

Make every effort to keep your hands positioned as described to prevent injuries to the liver, ribs, or other vital organs. No matter how carefully the treatment has been applied, possible breastbone damage may occur. If the patient is to be moved, it should be done very carefully to prevent possible injury to internal organs by unrecognized bone fractures.

2-23. The mouth-to-mouth or mouth-to-nose technique of artificial respiration is the most effective of all the resuscitation techniques. It is the most practical method for emergency ventilation of an individual of any age who has stopped breathing, in the absence of equipment or of help from a second person regardless of the cause of cessation of breathing. Persons who are trained in first aid do not usually have the experience, training, and essential equipment to determine whether lack of breathing is a result of disease or accident. Therefore, some form of artificial respiration should be started at the earliest possible moment. Any procedure that will obtain and maintain an open air passageway from the lungs to the mouth and provide for an alternate increase and decrease in the size of the chest, internally or externally, will move air in and out of a non-breathing person. The mouth-to-mouth or mouth-to-nose technique has the advantage of providing pressure to inflate the victim's lungs immediately. It also enables the rescuer to get more accurate information on the volume, pressure, and timing of efforts to inflate the victim's lungs than afforded by other methods.

RESPONSIBILITY FOR SAFETY

2-24. Safety is everyone's responsibility. It cannot be left to an individual or an office. Everyone must always be on the alert to avoid causing accidents. One of the basic rules of safety is the proper behavior of every individual. Practical jokes and horseplay cannot and will not be tolerated. The possible consequences are too high a price to pay for the small amount of humor derived. Another basic rule is cleanliness, or just plain good housekeeping practices. Decks must be kept clean and free of oil. Oil spots are very slippery and can cause accidents.

Chapter 3

BASIC PRINCIPLES OF ENGINES

INTRODUCTION

3-1. Before you can fully understand what takes place in an operating diesel engine, you must understand the basic principles and sciences. These include basic physics, engine performance and efficiency, engine mechanics, and the petroleum products used in diesel engines. This chapter covers this material in depth. It is intended to further understanding of diesel engines as well as help operate and troubleshoot diesel engines more effectively.

SECTION I. UNITS OF MEASUREMENT

Physical quantities can be expressed by one of three units—distance (length), force (weight or pressure), and time. In the English system used in the United States, the standard units are foot (ft.), pound (lb.) and second (sec.).

DERIVED UNITS

3-2. Derived units are formed and developed from the three standard English units and are used to measure quantities encountered in engineer-practice.

AREA

3-3. Area is a measure of surface expressed as the product of the length and width, or of two characteristic lengths of the surface. Areas are expressed in square units, such as square feet (sq. ft. or $ft.^2$).

VOLUME

3-4. Volume is a measure of space expressed as the product of area and depth, or of three characteristic lengths of the space. Volumes are measured in cubic units, such as cubic feet (cu. $ft.^3$) or cubic inches (cu. in. or in^3).

LINEAR MOTION

3-5. Linear motion is the length of the line along which a point or a body has moved from one position to another. Linear distance is measured in units of length (such as feet or inches).

ROTARY MOTION

3-6. Rotary motion (see Figure 3-1) is the movement of a point or body along a circular path. The position of a point or body rotating about a fixed point in a plane is expressed by the angle through which it has rotated. When point "a" has moved to "b", remaining all the time at a constant length "r" from the fixed point "o", the position of "b" is determined by the angle of rotation "c". Angles are measured in degrees, 360 degrees corresponding to one complete revolution.

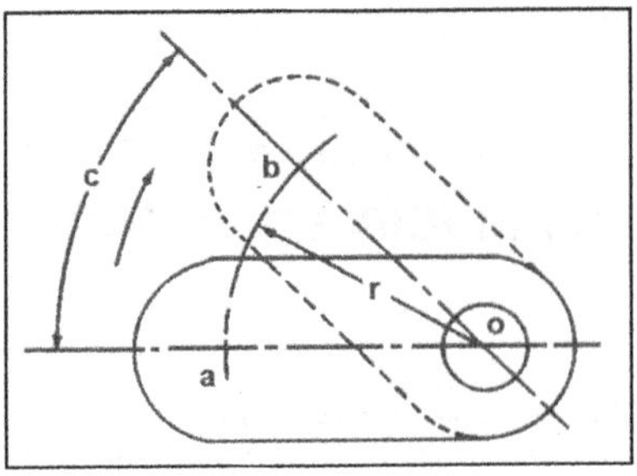

Figure 3-1. Sample of Rotary Motion

COMPARATIVE UNITS

VELOCITY

3-7. Velocity is the distance traveled by a moving point in a unit of time (as second, minute, or hour). Velocity is computed by dividing the distance traveled by the time used for the travel:

Velocity distance ÷ time

When distance is expressed in feet and time is expressed in minutes, velocity will be expressed in feet per minute (ft. per min. or ft./min.). If time is expressed in seconds, the velocity will be expressed in feet per second (ft. per sec. or ft./sec.).

Example: A point traveled 1,800 feet in 2.5 minutes. Find its velocity.

Velocity = distance ÷ time

or

1,800 divided by 2.5 = 720 ft./min.

Velocity may be uniform or varying. If the motion is uniform, that is, if the velocity is constant, the above formula will give the actual velocity. If the motion, and also the velocity, is not uniform, as in the reciprocating motion of a piston in an engine cylinder, then the above formula will give the average velocity. The average piston velocity is referred to as piston speed. The velocity of a moving vehicle or aircraft is generally called speed and is expressed in miles per hour (mph).

Example: Find the velocity in feet/minute of an automobile traveling at a speed of 50 mph. 1 mile = 5,280 ft.; 1 hour = 60 min.; speed = 50 mph; therefore—

50 × 5,280 ÷ 60 = 4,400 ft./min.

In reference to the flow of fluids (such as gas or water), the rate of flow is called velocity expressed in terms of feet per minute (ft./min.) or feet per second (ft./sec.). However, the term speed is applied in reference to the rotary motion of a mechanism. Therefore, engine speed is said to be so many revolutions of its crankshaft per minute, and is designated as rpm.

ACCELERATION

3-8. Acceleration is a change in the velocity of a moving body within a unit of time. Acceleration may be uniform or varying. Acceleration is positive when the velocity increases and negative when the velocity decreases. Negative acceleration is called deceleration. Acceleration is computed by dividing the change in velocity by the time during which this change takes place. If the acceleration is uniform, then this formula will give the actual acceleration. If the change in velocity is not uniform, then this formula will give the average acceleration. When velocity is expressed in feet/minute, acceleration will be expressed in feet/minute per minute or feet/minute^2. If the velocity is expressed in feet/second, the acceleration will be expressed in feet/second per second or feet/second^2. An example of constant acceleration is the acceleration of the earth's force of gravity which, for all technical calculations, can be set at 32.2 ft./sec.^{2}.

Example: A ship accelerates from dead in the water to 20 knots in 5 minutes. A speed of 20 knots is considered to be 2,000 feet per minute. What is the average acceleration in ft./sec.^{2}? First, convert all factors into seconds—

2,000 ft/min. = 2,000 ÷ 60 = 33.33 ft./sec.

5minutes = 60 × 5 = 300 seconds

$$\text{Average acceleration} = \frac{\text{change in velocity}}{\text{change in time}}$$

or

$$\frac{33.33}{300} - 0 = 0.11 \text{ ft./sec.}^2$$

PRESSURE

3-9. Pressure may be defined as force acting on a unit of area. Pressure may be exerted either by a solid or by a fluid.

Example: Determine the pressure on the subbase of a 1,800-pound engine whose contact area between the engine and the subbase consists of two strips, each of which is 2 inches wide and 40 inches long. In this case, the weight of the engine is a force pressing the engine against the subbase. The pressure (p) will be equal to the force (F) divided by the area (A)—

$$p = F \div A$$

or

$$p = \frac{1{,}800}{2(2 \times 40)} = 11.25 \text{ psi}$$

The force may be in pounds, and the area upon which the force is acting may be expressed in square inches or square feet. Accordingly, pressure may be in pounds per square foot (psf) or in pounds per square inch (psi). In the case of contact between two solid bodies, the surfaces have a perfect uniform contact only in exceptional cases. The presence of uneven areas will give higher pressures at the high spots, and lower pressures, if any, at the places of depression. In such a case, the pressure determined, as in the above example, will give only the average value. However, when a force is transmitted by a fluid (either liquid or gas), the pressure between the fluid and the walls of the container will be uniform and equal in all directions regardless of the shape of the walls.

Example: Determine the pressure in an air compressor if the force acting upon the piston is 750 pounds and the piston diameter is 3 inches (the area of a 3-inch circle)—

$$A = \pi r^2 = \pi (d \div 2)^2 = (\pi d^2) \div 4$$

$$= 3.14 \times 3^2 \div 4 = 7.07 \text{ sq. in.}$$

and the pressure will be according to the equation—

$$p = 750 \div 7.07 = 106 \text{ psi}$$

SPECIFIC GRAVITY

3-10. The ratio of the weight of a certain volume of a liquid to the weight of an equal volume of water is called specific gravity. For practical use, it is well to remember that:

1,732 cu. in. = 1cu. ft.

1 cu. ft. of freshwater weighs 62.4 lb.

1 gal. = 231 cu. in.

therefore, 1 gal. weighs—

62.4 × 231 ÷ 1,732 = 8.34 lb./gal.

Example: Determine the weight of 1 gallon of fuel oil which has a specific gravity of 0.84. The weight of the oil will be equal to the weight of the water times the specific gravity.

8.34 × 0.84 = 7 lb./gal.

WORK

3-11. Work is done when a force is moving a body through a certain distance. Work (W) is measured by the product of the force (F) multiplied by the distance (d) moved in the direction of the force.

Work = force × distance

Work is expressed in ft.-lb. or in in.-lb.

Example: Find the work necessary to raise the weight of 100 pounds a distance of 2 3/4 feet.

100 × 2.75 = 275 ft./lb.

POWER

3-12. Power is the rate at which work is performed, or the number of units of work performed in one unit of time. Power is measured in foot-pounds/minute (ft.-lb./min.). In engineering calculations, 550 foot-pounds/second (ft.-lb./sec.) = 33,000 foot-pounds/minute and is called a horsepower.

Power = work ÷ time

or

P = W ÷ t

Example: Determine the power required to do the work of the previous example if the work is to be performed (a) in 5 sec. or (b) in 25 sec.

FOR CASE (a): The rate of performing the work is—

275ft.-lb. ÷ 5 sec. = 55 ft.-lb./sec.

Expressed in horsepower this is

55ft.-lb./sec. = 0.1 hp
550 ft.-lb./sec.

or since—

55 ft.-lb./sec. × 60 sec./min. = 3,300 ft.-lb./min.

then—

3,300 ft.-lb./min. = 0.1 hp
33,000 ft.-lb./min.

FOR CASE (b): The rate of performing the work is—

275ft.-lb. ÷ 25 sec. = 11 ft.-lb./sec.

so—

11 ft.-lb./sec. = 0.02 hp
550 ft.-lb./sec.

ELECTRIC POWER

3-13. Electric power is measured in units called watts (1,000 watts equals 1 kilowatt (kw)). The conversion factors between hp and kw are—

1 hp = 0.746 kw or 1 kw = 1.341 hp

TEMPERATURE

3-14. The temperature of a body is a characteristic which can be determined only by comparison with another body. When two bodies are placed in close contact, the hotter one will begin to heat the other and is said to have a higher temperature. The scales of temperature are set arbitrarily. The two scales in general use are the Celsius (C) or centigrade scale and the Fahrenheit (F) scale. In the Fahrenheit scale, the two reference points are the temperature of melting ice, designated as 32° F and the temperature of steam with the water boiling under normal barometric pressure, designated as 212° F. The distance on the scale between these two points is divided into 180 equal parts called degrees. The scale is continued in both directions, above 212° F and below 32° F. Below 0° F the temperatures are designated by a minus (-) sign. In theoretical calculations pertaining to gases, another scale, the absolute or rankine (R) is used. The unit of the rankine scale is the degree, the same as in the Fahrenheit scale, but the absolute zero is placed at -460^{0} F. Therefore, the relation between the absolute temperature (T) and the corresponding Fahrenheit temperature (t) is--

degree rankine - degree Fahrenheit = 460

or

$T = t + 460$

In technical calculations pertaining to gases, 60^{0} F is the normal or standard temperature.

GAS PRESSURE RELATIONSHIPS

3-15. Pressure has already been discussed. Also, as applied to gas, the pressure of a gas is often expressed by the height of the column of either water or mercury which balances the gas pressure in the space under consideration. The relationship between the gas and water or between the gas and mercury can be established as follows:

1 cu. ft. of freshwater at room temperature weighs 62.4 lb.

or

a column of water 1-foot high acting upon an area of 1 square foot is equal to a pressure of 62.4 psf.

since—

1 cu. ft. contains 1,728 cu. in., the weight of 1 cu. in. of water is 62.4 ÷ 1,728 = 0.0361 lb.

therefore—

a column of water 1 in. high, acting upon 1 sq. in. will produce a pressure of 0.0361 psi.

To obtain a pressure of 1 psi, the column must be higher in the proportion of:

1 ÷ 0.0361 = 27.70

or

27.70 in. or also 27.70 ÷ 12 = 2.309 ft.

Mercury is 13.6 times heavier than water. Therefore, a column of mercury must be shorter in this proportion

or

1 psi = 27.70 ÷ 12.6 = 2.036 in. mercury

conversely—

1 in. mercury = 1 ÷ 2.036 = 0.491 psi.

Gauge and Absolute Pressures

3-16. Instruments measure the pressure of gases in respect to the pressure of atmospheric air, also called barometric pressure. Pressures measured thusly are called gauge pressures which usually indicate pounds per square inch gauge (psig). The actual pressure exerted on the gas can be obtained by adding the barometric pressure to the gauge pressure. This pressure is called absolute pressure and is indicated as pounds per square inch absolute (psia) and pounds per square foot absolute (psfa). If absolute pressure is designated P_a, gauge pressure P_g, and barometric pressure b, then the relation can be written as:

absolute pressure = gauge pressure + barometric pressure

or

$P_a = P_g + b$

Barometric pressure (b) is not constant because it changes with altitude and weather. Normal or standard barometric pressure at sea level is 29.92 inches of mercury

or

29.92 ÷ 2.306 = 14.0 psia

3-17. Volume is the space occupied by a body whether solid, liquid, or gas. If the body is a vapor or gas, its volume must be confined on all sides. In engines, the volume of gas is usually confined by a cylinder that has one end closed by a stationary cylinder head and the other end closed by the head of a piston. The piston provides a gastight seal. When the piston changes position, the volume of the gas also changes. When the piston approaches the cylinder head, the volume is being decreased and the gas is compressed. When the piston moves away from the cylinder head, the volume increases and the gas expands.

Gas Pressure, Volume, and Temperature

3-18. In gases, the three measurable quantities--pressure, volume, and temperature--are called gas properties or characteristics. The three characteristics are connected by a simple relation, which for any gas can be written as:

$$pV = WRT$$

where p is the absolute pressure in pounds per square foot absolute, V is the volume in cubic feet, W is the weight of the gas in pounds, T is the absolute temperature in degrees rankine, and R is a constant called the gas constant. The numerical value of R is known for all gases. It is expressed in foot-pounds per pound per degree rankine. The equation shows that if the three characteristics of a certain amount of gas are known, the weight can be found; or, if the weight is known, any one of the three characteristics can be found if the two others are measured.

Example: Find the weight of air contained in a 2-cu.-ft. cylinder at a pressure of 100 psig with a temperature of 72° F and the gas constant of air of R = 53.3.

Find the absolute pressure, assuming a normal barometric pressure (b) of 14.7 psia—

$$\text{pressure} = 100 + 14.7 = 114.7 \text{ psia}$$

or

$$p = 114.7 \times 144 = 16{,}500 \text{ psfa}$$

Find the absolute temperature by equation—

$$T = 72 + 460 = 532^{\circ} R$$

Solving the equation for W and substituting the corresponding numerical values gives—

$$W = pV \div RT = 16{,}500 \times 2 \div (53.3 \times 532) = 1.16 \text{ lb.}$$

Example: Determine what will happen if the air in the example is heated to 150° F. The new absolute temperature is $T = 150 + 460 = 610^{\circ}$ R. The characteristic which will change is p. Solving for p, and substituting the corresponding values gives—

$$p = WRT \div V = 1.16 \times 53.3 \times 610 \div 2 = 18{,}900 \text{ psfa}$$

Converting 18,900 psfa to psia gives 18,900 ÷ 144 = 131.2 psia. The gauge pressure will be—

$$pg = 131.2 - 14.7 = 116.5 \text{ psig}$$

or

an increase of 16.5 psi over the original ; pressure of 100 psig.

ENERGY

3-19. Energy of a body is the amount of work it can do. Energy exists in several different forms. A body may possess energy through its position, motion, or condition. Energy due to a position occupied by a body is called mechanical potential energy. An example of mechanical potential energy is a body located at a higher level, such as water behind a dam. When a body is moving with some velocity, it is said to possess energy of motion or kinetic energy (such as a ball rolling upon a level floor). A third form of energy is internal energy or energy stored within a body. It can be a gas, liquid, or solid, due to the forces between the molecules or atoms composing the body (such as in steam or gas under pressure). Chemical energy in fuel or in a charged storage battery is also classified as internal energy. These three forms of energy (mechanical potential, kinetic, and internal) have in common the characteristic of being forms in which energy may be stored for future use. Work can be classified as mechanical or electrical energy in the state of transformation or transfer. Work done by raising a body stores mechanical potential energy in the body due to the force of gravity. Work done to set a body in motion stores kinetic energy. Work done in compressing a gas stores internal energy in the gas. Electrical work can be transformed into mechanical work by means of an electric motor and, after that, it may undergo other changes, the same as mechanical work. Heat, like work, is energy in the state of transfer from one body to another, due to a difference in temperature of the bodies.

Units of Energy

3-20. Two basic independent units of energy are the foot-pound (ft.-lb.) and the British thermal unit (Btu). The ft.-lb. is the amount of energy as shown by work and is equivalent to the action of a force of 1 pound through a distance of 1 foot. The Btu is the energy required to raise the temperature of 1 pound of pure water by 1° F at a standard atmospheric pressure of 14.70 psia. The conversion factor from ft.-lb. to Btu, often called the mechanical equivalent of heat, is—

1 Btu = 778 ft.-lb.

Two other energy units are used in engineering calculations derived from the basic unit of foot-pounds—

The horsepower-hour (hp-hr.) and the kilowatt-hour (kw-hr.) is the transfer of energy at the rate of 33,000 ft.-lb. per min. during 1 hr., or a total of 1,980,000 ft.-lb. or, using the factor 778—

1 hp-hr. = 1,980,000 ÷ 778 = 2,544 Btu

The kw-hr. is the transfer of energy at the rate of 1,000 watts per hour or 1.341 hp per hour which is equivalent to 44,253 ft.-lb. per min. during 1 hr., or a total of 2,655,180 ft.-lb.

or

1 kw-hr. = 2,655,180-778 ÷ 778 = 3,412 Btu

Kinetic Energy

3-21. Kinetic energy of a body is computed by—

$$KE = 1/2(W \div g)v^2$$

where g is the acceleration due to gravity, 32.2 ft./sec.^{2} and v is the velocity of the body, ft./sec. Work done by kinetic energy is due to a change of velocity from an initial value of v_1 to a final value v_2.

Example: Find the work done by 500 pounds of exhaust gases discharged upon the blades of a supercharger turbine if the initial velocity of the gases were 9,000 ft./min. and the exit velocity were 5,400 ft./min.

First the velocity must be changed to ft./sec.—

v_1 = 9,000 ÷ 60 = 150 ft./sec.

v_2 = 5,400 ÷ 60 = 90 ft./sec.

The kinetic energies before and after the turbine are—

$KE_1 = 1/2(500 \div 32.2) \times 150^2$ = 175,000 ft.-lb.

$KE_2 = 1/2(500 \div 32.2) \times 90^2$ = 63,000 ft-lb.

and the work done—

175,000 – 63,000 = 112,000 ft.-lb.

Energy conservation

3-22. The principle of energy conservation is that energy may exist in many varied and interchangeable forms but may not be quantitatively destroyed or created. Therefore, mechanical energy may be transformed into heat, or vice versa, but only in a definite relation as given before (1 Btu = 778 ft.-lb.). Potential or internal energy may be changed to kinetic energy, and so on.

HEAT FLOW

3-23. As previously stated, heat is a form of energy in a state of change; it is expressed in Btu. Quantitatively, a flow of heat is determined by the change of temperature of a body. Heat is conveyed if the temperature of the body rises and it is taken away if the temperature drops. A quantitative measurement of heat is possible only by comparison with the behavior of some other body selected as a standard. Since the Btu (heat unit) is determined with the aid of water, water is used as the standard to determine the behavior of all other substances in respect to a change of heat.

SPECIFIC HEAT

3-24. The heat of a substance is the ratio of heat flow required to rise by 1° F the temperature of a certain weight of the substance to the heat flow required to rise by 1° F the temperature of an equal weight of water. Due to the definition of 1 Btu, the specific heat of water is unity, or 1 Btu/lb.-deg. F. Numerically, the specific heat of a substance is equal to the heat flow, in Btu, required to rise by 1° F the temperature of 1 pound of the substance. Denoting the specific heat by c, the heat flow Q required in raising the temperature of W lb. of a substance from t_1 to t_2 degrees F is—

Heat = Weight (of body) × specific heat × temperature difference

or

$$Q = Wc(t_2 - t_1)$$

In general, specific heat varies with the temperature. For gases it depends also upon conditions of pressure and volume. For many calculations, a mean value of specific heat can be used.

Example: Find the heat which is transferred to 53 gallons of lubricating oil when its temperature rises from 70° to 165° F. The specific heat of the oil is 0.5 Btu/lb.-deg. F and its specific gravity is 0.925.

1 gal. of oil weighs 8.34 × 0.925 = 7.71 lb.

53 gal. weighs 53 × 7.71 = 408 lb.

Therefore, by equation, the heat transferred is—

$$Q = 408 \times 0.5(165 - 70) = 19{,}380 \text{ Btu}$$

HEAT TRANSFER

3-25. Generally speaking, heat is transferred by three methods: conduction, radiation, and convection.

- *Conduction.* Conduction is energy transfer by actual contact from one part of a body having a higher temperature to another part of it or to a second body having a lower temperature.
- *Radiation.* Radiation is energy transfer through space from a hotter body to a colder body.
- *Convection.* Convection is not a form of energy transfer. Convection designates a process in which a body and the energy in it are moved from one position to another without change of state. An example of convection is the movement of heated air from one part of a room to another. The basic principle of heat flow is that heat can flow from one body to a second body only if the temperature of the first body is higher than the temperature of the second body.

SECTION II. ENGINE MECHANICS

In this manual, the term "engine mechanics" covers such items as piston and crankshaft travel, piston speed, inertia, torque, and speed. These factors are important because they play a large part in limiting the capabilities of a diesel engine.

PISTON AND CRANKSHAFT TRAVEL

3-26. The movements of the piston (see Figure 3-2) are transmitted to the crankshaft by the connecting rod, the crankshaft pin, and the crankshaft throw. By these members, the motion of the piston (called reciprocating motion) is transformed into rotary motion. The travel of the crankshaft pin can be considered as a uniform motion along a circle described by the radius (R) equal to the length of crankshaft throw. The total piston travel or stroke m-n = o-r = 2R.

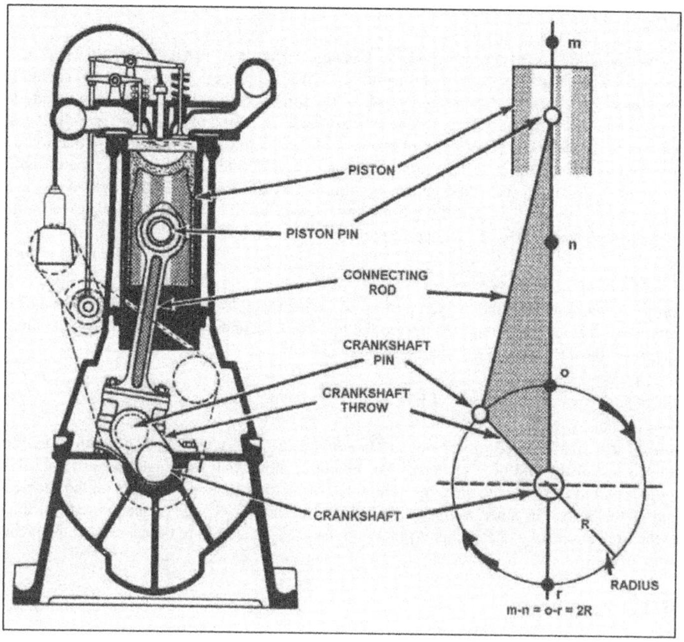

Figure 3-2. Piston and Crankshaft Travel

PISTON SPEED

3-27. While the crankpin travel is uniform and has a constant velocity, the piston travel is not uniform and the piston speed constantly varies. At each dead center, the piston comes to a standstill, and its speed becomes zero. As the piston begins to move, the speed gradually increases and reaches a maximum when the angle formed by the crankshaft is equal to 90°. After this position, the piston speed begins to decrease and at the dead center again becomes zero. For many calculations, the average or mean piston speed must be known or the constant speed at which the piston must move to travel the same distance in the same time as it travels at the actual variable speed must be known. Using our definition of velocity, this speed is expressed as the distance traveled divided by the time used for the travel. Piston speed is expressed in ft./min. The distance traveled by the piston in one revolution is two strokes. The piston stroke is usually measured in inches. Divide the strokes by 12 to convert to feet. Multiply by the number of rpm and divide 12 by 2 because the piston travels down as well as up. Therefore, stroke × rpm ÷ 6 = the distance traveled in 1 minute. At the same time, according to the definition of velocity, this will be the mean piston speed, usually simply called piston speed

or

Piston speed = stroke × rpm ÷ 6

Example: Find the piston speed of an 8 1/2 × 10 1/2 engine running at 750 rpm.

Piston speed = 10.5 × 750 ÷ 6 = 1,313 ft./min.

In diesel engines now in use, the piston speed at rated rpm varies from 1,000 to about 2,000 ft./min.

INERTIA

3-28. Inertia is the resistance of a body to a change of motion. That is, the tendency of an object to remain at rest if it is stationary or to continue to move if it is moving. Inertia, as such, cannot be measured directly. However, it can be expressed in terms of the force which must be applied to a body to change its velocity. As with any force, inertia forces are expressed in pounds. A change in velocity is defined as acceleration or deceleration. Therefore, inertia may also be defined as being equal to that force which must be applied to a body to impart to it a certain acceleration or deceleration, either to speed it up or to slow it down. Numerically, the force of inertia F is equal to the weight (W) of a body, divided by the acceleration of the force of gravity (g) = 32.2, and multiplied by the acceleration (a), which is imparted to the body:

Inertia force = (weight ÷ 32.2) × acceleration

or

$F = (W \div g)a$

Example: Determine the force of inertia of a body which weighs 12 pounds and moves uniformly with a velocity of 15 ft./sec. that is required to stop the body in 2 seconds. The acceleration, or since in this case it is negative, the deceleration per second is—

$a = 15 \div 2 = 7.5$ ft./sec.

and the negative acceleration force of F is—

$F = (12 \div 32.2) \times 7.5 = 2.8$ lb.

Therefore, if a steady force of 2.8 pounds is applied to the body against its motion, the body will be brought to rest at the end of 2 seconds. The example shows that the force of inertia of a body is not a fixed quantity but is a variable quantity. The force of inertia of a body depends on the acceleration that is applied to the body. In other words, it depends on the rate of change of the velocity of the body. The lesser the time in which a change takes place and the higher the required acceleration, the greater becomes the force of inertia.

TORQUE

3-29. Torque (T) (see Figure 3-3) is the effect which rotates or tends to rotate a body. To produce rotation of a free body, two equal and opposite forces must act along parallel lines but at separate points of the body. These two forces form a "couple." The perpendicular, or shortest, distance between the lines of action of the forces is called the arm of the couple. The magnitude, or moment, of the couple is expressed as the product of one of the forces multiplied by the length (L) of the arm of the couple. When a body rotates about the point of application of one of the forces of a couple, as on a fixed pivot, the arm of the couple is known as the lever and the turning moment is called the torque. If the lever is fastened to a rotating shaft, it is called a crank. Figure 3-3 shows a force (F) acting perpendicularly to a crank having a length from the center of rotation 0 of a rotating shaft. The torque acting on this shaft is—

Torque = Force × lever

or

Torque is measured in pound-foot (lb.-ft.) or pound-inch (lb.-in.).

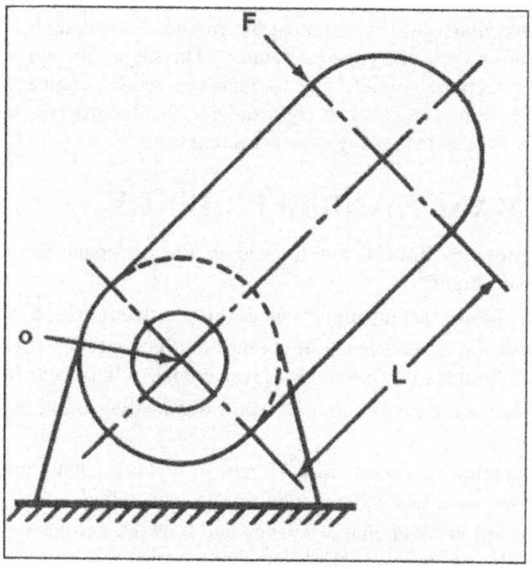

Figure 3-3. Torque

SPEED

3-30. Engines are often classified according to their speed capacity. Some are classified as low-speed, others as medium-speed, and still others as high-speed engines. However, unless a definite yardstick is used, the designations remain vague. There have been attempts to use either the engine speed (rpm) or its piston speed (ft./min.) as a measure of speed capacity. Neither of these two methods can give correct indications. Rotative speed, as such, is not suitable as a speed characteristic because it does not take into consideration the size of the engine. A 6-cylinder × 3 1/2 × 4 1/2 × 900-rpm engine is not a high-speed engine because this type engine normally operates at speeds up to 2,000 rpm and higher. On the other hand, 8 1/2 × 10 1/2–inch diesel engines usually operate at speeds not exceeding 750 rpm. Even at this lower speed, they have many features in common with high-speed engines. The same is true, only in reverse, with respect to piston speeds. In a large engine, a relatively high piston speed (1,800 ft./min. or more) may be obtained with a relatively low rpm. In a small, high-speed engine the piston speed is not high. A good speed characteristic, called speed factor is obtained as a product of rpm and piston speed. For the sake of obtaining smaller, more easily remembered figures, the product is divided by 100,000, for example:

$$\text{Speed factor} = \frac{\text{rpm} \times \text{piston speed (ft./min.)}}{100{,}000}$$

The figures obtained for various existing diesel engines lie between the limits of 1 and slightly less than 81. According to this data, all engines can be divided into four classes. In each class, the high limit is obtained by multiplying the low limit by 3:

- Low-speed engines with a speed of 1 to3.
- Medium-speed engines with a speed factor of 3 to 9.
- High-speed engines with a speed factor of 9 to 27.
- Super high-speed engines with a speed factor of 27 to 81.

Example: Find the speed factor and speed classification of a 16- × 8 1/2 × 10 1/2 × 750-rpm engine. As found in the previous example, the piston speed is 1,313 and therefore, by equation—

$$\text{Speed factor} = \frac{750 \times 1{,}313}{100{,}000} = 9.85$$

According to the four classes of engines, an engine having a speed factor over 9, but under 27, is classified as a high-speed engine. Classifying an engine in one of the above-named groups according to its speed factor has a particular value in designing the engine. Knowledge of the speed group to which an engine belongs is also of value to the engine operator. The higher the speed classification of an engine, the more attention the operator should give to maintaining the engine in its best possible running condition. This can be done by observing every detail given in the manufacturer's instruction book. Also more care should be given when inspecting or overhauling the engine.

ENGINE PERFORMANCE AND EFFICIENCY

3-31. The following provides information to help in the understanding of engine performance and efficiency. The engineman should--

- Know how the power that an engine can develop is determined by limiting factors.
- Learn how heat losses, efficiency of combustion, volumetric efficiency, and the proper mixing of fuel and air, limit the power which a given engine cylinder can develop.
- Become familiar with the factors that cause overloading of an engine and unbalance between engine cylinders.
- Know the symptoms, causes, and effects of cylinder load unbalance and know what is necessary for an equal load to be maintained on each cylinder.
- Know what is meant by engine efficiency and how the various types of efficiencies and losses are used in analyzing the internal-combustion process.
- Be familiar with what causes various efficiencies to increase or decrease and the way in which these variations affect engine performance.

POWER LIMITATIONS

3-32. The amount of power that an engine can develop is limited chiefly by design. These limiting factors are mean effective pressure, length of piston stroke, cylinder bore, and piston speed. This latter is itself limited by the frictional heat generated and by the inertia of moving parts.

Mean Effective Pressure

3-33. The average pressure exerted on the piston during power stroke is referred to as mean effective pressure (mep). It is determined from a formula or by a planimeter. There are two kinds of mean effective pressure. One is indicated mean effective pressure (imep), which is developed in the cylinder and can be measured. The other is brake mean effective pressure (bmep), which is computed from the brake horsepower (bhp) delivered by the engine.

Length of Stroke

3-34. The distance a piston travels between top dead center (TDC) and bottom dead center (BDC) is known as the length of stroke. It is one of the factors that determine the piston speed. The shorter the length of stroke, the faster the engine can turn without placing excessive strain on rods and bearings. On the other hand, a slower engine can develop more power if it has a longer stroke.

Cylinder Bore

3-35. Bore is used to identify the diameter of the cylinder. The diameter must be known to compute the area of the piston crown. It is upon this area that the pressure acts to create the driving force. This pressure is calculated and expressed for an area of 1 square inch. The ratio of length of stroke to cylinder bore is fixed in engine design. In most slow-speed engines the stroke is greater than the bore.

Piston Speed (Revolutions Per Minute)

3-36. The speed at which the crankshaft rotates is measured in revolutions per minute (rpm). Since the piston is connected to the shaft, the rpm, along with the length of the stroke, determines piston speed. During each revolution, the piston completes one upstroke and one downstroke. Therefore, piston speed is equal to rpm times twice the length of the stroke. As described previously, this speed is usually expressed in feet per minute.

HORSEPOWER COMPUTATION

3-37. The power developed by an engine depends on the type of engine as well as the engine's speed. Remember that a cylinder of a single-acting, 4-stroke engine will produce one power stroke for every two crankshaft revolutions. A single-acting, 2-stroke cycle engine produces one power stroke for each revolution.

Indicated Horsepower (ihp)

3-38. The power developed within a cylinder is calculated by measuring the imep and engine speed. The rpm of the engine is converted to the number of power strokes per minute. With the bore and stroke known (available in engine manufacturer's technical manuals [TMs]), the horsepower can be computed for the type of engine involved. This power is called ihp because it is obtained from the pressure measured with an engine indicator. Power loss due to friction is not considered in computing ihp. Using the factors which influence the engine's capacity to develop power where:

P = Mean indicated pressure, in psi

L = Length of stroke, in feet

A = Effective area of the piston, in square inches

N = Number of power strokes per minute

33,000 = Unit of power (1 horsepower), or foot-pounds per minute

The general or standard equation for calculating ihp is—

$$\text{ihp} = \frac{P \times L \times A \times N}{33{,}000}$$

Example: Assume that a 12-cylinder, 2-stroke cycle, single-acting engine has a bore of 8 1/2 inches and a stroke of 10 inches. Its rated speed is 744 rpm. With the engine running at full load and speed, the imep is measured and found to be 105 psi. What is the ihp developed by one cylinder of the engine?

In this case:

$$P = 105;\ L = \frac{10}{12};\ A = 3.1416 \left(\frac{8.5}{2}\right)^2 : N = 744$$

or for one cylinder:

$$105 \times \frac{10}{12} \times 3{,}1416 \left(\frac{8.5}{2}\right)^2 \times 744$$

or for a 12-cylinder engine:

ihp = 12 × 111.9 = 1,343

Brake Horsepower (bhp)

3-39. Sometimes called shaft horsepower, bhp is the amount available for useful work. Because of the various power losses that occur during engine operation, bhp is less than ihp. To obtain the brake or shaft horsepower developed by an engine and delivered as useful work, deduct the total of all mechanical losses from the total ihp.

Cylinder Performance Limitations

3-40. The factors that limit the power which a given cylinder can develop are speed and mep. As stated earlier, the piston speed is limited by the inertial forces set up by the moving parts and frictional heat. In the case of mep, the limiting factors are heat losses and efficiency of combustion; volumetric efficiency, or the amount of air charged into the cylinder and the degree of scavenging and mixing of the fuel and air. The manufacturer prescribes the limiting mep (both brake and indicated) which should never be exceeded. In a direct-drive ship, the mep developed is determined by the rpm of the power shaft. In electric-drive ships, the horsepower and bmep can be determined by a computation based on readings from electrical instruments and on generator efficiency.

Cylinder Load Balance

3-41. To ensure a balanced, smooth-operating engine, load balance must be properly maintained so that the power output of the individual cylinders is within the prescribed limits at all loads and speeds. To have a balanced load on the engine, each cylinder must produce its share of the total power developed. If the engine is developing its rated full power, or nearly so, and one cylinder or more is producing less than its share, the remainder of the cylinders will become overloaded. Using the rated speed and bhp, it is possible to determine for each individual cylinder a rated bmep which cannot be exceeded without overloading the cylinder. If engine rpm drops below the rated speed, then the cylinder bmep generally drops to a lower value. The bmep should never exceed the normal mep at lower engine speed. Usually, it should be somewhat lower if the engine speed is decreased. Some engine manufacturers design the fuel systems so that rated bmep cannot be exceeded. By installing a positive stop to limit the maximum throttle or fuel control, the maximum amount of fuel that can enter the cylinder is regulated. Therefore, the maximum load of the cylinder is kept within the rated bmep. Engines used aboard ship are generally rated lower than those for industrial use in order to meet emergency situations. The economical speed for most diesel engines is approximately 90 percent of the rated speed. For this speed the best load conditions have been found to be from 70 percent to 80 percent of the rated load or output. On this basis, if an engine is operated at an 80 to 90 combination (80 percent of rated load at 90 percent rated speed), the parts will last longer and the engine will stay cleaner and in better operating condition. Diesel engines do not operate well at exceedingly low bmep, such as that occurring at idling speeds. Idling an engine tends to gum up parts associated with the combustion spaces. Operating an engine at idling speed for long periods will require cleaning and overhauling much sooner than operating at 50 percent to 100 percent of load.

SYMPTOMS OF UNBALANCE

3-42. An unbalanced condition existing between the cylinders of an engine may be indicated by several factors. These include black exhaust smoke, high exhaust temperatures, high temperatures in lubricating oil and cooling water, excessive heat, or excessive vibration or unusual sound.

Black Exhaust Smoke

3-43. When the exhaust emits black smoke, it is not possible to determine immediately whether the entire engine or just one of the cylinders overloaded. To determine which cylinder is overloaded, open the indicator cock on individual cylinders and check the color of the exhaust.

HIGH EXHAUST TEMPERATURES

3-44. When the temperatures of exhaust gases from individual cylinders become higher than normal, an overload within the cylinder is indicated. Higher than usual temperatures of the gases in the exhaust header indicates that all cylinders are probably overloaded. A frequent check on the pyrometer will indicate whether each cylinder is firing properly and carrying its share of the load. Any sudden change in the exhaust temperature of any cylinder should be investigated immediately. The difference in exhaust temperatures between any two cylinders should not exceed the limits prescribed in the engine manufacturer's TM.

HIGH LUBRICATING OIL AND COOLING WATER TEMPERATURES

3-45. If the temperature gauges for the lubricating oil and cooling water systems show an abnormal rise, an overloaded condition may exist. The causes of abnormal temperatures in these systems should be determined and corrected immediately if engine efficiency is to be maintained.

EXCESSIVE HEAT

3-46. In general, excessive heat in any part of the engine may indicate overloading. An overheated bearing may be the result of overloading the engine as a whole.

EXCESSIVE VIBRATION OR UNUSUAL SOUND

3-47. If all cylinders are not developing an equal amount of power, the forces exerted by individual pistons will be unequal. When this occurs, the unequal forces may cause an uneven turning moment to be exerted on the crankshaft and vibrations will be set up. Through experience, the vibrations and sound of an engine will tell you when a poor distribution of load exists. Use every opportunity possible to observe engines running under all conditions of loading and performance.

CAUSES OF UNBALANCE

3-48. To obtain equal load distribution among individual cylinders, the clearances, tolerances, and general condition of all parts that affect the cycle must be maintained so that very little, if any, vibration exists. In this connection, unbalance will occur unless the following are as nearly alike as possible for all cylinders:

- Compression pressures.
- Fuel injection timing.
- Quantity and quality of fuel injected.
- Firing pressures.
- Valve timing and lift.

3-49. When unbalance occurs, correction usually involves repair, replacement, or adjustment of the affected part or system. Before any adjustments are made to eliminate unbalance, it must be determined beyond all doubt that the engine is in proper mechanical condition. When an engine is in good mechanical condition, few, if any, adjustments will be required. However, after an overhaul in which piston rings or cylinder liners have been renewed, considerable adjustment may be necessary. Until the rings become properly seated, some lubricating oil will leak past the rings into the combustion space. This excess oil will burn in the cylinder, giving an incorrect indication of fuel oil combustion. If the fuel pump is set for normal compression and the rings have not seated properly, the engine will become overloaded. As the compression rises to normal pressures, an increase in the power develops, as well as in the pressure and temperature under which the combustion takes place. Therefore, when an overhaul has been completed, the engine instruments must be carefully watched until the rings are seated and any necessary adjustments are made. Frequent compression tests will help in making the necessary adjustments. Unless an engine is equipped so that compression can be readily varied, the engine should be operated under light load until the rings are properly seated.

EFFECT OF UNBALANCE

3-50. In general, the result of unbalance will be overheating of the engine. The clearances established by the engine designer allows for sufficient expansion of the moving parts when the engine is operating at the designed temperatures. An engine operating at temperatures in excess of those for which it was designed is subject to many casualties. Excessive expansion soon leads to seizure and burning of the engine parts. If the temperatures rise above the flashpoint of the lubricating oil vapors in the crankcase, an explosion may occur. High temperature may destroy the oil film between adjacent parts and the resulting increased friction will further increase the temperature. Since power is directly proportional to the mep developed in a cylinder, any increase in mep will cause a corresponding increase in power. If the mep in the individual cylinders varies, power will not be evenly distributed among the cylinders. The quality of combustion obtained depends on the heat content of the fuel. The amount of heat available for power depends on temperature. Temperature varies directly with the pressure. Therefore, a decrease in pressure will result in a decrease in temperature and in poor combustion. The results of poor combustion will be lowered thermal efficiency and reduced engine output. Cylinder load balance is essential for efficient performance of an engine. To avoid the harmful effects of overloading and unbalancing of load, the load on an engine should be properly distributed among the working cylinders. No cylinder, or the engine itself, should ever be overloaded. In general, load balance in an engine can be maintained if the following procedures are observed:

- Maintain the engine in proper mechanical condition.
- Adjust the fuel system according to the manufacturer's instructions.
- Operate the engine within the temperature limits specified in appropriate instructions.
- Keep cylinder temperatures and pressures as evenly distributed as possible.
- Train yourself to recognize the symptoms of serious engine conditions.

ENGINE EFFICIENCY

3-51. Engine efficiency is the amount of power developed compared to the energy input, which is measured by the heating value of the fuel consumed. In other words, the term "efficiency" is used to designate the relationship between the result obtained and the effort expended to produce the result. The term "compression ratio" is frequently used in connection with engine performance and the various types of efficiencies. Remember that compression ratio is the ratio of the volume of air above the piston when it is at the BDC position to the volume of air above the piston when it is at the TDC position. The principal efficiencies in the internal-combustion process are cycle, thermal, mechanical, and volumetric.

CYCLE EFFICIENCY

3-52. The efficiency of any cycle is equal to the output divided by the input. The efficiency of the Diesel cycle is considerably higher than the Otto or Constant-Volume cycle because of the higher compression ratio and because combustion starts at a higher temperature. In other words, the heat input is at a higher average temperature. Theoretically, the gasoline engine using the Otto cycle could be more efficient than the diesel engine if equivalent compression ratios could be used. However, in practice, engines operating on the Otto cycle cannot use a compression ratio comparable to those of diesel engines. This is because the fuel and air are drawn into the cylinder together and compressed. If comparable compression ratios were used, the fuel would fire or detonate before the piston reached the correct firing position. Since temperature and amount of heat content available for power are proportional to each other, the cycle efficiency is actually computed from measurements of temperature. The specific heat of the mixture in the cylinder is either known or assumed. When combined with the temperature, the heat can be calculated at any instant.

THERMAL EFFICIENCY

3-53. Thermal efficiency may be regarded as a measure of the efficiency and completeness of combustion of the fuel. More specifically, it is the ratio of the output or work done by the working substance in the cylinder in a given time to the input or heat energy of the fuel supplied during the same time. Generally, two kinds of thermal efficiency are considered for an engine: indicated thermal and overall thermal.

Indicated Thermal Efficiency

3-54. The work done by the gases in the cylinder is called indicated work. The thermal efficiency determined by its use is therefore often called indicated thermal efficiency (ite). If all the potential heat in the fuel could be delivered as work, the thermal efficiency would be 100 percent. Because of the various losses, this percentage is not possible in actual installations. If the amount of fuel injected is known, the total heat content of the injected fuel can be determined from the heating value or Btu per pound of the fuel. The thermal efficiencies for the engine can then be calculated from the mechanical equivalent of heat (778 ft.-lb. = 1 Btu and 2,545 Btu = 1 hp-hr.) and the number of ft.-lb. of work contained in the fuel can be computed. If the amount of fuel injected is measured over a given time, the rate at which the heat is put into the engine can be converted into potential power. Therefore, if the ihp developed by the engine is calculated as previously explained, the ite can be computed as:

$$\text{Indicated thermal efficiency (ite)} = \frac{\text{indicated hp} \times 2{,}545 \text{ Btu per hr. per hp}}{\text{rate of heat input of fuel in Btu per hr.}} \times 100$$

Example: Assume that the same engine used as an example in computing ihp consumes 360 lb. (approximately 50 gallons) of fuel per hour, and the fuel has a value of 19,200 Btu per pound. What is the ite of the engine?

The work done per hour when 1,343 ihp are developed is 1,343 × 2,545 or 3,417,935 Btu. The heat input for the same time is 360 × 19,200, or 6,912,000 Btu. Then, by equation, the ite is—

$$\text{ite} = \frac{1{,}343 \times 2{,}545}{360 \times 19{,}200} \times 100$$

$$= \frac{3{,}417{,}935}{6{,}912{,}000} \times 100 = 49.4 \text{ percent}$$

Overall Thermal Efficiency

3-55. The other type of thermal efficiency (overall thermal efficiency) for an engine is determined by a ratio similar to ite, except that the useful or shaft work (bhp) is used. Therefore, overall efficiency (often called brake thermal efficiency) is computed as:

$$\text{overall thermal efficiency} = \frac{\text{brake horsepower}}{\text{heat input of fuel}} \times 100$$

Converting these factors into the same units (Btu), the equation is written as power output in Btu divided by fuel input in Btu.

Example: If the engine used in the previous example delivers 900 bhp (determined by the manufacturer), what is the overall thermal efficiency of the engine?

1 hp-hr. = 2,545 Btu

900 bhp × 2,545 Btu per hp-hr. = 2,290,500 Btu output per hr.

Substituting factors already known, overall thermal efficiency is computed as follows:

$$\text{overall thermal efficiency} = \frac{2{,}290{,}500}{6{,}912{,}000} = 0.331, \text{ or } 33.1 \text{ percent}$$

Compression ratio influences the thermal efficiency of an engine. Theoretically, thermal efficiency increases as compression ratio increases. The minimum value of a diesel engine compression ratio is determined by the compression required for starting. This compression, to a large extent, depends on the type of fuel used. The maximum value of the compression ratio is not limited by the fuel used. It is limited by the strength of the engine parts and the allowable engine weight per bhp output.

MECHANICAL EFFICIENCY

3-56. Mechanical efficiency is the rating of how much of the power developed by expansion of gases in the cylinder is actually delivered as useful power. The factor that has the greatest effect on mechanical efficiency is friction within the engine. The friction among moving parts in an engine remains practically constant throughout the engine's speed range. Therefore, the mechanical efficiency of an engine will be highest when the engine is running at the speed at which bhp is developed. Since power output is bhp and the maximum horsepower available is ihp, then:

$$\text{mechanical efficiency} = \frac{\text{brake hp}}{\text{indicated hp}} \times 100$$

During the transmission of ihp through the piston and connecting rod to the crankshaft, the mechanical losses which occur may be due to friction. They may also be due to power absorbed. Friction losses occur because of friction in the various bearings or between piston and piston rings and the cylinder walls. Power is also absorbed by valve and injection mechanisms and by various auxiliaries such as the lubricating oil and water circulating pumps and the scavenge and supercharge blowers. As a result, the power available for doing useful work (bhp) is less than indicated power. Mechanical losses which affect the efficiency of an engine may be called frictional horsepower (fhp) or the difference between ihp and bhp. The fhp of the engine used in the previous examples, then would be 1,343 ihp - 900 bhp = 443 fhp, or 33 percent of the ihp developed in the cylinders. Using the equation for mechanical efficiency, the percentage of power available at the shaft is computed as:

$$\text{mechanical efficiency} = \frac{900}{1{,}343} = 0.67 \text{ or } 67 \text{ percent}$$

When an engine is operating under part load, it has a lower mechanical efficiency than when operating at full load. The reason for this is that most mechanical losses are almost independent of the load. Therefore, when the load decreases, ihp decreases relatively less than bhp. Mechanical efficiency becomes zero when an engine operates at no load because then bhp = 0, but ihp is not zero. In fact, if bhp is zero and the expression for fhp is used, ihp is equal to fhp. To show how mechanical efficiency is lower at part load, assume that the engine used in previous examples is operating at 3/4 load. Brake horsepower at 3/4 load is 900 × 0.75 or 675. Assuming that fhp does not change with load, fhp = 443. The ihp is, by expression, the sum of bhp and fhp.

$$\text{ihp} = 675 + 443 = 1{,}118$$

$$\text{mechanical efficiency} = 675 \div 1{,}118 = 0.60 \text{ or } 60 \text{ percent}$$

This is appreciably lower than the 67 percent indicated for the engine at full load.

VOLUMETRIC EFFICIENCY

3-57. The term "volumetric efficiency" applies to 4-stroke engines. It is an indication of the efficiency of the air intake system. An engine would have 100 percent volumetric efficiency if an amount of air exactly equal to piston displacement could be drawn into the cylinder. This is not possible, except by supercharging. This is because the passages through which the air must flow offer a resistance, the force pushing the air into the cylinder is only atmospheric, and the air absorbs heat during the process. Therefore, volumetric efficiency is determined by measuring (with an orifice or venturi-type meter) the amount of air taken in by the engine, converting the amount to volume, and comparing this volume to the piston displacement.

$$\text{volumetric efficiency} = \frac{\text{volume of air admitted to cylinder}}{\text{volume of air equal to piston displacement}} \times 100$$

The concept of volumetric efficiency does not apply to 2-stroke cycle engines. Instead, the term "scavenge efficiency" is used, which shows how thoroughly the burned gases are removed and the cylinder filled with fresh air. As in the case of a 4-stroke cycle engine, the air supply should be sufficiently cool. Scavenge efficiency depends largely on the arrangement of the exhaust, scavenge air ports, and valves.

ENGINE LOSSES

3-58. As the heat content of a fuel is transformed into useful work during the combustion process, many different losses take place. These losses can be classed as thermodynamic and mechanical. The net useful work delivered by an engine is the result obtained by deducting the total losses from the heat energy input.

Thermodynamic Losses

3-59. Thermodynamic losses are a result of loss to the following:

- Cooling and lubricating systems.
- Surrounding air.
- Exhaust.
- Lack of perfect combustion.

Heat energy losses from the cooling water system and the lubricating oil system are always present. Some heat is conducted through the engine parts and radiated to the atmosphere or picked up by the surrounding air by convection. The effect of these losses varies according to the part of the cycle in which they occur. The heat of the jacket cooling water cannot be taken as a true measure of heat loss since all this heat is not absorbed by the water. Heat is also lost by:

- The jackets during the compression, combustion, and expansion phases of the cycle.
- To the atmosphere during the exhaust stroke.
- Absorbed by the walls of the exhaust passages.

Heat losses to the atmosphere through the exhaust are unavoidable because the engine must be cleared of the hot exhaust gases before the next air intake charge can be made. The heat lost to the exhaust is determined by the temperature within the cylinder where exhaust begins. The amount of fuel injected and the weight of air compressed within cylinder are controlling factors. Improper timing of the exhaust valves, whether early or late, will result in increased heat losses. If early, the valve releases the pressure in the cylinder before all the available work is obtained. If late, the necessary amount of air for complete combustion of the next charge cannot be realized although a. small amount of additional work may be obtained. Proper timing and seating of the valves is essential to keep to a minimum the heat loss to the exhaust. Heat losses due to imperfect or incomplete combustion have a serious effect on the power that can be developed in the cylinder. Because of the short interval necessary for the cycle in modern engines, complete combustion is not possible. But heat losses can be kept to a minimum if the engine is kept in proper adjustment. It is often possible to detect incomplete combustion by watching for abnormal exhaust temperatures and changes in the exhaust color and by being alert for unusual noises in the engine.

Mechanical Losses

3-60. There are several kinds of mechanical losses, but all are not present in every engine. The mechanical or friction losses of an engine include the following:

- Bearing friction.
- Piston and piston ring friction.
- Pumping losses caused by operation of water pumps, lubricating pumps, and scavenging air blower.
- Power required to operate valves.

Friction losses cannot be eliminated but they can be kept to a minimum if the engine is kept in good mechanical condition. Bearings, pistons, and piston rings should be properly installed and fitted, shafts must be in alignment, and lubricating and cooling systems should be at their highest operating efficiency. Remember that the total of these mechanical losses must be deducted from ihp of the engine to determine actual bhp.

SECTION III. PETROLEUM PRODUCTS

Except in an emergency, fuels burned in internal-combustion engines used by the Navy must meet the specifications prescribed by the Naval Sea Systems Command. Therefore, selecting a fuel which has the required properties is not your responsibility. Your primary responsibility is to follow the rules and regulations dealing with the proper use of fuels. Strict adherence to prescribed safety precautions is required. Every possible precaution must be taken to keep fuel free from impurities.

FUEL HANDLING

3-61. Fuels are generally delivered clean and free from impurities. However, the transfer and handling of fuel increases the danger of it becoming contaminated with foreign material which interferes with engine performance. Foreign substances such as sediment and water cause wear, gumming, and corrosion in the fuel system. They also cause an engine to operate erratically with a power loss. For these reasons, periodic inspection, cleaning, and maintenance of fuel handling and filtering equipment are necessary. Even though proper handling and proper use of fuel are your prime responsibility, knowledge of fuels and their characteristics will make problems encountered in engine operation and maintenance more readily understood.

DIESEL ENGINE FUEL OIL

3-62. Fuel normally used in diesel engines is diesel fuel oil. However, others such as JP-5 and Navy distillate fuels have been authorized for use in diesel engines when it would be a logistic advantage. Diesel engines require a fuel which is particularly clean. Otherwise, the closely fitted parts of the injection equipment will wear rapidly and the small passages which create the fuel spray within the cylinders will become clogged. The composition of diesel fuel oil must be such that it can be injected into the cylinders in a fine mist of fog. Ignition qualities must be such that the fuel will ignite properly and burn rapidly when it is injected into cylinders. The self-ignition point of a fuel is a function of temperature, pressure, and time. In a properly operating diesel engine, the intake air is compressed to a high pressure (thereby increasing the temperature), and the injection of fuel starts a few degrees before the piston reaches TDC. The fuel is ignited by the heat of compression shortly after the fuel injection starts and combustion continues throughout the injection period. Combustion is much slower than in a gasoline engine and the rate of pressure rise is relatively small. After injection, the first effect on the fuel is a partial evaporation with a resultant chilling of the air in the immediate vicinity of each fuel particle. However, the extreme heat of compression rapidly heats the fuel particles to the self-ignition point and combustion begins. The fuel particles burn as they mix with the air. The smaller particles burn rapidly. However, the larger particles take more time to ignite because heat must be transferred into them to bring them to the self-ignition point. There is always some delay between the time that is injected and the time that it reaches the self-ignition point. This delay is commonly referred to as "ignition delay," or "lag." The duration of the ignition delay depends on the following factors:

- Characteristics of the fuel.
- Temperature and pressure of the compressed air in the combustion space.
- Average size of the fuel particles.
- Amount of turbulence present in the space.

3-63. As combustion progresses, the temperature and pressure within the space rise. Therefore, the ignition delay of fuel particles injected late in the combustion process is less than in those injected earlier. The delay period between the start of injection and the start of self-ignition is sometimes referred to as the first phase of combustion in a diesel engine. The second phase includes ignition of the fuel injected during the first phase and the spread of the flame through the combustion space injection continues. The resulting increase in temperature and pressure reduces the ignition lag for the fuel particles entering the combustion space during the remainder of the injection period. Remember that only a portion of the fuel has been injected during the first and second phases. As the remainder of the fuel is injected, the third and final phase of combustion takes place. The increase in both temperature and pressure during the second phase and progression into the third phase are sufficient to cause most of the remaining fuel particles to ignite with practically no delay as they come from the injection equipment. The rapid burning during the final phase of combustion causes an additional rapid increase in pressure which is accompanied by a distinct and audible knock. A knock so caused is characteristic of normal diesel operation, particularly at light loads. The knock that occurs during the normal operation of a diesel engine should not be confused with "detonation." Detonation in a diesel engine is generally an instantaneous explosion of a greater than normal quantity of fuel in the cylinder, instead of only a portion of the fuel charge, as in a gasoline engine. Whether combustion is normal or whether detonation occurs is determined by the amount of fuel that ignites at one time. The greater the pressure rise, the more severe the knock. Detonation in a diesel engine is generally caused by too much delay in ignition. The delay can be the result of poor injector timing or cold combustion spaces. The greater the delay, the greater the amount of fuel that accumulates in the cylinder before ignition. When the ignition point of the excess fuel is reached, all of this fuel ignites simultaneously, causing extremely high pressures in the cylinder and an undesirable knock. Therefore, detonation in a diesel engine generally occurs at what is normally considered the start of the second phase of combustion instead of during the final phase, as in a gasoline engine. Detonation in a diesel engine may occur at the following times:

- When the engine is not warmed up sufficiently.
- When fuel injection is timed too late or too early.
- When injection valves permit excessive fuel to accumulate in the cylinder.

3-64. Even though diesel fuel must have the ability to resist detonation, it must ignite spontaneously at the proper time under the pressure and temperature existing in the cylinder. The ease with which a diesel fuel oil will ignite and the manner in which it burns determines the ignition quality of the fuel. This quality is determined by the fuel cetane rating, or cetane value. The cetane value of a fuel is a measure of the ease with which the fuel will ignite. The cetane rating of any given fuel is identified by its cetane number. The higher the cetane number, the less the lag between the time the fuel enters the cylinder and the time it begins to burn. The cetane rating of a diesel fuel is determined in a manner similar to that used to determine the octane value of gasoline. However, the hydrocarbons used for the reference fuel are cetane and alpha-methyl-napthalene. Cetane has an excellent ignition quality (100) and alpha-methyl-napthalene has a very poor ignition quality (0). By comparing the performance of a reference fuel with that of a fuel whose ignition quality is unknown, the unknown cetane rating or number can be determined. The cetane number represents the percentage of pure cetane in a reference fuel which will just match the ignition quality of the fuel being tested. The higher the cetane number, the quicker burning the fuel and the better the fuel from the standpoint of ignition and combustion. More information on fuel injection and engine controls is given in Chapter 8.

ENGINE LUBRICANTS

3-65. Lubrication is as important to successful engine operation as air, fuel, and heat are to combustion. Lubrication is frequently considered one of the most important factors in efficient engine operation. The proper type of lubricant must be used. The lubricant must also be supplied to the engine parts in the proper quantities, at the proper temperature, and any impurities that enter the system must be removed. The contacting surfaces of all moving parts of an engine must be kept free from abrasion and friction and wear must be kept to a minimum. If sliding contact is made by two dry metal surfaces under pressure; excessive friction, heat, and wear will result. Friction, heat, and wear can be greatly reduced if metal-to-metal contact is prevented through the use of a clean film of lubricant between the metal surfaces. The film between the bearing surfaces in naval machinery is provided by either a specified oil or grease.

OILS

3-66. A lubricating oil with the necessary properties and characteristics will provide the following:

- A film of proper thickness between the bearing surfaces under all operating conditions.
- It will remain stable under changing temperature conditions.
- It will prevent corrosion of the metal surfaces.

If the lubricating oil is to meet these requirements, the engine temperature during operation must not be allowed to exceed a specified limit. In addition to preventing metal-to-metal contact, the lubricating oil is required to form a seal between the piston rings and the cylinder wall, to aid in engine cooling, and to help to keep the inside of the engine free of sludge. A direct metal-to-metal moving contact has an action comparable to a filing action. The filing action is due to minute irregularities in the surfaces. The severity of the action depends on the finish of the surfaces and the force with which the surfaces are brought into contact as well as on the relative hardiness of the materials used. Lubricating oil fills the minute cavities in bearing surfaces, thereby preventing high friction losses, rapid wear of engine parts, and many other difficulties. Lack of a proper oil film results in seized or frozen pistons, warped bearings, and stuck piston rings. Lubricating oil assists in cooling the engine by transferring or carrying away heat from localized hot spots in the engine. The principal parts from which oil absorbs heat are the bearings, the journal surfaces, and the pistons. In some engines, the oil carries heat to the sump where the heat dissipates in the mass of oil. However, most modern internal-combustion engines use a centralized pressure-feed lubrication system. This type of system incorporates an oil cooler or heat exchanger where the heat from the oil is transferred to the circulating water of the cooling system.

Causes of Sludge

3-67. Almost any type of gummy or carbonaceous material that accumulates in lubricating oil is called sludge. Most engine lubricating oils have some natural ability for preventing sludge to form and for carrying sludge that does form in a finely suspended state until it is removed by filtering equipment. Chemicals are added to some oils to improve their ability to prevent and to remove sludge. The formation of sludge is greatly reduced if the lubricating oil has the proper stability. Such stability is essential if a strong oil film, or body of oil, is to be maintained under varying temperatures. Stability of the oil should be such that a proper oil film is maintained throughout the entire operating temperature range of engine. Such a film will ensure sufficient oiliness, or film strength, between the piston and the cylinder wall so that partly burned and exhaust gases cannot get by the piston rings to form sludge. Various factors may tend to cause sludge to form in an engine. Carbon from the combustion chambers or from the evaporation of oil on a hot surface (such as on the underside of a piston), will cause the formation of sludge. Gummy, partially burned fuel, which gets past the piston rings, or an emulsion of lubricating oil and water, which may enter the lubricating oil system, will also tend to cause sludge.

Effects of Sludge

3-68. Sludge in the lubricating oil system of an engine is harmful for several reasons. In addition to carbon and gummy material, sludge may contain abrasive ingredients. These may be dust from the atmosphere, rust caused by water condensation in the engine, and metallic particles resulting from wear of engine parts. Sludge in engine lubricating oil causes premature wear of parts and breakdown of the engine. Sludge may clog the oil pump screen or collect at the end of the oil passage leading to a bearing. This prevents sufficient oil from reaching the parts to be lubricated. Sludge will also:

- Coat the inside of the crankcase.
- Act as insulation.
- Blanket the heat inside the engine.
- Raise the oil temperature.
- Induce oxidation.

Sludge will accumulate on the underside of the pistons and prevent proper heat transfer, thereby raising piston temperatures. Sludge in lubricating oil also contributes to piston ring sticking.

Oil Characteristics and Tests

3-69. Lubricants are tested for such characteristics as viscosity, pour point, flashpoint, fire point, autogenous ignition point, neutralization number, demulsibility, and precipitation number. The lubricants must meet the following requirements. They must–

- Have a suitable viscosity at the operating temperature of the bearing being lubricated.
- Form durable boundary films on the metal rubbing surfaces.
- Not chemically attack the journal or the bearing metals.
- Not change chemical composition with use.

GREASES

3-70. Greases are used in certain places where oil will not provide proper lubrication. Operating temperatures, the rate at which lubrication must be supplied, and the design of the equipment, may make the use of oil impractical. At points where oil will not provide proper lubrication, machinery manufacturers have installed fittings (either pressure type or cup type) for applying grease of the proper type and grade. The location of grease fittings and the type of grease required are shown on machinery lubrication charts and in manufacturer's TMs. You must follow lubrication instructions, since some greases are developed for general use and others are developed for special purposes. Maintenance problems involving lubrication may be more readily understood if familiar with the composition and classification of greases.

Chapter 4

RECIPROCATING INTERNAL-COMBUSTION ENGINES

INTRODUCTION

4-1. This chapter discusses the specific functions that occur in a diesel engine. Three basic engine designs are presently used. They are the in-line type, the V-type, and the opposed-piston type. Engine designation and cycles of operation are also covered in this chapter.

IN-LINE DIESEL ENGINE

4-2. The in-line diesel engine is the simplest arrangement with all cylinders parallel and in line as shown in Figure 4-1. This construction is limited in respect to its length. The longer the engine, the more difficult it is to make a sufficiently rigid frame and crankshaft. In-line diesel engines rarely have more than six cylinders. Those engines which have more require special designs which usually add to the weight of the engine.

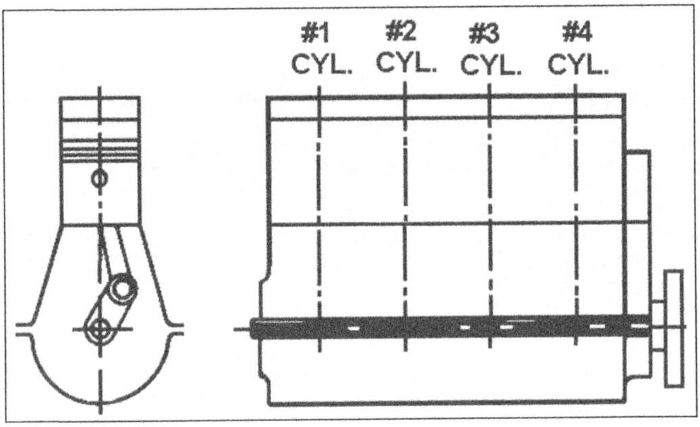

Figure 4-1. In-line Diesel Engine

V-TYPE ENGINE

4-3. The V-type engine, with two connecting rods attached to each crankpin, permits the entire length to be reduced by one-half, thereby making it much more rigid with a stiff crankshaft. This is a common arrangement for engines with 8 to 16 cylinders. Cylinders lying in one plane are called a bank. The angle "a" between the banks (see Figure 4-2) may vary from 30 inches to 120 inches, the most common angle being between 40 inches and 75 inches.

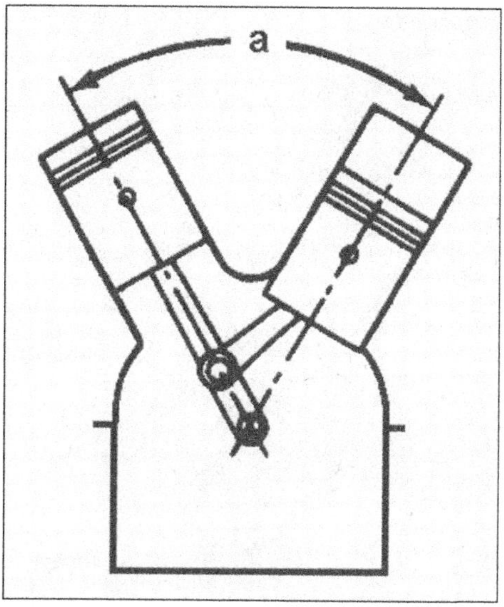

Figure 4-2. V-type Diesel Engine

OPPOSED-PISTON ENGINE

4-4. Engines of the opposed-piston type (see Figure 4-3) are not to be confused with engines of the "flat" or 180° V-type. Flat engines have two rows of cylinders in a horizontal plane with one crankshaft. The crankshaft, located between the rows, serves both rows of cylinders. This type engine is single-acting and is sometimes referred to as a horizontal-opposed engine (the term is based on cylinder arrangement). With respect to combustion-gas action, the term opposed-piston identifies those engines which have two pistons and one combustion space in each cylinder. The pistons are arranged in "opposed" positions; that is, crown to crown, with the combustion space in between. When combustion takes place, the gases act against the crowns of both pistons, driving them in opposite directions. Therefore, the term "opposed" not only signifies that with respect to pressure and piston surfaces the gases act in "opposite" directions but also classifies the piston arrangement within the cylinder. In modern engines that have the opposed-piston arrangement, two crankshafts (upper and lower) are required for transmission of power. Both shafts contribute to the power output of the engine. In most opposed-piston engines common to Navy service, the crankshafts are connected by a vertical drive.

ENGINE DESIGNATION

4-5. Diesel engines are designated by the number of cylinders, bore, stroke, and speed (if the engine operates at a definite speed) in the order named. Therefore, a 6-cylinder × 3 3/4-inch × 5-inch × 1,500-rpm (often written simply as 6 × 3 3/4 × 5 × 1,500) designation means an engine with 6 cylinders, a 3 3/4-inch bore, and a 5-inch stroke, which is normally operated at 1,500 rpm. Another designation is by piston displacement or, for short, by displacement. To find the engine displacement, multiply the piston area times the piston stroke times the number of cylinders. Engine displacement is not concerned with engine speed. In English units, displacement is expressed in cubic inches. Sometimes the displacement is indicated as the number of cylinders and the displacement of only one cylinder.

Example: Find the piston displacement of an 8 × 8 × 10 × 720 naval engine.

Piston area:

$$\pi r^2 = 3.14159 \times \left(\frac{8}{2}\right)^2 = 50.27 \text{ sq. in.}$$

Stroke: 10 in.

Displacement of one cylinder: 50.27 × 10 = 502.7 cu. in.

Number of cylinders: 8

therefore—

Total displacement = 50.27 × 10 × 8 = 4,021.6 cu. in., or the engine may be designated as 8-503 to avoid decimal fractions in the designation.

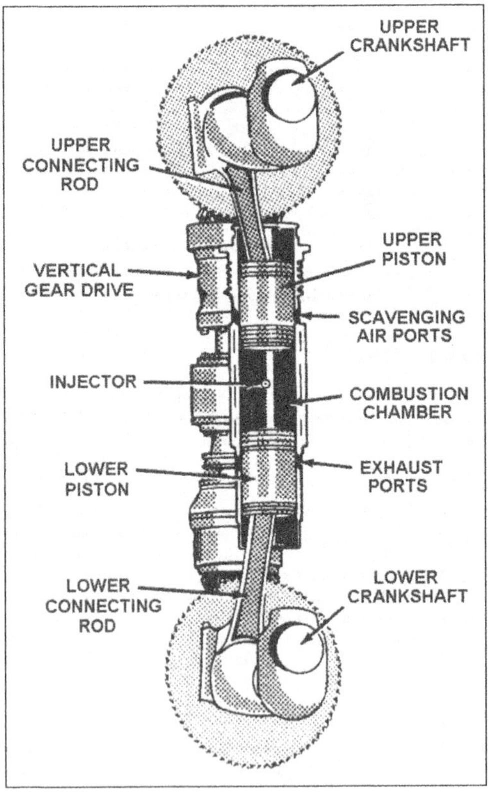

Figure 4-3. Opposed-piston Diesel Engine

CYCLES OF OPERATION

4-6. The operation of an engine involves the admission of fuel and air into a combustion space, and the compression and ignition of the charge. The resulting combustion releases gases and increases the temperature within the space. As temperature increases, pressure increases and forces the piston to move. The piston movement is transmitted to a rotating shaft. The rotary motion of the shaft is used to perform work. Therefore, heat energy is transformed into useful mechanical energy. For the process to be continuous, the expanded gases must be removed from the combustion space, a new charge admitted, and then the process repeated. In the process of engine operation, beginning with the admission of air and fuel and following through to the removal of the expanded gases, a series of events or phases take place. The term "cycle" identifies the sequence of events that take place in the cylinder of an engine for each power impulse transmitted to the crankshaft. These events always occur in the same order each time the cycle is repeated. The mechanics of engine operation is sometimes referred to as the Mechanical (or Operating) cycle of an engine. The heat process, which produces the forces that move engine parts, may be referred to as the Combustion cycle. A cycle of each type is included in a cycle of engine operation.

MECHANICAL CYCLES

4-7. The previous paragraphs discuss events taking place in a cycle of engine operation. Little is said about piston strokes except that a complete sequence of events will occur during a cycle regardless of the number of strokes made by the piston. The number of piston strokes occurring during any one series of events is limited to either two or four, depending on the design of the engine. Therefore, we have a 4-stroke cycle and a 2-stroke cycle. These cycles are known as the mechanical cycles of operation. The terms "4-stroke" and "2-stroke" identify a cycle of events. Both types of mechanical cycles are also used in 4- and 2-stroke cycle reciprocating engines. Familiarization with the principal differences in these cycles is important. The relationship of the events and piston strokes occurring in a cycle of operation involves some of these differences.

Relationship of Events and Strokes in a Cycle

4-8. Remember that a piston stroke is the movement of a piston between its limits of travel. We will discuss the cycles of engine operation. An engine operating on a 4-stroke cycle involves four piston strokes (intake, compression, power, and exhaust). An engine operating on a 2-stroke cycle involves two piston strokes (power and compression). The strokes are named to correspond to the events which occur in particular strokes. However, six events are listed for diesel engines (there are five events for gasoline engines). It is therefore evident that more than one event takes place during some of the strokes, especially in the 2-stroke cycle. Therefore, it is common practice to identify some of the events as strokes of the piston. This is because such events as intake, compression, power, and exhaust in a 4-stroke cycle involve at least a major portion of a stroke and in some cases more than one stroke. The same is true of power and compression events and strokes in a 2-stroke cycle. Such relationship between events and strokes ignores other events which are also taking place during the cycle of operation. This feature of the event-stroke relationship sometimes leads to confusion when studying the operation of an engine or dealing with maintenance problems involving the timing of fuel injection systems.

4-stroke Cycle

4-9. To point out the relation between events and strokes, this discussion covers the events which occur during a specific stroke, the duration of an event with respect to a piston stroke, and the cases where one event overlaps another. The relationship of events and strokes can be shown best by a graphic presentation of the changing situation occurring in a cylinder during a cycle of operation. Figure 4-4 illustrates these changes for a stroke cycle diesel engine. The relationship of events and strokes is more readily understood if the movements of a piston and its crankshaft are considered first. Looking at A in the figure, the travel of a piston during two piston strokes is shown along with the rotary motion of the crank. The positions of the piston and crank at the start and end of a stroke are marked "top" and "bottom," respectively. If these positions and movements are marked on a circle (B in figure), the piston position at the top of a stroke is located at the top of the circle. When the piston is at the bottom of a stroke, the piston position is located at the bottom center of the circle. Top center and bottom center are two terms encountered frequently when discussing the timing of fuel injection systems. Note in both A and B that top center and bottom center identify points where the piston changes its direction of motion. In other words, when the piston is at top center, upward motion has stopped and downward motion is ready to start or, with respect to motion, the piston is "dead." The points which designate changes in direction of motion for a piston and crank are frequently called top dead center (TDC) and bottom dead center (BDC). If the circle illustrated in C is broken at various points and "spread out," the events of a cycle and their relationship to the strokes can be shown, including how some of the events of the cycle overlap. TDC and BDC should be kept in mind since they identify the start and end of a stroke and they are the points from which the start and end of events are established. By following the strokes and events as illustrated, note that the intake event starts before TDC, or before the actual down stroke (intake) starts, and continues on past BDC, or beyond the end of the stroke. The compression event starts when the intake event ends, but the upstroke (compression) has been in process since BDC. The injection and ignition events overlap with the latter part of the compression event, which ends at TDC. The fuel continues to burn until a few degrees past TDC. The power event or expansion of gases ends several degrees before the down (power) stroke ends at BDC. The exhaust event starts when the power event ends and continues through the complete upstroke (exhaust) and past TDC. Note that the exhaust event overlaps with the intake event of the next cycle. The details on why certain events overlap and why some events are shorter or longer with respect to strokes are given later in this manual. From the previous discussion, you should see why the term "stroke" is sometimes used to identify an event which occurs in a cycle of operation. Remember, that a stroke involves 180° of crankshaft rotation (or piston movement between dead centers) while the corresponding event may take place during a greater or lesser number of degrees of shaft rotation (see Figure 4-4).

2-stroke Cycle

4-10. The relationship of events to strokes in a 2-stroke cycle diesel engine is shown in Figure 4-5. Comparison of the two illustrations shows a number of differences between the two types of mechanical or operating cycles. These differences are not too difficult to understand if you remember that four piston strokes and 720° (180° per stroke) of crankshaft rotation are involved in the 4-stroke cycle, while only half as many strokes and degrees are involved in a 2-stroke cycle. The five cross-sectional illustrations shown (see Figure 4-5) help to associate the event with the relative position of the piston. Even though the two piston strokes are frequently referred to as power and compression, we shall refer to them as the "downstroke" (TDC to BDC) and "upstroke" (BDC to TDC) to avoid confusion when reference is made to an event. Starting with the admission of air (1) in the circle (see Figure 4-5), we find that the piston is in the lower half of the downstroke and that the exhaust event (6) is in process. The exhaust event (6) started a number of degrees before scavenging. Both exhaust and scavenging start several degrees before the piston reaches BDC. These events overlap so that the incoming air (1) can aid in clearing the cylinder of exhaust gases. Note that the exhaust event stops a few degrees before the intake event stops, but several degrees after upstroke of the piston has started. The exhaust event in some 2-stroke cycle diesel engines ends a few degrees after the intake event ends. When the scavenging event ends, the cylinder is charged with the air which is to be compressed. The compression event (2) takes place during the major portion of the upstroke. The injection event and ignition (3) and (4) occur during the latter part of the upstroke. The point at which injection ends varies with engines. In some engines, it ends before TDC; in others, a few degrees after TDC. The intense heat (approximately 1,000° F) generated during the compression of the air ignites the fuel/air mixture and the pressure resulting from combustion forces the piston down. The expansion (5) of the gases continues through a major portion of the downstroke. After the force of the gases has been expended, the exhaust valve opens (6) and permits the burned gases to enter the exhaust manifold. As the piston moves downward, the intake ports are uncovered (1), and the incoming air clears the cylinder of the remaining exhaust gases and fills the cylinder with a fresh air charge (1). The cycle of operation will now start again. Therefore, from the standpoint of the mechanics of operation, the principal difference between the 2- and 4-stroke cycles is the number of piston strokes taking place during the cycle of events. A more significant difference is that a 2-stroke cycle engine delivers twice as many power impulses to the crankshaft for every 720° of shaft rotation.

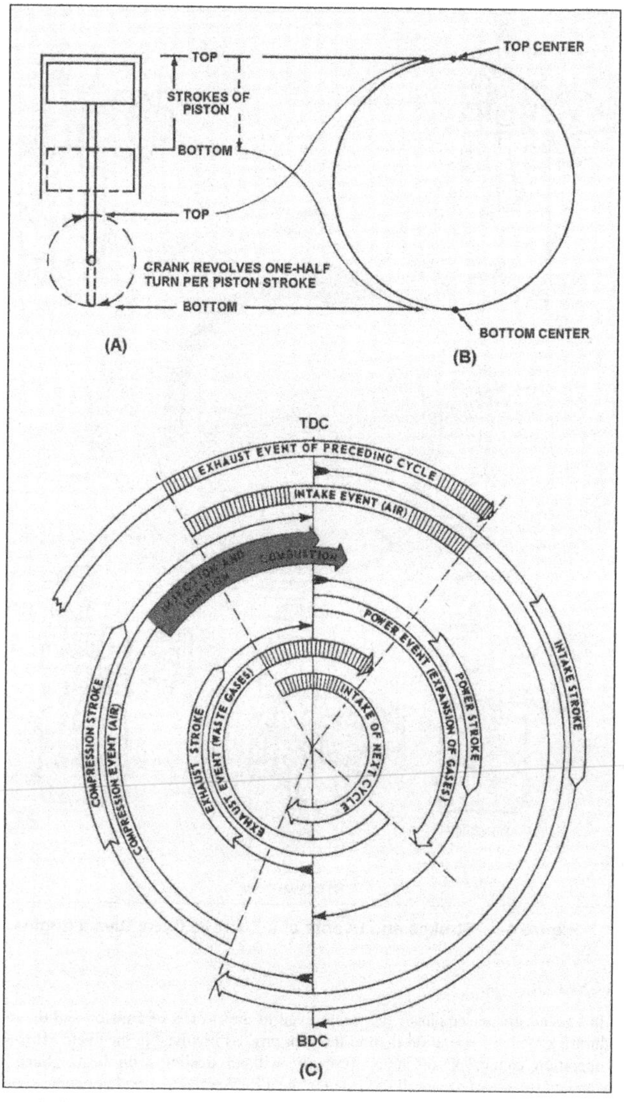

Figure 4-4. Relationship of Events and Strokes in a 4-stroke Cycle Diesel Engine

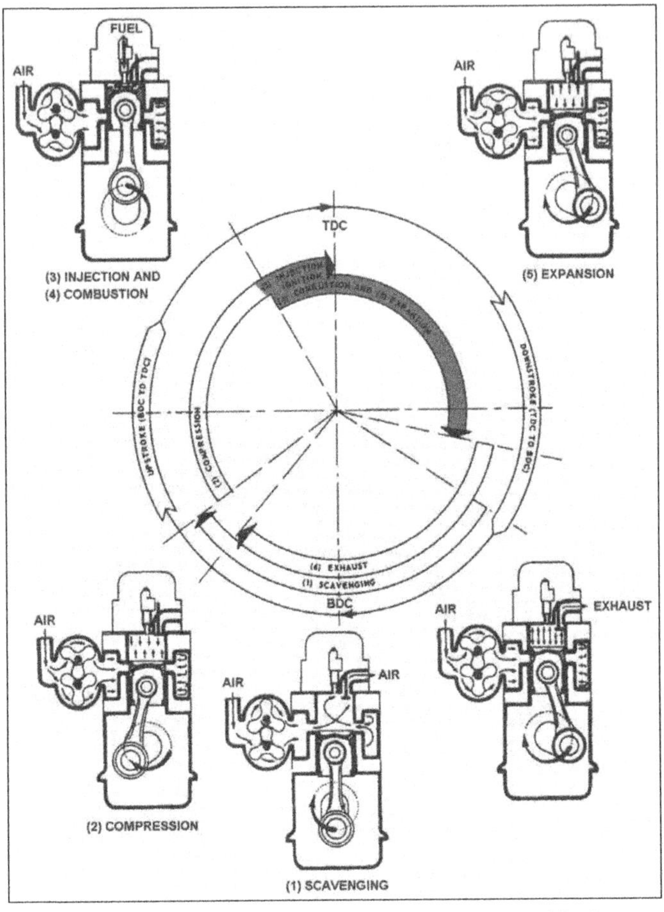

Figure 4-5. Strokes and Events of a 2-stroke Cycle Diesel Engine

COMBUSTION CYCLES

4-11. To this point, greater emphasis has been given to the strokes of a piston and the related events which take place during a cycle of operation than to the heat process involved in the cycle. However, the mechanics of engine operation cannot be discussed properly without dealing with heat. Such terms as ignition, combustion, and expansion of gases, all indicate that heat is essential to a cycle of engine operation.

Relationship of Temperature, Pressure, and Volume

4-12. To illustrate the relationship of temperature, pressure, and volume in an engine; consider what takes place in a cylinder fitted with a reciprocating piston (see Figure 4-6). Instruments indicate the pressure within the cylinder as well as the temperature inside and outside the cylinder. Assume that the air in the cylinder is at atmospheric pressure and that the temperatures inside and outside the cylinder are approximately 70° (as shown in A in Figure 4-6). If the cylinder is an airtight container and a force pushes the piston toward the top of the cylinder, the entrapped charge will be compressed. As the compression progresses, the volume of the air decreases, the pressure increases, and the temperature rises (as shown in B and C of Figure 4-6). These changing conditions continue as the piston moves. As the piston nears TDC (see D in Figure 4-6), there is a marked decrease in volume, and both pressure and temperature are much greater than at the beginning of compression. Note that temperature has gone from 70° F to approximately 1,000° F. These changing conditions indicate that mechanical energy, in the form of force applied to the piston, have been transformed into heat energy in the compressed air. The temperature of the air has been raised sufficiently to cause ignition of fuel when the fuel is injected into the cylinder. Further changes take place after ignition. Since ignition occurs shortly before TDC, there is little change in volume until the piston passes TDC. However, there is a sharp increase in pressure and temperature shortly after ignition takes place. The increased pressure forces the piston downward. As the piston moves downward, the gases expand, or increase in volume, and pressure and temperature decrease rapidly. The changes in volume, pressure, and temperature (as described and illustrated) represent the changing conditions in the cylinder of a modern diesel engine. The changes in volume and pressure in an engine cylinder can be illustrated by diagram. Such a diagram is made by devices which measure and record the pressures at various piston positions during a cycle of engine operation. Diagrams which show the relationship between pressures and corresponding piston positions are called pressure-volume diagrams (see Figure 4-7) or indicator cards. On diagrams which provide a graphic representation of cylinder pressure, as related to volume, the vertical line P on the diagram represents pressure and the horizontal line V represents volume. When a diagram is used as an indicator card, the pressure line is marked off in units of pressure and the volume line is marked off in inches. Therefore, the volume line shows the length of the piston stroke which is proportional to volume. The distance between adjacent letters on each of the diagrams represents an event of a combustion cycle; that is, compression of air, burning of the charge, expansion of gas, and removal of gases.

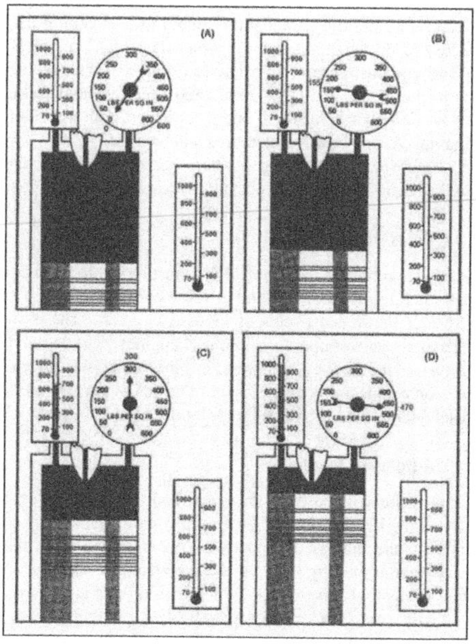

Figure 4-6. Volume, Temperature, and Pressure Relationship in a Cylinder

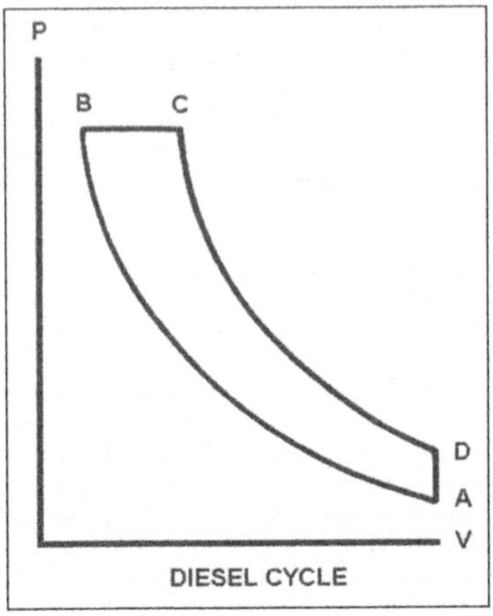

Figure 4-7. Pressure-volume Diagram for Theoretical Combustion Cycles

True Diesel (Constant-pressure) Cycle

4-13. The true diesel cycle may be defined as one in which combustion, induced by compression ignition, theoretically occurs at a constant pressure. Adiabatic (without loss or gain of heat) compression (see line AB, Figure 4-7) of the air increases its temperature to a point when ignition occurs automatically as the fuel is injected. Fuel injection and combustion are so controlled that constant-pressure combustion (line CD) and constant volume rejection of the gases (line DA) are assured. In the true diesel cycle, the mixture of fuel and compressed air burns in a process that is relatively slow compared with the quick, explosive-type combustion process of the gasoline engine. As injected fuel penetrates the compressed air, some ignites, and the rest of the fuel charge feeds the fire. The expansion of the gases keeps pace with the change in volume caused by piston travel. Therefore, combustion is said to occur at constant pressure (see line BC, Figure 4-7).

Actual Diesel Combustion Cycles

4-14. The previous discussion covers the theoretical combustion cycle for modern diesel engines. In actual operation, modern engines operate on modifications of the theoretical cycle. However, characteristics of the true cycles are incorporated in the actual cycles of modern engines. These are discussed in the following examples representing the actual cycle of operation in diesel engines. The actual diesel combustion cycle is one in which the combustion phase, induced by compression ignition, begins on a constant-volume basis, where pressure increases and ends on a constant-pressure basis. The actual cycle is used as the basis for the design of practically all modern diesel engines and is referred to as modified diesel cycle, or semi-diesel cycle.

Modified 4-stroke Diesel Combustion Cycle

4-15. An example of a pressure-volume diagram for a modified 4-stroke cycle diesel engine is shown in Figure 4-8. Note that the volume line is divided into 16 units, indicating a 15:1 compression ratio. The higher compression ratio accounts for the increased temperature necessary to ignite the charge. Fuel is injected at point "c" and combustion is represented by line "cd". Combustion in the actual diesel cycle takes place with volume practically constant for a short time, during which period there is a sharp increase in pressure, until the piston reaches a point slightly past TDC. Then, combustion continues at a relatively constant pressure, dropping slightly as combustion ends at d.

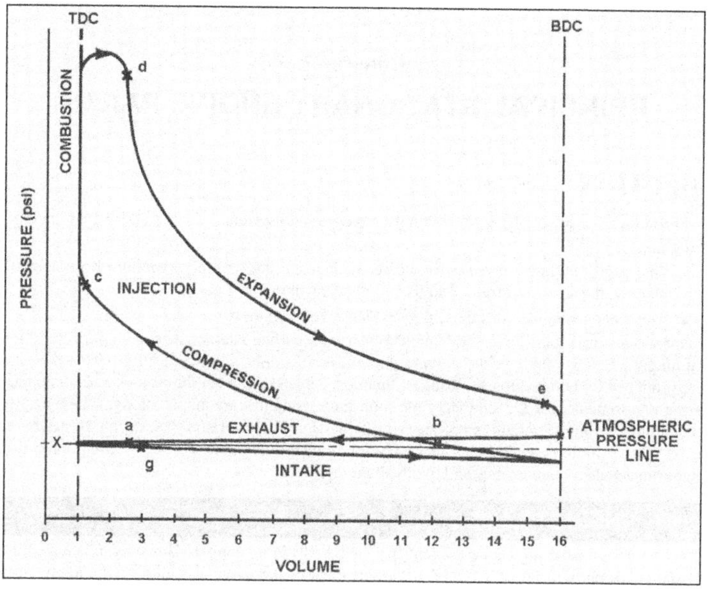

Figure 4-8. Pressure-volume Diagram, Diesel 4-stroke Cycle

Modified 2-stroke Diesel Combustion Cycle

4-16. Pressure-volume diagrams for diesel engines that operate on the 2-stroke cycle would be similar to those just discussed. The exception is that separate exhaust and intake curves will not exist. This is because intake and exhaust occur during a relatively short interval near BDC and do not involve full strokes of the piston as in the 4-stroke cycle. Therefore, a pressure-volume diagram for a 2-stroke modified diesel cycle will be similar to a diagram formed by f-b-c-d-e-f (see Figure 4-8). The exhaust and intake phases will take place between e and b with some overlap of the events. Remember, some of the main differences among diesel engines involve engine designs, designations, and principles.

Chapter 5

PRINCIPAL STATIONARY ENGINE PARTS

INTRODUCTION

5-1. The main parts of an engine, excluding accessories and systems, may be divided into the following two principal groups.

- One group includes those parts which, with respect to engine operation, do not involve motion; namely, the structural frame and its components and related parts.
- The other group includes those parts which involve motion.

This chapter deals principally with the main stationary parts of an engine. The main purpose of the stationary parts of an engine is to maintain the moving parts in their proper relative position. This is to allow the gas pressure produced by combustion to fulfill its function, which is to push the pistons and rotate the crankshaft. The prime requirements for the stationary parts of marine engines are ample strength, low weight, minimum size, and simple design. Strength is necessary if the parts are to withstand the extreme forces developed in an engine. Space limitations aboard ship make minimum weight and size essential. Simplicity of design is of great importance when maintenance and overhaul are involved.

SECTION I. ENGINE FRAMES

The term "frame" is sometimes used to identify a single part of an engine. However, it is also used to identify several stationary parts fastened together to support most of the moving engine parts and engine accessories. In this chapter, the latter definition is used. Designs of modern engine frames differ somewhat from earlier designs. Some of the earlier frames types were referred to as the following:

- A-frame.
- Crankcase.
- Trestle.
- Staybolt or tie-rod.

These old frames were named according to their shape or the manner in which the parts were fastened together. Many of the features common to early engine frames have been incorporated in frames of more recent design. As the load-carrying part of the engine, the frame of the modern engine may include such parts as the cylinder block, crankcase, bedplate or base, sump or oil pan, and end plates. Engines also have access openings and covers and bearings.

CYLINDER BLOCK

5-2. The part of the engine frame which supports the engine's cylinder liners and head or heads is generally referred to as the cylinder block. The blocks for most large engines are of welded steel construction. In this type construction, the block is of welded steel with plates located at the places where loads occur. Deck plates are generally fashioned to house and hold the cylinder liners. The uprights and other members are welded with the deck plates into one rigid unit. Blocks of small, high-speed engines may be of the en-bloc construction, in which the block is one piece of cast iron. A cylinder block may contain passages to allow circulation of cooling water around the liners. If the liner is constructed with integral cooling passages, the cylinder block generally does not have cooling passages. Most 2-stroke cycle engines have air passages in the block. In other words, a passage which is an integral part of an engine block may serve as a part of the engine's cooling system, lubricating system, or air system. One cylinder block is generally thought of in connection with all cylinders of an engine. However, some engines have one cylinder block for each cylinder or for each pair of cylinders. Engines that have V-type cylinder arrangements may have a separate block for each bank of cylinders. The cylinder block shown in Figure 5-1 is somewhat larger than the one just described. It is constructed of welded steel forgings and steel plate. This type of block is secured to a separate engine base. When the two parts are bolted together, they form the frame for the main bearings which carry the crankshaft. Note that the camshaft bearing supports, consisting of forged transverse members, are an integral part of the block. Pads are welded to the block and are machined to carry engine parts and accessories. The block shown has no water passages because each cylinder liner and cylinder head has its own water jacket. The block discussed so far is from an engine with in-line cylinder arrangement. The block illustrated in Figure 5-2 represents blocks constructed for some engines with the V-type cylinder arrangement. Blocks of this type are usually constructed of forgings and steel plates welded together. In the V-type construction, the upper and lower deck plates of each side of the V are bored to the cylinder liners. The space between the decks and the space between the two banks form the scavenging air chamber or air box. In some V-type blocks, the liner bore in the lower deck plate is made with a groove that serves as a cooling water inlet for the liner. Some V-type blocks are constructed with the mounting pads for the main bearing seats as integral parts of the forged transverse members at the bottom of the block. In some blocks, the lower bearing seats for the camshaft are located in a pocket which is an integral part of the block. Note the camshaft pocket in Figure 5-2.

CRANKCASE

5-3. The engine frame part, which serves as a housing for the crankshaft is commonly called the crankcase. In some engines, the crankcase is an integral part of the cylinder block (see Figure 5-1). It requires an oil pan, sump, or base to complete the housing. In others, the crankcase is a separate part and is bolted to the block. This crankcase, an all-steel welded structure, incorporates the bolting flanges, the main bearing saddles, and an oil trough. This type crankcase is sometimes called the main engine base.

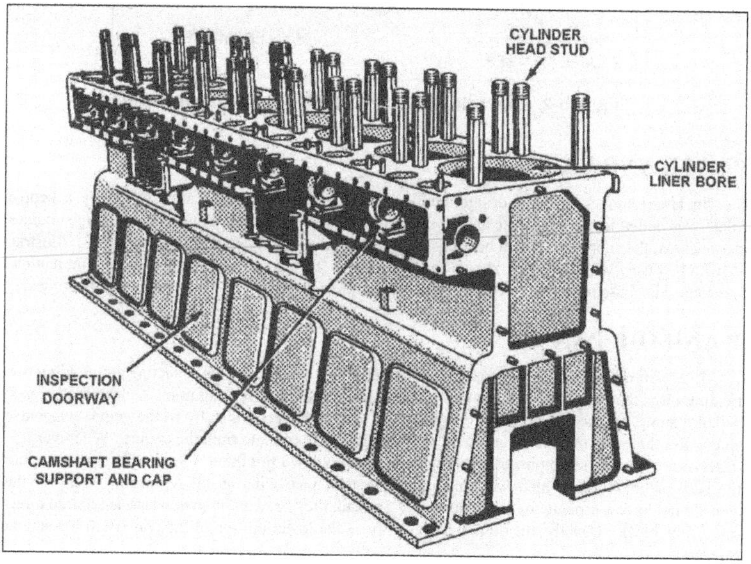

Figure 5-1. Cylinder Block

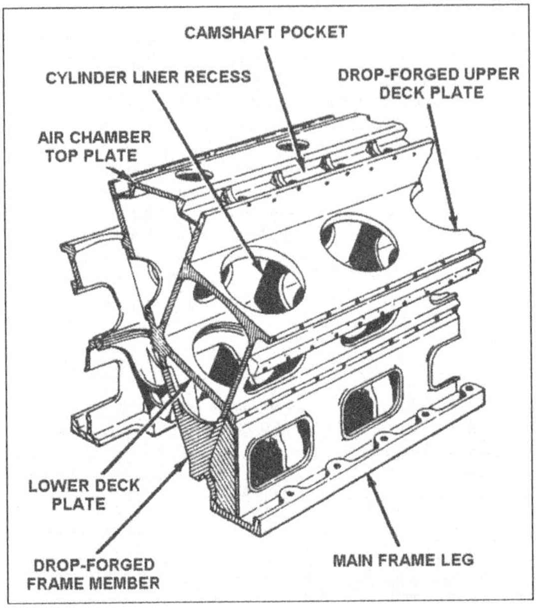

Figure 5-2. Example of V-type Cylinder Block Construction

BEDPLATE AND BASE

5-4. In large engines of early design, support for the main bearings was provided by a bedplate. The bedplate was bolted to the crankcase and an oil pan was bolted to the bedplate. In some large engines of more modern design, the support for main bearings is provided by a part called the base. Figure 5-3 illustrates such a base (which is used with the block shown in Figure 5-1). This type of base serves as a combination bedplate and oil pan. This base requires the engine block to complete the frame for the main engine bearings.

SUMP AND OIL PAN

5-5. Lubrication is essential for proper engine operation. A reservoir for collecting and holding the engine's lubricating oil is therefore a necessary part of the engine structure. Depending on its design, the reservoir may be called a sump, an oil trough, or an oil pan. It is usually attached directly to the engine. However, in dry sump engines the sump may be located at some point relatively remote from the engine. Wherever it is located, the reservoir serves the same purpose. In the engine base shown in Figure 5-3, the oil sump is an integral part of the base or crankcase, which has functions other than just being an oil reservoir. Many of the smaller engines do not have a separate base or crankcase. Instead, they have an oil pan, which is secured directly to the bottom of the block. Usually, the oil pan serves only as the lower portion of the crankshaft housing and as the oil reservoir.

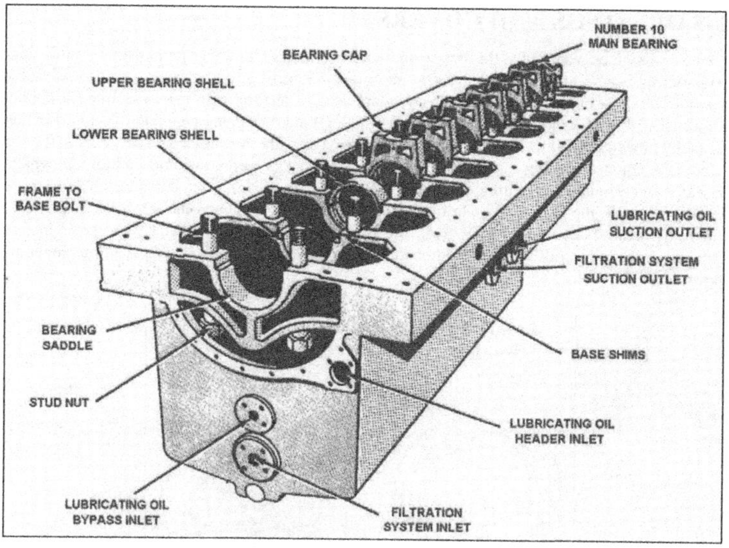

Figure 5-3. Engine Base

END PLATES

5-6. Some engines have flat steel plates attached to each end of the cylinder block. End plates add rigidity to the block. They also provide a surface to which housings may be bolted for such parts as gears, blowers, pumps, and generators. An end plate and gasket for the block in Figure 5-1 are shown in Figure 5-4.

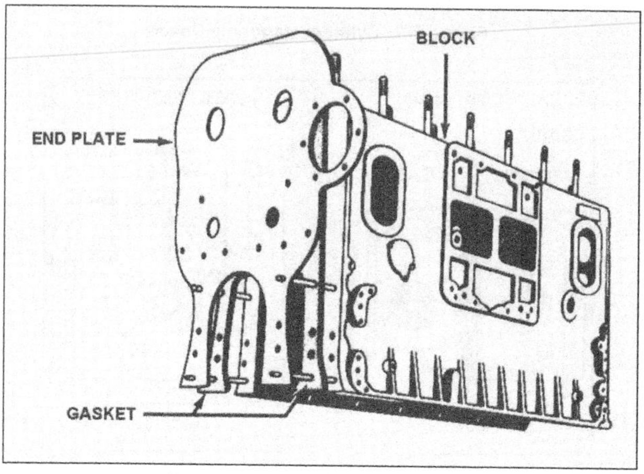

Figure 5-4. Front End Plate and Block

ACCESS OPENINGS AND COVERS

5-7. Many engines, especially the larger ones, have openings in some part of the engine frame. These openings permit access to the cylinder liners, main and connecting rod bearings, injector control shafts, and various other internal engine parts. Access doors are usually secured with handwheel or nut-operated clamps and are fitted with gaskets to keep dirt and foreign material out of the engine's interior. On some engines, the covers (sometimes called doors or plates) to access openings are constructed to serve as safety devices. A safety cover is equipped with a spring-loaded pressure plate. The spring maintains a pressure which keeps the cover sealed under normal operating conditions. In the event of a crankcase explosion or extreme pressure within the crankcase, the excess pressure overcomes the spring tension and the safety cover permits the access opening to act as an escape vent. The release of excess pressure prevents damage to the engine. The access opening and cover for one of the blocks discussed earlier is shown in Figure 5-5. A safety cover and its parts are shown in Figure 5-6.

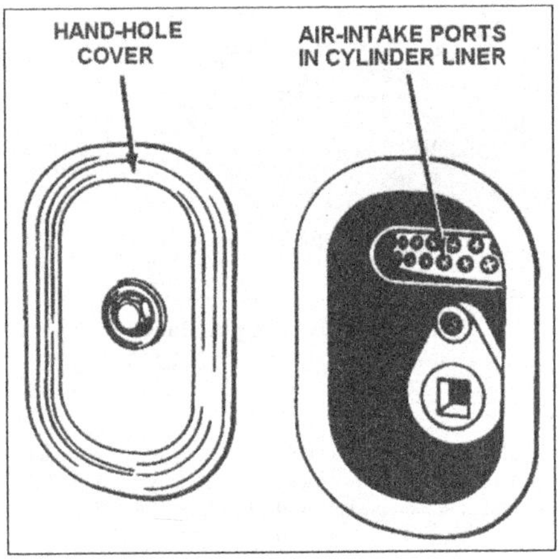

Figure 5-5. Cylinder Handhole Cover

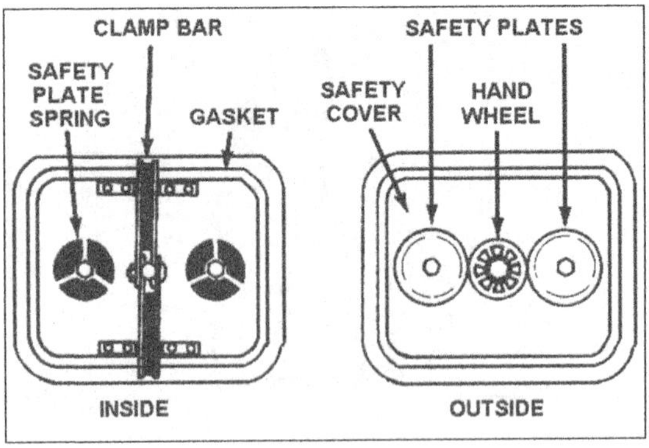

Figure 5-6. Safety-type Crankcase Handhole Cover

BEARINGS

5-8. The bearings of an engine are an important group of parts. Some bearings remain stationary in performing their function while others move. One principal group of stationary bearings, generally called main bearings, supports the crankshaft. Additional information on these and other bearings is given in later chapters in connection with related moving parts.

SECTION II. CYLINDER ASSEMBLIES

One of the main stationary parts, the cylinder assembly (see Figure 5-7), completes the structural framework of an engine. The cylinder assembly, along with various related working parts, serves to confine and release the gases. For the purpose of this discussion, we shall consider the cylinder assembly to consist of the liner, the head, the studs, and the gasket. The other engine parts, many of which involve motion, are covered later in this manual. The design of the parts of the cylinder assembly varies considerably from one type of engine to another. However, regardless of differences in design, the basic components of all cylinder assemblies function, along with related moving parts, to provide a gas- and liquid-tight space. Differences other than in design will be found in cylinder assemblies. For example, a gasket is necessary between the head and block of most cylinder assemblies. However, such gaskets are not used on all engines. When a gasket is not a part of the assembly, the mating surfaces of the head and block are accurately machined to form a seal between the two parts. Other differences in cylinder assemblies exist, some of which are pointed out in the discussion that follows.

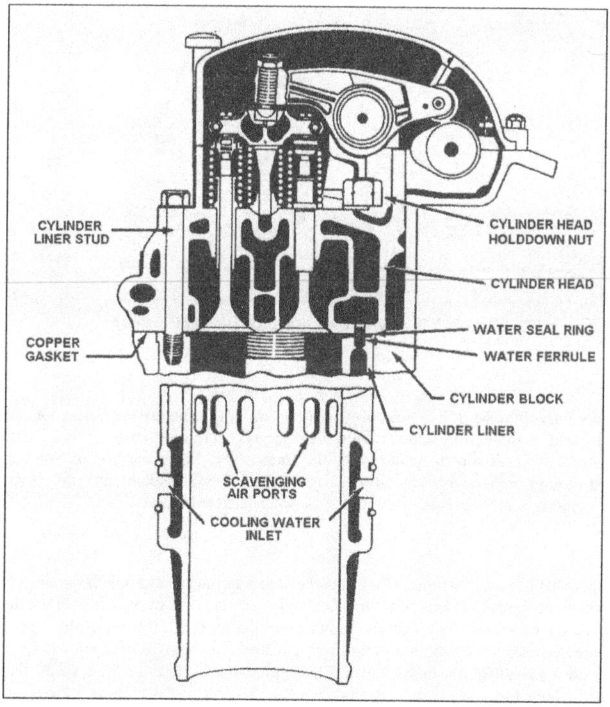

Figure 5-7. Principal Stationary Parts of a Cylinder Assembly

CYLINDER LINERS

5-9. The barrel or bore in which an engine piston moves back and forth may be an integral part of the cylinder block or it may be a separate sleeve or liner. The former, common in gasoline engines, has the disadvantage of not being replaceable. When excessive wear occurs in a barrel of this type, the barrel must be rebored and honed. Reconditioning cannot be repeated indefinitely; in time, the entire block must be replaced. Another disadvantage is the inconvenience, especially in large engines, of removing the entire cylinder block from a ship to recondition the cylinders. For these reasons, practically all diesel engines are constructed with replaceable cylinder liners. The material of a liner must withstand the extreme heat and pressure developed within the cylinder. At the same time, the material of a liner must permit the piston and rings to move with a minimum of friction. Close-grained cast iron is the material most commonly used for liner construction. However, steel is sometimes used. Some liners are plated on the wearing surface with porous chromium. Chromium has greater wear-resistant qualities than other materials used. The pores in the chromium plating also tend to hold the lubricating oil. This helps to maintain the lubrication film necessary to reduce friction and wear. Five replaceable cylinder liners are shown in Figure 5-8. These liners show some of the differences in the design of liners and the relative size of the engines represented.

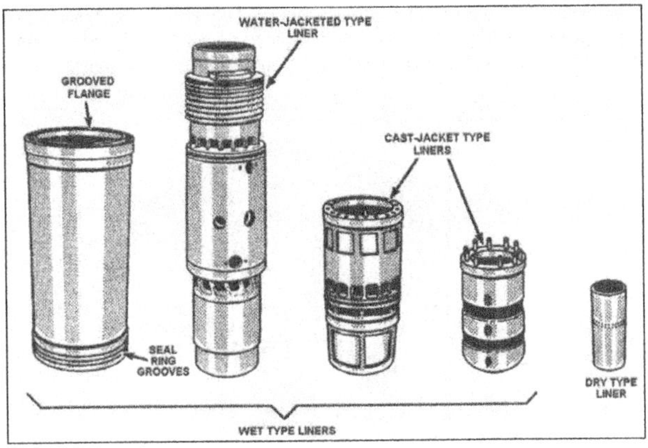

Figure 5-8. Cylinder Liners of Diesel Engines

TYPES OF LINERS

5-10. Cylinder liners may be of two general types (dry and wet). The dry liner does not come in contact with the coolant. Instead, it fits closely against the wall of the cooling jacket in the cylinder block. With the wet liner, the coolant comes in direct contact with the liner. Wet liners may be of the type which requires circumferential sealing devices, or they may be of the type which contains integral passages. Liners with integral cooling passages are sometimes referred to as water-jacketed liners.

Dry Liners

5-11. Dry liners have relatively thin walls compared with wet liners. The smaller liner in Figure 5-8 is a dry type. The coolant circulates in passages in the block and does not come in contact with the liner. Dry liners are installed in some engines with a press fit and in others with a loose fit. Manufacturers recommend that when replacements are necessary, liners with a press fit be replaced with those having a loose fit. In such cases, the liners for both press and loose fits are identical. Therefore, when replacements are made, the cylinder bores in the block must be honed to a larger diameter to permit the loose fit. All liners in a block must have the same type fit. Liners with a press fit require special tools for installation and removal. In small engines, liners with a loose fit can usually be removed by hand after the liner has been loosened.

Wet Liners

5-12. In wet liners, which do not have integral cooling passages, the water jacket formed by the liner and a separate jacket fits within the block or frame. A seal must be provided at both the combustion and crankshaft ends of the cylinders to prevent leakage. Generally, the seal at the combustion end of a liner consists of either a gasket under a flange or a machined fit. Rubber or neoprene rings generally form the seal at the crankshaft end of the liner. Liners of this type are constructed to permit lengthwise expansion and contraction. The walls are also strong enough to withstand the full working pressure of the combustion gases. In Figure 5-8, the liner with the largest diameter is an example of this type of wet liner. Note the grooved flange at the top and the seal ring grooves at the lower end. The groove in the flange and the tongue of the cylinder head make a metal-to-metal joint and seal. The joint between the flange and the cooling jacket is sealed with a nonhardening sealing compound. A cross section of a wet liner is shown in Figure 5-9.

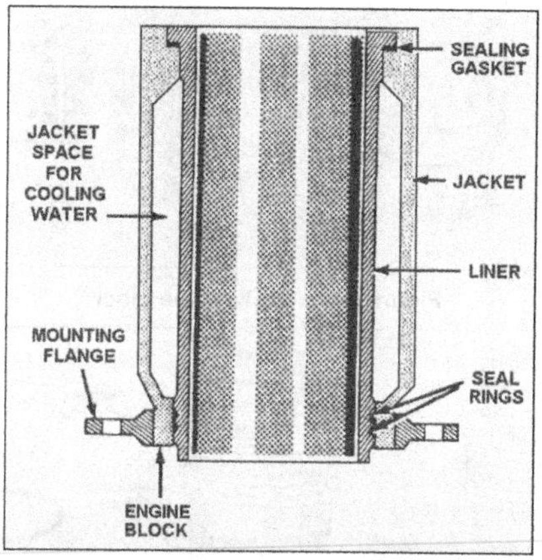

Figure 5-9. Cross Section of a Wet Liner

Water Jacket Liners

5-13. This type cylinder sleeve has its own coolant jacket as an integral part of the liner assembly. The jacket may be cast on, shrunk on, or sealed on the liner. Water is admitted into the lower section of the jacket and it leaves through the top (see Figure 5-10). The liner shown is that of a 4-stroke cycle engine. Most 2-stroke cycle engines are equipped with jacket-type liners because such liners provide the most effective watertight seal around the ports. Another feature of some liners is the counter-bored area. Such an area is identified in B of Figure 5-10. The counterbore extends down to the top point of travel of the top piston ring. The diameter of the liner in the counterbored area is slightly larger than the diameter in the area of piston ring travel. The counterbore prevents the formation of a ridge or lip on the liner surface at the upper point of ring travel. After extensive engine operation, the liner surface may wear where the rings make contact. If the diameter of the liner is the same throughout its length, the increase in diameter in the ring contact area results in the formation of a ridge at the upper limit of piston ring travel. A counterbore of larger diameter above this point prevents the formation of such a ridge or lip caused by wear of the liner. A lip on the surface of a liner may cause broken piston rings and possibly extensive damage to an engine. The liner in Figure 5-10 and two of the liners in Figure 5-8 have cast-on jackets. In other words, the jacket is cast as an integral part of the liner. In the longest liner shown in Figure 5-8, instead of being cast as an integral part of the liner, the water jacket is sealed on the liner with rubber seal rings. Details of a liner assembly, with a sealed-on jacket, are shown in Figure 5-11. Of the four jacket-type liners discussed and shown in Figure 5-8, Figure 5-10, and Figure 5-11, note that three have ports. In the largest of these three, the location of the ports (see Figure 5-11) eliminates the problems of the seal of the water jacket with respect to the air ports. In the other two liners (see Figure 5-8), there is no problem of leakage since the ports are cast as a part of the assembly. In the cast-jacket liner, the water jacket is formed by the inner and outer walls of the liner. The ports divide the water space into lower and upper spaces. These are connected by vertical passages between the ports.

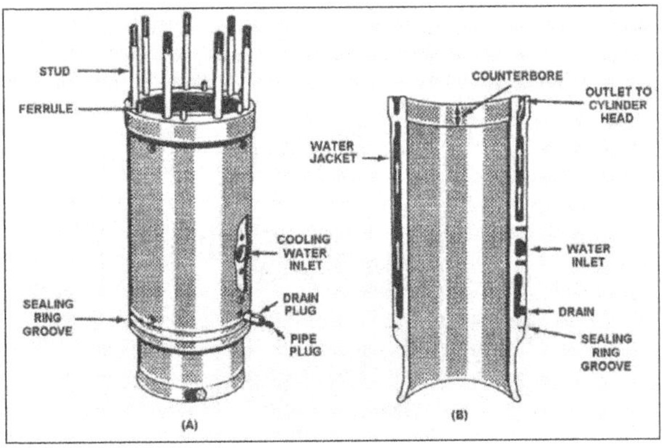

Figure 5-10. Jacket-type Liner

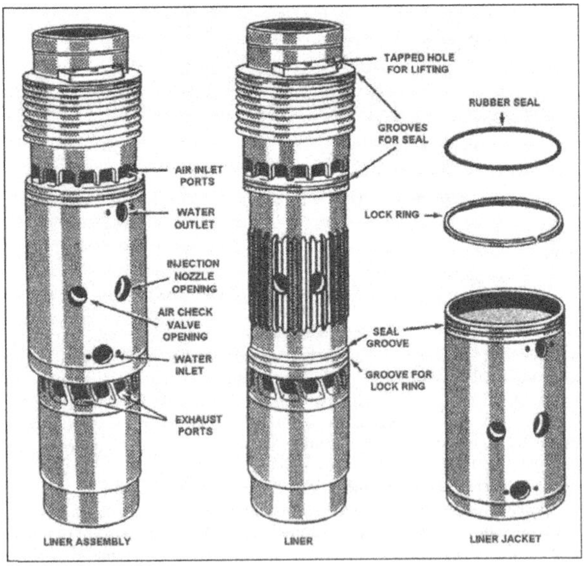

Figure 5-11. Cylinder Liner Assembly With Sealed-on Jacket

REPAIR OF LINERS

5-14. If an inspection or test of a cylinder liner indicates that it must be replaced, the liner must be removed. Specific instructions, including a list of tools required, will be found in the manufacturer's technical manual (TM) for the engine. Cylinder liner removal is a major operation and may mean laying up the engine for one or more days. All coolant must be removed from the engine and the cylinder head removed. On large engines, only the particular cylinder head needs to be removed. Usually these steps must be taken before the liner can be inspected. The piston must be removed, which means opening the crankcase to disconnect the connecting rod. Again, on large engines this may be a more simple operation because a handhole or door may be opened instead of disassembling the entire crankcase or sump. A lot of force must be applied to break the liner free from its seat. Usually, a liner puller attachment is fastened to the liner and the necessary lifting force is applied with a chain fall mounted over the engine. A block of wood can also be placed under the liner, and force can be applied to it by turning the crankshaft with the turning gear. This will force the crank throw against the block. If the liner can be repaired, it must be handled carefully to prevent further damage.

Cracked, Broken, and Distorted Liners

5-15. Cracks in liners may be suspected when any one of the following occurs:

- Excessive water in the lubricating oil.
- An accumulation of water in a cylinder of a secured engine.
- An abnormal decline in the water level of the expansion tank.
- Excessive temperature or fluctuating pressure of the cooling water resulting when the combustion gases blow into the cooling passages.
- Gas in the cooling water of an operating.

If cracks cannot be located by visual inspection, other methods must be used. After removal from the engine, liners with integral cooling passages may be checked by plugging the outlets and filling the passage with glycol-type antifreeze. This liquid will leak from even the smallest cracks. Cracks in dry liners may be more difficult to locate because no leakage of water will occur as a result of such cracks. The location of cracks in dry liners may require the use of magnaflux equipment. Cracking of cylinder liners may result from the following:

- Poor cooling.
- Improper fit of piston or pistons.
- Incorrect installation.
- Foreign bodies in the combustion space.
- Erosion and corrosion.

Improper cooling, which generally results from restricted cooling passages, may cause uneven heating of liners. Liner failure then occurs as a result of thermal stress. Uneven heating of liners may also result from the formation of scale on the cooling passage surfaces. Wet liners are subject to scale formation. Proper cooling of dry liners requires clean contact surfaces between the liner and cylinder block. Particles of dirt between these surfaces cause air spaces which are poor conductors of heat. Films of oil or grease on these mating surfaces also offer resistance to the flow of heat. Distortion, wear, or breakage may result if a liner is not properly seated. Causes of improper liner seating may be metal chips, picks, or burrs, or improper fillets. In Figure 5-12 an improper fillet on the cylinder deck proper liner seating. The fillet should be ground down until the lower surface of the flange seats properly on the mating surface of the cylinder deck. An oversize sealing ring may cause improper positioning of the liner. As the sealing ring is over compressed, the rubber loses its elasticity and becomes hard. This may result in distortion of the cylinder. The clearance between the mating surfaces can be checked with feeler gauges. If the manufacturer's TM specifies the distance from the cylinder deck to the upper surface of the liner flange, this dimension can be used to check the seating of the liner. Obstructions in the combustion chamber may be destructive not only to the liner but also to the cylinder head and other parts. Erosion and corrosion may take place in a few isolated spots and weaken a liner enough to cause cracks. Replacement is the only satisfactory means of correcting cracked, broken, or badly distorted cylinder liners.

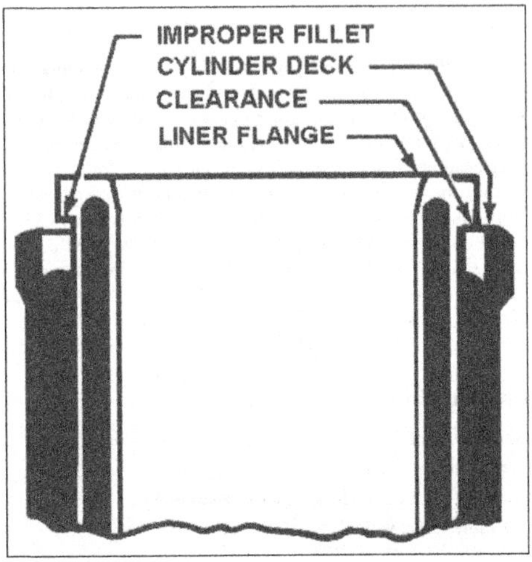

Figure 5-12. Improperly Seated Cylinder Liner

Scored Clinder Lners

5-16. Scoring may be in the form of deep or shallow scratches in the liner surface. With most liner scoring, corresponding scratches will be found on the piston and piston rings. The symptoms of scoring may be low firing or compression pressure and rapid wear of piston rings. Visual inspection is the best method of detecting scoring and may be done through liner ports, through the crankcase housing with piston in top position, or when the engine is disassembled. Scored cylinder liners may be caused by broken piston rings, defective piston, improper cooling, improper lubrication, or the presence of foreign particles. Dust particles drawn into an engine cylinder will mix with the oil and become an effective but undesirable lapping compound that may cause extensive damage. Keeping the intake air clean is of very importance. When an engine is being assembled, be careful that no metal chips, nuts, bolts, screws, or tools remain in the cylinder when the head is replaced. Pistons and rings which are badly worn permit blowby of combustion gases. Not only is operating efficiency reduced, but there is also a greater tendency for scoring. This is because of the increased temperature and because blowby may reduce the oil film until metal-to-metal contact takes place. Inspect the pistons and rings carefully. A piston with a rough surface (such as one that has seized) will cause scoring of the liner. Scoring as a result of insufficient lubrication or dirt in the lubricating oil can be prevented if lubricating equipment (filters, strainers, and centrifuges) is maintained properly. Lube oil must be purified according to required procedures. Repair of scored liners is not generally undertaken by the ship's force. Spare liners are installed. When necessary, liners with minor scoring may be kept in service if the cause of scoring is eliminated and the minor defects can be corrected. The surface of the liner must be inspected carefully, especially in the area near the ports. Look for any burrs, projections, or sharp edges that will interfere with piston and ring travel. Most projections can be removed by hand-stoning, using a fine stone.

Excessively Worn Liners

5-17. The best method of determining whether excessive wear has taken place is to measure the cylinder liner with inside micrometer calipers (see Figure 5-13). Clearance between a piston and a liner is generally checked by micrometer measurements of both parts. On small engines, a feeler gauge can be used. Clearance in excess of that specified by the manufacturer is generally due to liner wear, which normally is greater than piston wear. Measurements for determining liner wear should be taken at the following three levels:

- The first measurement slightly below the highest point to which the top ring travels.
- The second measurement slightly above the lowest point of compression ring travel.
- The third measurement at a point about midway between the first two.

Note: All readings should be recorded so that rapid wear of any particular cylinder liner will be evident to the operator.

If excessive wear or out-of-roundness exists beyond specified limits, the liner should be replaced. Shown are two examples of taking inside measurements. The liner shown in Figure 5-13 is for an opposed-piston engine and requires at least twice as many measurements as other types of liners. Accurate measurements cannot be obtained unless the caliper or gauge is properly positioned in the liner. Common errors in positioning are illustrated in Figure 5-14. One end of the caliper should be held firmly against the liner wall (as shown in A of Figure 5-14). The free end can then be moved back and forth and up and down until the true diameter of the liner is established. Considerable experience in using an inside micrometer or cylinder gauge is necessary to ensure accuracy. As a precaution against error, it is a good practice for two persons to take the liner measurement. Therefore, any discrepancy between the two sets of readings can be rechecked. Excessive or abnormal wear of cylinder liners may be caused by cooling water temperature being too low, insufficient lubrication, dirt, or improper starting procedures. The cooling water of an engine should always be maintained within the specified temperature ranges. If the temperature is allowed to drop too low, corrosive vapors will condense on the liner walls. The lubricating systems must be maintained in proper working order. The method of cylinder liner lubrication varies with different engines. The proper grade of oil, in accordance with engine specifications, should be used. A dirty engine must not be operated. The air box, crankcase, and manifold should be cleaned and kept clean to avoid cylinder wear and scoring. Attention to the air cleaner, oil filters, and oil centrifuge are the best precautions against the entrance of dirt into the engine. Improper starting procedures will cause excessive wear on the liners and pistons. When an engine is first started, some time may elapse before the flow of lubricating oil is complete. Also, the parts are cold and corrosive vapors tend to condense more rapidly. These two factors (lack of lubrication and condensation of corrosive vapors) make the period immediately after starting critical for cylinder liners. If an independently driven oil pump is installed, it must be used to prime the lube oil system and to build up oil pressure before the engine is started. The engine should not be subjected to high load during the warm up period. In fact, it is best to warm up the engine to the operating temperature before applying any load. Cylinder liners worn beyond the maximum allowable limit should be replaced. Maximum allowable wear limits for engines may be found in the appropriate TM. In the absence of such specific information, the following wear limits apply in general:

- Two-stroke cycle engines with aluminum pistons: 0.0025 inch per diameter.
- Slow-speed engines over 18-inch bore: 0.005 inch per inch diameter.
- All other engines: 0.003 inch per inch diameter.

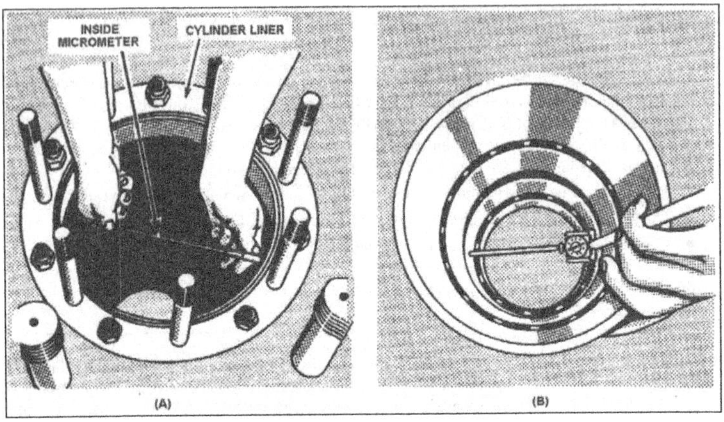

Figure 5-13. Measuring the Inside of a Cylinder Liner

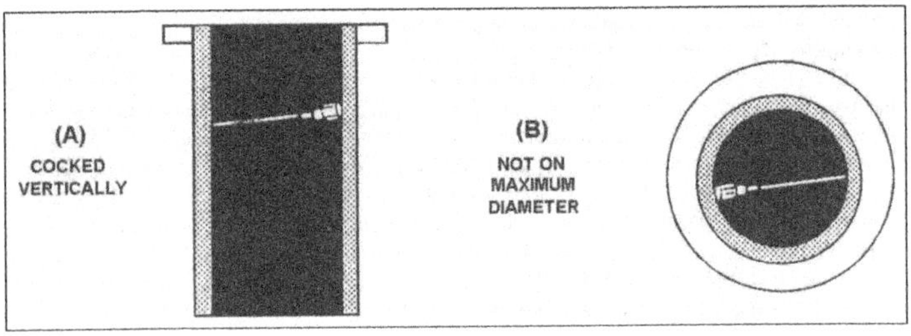

Figure 5-14. Errors to Avoid When Taking Liner Measurements

CYLINDER HEADS

5-18. The liners or bores of an internal-combustion engine must be sealed tightly to form the combustion chambers. In most engines, except for engines of the opposed-piston type, the space at the combustion end of a cylinder is formed and sealed by a cylinder head (see Figure 5-15) which is usually a separate unit from the block or liner. A number of engine parts which are essential to engine operation may be found in, or attached to, the cylinder head. The cylinder head may house intake and exhaust valves, valve guides, valve seats, or only exhaust valves and related parts. Rocker arm assemblies are frequently attached to the cylinder head. In a diesel engine, the fuel injection device is almost universally in the cylinder head or heads, while the spark plugs are always in the cylinder head of gasoline engines. Cylinder heads of a diesel engine may also be fitted with air-starting valves, indicator cocks, and safety valves. The parts which may be attached to or housed in the cylinder head are covered in more detail in later chapters. The design and material of a cylinder head must be such that it can withstand the rapid changes of temperature and pressure taking place in the combustion space. It must also withstand the stress resulting from the head's being bolted securely to the block. Cylinder heads are almost universally made of heat-resisting, alloy cast iron. The number of cylinder heads found on engines varies considerably. Small engines of the in-line cylinder arrangement use one head for all cylinders. A single head serves for all cylinders in each bank of some V-type engines. Large diesel engines generally have one cylinder head for each cylinder. Some engines use one head for each pair of cylinders. A cylinder head of the type used to seal all cylinders of a block is shown in Figure 5-15.

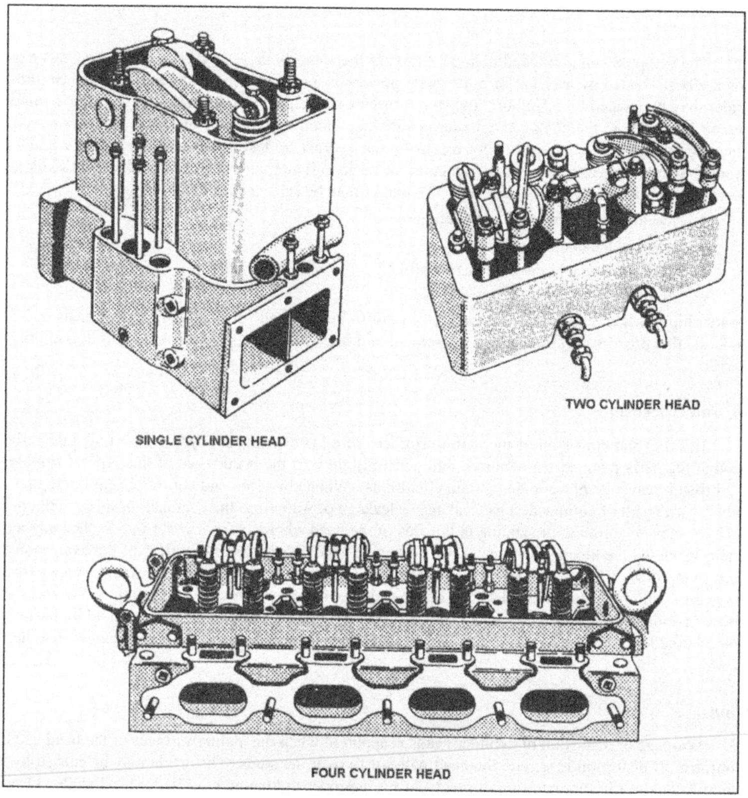

Figure 5-15. Types of Cylinder Heads

Cooling of Cylinder Heads

5-19. Cooling passages are common to most cylinder heads. The coolant enters the head from the cylinder block or liner and cools the head and attached parts. The connection will vary between the cooling passages in block or liners and the head. Such connections may consist of ferrules or similar connections, or of outside jumper lines.

Repair of Cylinder Heads

5-20. Conditions requiring repair of a cylinder head are, in many ways, similar to those encountered in cylinder liners. They can be grouped generally as cracks, corrosion, distortion, and fouling.

Cracks

5-21. The symptoms of a cracked cylinder head are the same as those of a cracked liner. Cracks in cylinder heads are best located by a visual or a magnetic powder inspection. On some types of engines, the defective cylinder may be located by bringing the piston of each cylinder, in turn, to TDC and applying compressed air. When air is applied to the damaged cylinder, a bubbling sound will indicate leakage. When removed from the engine, the cylinder may be checked for cracks by the hydrostatic test that is used on cylinder liners equipped with integral cooling passages. Cracks generally occur in cylinder heads on the narrow metal sections existing between such parts as valves and injectors. The cracks may be caused by the following:

- Addition of cold water to a hot engine
- By restricted cooling passages.
- By obstructions in the combustion space.
- By improper tightening of studs.

Aboard ship, cracked cylinder heads generally have to be replaced. It is possible to repair them by welding. However, this process requires special equipment and highly skilled personnel generally found only at repair activities.

Burning and Corrosion

5-22. Burning and corrosion of the mating surfaces of a cylinder head may be caused by a defective gasket. Although regularly planned maintenance will generally prevent the occurrence of this type of trouble, burning and corrosion may take place under certain conditions. When corrosion and burning occur, there may be a loss of power as a result of combustion gas leakage or leakage of water into the combustion space. Other symptoms of leakage may be a hissing or sizzling in the area of the head where gases or water may be leaking between the cylinder head and the block. Bubbles in the expansion tank sight glass or overflow of the expansion tank may also be symptoms of leakage. Gaskets and grommets which seal combustion spaces and water passages must be in good condition. Otherwise, leakage of the fluids will cause corrosion or burning of the areas contacted. Improper cooling water treatment may also accelerate the rate of corrosion. In general, cylinder heads which become burned or corroded as a result of gas or water leakage are damaged to such an extent that they must be replaced.

Distortion

5-23. Warpage or distortion of cylinder heads is apparent when the mating surfaces of the head and block fail to conform. If distortion is severe, the head will not fit over the studs. Distortion may be caused by improper welding techniques in the repairing of cracks or by improper tightening of cylinder head studs. Occasionally, new heads may be warped because of improper casting or machining processes. Repair of distorted or damaged cylinder heads is often impracticable. They should be replaced as soon as possible and turned in to the nearest supply activity. This activity will determine the extent of damage and the method of repair.

Fouling

5-24. If the combustion spaces become fouled, the efficiency of combustion will decrease. Combustion chambers are designed to create the desired turbulence of mixing the fuel and air. Any accumulation of carbon deposits in the space will impair both turbulence and combustion by altering the shape and decreasing the volume of the combustion chamber. Symptoms of fouling in the combustion spaces are smoky exhaust, loss of power, or high compression. Such symptoms may indicate the existence of extensive carbon formation or clogged passages. In some engines, these symptoms indicate that the shutoff valves for the auxiliary combustion chamber are stuck. Combustion chambers may also become fouled because of the following:

- Faulty injection equipment.
- Improper assembly procedures.
- Excessive lubricating oil getting past the piston.

Cleaning of fouled combustion spaces generally involves removing the carbon accumulation. The best method is to soak the dirty parts in an approved solvent and then wipe off all traces of carbon. Use a scraper to remove carbon, but be extremely careful not to damage the surfaces. If lubricating oil is causing carbon formation, check the wear of the rings, bearings, pistons, and liners. Replace or recondition excessively worn parts. Carbon formation resulting from improperly assembled parts can be avoided by following the procedures described in the manufacturer's TM.

CYLINDER HEAD STUDS AND GASKETS

5-25. In most installations, the seal between the cylinder head and the block depends principally upon the studs and gaskets. The studs, or stud bolts, secure the cylinder head to the cylinder block. A gasket between the head and the block is compressed to form a seal when the head is properly tightened.

STUDS

5-26. A round rod, generally of alloy steel, is used for cylinder head studs. Threads are cut on both ends. Those which screw into the block generally have a much tighter fit than those on the nut end. The tighter fit in the block aids in preventing the stud's unscrewing when the stud nuts are removed. Studs in good condition should normally not be removed from a cylinder block. Sets of cylinder head studs in position in a cylinder block are shown in Figure 5-16. All stud nuts should be tightened equally and in accordance with specifications given in the manufacturer's TM. Overtightening is as undesirable as undertightening. Sometimes studs that are relatively inaccessible are neglected during periodic checks for tightness. Such an oversight may result in the studs coming loose and failing. When installing stud nuts, carefully clean the threads of the studs and the nuts by wire brushing and applying an approved solvent. Cleaning will reduce wear and distortion of threads resulting from dirt as well as increase the accuracy of the torque wrench readings. A higher torque wrench reading will be needed to reach required tension when threads are dirty than when they are clean and well lubricated. Figure 5-16 illustrates a sequence for tightening studs for two types of cylinder heads. This order is not a hard-and-fast rule but can be followed in the absence of more specific information. Studs are generally tightened sufficiently to seat the cylinder head lightly-finger-tight. At least two or three rounds of tightening should be made before bringing all studs up to the specified torque. Cylinder heads should be retorqued after 100 hours of operating time.

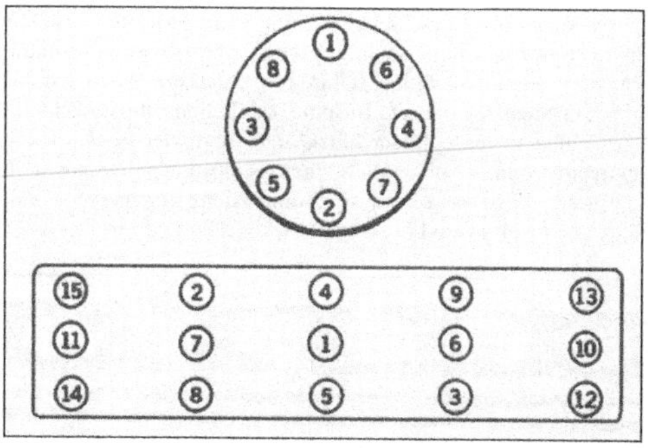

Figure 5-16. Order of Sequence for Tightening Cylinder Head Studs

GASKETS

5-27. Even though gasket design varies, all gaskets have "compressibility" as a common property. The mating surfaces of a cylinder block and head appear to be quite smooth. However, if the mating surfaces are highly magnified, existing irregularities can be seen in the surfaces. Such irregularities, though slight, are sufficient to allow leakage of the combustion gases, oil, or coolant unless compressible material is used between the mating surfaces. The compressible gasket material fills the openings caused by the irregularities.

Figure 5-17 shows two types of gaskets. The material used in the manufacture of gaskets varies as widely as does gasket design. Gaskets can be made from the following:

- Copper and other relatively soft metals.
- Laminated steel sheets.
- Fiber.
- Cork.
- Rubber.
- Synthetic rubber.
- A combination of materials such as copper and asbestos.

5-28. Combinations of gaskets, seal rings, and grommets or similar devices may be used to prevent leakage of oil, water, and gas between a cylinder block and head. Leakage is the chief trouble encountered with gaskets. Leakage becomes apparent when compression pressure becomes low; resulting in starting difficulties, or when fluid escapes between the head and block. Sometimes there may be no external leakage from a defective gasket. The only way to positively determine internal leakage is to remove the cylinder head. Leaky gaskets may be the result of the following:

- A permanently compressed gasket.
- Improper tightening of cylinder head stud nuts.
- Careless installation of the gasket.
- Installation of a damaged or defective gasket.

5-29. After prolonged use, gaskets become permanently compressed and lose their ability to conform to irregularities in the machine surfaces. If leakage occurs past the gasket, the operator may attempt to stop the leakage by tightening the holddown or cylinder head nuts. A gasket in very poor condition may not stop leakage even though the nuts are pulled up to the breaking point. If a gasket is torn or burned across an area that must be sealed, it should be replaced. Gaskets should be removed and replaced in accordance with the engine manufacturer's instructions. Gaskets may also be damaged by tightening the head nuts unevenly or insufficiently. That portion of the gasket below the tight nuts may become pinched or cut, particularly if the mating surfaces or the head and cylinder are not perfect planes. If the nuts are not tightened to the specified torque wrench reading, the gasket may not be compressed sufficiently to cause it to the irregularities in the machined surfaces. This will prevent proper sealing and allow blowby of the combustion gases and burning of the gasket material. This may result in complete failure of the gasket. When installing a gasket over the studs, be careful not to bend, tear, or break the gasket. In some installations, the gaskets do not surround the studs but can be placed in recesses either in the cylinder block or the cylinder head. The gasket must be positioned properly in the recess to avoid cutting or pinching it when the head is pulled down. It is also important that the correct type of gasket be used. A gasket of improper dimensions may prevent proper sealing of the cylinder head on the cylinder block and result in stud breakage as well as burned surfaces.

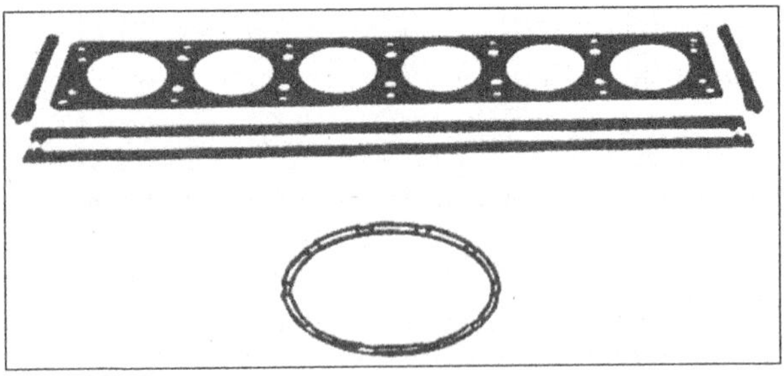

Figure 5-17. Types of Gaskets

SECTION III. ENGINE MOUNTINGS

The devices used to secure an engine in place are not an actual part of the engine. However, these devices are discussed because they are essential for installation purposes and because they serve an important part in reducing the possibility of damage to an engine and the mechanism it drives. Different terms are used to identify the devices that secure an engine to a ship. Such terms as base, subbase, bed, frame, rails, mountings, and securing devices appear in various engine TMs. To avoid confusion with the engine base already discussed in this chapter, the supporting connecting pedestal between an engine and the structure of a ship are referred to in this discussion as a subbase. Also, the devices used to fasten the subbase to the ship are referred to as securing devices.

SUBBASE

5-30. The size and design of the subbase depend on the engine involved and its use. In many installations, the engine and the mechanism which it drives are mounted on a common subbase. One advantage of mounting both units on a common subbase is that misalignment is less likely than when the units are mounted separately. Diesel engine generator sets are usually secured to a common subbase. A different type of mounting involves the use of handfitted chocks or blocks between the engine and the structure of the ship. Bolts are used to secure the engine rigidly in place and maintain alignment. A block mounting is used with the engine base shown in Figure 5-3.

SECURING DEVICES

5-31. The securing devices used to fasten a subbase to the structure of a ship may be generally classed as rigid or flexible. Propulsion engines are secured rigidly to avoid misalignment between the engine, reduction gear (or other driven mechanisms), and propeller shaft. Engines which drive auxiliary equipment may be secured by either rigid or flexible devices. In installations where rigidity is of prime importance, bolts are used as the securing devices. Flexible securing devices are generally between the subbase of a generator set and the ship's structure. Flexible devices may be placed between the engine and the subbase. Although flexible devices are not necessary for every type of generator set, they are desirable for generator sets mounted near the side of the hull to reduce vibration. Flexible devices also aid in preventing damage from shock loads imposed by external forces. Two general types of flexible securing devices are the vibration isolator and the shock absorber. Both types may be incorporated in one device.

VIBRATION ISOLATOR

5-32. The vibration isolator is designed to absorb the forces of relatively minor vibrations common to an operating engine. Such vibrations are referred to as high-frequency, small-amplitude vibrations. They result from an unbalanced condition created by the movement of operating engine parts. Isolators may be equipped with coil springs or flexible pads to absorb the energy of vibration. An isolator reacts in the same manner, whether it is the spring-type or the flexible-pad type. Examples of both types of isolators are shown in Figure 5-18. Four or more spring type isolators, shown in A, are used to support a generator set. The flexible pad or "rubber sandwich" type isolator, shown in B, is used for mounting small engines. The rubber block in the isolator shown is bonded to steel plates which are fitted for attachment to the engine and subbase.

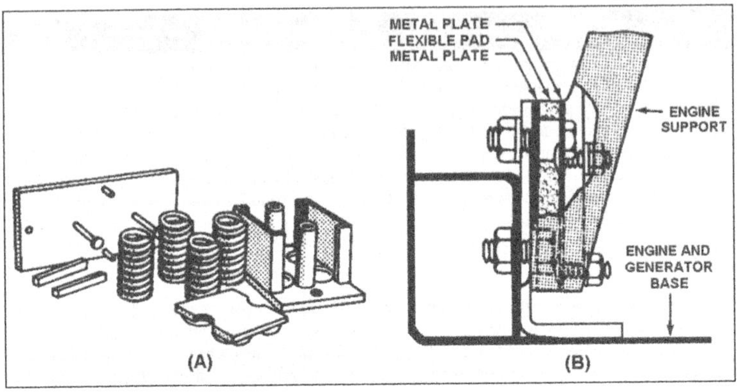

Figure 5-18. Vibration Isolators

SHOCK ABSORBERS

5-33. Shock absorbers are used to absorb vibration forces greater than those originating in the engine. Such forces or shock loads may be induced by the detonation of shells, torpedoes, and bombs. The shock absorber operates on the same principle as the vibration isolator but incorporates an additional device to protect the engine against severe shock loads. Figure 5-19 shows a commonly used type of shock absorber.

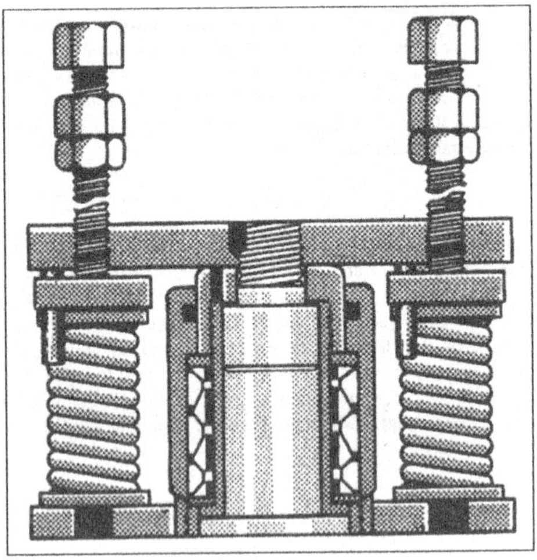

Figure 5-19. Shock Absorber

Chapter 6

PRINCIPAL MOVING AND RELATED PARTS

INTRODUCTION

6-1. Many of the principal parts within the main structure of an engine are moving parts. They convert the power developed by combustion in the cylinder into mechanical energy. This energy is available for useful work at the output shaft. This chapter discusses the moving and related parts that seal and compress gases in the cylinder and transmit the power developed in the cylinder. Additional parts required for the development and transmission of power (such as timing gears and gear trains) are discussed in subsequent chapters of this manual.

SECTION I. TYPES OF VALVES AND VALVE ACTUATING MECHANISMS

INTAKE VALVES

6-2. Intake valves and exhaust valves used in internal-combustion engines are of the poppet type. Poppet valves have heads with conically shaped or beveled edges and a beveled seat. This gives the valves a self-centering action. Intake valves are generally made of low-alloy steels since these valves are not subjected to the corrosive action of the hot exhaust gases. Intake valves generally give longer periods of trouble-free operation than exhaust valves. This is because the operating temperature of the intake valve is lowered by the air which enters the cylinder.

EXHAUST VALVES

6-3. Exhaust valves are usually made of silicon-chromium steel or of steel alloys. A high content of nickel and chromium is usually included in the steel or steel alloy to resist corrosion caused by high-temperature gases. A hard alloy (such as Stellite) is often welded to the seating surface of the valve head and to the tip of the valve stem. The hard alloy increases wearing qualities of the surfaces which make contact with other metal parts. Sodium is used in the exhaust valves of some engines to aid in cooling the valves. The hollow valve stems are partially filled with sodium. At engine operating temperatures, the sodium melts and splashes up and down inside the valve. Sodium offers a very effective means of transferring heat from the hot exhaust valve head through the stem and valve guides to the engine cooling system. Although valve assemblies may differ, depending on engine construction, certain troubles are common to all applicable parts of valve assemblies. These common casualties include:

- Sticking, bent, and burned valves.
- Cracked and loose valve seats.
- Weak and broken valve springs.
- Worn valve keepers and retaining washers.
- Broken valve heads.

STICKING VALVES

6-4. Sticking valves cause the engine to misfire and produce unusual noises at the cam follower, push rod, and rocker arm. Improper lube oil or improper fuel will leave resinous deposits which, if allowed to accumulate, may cause valves to stick. If the resinous formation is slight, it can be removed by applying a half-and-half mixture of kerosene and lube oil to the valve stem and guide. Since any appreciable amounts of mixture settling in the cylinder can cause a serious explosion, caution must be observed when using this mixture. There are a number of approved commercial products that can be used when the gum deposits are slight or when a complete disassembly is impracticable.

BENT VALVES

6-5. A valve that hangs open not only prevents the cylinder's firing, but it is likely to be struck by the piston and bent so that it cannot seat properly. Symptoms of warped or slightly bent valves usually show up as damage to the surface of the valve head. To lessen the possibility of bending or otherwise damaging the cylinder head valves during overhaul, never place a cylinder head directly on a steel deck or grating. Use a protective material such as wood or cardboard and never pry a valve open with a screwdriver or other tool.

BURNED VALVES

6-6. Burned valves are indicated by irregular exhaust gas temperatures and sometimes by an excessive noise. The following are main causes of burned valves:

- Carbon deposits.
- Insufficient tappet clearance.
- Defective valve seats.
- Valve heads that have been excessively reground.

The most frequent cause of burned exhaust valves is small particles of carbon that lodge between the valve head and the valve seat. These particles come from the engine cylinder and head when carbon deposits on those parts become excessive. The particles hold the valve open just enough to allow the combustion gases to pass, or leak, at velocities and temperatures high enough to cause the valve head to erode and burn. The valve seat seldom burns because the jackets surrounding the seat usually provide enough cooling to keep the temperature below the danger point. The valve is generally cooled by several factors, including its contact with the valve seat. When carbon particles prohibit contact, the heat normally transferred from the valve head to the seat remains in the valve head. When cleaning carbon from cylinder heads, be sure that ail loose particles are removed from the crevices. Be extremely careful to prevent nicking or scratching of the valve or seat. It will be easier to clean the passages and to remove the carbon deposits from the underside of the valve heads if the valves are removed from the engine. Tappet clearance adjustments should be checked at frequent intervals for accuracy and for ensuring that the locking devices are secure. Adjustment of valve clearances is discussed later in this chapter.

CRACKED VALVE SEATS

6-7. Most diesel engines are equipped with valve seat inserts made of hard, heat-resisting, alloyed steel. A seat will occasionally crack and allow the hot gases to leak, burning both the insert and the valve. Sometimes a poor contact between the valve seat insert and the counterbore will prevent the heat from being conducted away. This results in high temperatures which deform the insert. When this occurs, both the seat and the valve will burn, and the seat insert must be replaced.

LOOSE VALVE SEATS

6-8. Loose valve seats can be avoided only by proper installation. The counterbore must be thoroughly cleaned to remove all carbon before an insert is shrunk in. The valve seat is chilled with dry ice and the cylinder head is placed in boiling water for approximately 30 minutes. Then the insert is driven into the counterbore with a large drift (see Figure 6-1). Never strike a valve seat directly. The driving operation must be done quickly, before the insert reaches the temperature of the cylinder head. When replacing a damaged valve with a new one, inspect the valve guides for excessive wear. If there is a side-to-side movement of the valve as it seats, the guides must be replaced. If the valve seat is secured firmly in the counterbore and is free of cracks and burns, slight damage such as pitting may be removed by hand-grinding (see Figure 6-2). The valve and valve seat are generally checked by using Prussian blue. If this is not available, any thin dark oil paint can be used. Allow the valve to seat by dropping it on the valve seat from a short distance. If the surfaces fail to make complete contact, regrinding is necessary. In any valve reconditioning job, the valve seat must be concentric with the valve guide. Concentricity can be determined with a dial indicator (see Figure 6-3). Hand-grinding methods should be held to a minimum. This should never be used instead of machine-grinding when a grinding stone is used to refinish the seat (see Figure 6-4). In the latter case, the stone is placed on the pilot (see Figure 6-3), and the motor is engaged. The motor is run a few seconds at a time and the seat is checked after every cut until it is free of pits. The objection to hand-grinding the valve to the seat is that a groove or indentation may be formed in the valve face. Since the grinding is done with the valve cold, the position of the groove with respect to the seat is displaced when the engine is running. This is because the temperature of the valve head becomes greater than that of the valve seat. Such a condition is shown (though greatly exaggerated) in Figure 6-5. Note that when "hot," the ground surface of the valve does not make any contact at all with the ground surface of the seat. Therefore, hand-grinding should be used only to remove slight pitting or as the final and finishing operation in a valve reconditioning job. Some valves and seats are not pitted enough to require replacement but are so pitted that hand-grinding is unsatisfactory. Such valves may be on a lathe (see Figure 6-6) and the valve seats may be reseated by power-grinding equipment (see Figure 6-4). Normally, these operations are done at general support maintenance facilities. Valve heads which are reground excessively (to the extent that the edge is sharp or almost sharp) will soon burn. A sharp edge cannot conduct the heat away fast enough to prevent burning. This factor limits the extent to which a valve may be refaced.

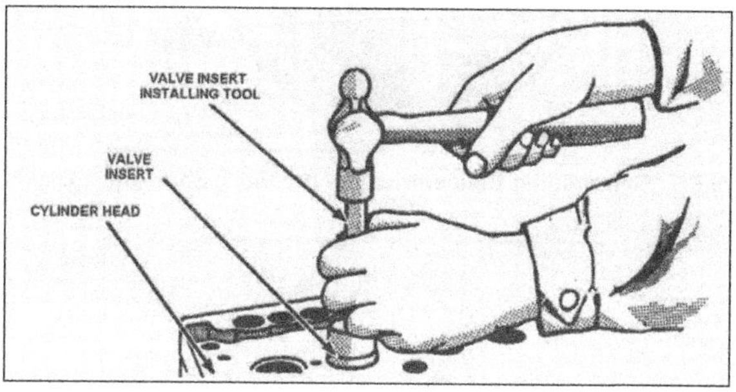

Figure 6-1. Driving a Valve Insert Into the Cylinder Head Counterbore

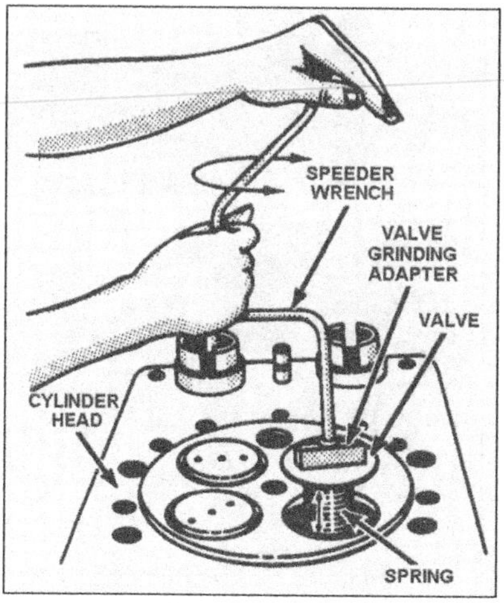

Figure 6-2. Hand-grinding a Valve and Valve Seat

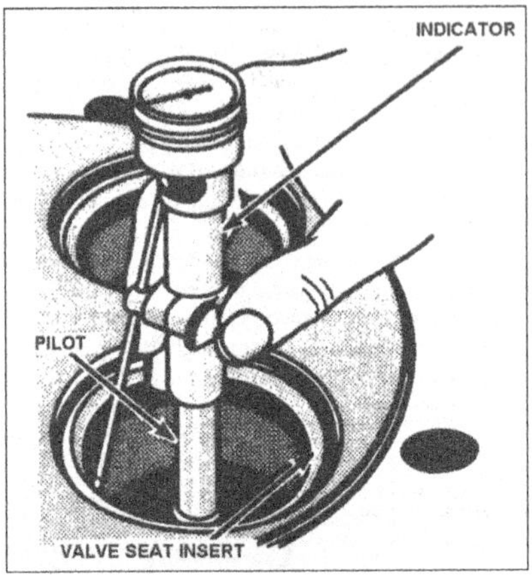

Figure 6-3. Determining Concentricity of the Valve Seat With a Dial Indicator

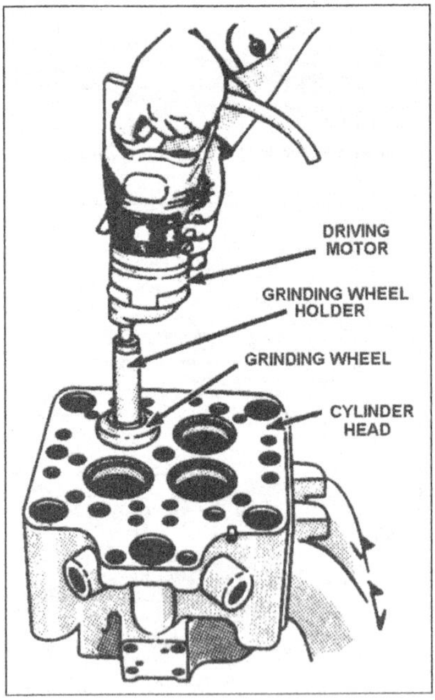

Figure 6-4. Machine-grinding a Valve Seat

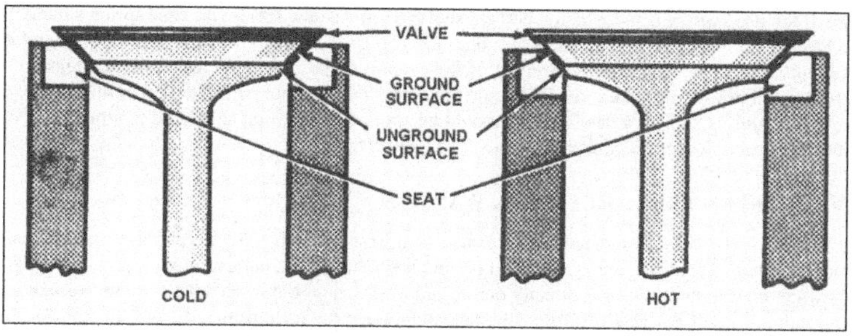

Figure 6-5. Excessively Hand-ground Valves

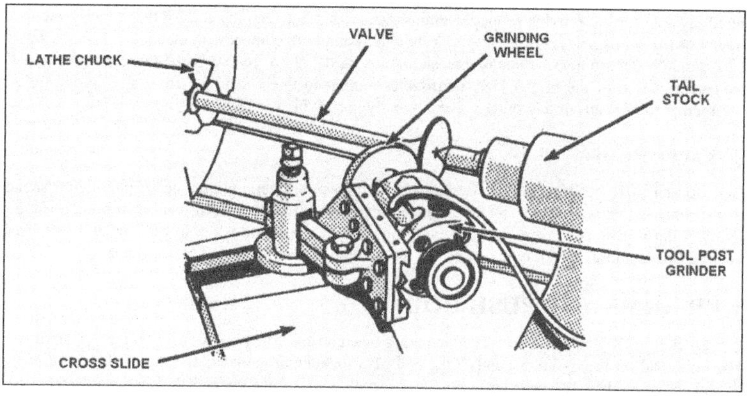

Figure 6-6. Facing a Valve On a Lathe

WEAK VALVE SPRINGS

6-9. Valves may close slowly or fail to close completely because of weak springs. At high speeds, valves may "float," therefore, reducing engine efficiency. Normal wear of valve springs is accelerated by excessive temperatures and by corrosion resulting from a combination of moisture and the sulfur present in the fuel.

BROKEN VALVE SPRINGS

6-10. Broken valve springs cause excessive valve noise and may cause erratic exhaust gas temperatures. Actual breakage of the valve springs is not always the most serious consequence. When a spring breaks, it may collapse sufficiently to allow the valve to drop into the cylinder. There it may be struck by the piston and extensive damage may occur. In addition, the valve stem locks or keepers may release the valve and allow it to drop into the cylinder. This causes severe damage to the piston, cylinder head, and other adjacent parts. A number of precautionary measures can prevent or minimize valve spring breakage (corrosion and metal fatigue). Be reasonably careful when assembling and disassembling. Before reassembling, thoroughly clean and inspect the valve spring. Kerosene or diesel fuel should be used for cleaning-never an alkaline solution because it will remove the protective coating. The best indication of impending valve spring failure is the condition of the surfaces. The use of magnaflux will greatly help in finding cracks which would otherwise be invisible. The free length of a spring should be within the limits specified in the manufacturer's technical manual (TM). If such information is not available, compare the length of a new spring with that of the used spring. If the used spring is more than 3 percent shorter than the new spring, the used spring should be replaced immediately. It must be noted, however, that loss of spring tension will not always be reflected in a proportional loss of overall length. Springs of the proper length may have lost sufficient tension to warrant replacement. Springs with nicks, cracks, or surface corrosion must not be reinstalled in the engine. When the protective coating is nicked, a new coating should be applied. Corrosion should be minimized by using clean lube oil, eliminating water leaks, and keeping vents open and clean.

WORN VALVE KEEPERS AND RETAINING WASHERS

6-11. Worn valve keepers and retaining washers may result if valve stem caps (used in some engines) are improperly fitted. Such caps are provided to protect and increase the service life of the valve stems. Trouble occurs when the cap does not bear directly on the end of the stem, but instead bears on the valve stem locks or spring retaining washer. This results in the actuating force being transmitted from the cap to the locks or retaining washer, and then to the stem. This causes excessive wear on the stem grooves and valve stem locks. As a result, the retaining washers will loosen and the valve stems may be broken. An improper fit on a valve stem cap may result from the omission of spacer shims or the use of improper parts. Steel spacer shims, required in some caps to provide proper clearance, are placed between the ends of the valve stems and the caps. Omission of the shims will result in the cap's shoulder coming in contact with the locks. When disassembling a valve assembly, determine whether or not shims are used. If so, record their location and exact thickness. Valve caps must be of the proper size, or troubles similar to those resulting from shim omission will occur. Never attempt to use caps or any other valve assembly parts which are worn.

BROKEN VALVE HEADS

6-12. Broken valve heads usually cause damage to the piston, liner, cylinder head, and other associated parts. They are generally repairable only by replacement of these parts. Whether the causes of broken valve heads are mechanical deformation or metal fatigue, every precaution should be taken to prevent their occurrence. If a valve head breaks loose, do not replace until all associated parts are thoroughly inspected.

ROCKER ARMS AND PUSH RODS

6-13. Rocker arms are part of the valve actuating mechanism. They pivot on a pivot pin or shaft secured to a bracket mounted on the cylinder head. One end of a rocker arm contacts the top of the valve stem, and the other end is actuated by the camshaft. In some installations where the camshaft is located near the cylinder head, the rocker arm may be actuated from the cam by hardened-steel rollers. In installations where the camshaft is located below the cylinder head, the rocker arms are actuated by push rods. One rocker arm and a bridge may be used in some installations to open two valves simultaneously. The principal trouble encountered with rocker arms and push rods is wear. This may occur in bushings or on the pads, end fittings, or tappet adjusting screws. The causes, prevention, and repair of worn rocker arm bushings are practically the same as for piston pin bushings. Any excessive wear of a bushing requires replacement of the part. Installation of a new bushing generally requires that a reamer be used for the final fit. Wear at the points of contact on a rocker arm is generally in the form of pitted, deformed, or scored surfaces. Insufficient lubrication or excessive tappet clearance will cause excessive wear on the rocker arm pads and end fittings. Push rods are usually positioned to the cam followers and rocker arms by end fittings. The pads are the rocker arm ends that bear on the valve stem or valve stem cap. When the tappet clearance is excessive, the rods shift around, increasing the rate of wear of both the rocker arm and the rod contact surfaces. Worn fittings necessitate the replacement of parts. Continued use of a poor fitting and a worn push rod is likely to result in further damage to the engine, especially if the rod should come loose. Worn tappet adjusting screws and locknuts usually make it difficult to maintain proper clearances and to keep the locknuts tight. Wear of the adjusting screws is usually caused by loose locknuts. This allows the adjusting screw to work up and down on the threads each time the valve is opened and closed. This type of wear can be prevented by tightening down the locknuts after each adjustment and by frequently checking for tightness. If the threads are worn, the entire rocker arm must be replaced. Repair, or the use a new tappet adjusting screw, should be resorted to only in cases of emergency. The adjustment of the rocker arm assembly consists chiefly of adjusting the tappets for proper running clearance. The valve clearance for both intake and exhaust valves should be readjusted whenever the cylinder has been removed. Some engines have tappet clearances specified for a cold engine, others are specified for hot. The manufacturer's manual or other source of tappet clearance specifications for an engine will indicate which condition should be met. Generally, clearances are set with the piston for the related valves at the TDC, firing position (on a 2-stroke cycle engine this is simply the TDC position). The clearance is measured between top of the valve stem and tappet bearing surface, using a feeler gauge. If adjustment is needed, the jam nut on the tappet screw is loosened and the screw is turned until the gauge will just slide freely in the gap. The jam nut, is then tightened. Since the jam nut tends to increase the clearance, the gap must be checked again after it is tightened. An overhauled engine will require a second adjustment it has run a few hours. See Figure 6-7 for adjusting valve clearance.

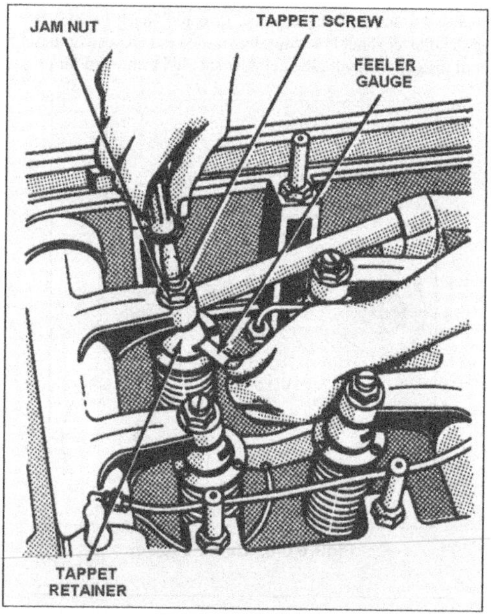

Figure 6-7. Adjusting Valve Clearance

CAM FOLLOWERS AND LASH ADJUSTERS

6-14. Cam followers are the part of the valve actuating mechanism that changes the rotary motion of the camshaft to reciprocating motion to open the valves. Cam followers ride the flat of the cam. They are raised, as the cam rotates, by the high side of the cam and lowered by tension from the valve spring. Figure 6-8 shows a cam follower used internal-combustion engines. Regardless of the type of cam follower, wear is the trouble most commonly experienced. Worn rollers usually are characterized by holes or pit marks in the roller surfaces. On the mushroom type, if the cam follower fails to revolve, the cams usually wipe the same surface each time the camshaft revolves. Normal use will cause surface disintegration, usually as a result of fatigue of the hardened surfaces. Nicks and dents on rollers will also start disintegration. A constant check must be maintained for defective rollers or surfaces and also for nicks, scratches, or dents in the camshaft. Whenever a defective cam follower is discovered, it should be replaced. In roller type cam followers, a worn cam follower body and guide or roller needle bearings (if used) must be replaced. Hydraulic valve lifters or lash adjusters provide a means of controlling valve gear clearances or lash. They may be installed on push rods, on rocker arms, or on the cam follower. Figure 6-9 shows a hydraulic lash adjuster. Hydraulic valve lifters or adjusters are usually adjusted with their internal parts in compression or with "zero valve clearance." Manufacturer's TMs must be followed closely to get normal life of valves and operating gear. Hydraulic lash adjusters may vary in design. However, they generally consist of such basic parts as a cylinder, a piston or plunger, a ball check valve, and a spring. As precision parts, hydraulic valve lifters or adjusters require special care in handling and they must be kept exceptionally clean. Hydraulic lash must be kept free of abrasive materials if they are to perform satisfactorily. Defective or poorly operating valve adjusters allow clearance or lash in the valve gear. Noisy operation of a lash adjuster indicates that there is insufficient oil in the cylinder of the unit. When a noisy lash adjuster is discovered and the oil supply or pressure is not the source of trouble, the unit should be removed from the engine and disassembled in accordance with the manufacturer's instructions. Since the parts are not interchangeable, only one unit should be disassembled at a time. Check for resinous deposits, abrasive particles, a stuck ball check valve, a scored check valve seat, and excessive leakage. All parts of the hydraulic lash adjuster should be carefully washed in kerosene or diesel fuel. Such parts as the cam follower body, plunger or piston, and hydraulic cylinder should be checked for proper fit.

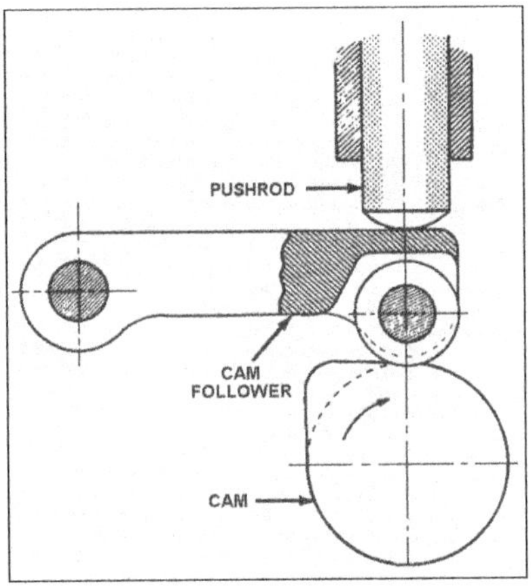

Figure 6-8. Cam Follower

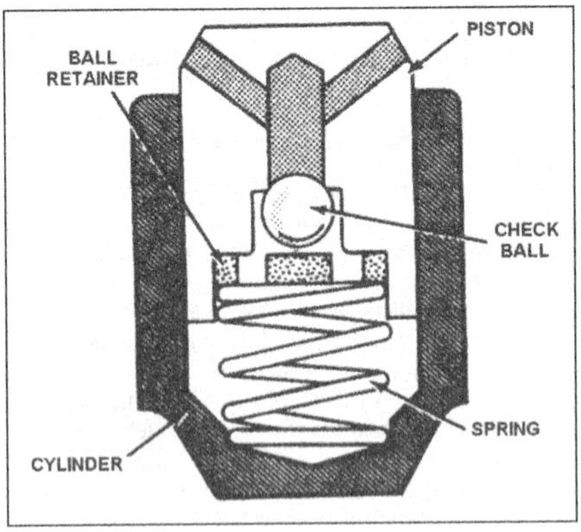

Figure 6-9. Hydraulic Lash Adjuster

SECTION II. PISTON AND ROD ASSEMBLIES

Piston and rod assemblies are composed of a piston, piston rings, piston pin, and connecting rod. A discussion of these components and their functions in engine operation follows.

PISTONS

6-15. The piston is one of the principal parts in the power-transmitting assembly. It must be so designed and of such materials that it can withstand the extreme heat and pressure of combustion. Pistons must also be light enough to keep inertia loads on related parts to a minimum. The piston aids in sealing the cylinder to prevent the escape of gas, and it transmits some of the heat through piston rings to the cylinder wall. Pistons are constructed of a variety of metals-cast iron, nickel cast iron, steel alloy, and aluminum alloy. Cast iron and aluminum are most commonly used at present. Cast iron gives longer service with little wear. It can be fitted to closer clearances, and it distorts less than aluminum. Lighter weight and higher conductivity are the principal advantages of aluminum. Cast iron is generally associated with the pistons of slow-speed engines and aluminum with those of high-speed engines. However, cast iron is also used for the pistons of some high-speed engines. The piston walls are of very thin construction and require additional cooling. Trunk-type pistons are used in single-acting and opposed-piston engines. Since most modern diesel engines use only the trunk piston, only that type is discussed here. Trunk pistons perform a number of functions. They serve as the unit which transmits the force of combustion to the connecting rod and conduct the heat of combustion to the cylinder wall. A trunk piston also serves as a valve opening and closing the ports of a 2-stroke cycle engine. Other functions of the piston and its parts are pointed out in the following discussion (see Figure 6-10).

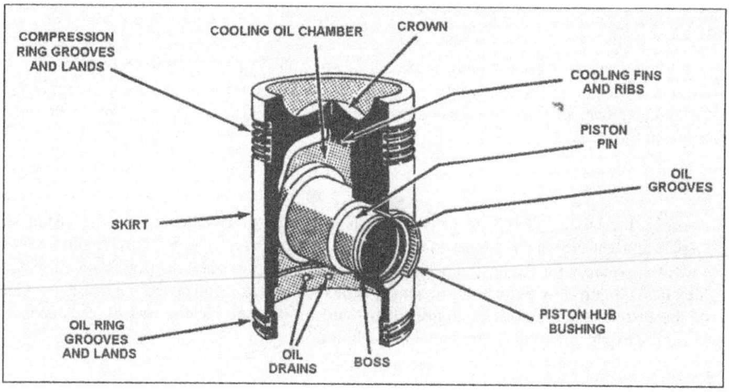

Figure 6-10. Trunk-type Piston

CROWN

6-16. The crown or head of a piston acts as the moving surface that changes the volume of the cylinder's content (compression), removes gases from the cylinder (exhaust), and transmits the force of combustion (power). Generally, the crown end of a piston is slightly smaller in diameter than the skirt end. The resulting slight taper allows for expansion of the metal at the combustion end. Even though slight, taper is sufficient so that, at normal operating temperatures, the diameter of the piston is the same throughout. Manufacturers have produced a variety of crown designs-truncated, cone, recessed, dome or convex, concave or cup, and flat. Concave piston crowns are common in marine engines. However, other types may be encountered. The concave shape has the advantage of assisting in creating air turbulence, which mixes the fuel with air during the last part of compression in diesel engines. Recesses are provided in the rim of some concave pistons to allow room for parts which protrude into the combustion space. Examples of such parts are the exhaust and intake valves, the air-starting valve, and the injection nozzle. In some 2-stroke cycle engines, piston crowns are shaped with irregular surfaces which deflect and direct the flow of gases.

SKIRTS

6-17. The skirt of a trunk piston receives the side thrust created by the movement of the crank and connecting rod. In turn, the piston transmits the thrust to the cylinder wall. In addition to receiving thrust, the skirt aids in keeping the piston in proper alignment within the cylinder. Piston skirts may be plain or smooth, slotted or split, or knurled (knurls are small beads on a metal surface). Some plain-skirt pistons have a smooth bearing surface throughout the length of the piston. In others, the diameter of the skirt near the bosses is slightly less than that of the rest of the piston. Pistons with slotted skirts are so constructed that the skirt can expand without increasing the piston diameter at heavy sections. The knurled skirt is of relatively recent design. One advantage claimed for knurled pistons is that longer service can be expected because of better lubrication afforded by the oil carried by the knurls. Most trunk pistons are of one-piece construction. However, some are made of two parts and of two metals. The trunk or skirt is made of cast iron or an alloy, and the crown or head is made of steel. In some pistons of this type construction, the crown is fitted to the trunk with a ground joint, while in others, the parts are welded together.

GROOVES AND LANDS

6-18. Without grooves and lands, the piston rings could not be properly spaced nor held in position. The number of grooves and lands found on a piston vary considerably, depending on such factors as size and type of the piston.

OIL DRAINS

6-19. Some pistons are constructed with oil drains (small holes) in the bottom of some of the grooves. In others, oil drains are located in the skirt of the piston. These holes serve as oil returns, permitting lubricating oil from the cylinder wall to pass through the piston into the crankcase. The location of the oil drains in a piston is shown in Figure 6-10.

BOSSES

6-20. Generally, the bosses (hubs) of a piston are heavily reinforced openings in the piston skirt. Some bosses are a part of an insert which is secured to the inside of the piston. The principal function of the bosses is to serve as mounting places for the bushings or bearings which support the wrist pin. They provide a means of attaching the connecting rod to the piston. In some pistons the bosses serve as the pin bearings. Generally, the diameter of the piston at the bosses is slightly less than the diameter of the rest of the piston. This is to compensate for the expansion of the extra metal in the bosses.

COOLING CHAMBER

6-21. Because of the intense heat generated in the combustion chamber, adequate cooling must be provided. The transmission of heat through the rings (approximately 30 percent of the heat absorbed by the piston) to the cylinder wall is not sufficient in many engines to keep the unit cooled within operating limits. Most pistons have fins or ribs and struts in a cooling chamber (see Figure 6-10) as an internal part of the piston. These parts provide additional surface for the dissipation of heat. Much of the heat is carried away by oil which may be pump-forced, splashed, or thrown by centrifugal force into the piston assembly. Oil is the principal means of cooling most piston assemblies. Intake air is also used in cooling hot engine parts. To exhaust or scavenge a cylinder of burned gases and to cool the engine parts, the intake and exhaust valves are so timed that both valves are open for a short interval at the end of the exhaust stroke. This allows the intake air to enter the cylinder, clean out the hot gases, and, at the same time, cool the parts.

PISTON RINGS

6-22. Piston rings are particularly vital to engine operation. They must effectively perform three functions: seal the cylinder, distribute and control lubricating oil on the cylinder wall, and transfer heat from the piston to the cylinder wall. All rings on a piston perform the latter function. However, two general types of rings (compression and oil) are required to perform the first two functions. There are many variations in the design of compression rings and oil rings. The types of rings discussed here and shown in Figure 6-11 are representative of many rings now in use. However, other variations exist. The number of rings on a piston varies with the type and size of the piston. For example, a piston of a small diesel engine may be fitted with three rings, while a piston in a large diesel may be fitted with as many as eight rings. The location of rings also varies considerably. The compression rings are located toward the crown or combustion end of the piston. The ring closest to the crown is sometimes referred to as the firing ring. In some pistons, both compression and oil rings are located toward the crown, above the pin bosses. In other pistons, the compression rings and one or two oil rings are above the bosses with one or two oil rings below the bosses.

Note: The terms "above" and "below" adequately identify ring location when the crown of the piston is at the top, as in engines of the vertical in-line type or even in some V-types. Where "above" and "below" may lead to confusion, such as when referring to ring location on the upper pistons of opposed-piston engines, piston ring location can be more accurately identified by referring to the crown or combustion end and the skirt or crankshaft end of the piston.

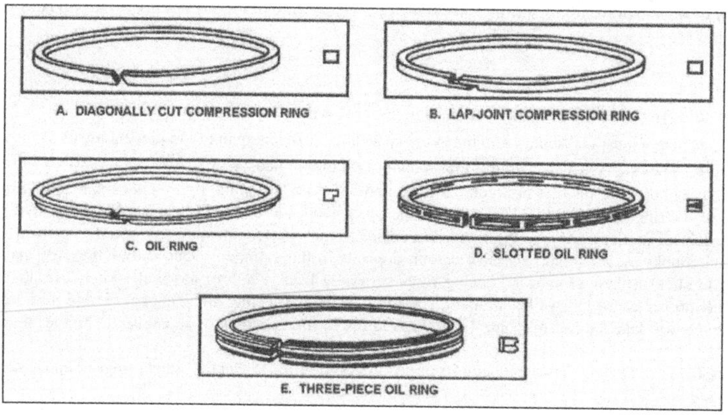

Figure 6-11. Types of Piston Rings

COMPRESSION RINGS

6-23. The principal function of compression rings is to seal the cylinder and combustion space. This is to prevent gases within the space from escaping until they have performed their function. Some oil is carried with the compression rings as they do their job. Most compression rings are made of gray cast iron. However, some types have special facing, such as a bronze insert in a slot cut circumferentially around the ring, or a treated surface. Rings with the bronze inserts are sometimes called gold seal rings while those with special facings are referred to as bimetal rings. The bimetal ring consists of two layers of metal bonded together, the inner layer being steel and the outer being cast iron. Compression ring designs come in a variety of cross sections. The rectangular cross section is the most common. Piston rings contribute as much as any other one thing toward maintaining pressure in a cylinder. They must therefore possess sufficient elasticity to press uniformly against the cylinder walls. The diameter of the ring, before installation, is slightly larger than the cylinder bore.

Because of the joint, the ring can be compressed to enter the cylinder. The tension created when the ring is compressed and placed in a cylinder, causes the ring to expand and produce a pressure against the cylinder wall. The pressure exerted by rings closer to the combustion space is increased by the action of the confined gases during compression and combustion. The gases enter behind the top ring through the clearance between the ring and groove and force the ring out against the cylinder and down against the bottom of the groove. The gas pressure on the second ring and other compression rings is progressively less. This is due to the gas reaching these rings being limited to that passing through the gap of the firing ring. When a piston assembly has been disassembled, visual inspection will reveal whether the compression rings have been functioning properly. If a ring has been working properly, the face (surface bearing against the cylinder wall) and the bottom of the ring will be bright and shiny because of contact with the cylinder wall and the groove. The top and back (inside surface) of the ring will be black since they are exposed to the hot combustion gases. The exposed sides and corresponding parts of the ring groove may be covered with deposits of carbon which must be removed during overhaul. Black areas on sealing surfaces will indicate that hot gases have been escaping. Under normal operating conditions and with engine parts functioning properly, there will be very little leakage of gas because of the excellent pressure. The oil which prevents metal-to-metal contact between the rings and cylinder wall also aids, to a degree, in making the seal. When a proper seal is established, the only point at which gas can normally leak is through the piston ring gaps. The gap of a piston ring is so small, compared to the total circumference of the ring, that the amount of leakage is negligible when rings are functioning properly. The time during which gas pressure is applied to the rings in a modern, high-speed engine is insufficient to cause any appreciable leakage through the joints. The leakage can be held to a minimum if the rings are placed so that the joints of successive rings are on alternate sides of the piston. Most pistons provide no means to prevent the rings from shifting or turning around in the grooves. However, some manufacturers provide metal pins or dowels to prevent this shifting. On some pistons, the ring or groove pin fits in the gap between the ends of the ring.

OIL RINGS

6-24. In performing their function, oil must do the following two things:

- They must distribute oil to the cylinder wall in sufficient quantity to prevent metal-to-metal contact.
- The rings must also control the amount of oil distributed.

Without an adequate oil film between the rings and the cylinder, undue friction occurs, resulting in excessive wear of the rings and the cylinder wall. On the other hand, too much oil is as undesirable as not enough. If too much oil is distributed by the rings, the oil may reach the combustion space and burn, wasting oil and causing smoky exhaust and excessive carbon deposits in the cylinder. Such carbon deposits may cause the rings to stick in their grooves. Sticking rings generally lead to a poor gas seal. Therefore, the rings must provide proper control as well as proper distribution of the lubricating oil. Rings of various designs have been constructed to take care of oil control and distribution in the cylinders of an engine. Three of these rings are shown in Figure 6-11 (C, D, and E). Manufacturers use a variety of terms to identify the oil rings of an engine. Some of these are—

- Oil control.
- Oil scraper.
- Oil wiper.
- Oil cutter.
- Oil drain.
- Oil regulating.

Regardless of the identifying terms used, such rings are to control and distribute lubricating oil within the cylinder of an engine. If a distinction is to be made between types of oil rings, perhaps the terms, "oil control" and "oil scraper" should be used. When this distinction is made, the oil control rings are considered to be those closest to the compression rings. The oil scraper or wiper rings are those farthest from the combustion end of the piston. Oil control rings prevent the flow of excessive amounts of oil to the compression rings and subsequent entry into the combustion space. Oil scraper rings regulate the amount of oil passing between the piston skirt and cylinder wall by wiping off the excess oil thrown into the cylinder bore by the crankshaft and connecting rod. In performing their function, the oil rings must permit enough oil to be carried to the upper part of the cylinder wall so that the piston and compression rings receive proper lubrication. In general, manufacturers name their oil rings according to the description of the function performed by the ring of their design. These terms, as well as design, vary with respect to location on any given piston. For example, a piston of a General Motors (GM) 6-71 has two oil control rings placed on the skirt below the pin. Both rings are identical, each consisting of three pieces (two rings and an expander). The ring shown in Figure 6-11, represents a three-piece oil ring. In these rings, the two "scraping" pieces have very narrow faces bearing on the cylinder wall. This permits the ring assembly to conform rapidly to the shape of the cylinder wall. Since the ring tension is concentrated on a small area, the rings will cut through the oil film easily and will efficiently remove excess oil. The bevel on the upper edge of each ring face causes the ring to ride over the oil film as the piston moves toward TDC. But as the piston moves on intake and power, the sharp, hook-like lip of each ring scrapes or wipes the oil from the cylinder wall. Another example of differences in terminology and location is found in the Fairbanks-Morse (FM) 38D8 1/8. A piston in this type of engine has three oil rings, all located on the skirt end. The two nearest the crankshaft end of the piston are called oil drain rings, while the ring nearest the pin bosses is called the scraper. The drain rings are slotted to permit oil to pass through the ring and to continue on through the holes drilled in the ring grooves. Figure 6-11D shows one type of slotted oil ring.

REPLACEMENT OF PISTON RINGS

6-25. While stuck rings may be freed and made serviceable, excessively worn or broken rings must be replaced with new ones. Installing a new set of rings in an engine is a job that requires great care. Damage generally occurs when the rings are being placed in the grooves of a piston or when the piston is being inserted in the cylinder bore. Be careful when removing the piston and connecting rod from the cylinder. In most engines, the piston should not be removed from a cylinder until the cylinder surface above the ring travel areas has been scraped. In addition to removing all carbon, the lip of any appreciable ridge must be removed before removing the piston. Do not remove a ridge by grinding because small abrasive particles from the stone may get into the engine. A metal scraper should be used, and a cloth should be placed in the cylinder to catch all metal cuttings. The lip of a cylinder ridge can generally be removed by scraping sufficiently to allow the piston assembly to slide out of the liner. After the piston has been removed, a more detailed inspection of the ridge can be made. Finish scraping the remaining ridge, but be careful not to go too deep. Finish the surface with a handstone. For removing large ridges, you may have to remove the liner and use a small power grinder. After the piston and connecting rod have been removed, check the condition and wear of the piston pin and piston pin bushing, both in the piston and in the connecting rod. Piston rings can be removed and installed with a tool similar to those shown in Figure 6-12. These tools generally are provided with a device which limits the amount that the ring can be spread and prevents the ring's being deformed or broken. If a ring is securely stuck in the groove, additional work will be necessary. The piston may have to be soaked overnight in an approved cleaning solvent or in diesel oil. If soaking does not free a stuck ring, the rings will have to be driven out with a brass drift. The end of the drift should be shaped and grounded to permit its use without damaging the lands. After removing the rings, thoroughly clean the piston giving special attention to the ring grooves.

Note: Diesel oil and/or kerosene are satisfactory cleaning agents.

In addition, it may be necessary to clean excessive deposits from the oil return holes in the bottom of the oil control ring grooves. This is done with a twist drill of a diameter corresponding to the original size of the holes. Another complete inspection should follow the cleaning of the piston. Check all parts for any defects which will require replacement of the piston. Pay particular attention to the ring grooves, especially if the pistons have been in service for a long time. A certain amount of enlargement in the width of the grooves is normal and shouldering of the groove may occur. Shouldering, as shown in Figure 6-13, results because of the "hammering out" motion of the rings. The radial depth of thickness of the ring is much less than the groove depth. Also, while the ring wears away, an amount of metal corresponding to its own width, the metal at the bottom of the groove remains unchanged. Shouldering usually requires replacement of the piston since the shoulders prevent the proper fitting of new rings. After determining that a piston is serviceable, inspect the rings carefully to determine whether they can be reused. If they do not meet specifications, new rings must be installed. When rings are being installed, measure the gap with a feeler gauge. New rings are placed inside the cylinder liner (see Figure 6-14A) or in a ring gauge when the gap is to be measured. When the gap is measured with the ring in the liner (see Figure 6-14B), the following two measurements are necessary:

- One just below the upper limit of the ring travel.
- One within the lower limit of travel.

These measurements are necessary because the liner may have a slight of taper caused by wear. Ring gap must be within the limits specified in the manufacturer's TM. If the gap of a new ring is less than specified, the ends of the ring will have to be filed with a straight-cut mill file until the proper gap is obtained. If the gap is more than specified, oversize rings will have to be installed. Ring gap of used rings should be checked with the rings held in place on the piston with a ring compressing tool. When ring gap measurement is made with the ring on piston, first measure the piston for wear and out-of-roundness (see Figure 6-15). After the proper gap clearance has been determined, the piston pin and connecting rod can be reinstalled. During reassembly and installation of a piston and connecting rod assembly, all parts should be well lubricated. The rings can be installed on the piston with tools similar to those used for removal. When installing piston rings, spread them as little as possible to avoid breaking them. The lowest ring should be inserted first. When all the rings have been installed, check the ring-to-land clearance (see Figure 6-15). If the clearance is too small, the ring may bind or seize, allowing improper sealing and blowby to occur. If the clearance is excessive, the ring may flutter and break the piston lands or rings. After all the rings have been properly installed, coat the entire assembly with oil, then insert it in the cylinder bore. Rings should be positioned so that the gap of each successive ring is on an alternate side and the gaps are in line with the piston pin bosses. On large engines, a chain fall should be used to hold the piston assembly in position as it is being lowered into the cylinder. When a piston is being inserted into a cylinder,, piston rings must be compressed evenly. Special funnel-type tools are usually provided for this purpose. Other types of ring-compressing tools include a steel band that is placed around the ring and is so constructed that it can be tightened.

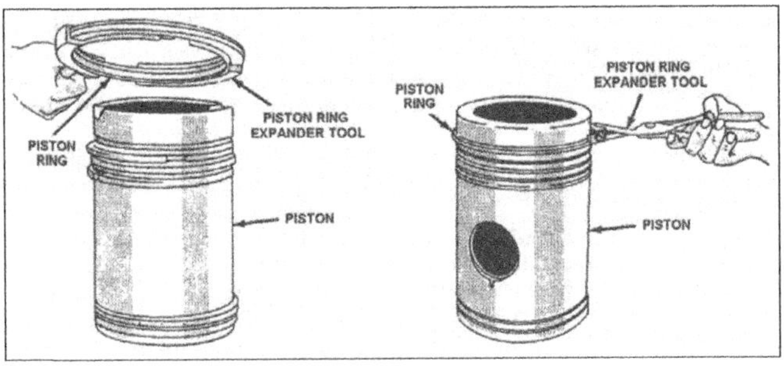

Figure 6-12. Piston Ring Tools Used for Removal or Installation

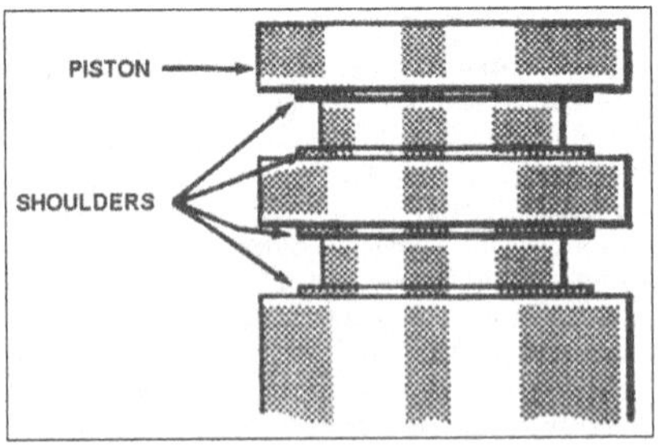

Figure 6-13. Ring Groove Shoulder Due to Wear

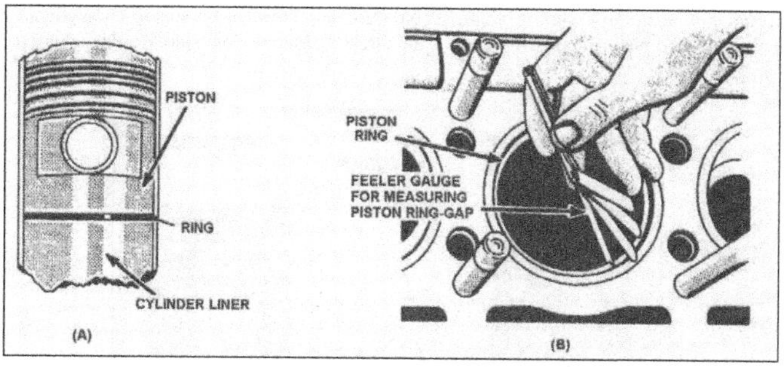

Figure 6-14. Leveling a Piston Ring in Cylinder Bore and Measuring Ring Gap Clearance

Figure 6-15. Checking Ring Groove Side Clearance

PISTON PINS AND PISTON BEARINGS

6-26. In trunk piston assemblies, the only connection between the piston and the connecting rod is the pin (sometimes referred to as the wrist pin) and its bearings. These parts must be of especially strong construction because the power developed in the cylinder is transmitted from the piston through the pin to the connecting rod. The pin is the pivot point where the straight-line or reciprocating motion of the piston changes to the reciprocating and rotating motion of the connecting rod. Therefore, the principal forces to which a pin is subjected are the forces created by change in direction of motion. Before discussing the pin further, consider the side thrust which occurs in a single-acting engine equipped with trunk pistons. Side thrust is exerted at all points during a stroke of a trunk piston, except at TDC and BDC. The side thrust is absorbed by the cylinder wall. Depending on the position of the piston and the rod and the direction of rotation of the crankshaft, thrust occurs first on one side of the cylinder and then on the other. Figure 6-16 (view A) indicates that gas pressure is forcing the piston downward (power). Since the crankshaft is rotating clockwise, the force of combustion and the resistance of the driven parts tend to push the piston to the left. The resulting side thrust is exerted on the cylinder wall. If the crankshaft were rotating counterclockwise, the situation would be reversed. Figure 6-16 (view B) shows the piston being pushed up (compression) by the crankshaft and connecting rod. This causes the side thrust to be exerted on the opposite side of the cylinder. Therefore, the side thrust alternates from side to side as the piston moves and down. Side thrust in an engine cylinder demands adequate lubrication and correct clearance. Without an oil film between the piston and the cylinder wall, metal-to-metal contact occurs and results in excessive wear. If the clearance between the piston and cylinder is excessive, a pounding noise (called piston slap) will occur as the thrust alternates from side to side. This can cause damage to the cylinder wall.

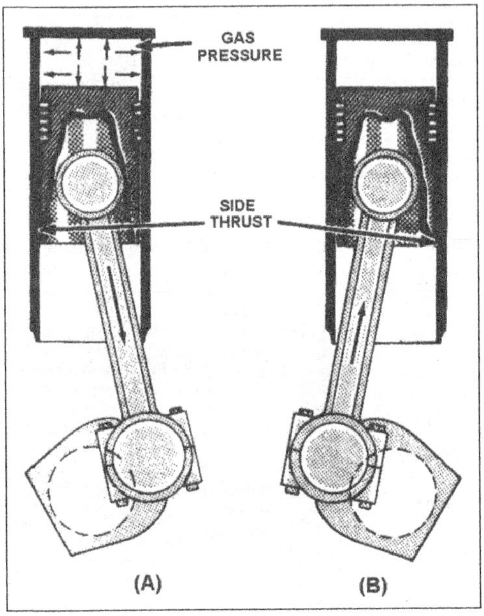

Figure 6-16. Side Thrust in a Trunk Piston in a Single-acting Engine

TYPES OF PISTON PINS

6-27. Pins are usually hollow and made of alloy steel, machined, hardened, and precision-ground and lapped to fit bearings. Some pins are chromium-plated to increase the wearing qualities. Their construction provides maximum strength with minimum weight. Lubrication is provided by splash from the crankcase or by oil forced through drilled passages in the connecting rods. Piston pins must be secured in position so that they do not protrude beyond the surface of the piston, or so they do not have end-to-end motion. Otherwise, the pin will tend to damage the cylinder wall. Piston pins may be secured in a piston rod assembly in one of three ways:

- Rigidly fastened into the piston bosses.
- Clamped to the end of the rod.
- Free to rotate in both piston and rod.

When piston pins are secured by these methods, the pins are identified as stationary (fixed), semi-floating, and full-floating, respectively.

Stationary (Fixed) Pin

6-28. The stationary (fixed) pin is secured to the piston at the bosses and the connecting rod oscillates on the pin. Since all movement is made by the connecting rod, there may be uneven wear on the contacting surfaces in this type of installation.

Semi-floating Pin

6-29. Semi-floating pins are secured in the middle of the connecting rod. The ends of the pin are free to move in the piston pin bearings in the bosses.

Full-floating Pin

6-30. The full-floating pin is the most common of the three types. It is not secured to either the piston or the connecting rod. Pins of this type may be held in place by caps, plugs, and snap rings or spring clips which are fitted in the bosses. The securing devices for a full-floating pin permit the pin to rotate in both the rod and the pin bosses.

TYPES OF PISTON PIN BEARINGS

6-31. Piston pin bearings are of three types:

- Integral bearing.
- Sleeve bearing or bushing.
- Needle-type roller bearing.

These bearings may be further identified according to location (the piston boss piston pin bearings and the connecting rod piston pin bearings). The bearings in the bosses (hubs) of most pistons are of the sleeve bushing type. Some boss hearings are an integral part of the piston (the bearing surface is precision-bored directly into the bosses). Pistons fitted with stationary piston pins require no bearing surfaces in the bosses. Sleeve bearings or bushings are made of bronze or similar material. The bushing material is a relatively hard-bearing metal and requires surface-hardened piston pins. The bore of the bushing is accurately ground in line for the close fit of the piston pin. Most bushings have a number of small grooves cut in their bore for lubrication purposes. Some sleeve bushings have a press fit, while others are "cold shrunk" into the bosses. Sleeve bushing bearings for both the bosses and the connecting rod are shown in Figure 6-17. Note that the bosses are a part of an insert. If the piston pin is secured in the bosses (stationary) of the piston or if it floats (full-floating) in both the connecting rod and the piston, the piston end of the rod may be fitted with a sleeve bushing. Pistons fitted with semi-floating pins require no bearing at the rod. Sleeve bushings used in the piston end of connecting rods are similar in design to those used in piston bosses as shown in Figure 6-17. Generally, bronze makes up the bearing surface. Some bearing surfaces are backed with a case-hardened steel sleeve, and the bushing has a shrink fit in the rod bore. In other bushings, the bushing fit is such that a gradual rotation (creep) takes place in the eye of the connecting rod. In another variation of the sleeve bushing, a cast bronze lining is pressed into a steel bushing in the connecting rod. In some engines, which use full-floating piston pins, the steel-backed bronze bushing rotates freely inside the piston end of the connecting rod.

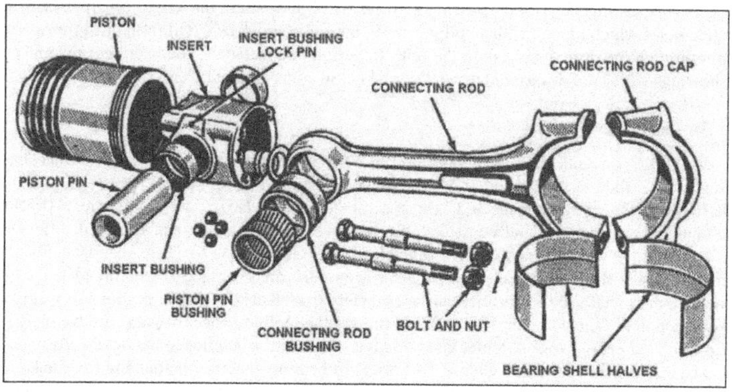

Figure 6-17. Piston and Rod Assembly With Sleeve Bushing Bearings

CARE OF PINS AND BEARINGS

6-32. Every time a piston assembly is removed from an engine, it should be inspected for wear. Piston pins and bushings should be measured with a micrometer to determine whether wear is excessive. Areas that do not make contact must be avoided during measuring. Such areas include those between the connecting rod and piston bosses, and the areas under the oil holes and grooves. Bushings may be pressed out of the rod with a mandrel and an arbor press or with special tools. It is also possible to remove bushings by first shrinking them with dry ice. The use of dry ice will also aid in the insertion of the new bearing. When new bushings are inserted, the bore into which they are compressed must be clean. Also, the oil holes in the bushing and the oil passages in the rod must be aligned. Sometimes a piston pin bushing will have to be reamed after it has been installed to obtain the proper clearance.

CONNECTING RODS

6-33. After a new bushing has been installed, alignment of the connecting rod should be checked. The manufacturer's TM should be consulted for details concerning clearance and alignment procedures. Connecting rods serve as the connecting link between the piston and crankshaft. To transmit the forces of combustion to the crankshaft, the rod changes the reciprocating motion of the piston to the rotating motion of the crankshaft. In general, the type of connecting rod used in an engine depends on cylinder arrangement and type of engine. Several types of connecting rods have been designed. However, only two of those likely to be found in marine engines are discussed here (the conventional rod and the fork and blade rod).

CONVENTIONAL RODS

6-34. The conventional rod is sometimes referred to as the "normal" or "standard" rod because of its extensive use by many manufacturers and its similarity to the rods used in many automobiles. An example of the conventional rod is shown in Figure 6-17. These rods are typical of those used in single-acting, in-line engines. Rods of this type are also used in opposed-piston engines and in some V-type engines. When used in V-type engines, two rods are mounted on a single crankpin. The two cylinders served by these two rods are offset so that the rods can be operated side by side. Rods are generally made of drop-forged, heat-treated, carbon steel (alloy steel forging). Most rods have an I- or H-shape cross section which provides maximum strength with minimum weight. The bore (hub, eye) at the piston end of the rod is generally forged as an integral part of the rod. However, the use of semi-floating piston pins eliminates the need for the bore. The bore at the crankshaft end is formed by two parts, one an integral part of the rod and the other a removable cap. Rods are generally drilled or bored to provide an oil passage to the piston end of the rod. The bore at the crankshaft end of a conventional rod is generally fitted with a shell-type precision bearing. Connecting rod bearings of most engines are pressure-lubricated by oil from adjacent main bearings through drilled passages. The oil is evenly distributed over the bearing surfaces by oil grooves in the shells. Bearing shells are provided with drilled holes which line up with an oil groove in the rod bearing seat. Oil from this groove is forced to the piston pin through the drilled passage in the rod. In design and materials, rod bearings are similar to the main journal bearings which are discussed later in this chapter in connection with crankshafts.

FORK AND BLADE (PLAIN) RODS

6-35. While two conventional rods are used to serve two cylinders in some V-type engines, a single assembly consisting of two rods is used in other V-type engines. As the name implies, one rod is fork-shape at the crankshaft end to receive the blade rod. In general, fork and blade rods (see Figure 6-18) are similar to conventional rods in material and construction. However, design at the crankpin end differs from that of conventional rods. The bearings of fork and blade rods are similar to those of conventional rods. The exception is that the shells must have a bearing surface on the outer surface to accommodate the blade rod. In some models, the metal used for bearing surfaces differs from that used in the bearings of many conventional rods. For example, in some high-speed, gasoline engines the shells are steel-backed with the inner surface lined with lead-tinplated pure silver. A center band of silver (unplated) is applied to the outer surface of the shells to provide a bearing surface for the blade rod. A variety of bearing materials is found in the crankpin bearings of some V-type diesel engines. In one model, the upper shell is steel (lined inside and outside with lead-bronze bearing metal which is lead-tinplated). The lower shell is a solid chilled, cast lead bronze, lead-tinplated, bearing.

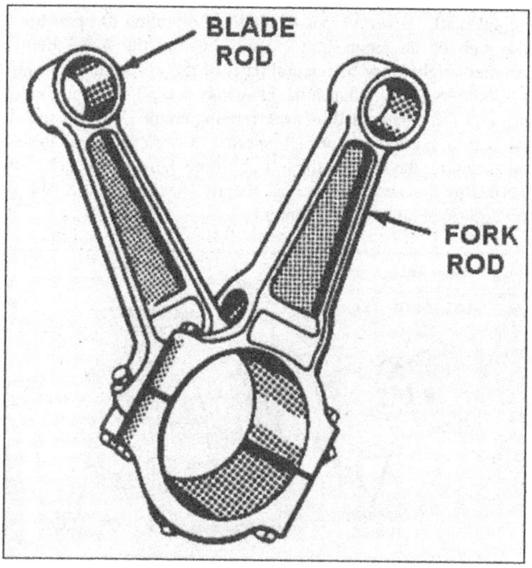

Figure 6-18. Fork and Blade Connecting Rod

SECTION III. SHAFTS

CRANKSHAFT

6-36. The crankshaft is one of the largest and most important moving parts in an engine. It changes the movement of the piston and the connecting rod into the rotating motion required to drive such items as reduction gears, propeller shafts, generators, and pumps. As the name implies, the crankshaft usually consists of a series of cranks (throws) formed as offsets in a shaft. As a result of its function, the crankshaft is subjected to all the forces developed in an engine. Because of this, the shaft must be of especially strong construction, usually machined from forged alloy or high-carbon steel. The shafts of some engines are made of cast iron alloy. Forged crankshafts are nitrided (heat-treated) to increase the strength of the shafts and to minimize wear. Crankshafts of a few large engines are of the built-up type (forged in separate sections and flanged together). However, the crankshafts of most modern engines are of one-piece construction. The parts of a crankshaft may be identified by a variety of terms. However, those shown in Figure 6-19 are commonly used for most of the engines used by the Army. The main bearing journals serve as the points of support and as the center of rotation for the shaft. As bearing surfaces, the journals (both crankpin and main) of most crankshafts are surface-hardened so that a longer-wearing, more durable bearing metal can be used without causing excessive wear of the shaft. As shown in Figure 6-19, crankshafts have a main journal at each end of the shaft. Usually, there is an intermediate main journal between the cranks. However, in small shafts, intermediate journals may not be used. Each crank (throw) of a shaft may be thought of as consisting of three parts (two webs and a pin). Crank webs are sometimes called cheeks or arms. The cranks or throws provide points of attachment for 'the connecting rods and serve as the connecting links between main journals. In many crankshafts, especially in large engines, the crankpins and main journals are of hollow construction. Hollow construction (see Figure 6-20) not only reduces weight considerably but also provides a passage for the flow of lubricating oil. The forces which turn the crankshaft of a diesel engine are produced and transmitted to the crankshaft in a pulsating manner. These pulsations create torsional vibrations which are capable of severely damaging an engine if they are not reduced, or damped, by opposing forces. To ensure satisfactory operation, many engines require an extra damping effect. This is usually provided in the form of torsional vibration dampers mounted on the free end of the crankshaft. Several types of torsional vibration dampers are currently used. On some crankshafts, part of the web of the crankshaft extends beyond the main journal to form or to support counterweights. The counterweights may be integral parts of the web or may be separate units attached to the web by studs and nuts. Counterweights balance the off-center weight of the individual crank and thereby hold in equilibrium the centrifugal force generated by each rotating crank. Without such balance, severe vibrations will be created by crank action, particularly at high speeds. Since excessive vibration causes rapid wear and leads to failure of metal structure, the shaft is likely to become damaged if such vibrations are not controlled. Counterweights use inertia, in the same manner as the flywheel, to reduce the pulsating effect of power impulses. Flywheels are described later in this chapter.

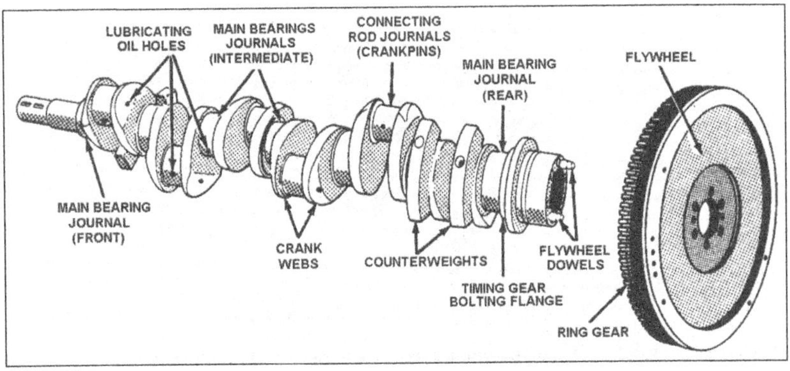

Figure 6-19. One-piece, 6-throw Crankshaft With Flywheel

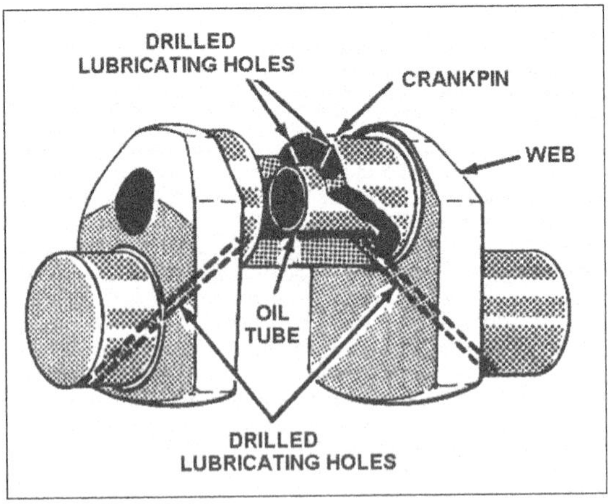

Figure 6-20. An Example of Hollow Crankpin Construction

Crankshafts and Lubrication

6-37. Whether a crankshaft is of solid or of hollow construction, the journals, pins, and webs of most shafts are drilled for the passage of lubricating oil. Two other variations in the interior arrangement of oil passages in crankshafts are shown in Figure 6-21. A study of the two oil passage arrangements will give you an idea of the part the crankshaft plays in engine lubrication. In the system shown in A, each oil passage is drilled through from a main bearing journal to a crankpin journal. The oil passages are in pairs which crisscross each other in such a way that the two oil holes for each journal are on opposite sides of the journal. These holes are in axial alignment with the oil grooves of the bearing shells when the shells are in place. Since the oil in a bearing goes at least halfway around the bearing, a part of the groove will always be aligned with at least one of the holes. Lubricating oil under pressure enters the main bearing oil grooves. The oil lubricates the main bearings and flows on through the drilled oil passages to the crankpin bearings. From the crankpin bearings, the oil is forced through the drilled passage in the connecting rod to lubricate the piston pin bearing and to be forced onto the interior surface of the piston crown for cooling. In the oil passage arrangement shown in Figure 6-21 (view B) (shaft is shown in Figure 6-19), the passage is drilled straight through the diameter of each main bearing and connecting rod journal. A single diagonal passage is drilled from the outside of a crankpin web to the center of the next main journal. The diagonal passage connects the oil passages in the two adjoining crankpins and main journals. The outer end of the diagonal passage is plugged. Lubricating oil under pressure enters the main bearing and is forced through the diagonal passage to lubricate the connecting rod bearing. From there it flows through the drilled connecting rod to lubricate the piston pin and cool the piston. In engines which use crankshaft oil passage arrangements like those just discussed and like that shown in Figure 6-20, the connecting rods are drilled to carry the lubricating oil to the piston pins and piston. In drilled V-type engines that use a fork and blade connecting rod, drilled passages supply oil to the main and connecting rod bearings. However, separate lines may supply oil for lubricating and cooling the piston assembly. Additional information on engine lubricating systems is given in Chapter 15 of this manual.

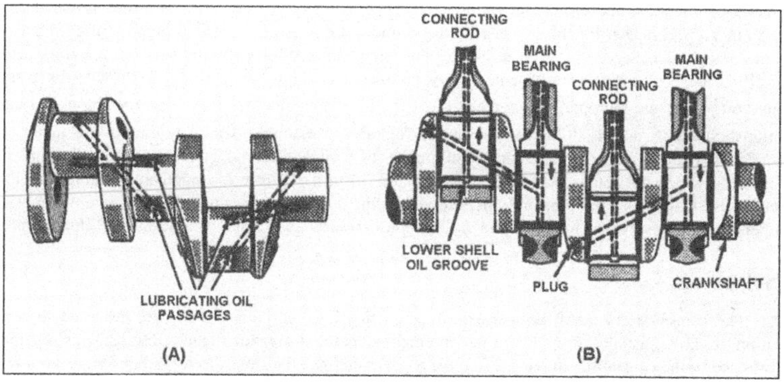

Figure 6-21. Examples of Crankshaft Oil Passage Arrangement

Crankshaft Throw Arrangements

6-38. The smooth operation of an engine and its steady production of power depend to a great extent on the arrangement of the cranks on the shaft and on the firing order of the cylinder. If the crankshaft in a multi-cylinder engine is to obtain uniform rotation, the power impulses must be equally spaced with respect to the angle of crankshaft rotation. They must also be spaced so that, when possible, successive explosions do not occur in adjacent cylinders. However, this arrangement is not always possible, especially in 2-, 3-, and 4-cylinder engines. Crankshafts may be classified according to the number of throws, such as 1-throw and 2-throw. The 6-throw shaft shown in Figure 6-19 is for a 6-cylinder, in-line, 2-stroke cycle engine. Shafts of similar design can be used in a 12-cylinder, V-type engine. The number of cranks and their arrangement on the shaft depend on a number of factors. These are the arrangement of the cylinders (in-line, V-type, flat) the number of cylinders, and the operating cycle of the engine. The arrangement of throws with respect to one another and with respect to the circumference of the main journals is generally expressed in degrees. In an in-line engine, the number of degrees between throws indicates the number of degrees the crankshaft must rotate to bring the pistons to TDC in firing order. This is not true in engines in which each throw serves more than one cylinder. See Figure 6-22 for examples of arrangements of throws with respect to cylinder arrangement, the numbers of cylinders served by each throw, and the firing order of the cylinders. (Note: The sketches in the right-hand column are not drawn to scale and therefore do not indicate relative size. The sketches are for illustrative purposes only.) In studying the examples in the table, remember that the crankshaft must make only one revolution (360°) in a 2-stroke cycle, where two revolutions are required in a 4-stroke cycle. In example (a) in the table, the shaft has the same number of throws, but factors are somewhat different. The 4-cylinder engine in example (a) operates on the 4-stroke cycle. Therefore, throws 1 and 3, 3 and 4, 4 and 2, and 2 and 1 (see firing order) must be 180° apart for the firing to be spaced evenly in 720° of crankshaft rotation. Note that in all the other examples, the throws are equally spaced regardless of cylinder arrangement, cycle of operation, or number of cylinders. In examples (b) and (c), the shaft design and the number of degrees between throws are the same. Yet the shaft in example (c) fires twice as many cylinders. This is possible because one throw, through a fork and blade rod, serves two cylinders, which are positioned in 60° banks. Therefore, even though both engines operate on the 4-stroke cycle, the 12-cylinder engine requires only 60° shaft rotation between power impulses. In examples (d) and (e), other variations in shaft throw arrangement and firing order are shown. Note that the differences are governed to a great extent by the following:

- Cylinder arrangement.
- The number of cylinders served by the shaft and by each throw.
- The operating cycle of the engine.

Notice how these factors influence throw arrangement and firing order by comparing some of the examples. For instance, there are six throws shown in examples (b) and (d), yet they are 120° apart in one and 60° apart in the other. The reason for this is because the cylinder arrangement, the total number of cylinders, and the number of cylinders served by each throw are the same. In examples (b) and (d), the operating cycle is the controlling factor in throw arrangement. This is not true in examples (d) and (e). Both shafts have six throws located 60° apart and the operating cycle is the same in both examples. However, the amount of crankshaft rotation between firings is 30° less in example (e) than in example (d). The controlling factors in these examples then are cylinder arrangement, total number of cylinders served, and the number of cylinders served by each throw. In the end views of examples (e) and (h) in the Throw Arrangement column, the numbers in parentheses identify the additional cylinders served by the throw. For instance, 1(7) in example (e) signify that number 1 crankpin serves cylinders 1 and 7. Examples (f) through (h) show variations in 8-throw crankshafts.

CAMSHAFTS

6-39. The camshaft is a shaft with eccentric projections, called cams. They are designed to control the operation of valves, usually through various intermediate parts. Originally, cams were made as separate pieces and fastened to the camshaft. However, in most modern engines the cams are forged or cast as an integral part of the camshaft. To reduce wear and to help withstand the shock action to which they are subjected, camshafts are made of low-carbon alloy steel with the cam and journal surfaces carburized before the final grinding is done. The cams are arranged on the shaft to provide the proper firing order of the cylinders they serve. The shape of the cam determines the point of opening and closing, the speed of opening and closing, and the amount of valve lift. If one cylinder is properly timed, the remaining cylinders are automatically in time. All cylinders are affected if there is a change in timing. The camshaft is driven by the crankshaft by various means. The most common method is by gears or by a chain and sprocket. The camshaft for a 4-stroke cycle engine must turn at one-half the crankshaft speed. In the 2-stroke cycle engine, it turns at the same speed as the crankshaft. The location of the crankshaft differs from engine to engine. Camshaft location depends on the arrangement of the valve mechanism. Location and operation of the camshaft in various types of engines are discussed and shown in Chapter 7. The speed of rotation of the crankshaft increases each time the shaft receives a power impulse from one of the pistons. The speed then gradually decreases until another power impulse is received. These fluctuations in speed (their number depends on the number of cylinders firing in one crankshaft revolution) result in an undesirable situation with respect to the driven mechanism as well as the engine. Therefore, some means must be provided to stabilize shaft rotation. In some engines, this is done by installing a flywheel on the crankshaft. In others, the motion of such engine parts as the crankpins, webs, and lower ends of connecting rods, and such driven units as the clutch and generator, serve the purpose. The need for a flywheel decreases as the number of cylinders firing in one revolution of the crankshaft and the mass of the moving parts attached to the crankshaft increases. Flywheels are further discussed at the end of the next section.

EXAMPLE	NUMBER OF CYLINDERS	CYLINDER ARRANGEMENT	CYCLE	NUMBER OF CYLINDERS SERVED BY EACH THROW	THROW ARRANGEMENT (SIDE VIEW)
(a)	4	IN-LINE	4-STROKE	1	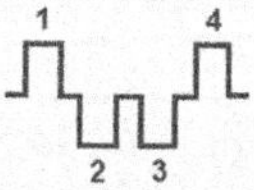
(b)	6	IN-LINE	4-STROKE	1	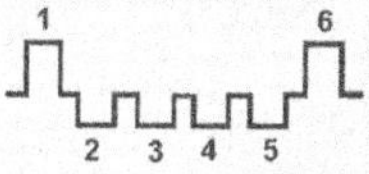
(c)	12	V	4-STROKE	2	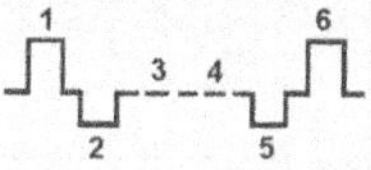
(d)	6	IN-LINE	2-STROKE	1	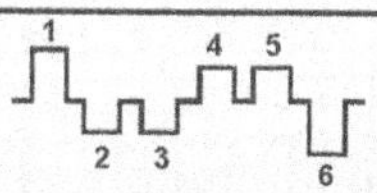
(e)	12	V	2-STROKE	2	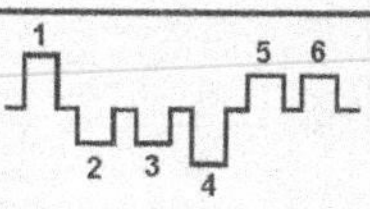
(f)	8	IN-LINE	4-STROKE	1	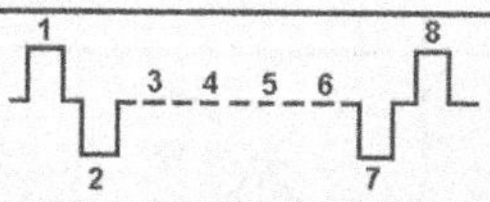
(g)	8	IN-LINE	2-STROKE	1	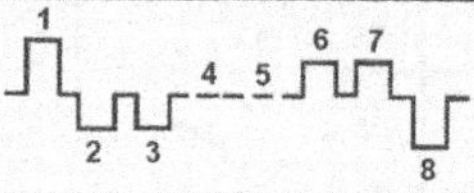
(h)	16	V	2-STROKE	2	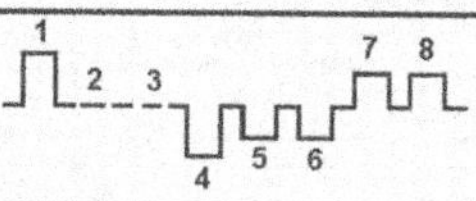

Figure 6-22. Examples of Crankshaft Throw Arrangement

EXAMPLE	THROW ARRANGEMENT (END VIEW)	FIRING ORDER	NUMBER OF DEGREES BETWEEN THROWS (SEE SKETCHES)	NUMBER OF DEGREES SHAFT ROTATION BETWEEN FIRINGS
(a)	 1&4, 2&3	1-3-4-2	4 THROWS 180° APART	180°
(b)	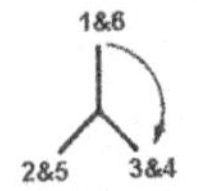 1&6, 2&5, 3&4	1-5-3-6-2-4	6 THROWS 120° APART	120°
(c)	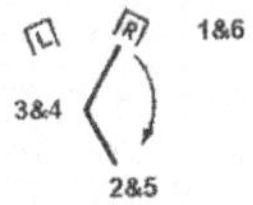 L, R, 1&6, 3&4, 2&5	L 1 2 3 4 5 6 R 1 2 3 4 5 6 1R, 3L, 4R, 5L, 2R, 1L, 6R, 4L, 3R, 2L, 5R, 6L	6 THROWS 120° APART	60°
(d)	 1, 5, 4, 3, 2, 6	1-5-3-6-2-4	6 THROWS 60° APART	60°
(e)	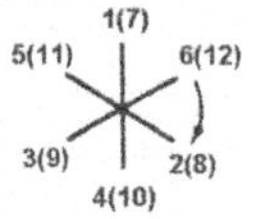 1(7), 5(11), 6(12), 3(9), 2(8), 4(10)	1 2 3 4 5 6 7 8 9 10 11 12 1-11-5-9-3-10-4-8-2-12-6-7	6 THROWS 60° APART	30°
(f)	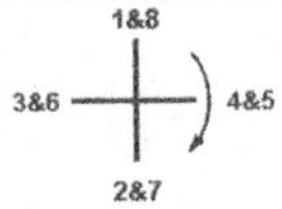 1&8, 3&6, 4&5, 2&7	1-3-2-5-8-6-7-4	8 THROWS 90° APART	90°
(g)	 1, 7, 6, 4, 5, 3, 2, 8	1-7-4-3-8-2-5-6	8 THROWS 45° APART	45°
(h)	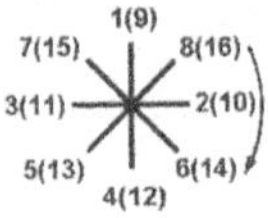 1(9), 7(15), 8(16), 3(11), 2(10), 5(13), 6(14), 4(12)	9 10 11 12 13 14 15 16 1 2 3 4 5 6 7 8 1-15-7-11-3-13-5-12-4-14-6-10-2-16-8-9	8 THROWS 45° APART	22 1/2°

Figure 6-22. Examples of Crankshaft Throw Arrangement (continued)

SECTION IV. BEARINGS

JOURNAL BEARINGS

6-40. In the past, journal bearings for internal-combustion engines were of the type known as poured babbit bearings. That is, the babbit lining was poured (cast) directly into the bearing housing and cap. Poured bearings have disadvantages in that they require hand-scraping to obtain a finished fit to the journal. Also, the bearing clearance must be adjusted by placing shims between the bearing housing and the cap. Poured babbit bearings are not commonly used in modern engines. Instead, replaceable precision bearings are used.

MAIN BEARINGS

6-41. Precision bearings, which act as supports and in which the main journal of a crankshaft revolves, are generally referred to as main bearings. In most engines, main bearings are of the sliding contact or plain type which consists of two half-shells (see Figure 6-23). The location of main engine bearings in one type of block is shown in Figure 6-24. The main journal bearings used in most Navy marine engines may be classed, according to the construction of the bearings and the materials used, into the following four principal groups:

Bronze-back Staco and Steel-back Staco Bearings

6-42. These bearings are sometimes referred to as the bimetal type. They consist of either a bronze or a steel back bonded with a bearing material of high lead content. The specifications for the back material are based on the type of bearing and the service for which it is intended. The bearing material (known as Staco) consists of approximately 98 percent lead and 1 percent tin.

Trimetal Bearings

6-43. The trimetal bearings have a steel back, bonded with an intermediate layer of bronze to which is bonded a layer of bearing material. The bearing material is either lead-base babbit or tin-base babbit.

Copper-lead Bearings

6-44. Copper-lead main journal bearings are usually constructed of a layer of copper-lead bonded to a steel back. Some of these bearings consist only of a copper-lead shell. Copper-lead bearings are sometimes plated with tin-lead or indium. The plating serves primarily as a protective coating against corrosion. Copper-lead bearings are relatively hard. Therefore, when they are used, the journal surfaces of the shaft must be harder than those required when other types of bearings are used. Remember this when bearings are being replaced.

Aluminum Alloy Bearings

6-45. Bearings made of aluminum alloys are becoming increasingly popular for use in diesel engines. These alloys may contain up to 6 percent tin. The bearings may be of either solid or bimetal construction.

6-46. Main precision bearings with shims are installed in some large engines. Shims provide a means of adjustment to compensate for wear. The bearings of medium and small engines have no shim adjustment. When non-adjustable bearings have worn the prescribed amount, they must be replaced. Main bearings, their housing, and caps are precision-machined with a tolerance sufficiently close so that, when properly installed, the bearings are in alignment with the journals and they fit with a predetermined clearance. The clearance provides space for the thin film of lubricating oil which is forced under pressure between the journals and the bearing surfaces. Under normal operating conditions, the film of oil surrounds the journals at all engine load pressures. Lubricating oil enters the bearing shells from the engine lubricating system, through oil inlet holes and oil grooves in the bearing shells. These inlets and grooves are located in the low-pressure area of the bearing. Main bearings are subjected to a fluctuating load, as are the crankpin bearings and the piston pin bearings.

However, the manner in which main journal bearings are loaded depends on the type of engine in which they are used. For example:

- In a 2-stroke cycle engine, a load is always placed on the lower half of the main bearings and the lower half of the piston pin bearings in the connecting rod. Also, the load is placed on the upper half of the connecting rod bearings at the crankshaft end of the rod. This is true because the forces of combustion are greater than the inertia forces created by the moving parts.
- In a 4-stroke cycle engine, the load is applied first on one bearing shell and then on the other. The reversal of pressure is the result of the large forces of inertia imposed during the intake and exhaust strokes. In other words, inertia tends to lift the crankshaft in its bearings during the intake and exhaust strokes.

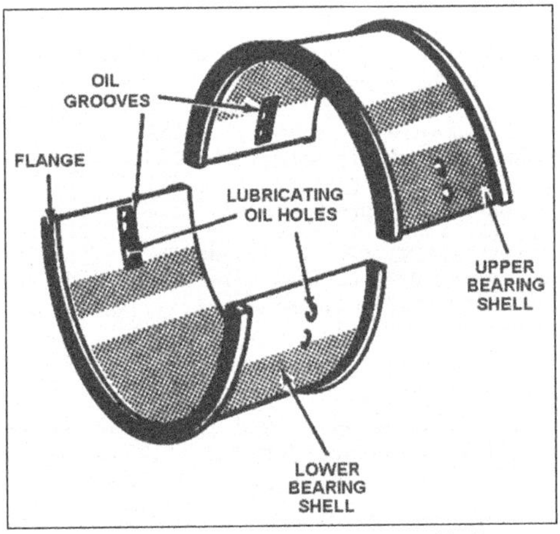

Figure 6-23. Main Journal Bearing Shells

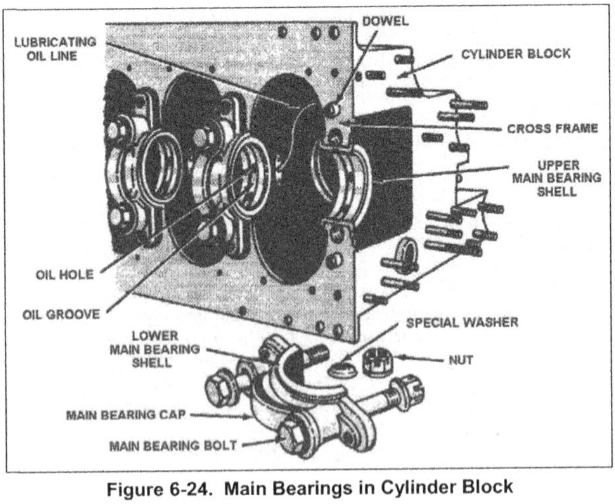

Figure 6-24. Main Bearings in Cylinder Block

6-47. Bearings can become a continual source of trouble unless engine operating personnel follow the recommended operation and maintenance procedures exactly. Severe bearing failures may be evidenced during engine operation by a pounding noise or by the presence of smoke near the crankcase. Impending failures may sometimes be detected by the following:

- A rise in the lubricating oil temperature.
- A lowering of the lubricating oil temperature.
- A lowering of the lubricating oil pressure.

Evidence of impending bearing failures may be detected during periodic maintenance checks or during engine overhauls by inspecting the bearing shells and backs for pits, grooves, scratches, or corrosion. The indication of an impending failure does not necessarily mean that the bearing has completed its useful life. Journal bearings may perform satisfactorily with as much as 10 percent of the load-carrying area removed by metal fatigue. Other minor repairs may be made so that a bearing will give additional hours of satisfactory service. Bearings should not be rejected or discarded because of minor pits, heavy scratches, or areas that indicate metallic contact between the bearing surface and the journal. Minute pits and raised surfaces may be smoothed by using crocus cloth or a bearing-scraping tool. After work has been performed on bearings, every effort must be made to ensure the cleanliness of the bearing surfaces. This also applies to the bearing back and the crank journal and crankpin. A thin film of clean lubricating oil should be placed on the journals and the bearing surfaces before they are reinstalled. The markings of the lower and upper bearing halves should always be checked so that they may be installed correctly. Many bearings are interchangeable when new. However, once they have become worn to fit a particular journal or crankpin, they must be reinstalled on that particular journal or pin. Each bearing half must be marked or stamped with its location (cylinder number) and the bearing position (upper or lower) to ensure correct installation. The connecting rod bearing cap nuts must be pulled down evenly on the connecting rod bolts. This is to prevent possible distortion of the lower bearing cap and consequent damage to bearing shells, cap, and bolts. A torque wrench should be used to measure the torque applied to each bolt and nut assembly. The same torque must be applied to each bolt. If a manufacturer recommends the use of a torque wrench, the specified torque may be obtained from the manufacturer's TM. After a bearing is reassembled, the engine should always be barred or jacked over by hand through several revolutions. This is to ensure that all reciprocating and rotating parts are functioning freely and that no binding exists among the main bearings and the connecting rod bearings and the crankshaft. Larger diesel engines must be turned over, first by the manual jacking gear provided and then by the engine starting system. The use of leads, shim stock, or other such devices is not recommended for determining the clearance of precision bearings. If they are used, there is danger that the soft bearing material may be seriously damaged. A micrometer especially fitted with a spherical seat should be used to determine the thickness of bearing shells. The spherical tip must be placed against the inside of the bearing shell to get an accurate reading and to prevent injury to the bearing material. Figure 6-25 shows a micrometer caliper specially fitted with a steel ball for measuring bearing thickness.

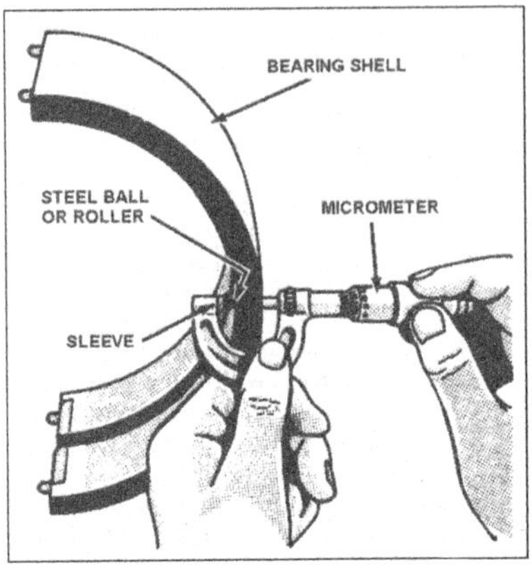

Figure 6-25. Measuring Bearing Shell Thickness With a Micrometer Caliper

6-48. An alternate method of determining clearance is with a Plastigage. The use of the Plastigage eliminates the possibility of leaving an impression in soft bearing metal due to the gauge being softer than the bearing. To use this method, place a length of the Plastigage or proper gauge across the bearing. Then, assemble the bearing cap and tighten it in place. Do not turn the crankshaft, because that will destroy the Plastigage. Next, remove the bearing cap. The width of the Plastigage, after crushing, gives the exact clearance, as indicated on the chart on the Plastigage package. Measurements must be taken at specified intervals, usually at every overhaul, to establish the amount of bearing wear. A number of crankshaft journal diameter measurements should also be taken at suitable points to determine possible out-of-roundness.

REPAIR OF SHAFTS AND JOURNAL BEARINGS

6-49. Repair of crankshafts, camshafts, and bearings varies with the part as well as with the extent of damage. There is no doubt of the necessity to replace such items as broken or bent crankshafts and camshafts with damaged integral cams and failed camshaft bearings. Out-of-round journals may be reground and undersize bearing shells may be installed, but this requires skilled personnel experienced in the use of precision tools. If available, a new shaft should be installed and the damaged shaft should be sent to a salvage reclamation center. Under certain conditions, scored crankshaft journals or damaged journal bearings may be kept in service if properly repaired.

SCORED JOURNALS

6-50. Repair of scored journals depends on the extent of scoring. If a crankshaft has been overheated, the effect of the original heat treatment will have been destroyed. It is advisable to replace the crankshaft. If journal scoring is only slight, an oilstone can be used for dressing purposes, provided that you are extra careful concerning abrasives. During the dressing operation, it is advisable to plug all oil passages within the journal as well as those connecting the main bearing journal and the adjacent crankpin. In the dressing procedure, a fine oilstone, followed with crocus cloth, should be used to polish the surface. After dressing, the journals must always be washed with diesel oil. This procedure must include washing of the internal oil passages as well as the outside journal surfaces. Some passages are large enough to accommodate a cleaning brush. Smaller passages can be cleaned by blowing out with compressed air. The passages should always be dried by blowing with compressed air. Never stow a crankshaft or bearing part on any metal surface. When a shaft is removed from an engine, it should be placed on a wooden plank with all journal surfaces protected. If the shaft is to be exposed for any length of time, protect each journal surface with a coating of heavy grease. Bearings should always be placed on wooden boards or clean cloths.

Crankshaft Overhaul

6-51. Crankshaft overhaul consists not only of an inspection and servicing for scoring and wear but also a determination of each crank web deflection. Crank web deflection readings should be taken whenever an overhauled crankshaft is installed in the engine. A strain gauge, often called a crank web deflection indicator, is used to take deflection readings. The gauge is merely a dial-reading inside micrometer used to measure variation in distance between adjacent crank webs as the engine shaft is barred over. When installing the gauge, or indicator, between the webs of a crank throw, see that the gauge is placed as far as possible from the axis of the crankpin. The ends of the indicator should rest in prick-punch marks in the crank webs. If these marks are not present, they must be made so that the indicator may be placed in its correct position. Consult the manufacturer's TM for the proper location of new marks. Readings are generally taken at the following four crank positions:

- TDC.
- Inboard.
- Near or at BDC.
- Outboard.

6-52. In some engines, it is possible to take at BDC. In others, the connecting rod may interfere, making it necessary to take the reading as near as possible to BDC without having the gauge come in contact with the connecting rod. The manufacturer's TM for the specific engine contains information concerning the proper position of the crank when readings are to be taken. When gauge is in its lowest position, the dial will be upside down. This necessitates the use of a mirror and flashlight so that a reading can be obtained. Once the indicator has been placed in position for the first deflection reading, the gauge should not be touched until all four readings have been taken and recorded. Variations in the readings obtained at the four crank positions will indicate distortion of the crank. Distortion may be caused by several factors, such as a bent crankshaft, worn bearings, or improper engine alignment. The maximum allowable deflection can be obtained from the manufacturer's TM. If the deflection exceeds the specified limit, the cause of the distortion must be determined and corrected. Deflection readings are also used in determining correct alignment between the engine and the generator, or between the engine and the coupling. However, when alignment is being determined, a set of deflection readings is usually taken at the crank nearest the generator or the coupling. In aligning an engine and generator, you may need to install new chocks between the generator and its base to bring the deflection within the allowable value. You may also need to shift the generator horizontally to get the proper alignment. When an engine and a coupling are to be aligned, the coupling must first be correctly aligned with the drive shaft. Then, the engine must be properly aligned to the coupling, rather than the coupling aligned to the engine.

Care of Camshafts

6-53. When a camshaft is removed from the engine, it must be thoroughly cleaned. Kerosene or diesel fuel may be used. After the shaft is cleaned, it should be dried with compressed air. After the cam and journal surfaces are cleaned, they should be inspected for any signs of scoring, pitting, or other damage. When a camshaft is being inserted or removed by way of the end of the camshaft recess, the shaft should be rotated slightly. Rotating the camshaft allows it to enter easily and reduces the possibility of damage to the cam lobes and the bearings.

FLYWHEELS

6-54. A flywheel stores up energy during the power events and releases it during the remaining events of the operating cycle. In other words, when the speed of the shaft tends to increase, the flywheel absorbs energy. Conversely, when the speed tends to decrease, the flywheel gives up energy to the shaft in an effort to keep shaft rotation uniform. In doing this, a flywheel keeps variations in speed within desired limits at all loads. It limits the increase or decrease in speed during sudden changes of load. It also aids in forcing the piston through the compression event when an engine is running at low or idling speed. It also helps to bring the engine up to speed when it is being cranked. Flywheels are generally made of cast iron, steel, or rolled steel. Strength of the material from which the flywheel is made is of prime importance because of the stresses created in the metal of the flywheel the engine is operating at maximum designed speed. In some engines, a flywheel is the point of attachment for such items as a starting ring gear, a turning ring gear, or an overspeed safety mechanism. The rim of a flywheel may be marked in degrees. With a stationary pointer attached to the engine, the degree markings can be used to determine the position of the crankshaft when the engine is being timed.

Chapter 7
OPERATING MECHANISMS FOR ENGINE PARTS

INTRODUCTION

7-1. To this point, the main engine parts (stationary and moving) have been discussed. At various points in the previous chapters, reference is made to the operation of other engine parts. However, little has been said about the source of power which causes these parts to operate. This chapter deals with this subject.

ENGINE PART OPERATING MECHANISMS

SOURCE OF POWER

7-2. The source of power for the operation of one engine part is also frequently the source of power for other parts and accessories of the engine. For example, the source of power for operation of engine valves may also be the source of power for the following:

- Operation of the governor.
- The fuel pump.
- The lubricating pump.
- The water pump.
- The overspeed trips.

Since mechanisms which transmit power to operate specific parts and accessories may be related to more than one engine system, they are explained before the engine systems are discussed. The parts which make up the operating mechanisms of an engine may be divided into two groups:

- The group which forms the drive mechanism.
- The group which forms the actuating mechanism.

The source of power for the operating mechanisms of an engine is the crankshaft.

Drive Mechanism

7-3. As used in this chapter, drive mechanism identifies that group of parts which takes power from the crankshaft and transmits it to various engine components and accessories. The drive mechanism does not change the type of motion in an engine, but it may change the direction of motion. For example, the impellers of a blower are driven or operated as a result of rotary motion which is taken from the crankshaft and transmitted to the impellers by the drive mechanism, an arrangement of gears and shafts. While the type of motion (rotary) remains the same, the direction of motion of one impeller is opposite to that of the other impeller as a result of the gear arrangement within the drive mechanism. A drive mechanism may be of the gear, the chain, or the belt type. The gear type is the most common, but some engines are equipped with chain assemblies. A combination of gears and chains is used as the drive mechanisms on small boat engines.

Accessory Drive

7-4. Some engines have a single-drive mechanism which transmits power for the operation of engine parts as well as accessories. In other engines, there may be two or more separate mechanisms. When separate assemblies are used, the one which transmits power for the operation of the accessories is called the accessory drive. Some engines have more than one accessory drive.

Camshaft Drive or Timing Mechanism

7-5. A separate drive mechanism which transmits power for the operation of engine valves is generally called the camshaft drive or timing mechanism. The camshaft drive, as the name implies, transmits power to the camshaft of the engine. The shaft, in turn, transmits the power through a combination of parts which causes the engine valves to operate. Since the valves of an engine must open and close at the proper moment (with respect to the position of the piston) and remain in the open and closed positions for definite periods, a fixed relationship must be maintained between the rotational speeds of the crankshaft and the camshaft. Camshaft drives are designed to maintain the proper relationship between the speeds of the two shafts. In maintaining this relationship, the drive causes the camshaft to rotate at crankshaft speed in a 2-stroke cycle engine and at one-half crankshaft speed in a 4-stroke cycle engine.

Actuating Mechanism

7-6. The actuating mechanism, as used in this chapter, identifies that combination of parts which receives power from the drive mechanism and transmits the power to the engine valves. In order for the valves (intake, exhaust, fuel injection, air start) to operate, there must be a change in the type of motion. In other words, the rotary motion of the crankshaft and drive mechanism must be changed to a reciprocating motion. The group of parts which, by changing the type of motion, causes the valves of an engine to operate is generally referred to as the valve-actuating gear or mechanism. A valve actuating mechanism may include the cams, cam followers, push rods, rocker arms, and valve springs. In some engines, the camshaft is so located that push rods are not needed. In such engines, the cam follower is a part of the rocker arm. Some actuating mechanisms are designed to transform reciprocating motion into rotary motion. However, in internal-combustion engines, most actuating mechanisms change rotary motion into reciprocating motion.

Design and Arrangement of Operating Mechanisms

7-7. Considerable variation exists in the design and arrangement of the parts of operating mechanisms found in different engines. The size of engine, cycle of operation, cylinder arrangement, and other factors govern the design and arrangement of the components as well as the design and arrangement of the operating mechanisms themselves. Some of the variations in operating mechanisms are described as follows. The arrangements of operating mechanisms described in this chapter represent those commonly found in marine engines used aboard Army vessels.

OPERATING MECHANISMS FOR A 2-STROKE CYCLE, IN-LINE DIESEL ENGINE

7-8. The operating mechanisms of some engines consist of a single-drive mechanism and the valve-actuating mechanism. The 6-cylinder, GM 6-71 engine is an example of such an engine. As a complete assembly, which transmits power from the driving part to the driven part, the operating mechanisms of the GM 6-71 consist of gears, shafts, couplings, and the parts of the valve-actuating mechanism.

Gears

7-9. When the driving mechanism of an engine consists only of gears, the mechanism is commonly referred to as a gear train. Figure 7-1 shows the gear train, or drive, for a GM 6-71 engine. The arrangement also shows the design for right-hand rotation. The gear train of a GM 6-71 functions as the camshaft drive as well as the accessory drive. The train consists of five helical gears, completely enclosed and located at the rear end of the engine. Note that all gears are driven by the crankshaft gear through an idler gear. The idler gear may be located on either the right or left side of the engine (see dummy hub, sometimes called spacer), depending on the direction of crankshaft rotation. Since the engine operates on the 2-stroke cycle, the camshaft and balancer gears are driven at the same speed as the crankshaft gear. Either the camshaft gear or the balancer gear may be driven by the crankshaft gear through the idler gear, the drive arrangement depending on the model (right- or left-hand rotation). The camshaft and balance shaft gears are counterweighted for balance purposes. The accessories of the GM 6-71 receive power from the blower drive gear which is driven by the camshaft gear.

Located on the blower side of the engine and supported by the rear end plate, the blower drive gear transmits power not only to the blower but also to the governor, water pump, and fuel pump.

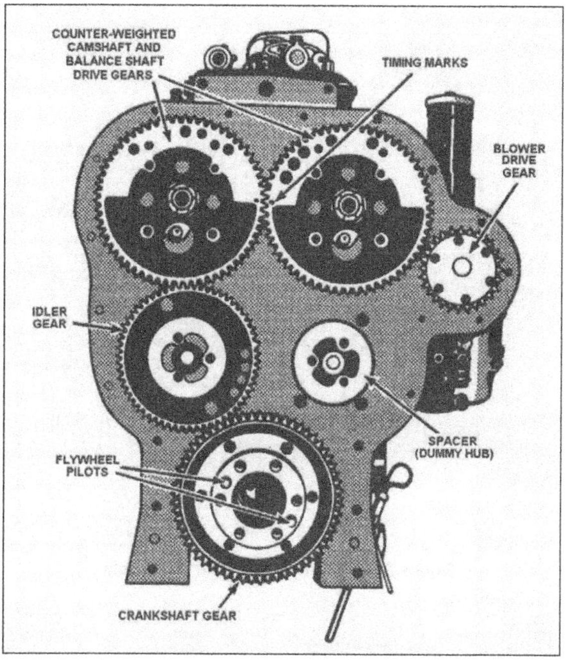

Figure 7-1. Driving Mechanism: Camshaft and Accessory Drive (GM 6-71) Showing Gear Train (Right-Hand Rotation)

Blower Drive Gear

7-10. Figure 7-2 shows the location of the various engine accessories and the shafts, gears, and couplings that transmit the power from the blower drive gear to each of the accessories. The blower end of the governor drive shaft is serrated or splined and it engages with corresponding serrations or splines inside the upper blower shaft. The fuel pump is bolted to the rear cover of the blower and is driven from the lower blower rotor shaft, through a device that acts as a universal joint. The water pump is mounted on the front end of the blower and is driven by the rotor shaft, through a coupling. The in-line engine discussed in the previously requires only one drive mechanism (gear train) to transmit power to the valve actuating gear and engine accessories. Our discussion will now cover and engine that uses two separate gear drives (gear trains), one at each end of the engine. The front gear train, as shown in Figure 7-3 consists of a crankshaft gear and two idler gears. The idler gears serve to drive the water pump (not shown) and balance the engines (see balance weights). The rear gear train (see Figure 7-3) consists of a crankshaft gear three idler gears, and two camshaft gears. The rear idler gears, like the front, also serve to balance the engine (see balance weights). The two other gears that are mounted on the rear of the engine (the accessory drive gear and the blower drive gear) are shown in Figure 7-4. The blower drive gear supplies the power that operates many of the same accessories as those on the GM 6-71 (see Figure 7-1). The correct relationship between the crankshaft and the two camshaft must be maintained so that the fuel injection, the opening and closing of exhaust valves, and the engine balance can be properly controlled. Since the camshaft must be in time with the crankshaft, timing marks are stamped on the face of the gears to facilitate correct gear train timing. The timing marks stamped on various gears are shown in Figure 7-1. When an engine is assembled, whether it is a 2-stroke or 4-stroke cycle engine, it is important that the appropriate timing marks be lined up on the gears as each gear is installed.

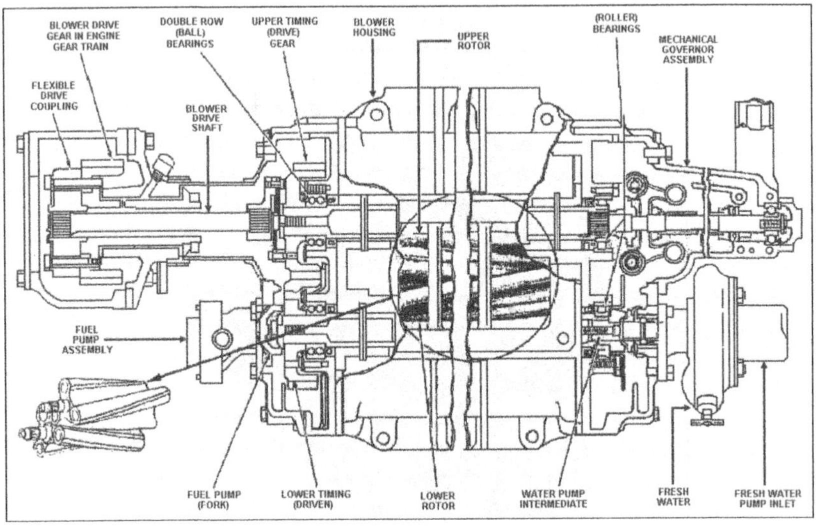

Figure 7-2. Blower Drive Gear Assembly

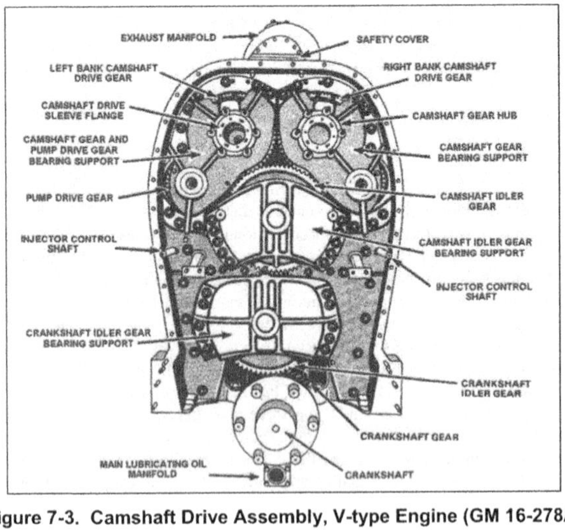

Figure 7-3. Camshaft Drive Assembly, V-type Engine (GM 16-278A)

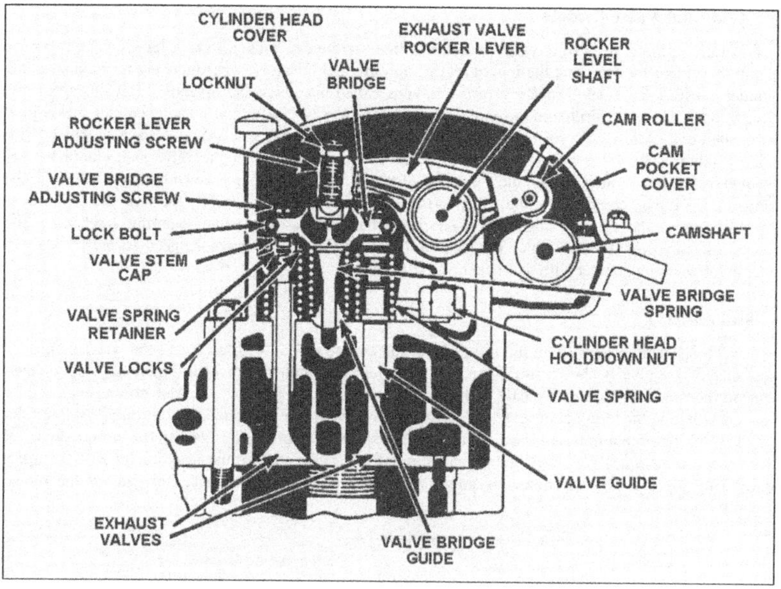

Figure 7-4. Valve-operating Mechanism With Valve Bridge (GM 16-278A)

Camshaft Drive and Valve-Actuating Gear

7-11. The camshaft gear train in the GM 16-278A is similar to that of the GM 6-71 except for two additional gears. Figure 7-3 shows an end view of the engine showing the camshaft drive assembly. Compare this assembly with that of the GM 6-71.

Gears and Couplings

7-12. Two camshaft gears are driven by the crankshaft gear at the power takeoff or coupling end of the shaft, through the crankshaft and the camshaft idler gears. Both camshaft gears mesh with the camshaft idler gear. The lubricating oil pump gears are driven by the camshaft gears. A camshaft gear also furnishes power to drive the governor and the tachometer. In some models, the governor drive shaft is driven by the camshaft gear through a flexible coupling. In other models, the shaft is flanged and bolted to a camshaft gear. The train of gears in the camshaft drive is enclosed in an oil-tight housing. A spring-loaded safety cover is fitted to a pressure relief opening in the top of the housing.

Camshafts

7-13. Each cylinder bank of the engine is fitted with a camshaft. These shafts rotate at the same speed as the crankshaft but in the opposite direction. Each shaft is made of two sections which are flanged and bolted together. The cams are case-hardened and are an integral part of each shaft section. As in the GM 6-71, three cams are provided for each cylinder. However, in the GM 16-278A, push rods are not required for the transmission of power.

Rocker Arms and Valve Bridges

7-14. Each cylinder head of the GM 16-278A is fitted with three rocker arms or levers. The two outer arms operate the exhaust valves, and the inner arm operates the fuel injector. Since there are four exhaust valves per cylinder, each exhaust valve rocker arm (sometimes called a lever) must operate a pair of valves through a valve bridge. The valve bridge in this engine is made of forged steel and has a hardened ball socket into which fits the ball end of the rocker lever adjusting screw. The valve bridge has two arms, each of which fits over an exhaust valve. The valve bridge spring keeps valve bridge tension off the valve stems until the bridge is actuated by the rocker arm. When the valve end of the rocker arm is forced down by the cam action, the valve bridge moves down, compressing the valve springs and opening the valves. By the time the action of the cam lobe has ceased, the valve springs will have closed the valves. The valve-operating mechanism shown in Figure 7-4 represents those in which the location of the camshaft eliminates the need for push rods. The cam lobes come in direct contact with the rocker arm cam rollers.

ACCESSORY DRIVE

7-15. The gear train located on the front end of the engine drives the blower and the water pumps. An end view of the assembly is shown in Figure 7-5. The gear train consists of helical gears of forged steel which transmit the rotation of the crankshaft to the blower and the water pumps. The assembly is enclosed in a case which is bolted to the blower housing. The drive gear is driven from the crankshaft through a splined shaft, one end of which fits into a hub that is bolted to the crankshaft. The other end fits into the blower drive gear hub. The drive gear operates the water pump gears through idler gears, and it meshes directly with the upper drive gear. The upper drive gear transmits power through a shaft to the rotor-driven gear of the blower gear assembly.

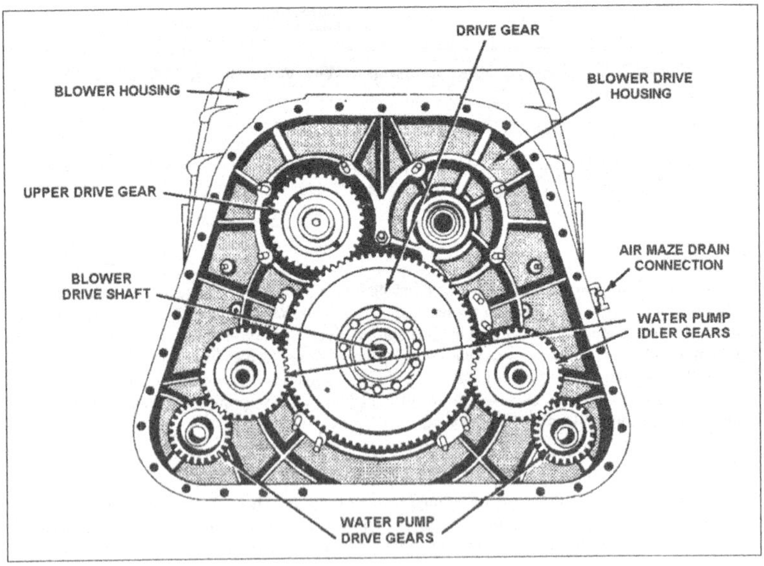

Figure 7-5. Accessory Drive Mechanism (GM 16-278A)

OPERATING MECHANISMS FOR AN OPPOSED-PISTON DIESEL ENGINE

7-16. The operating mechanisms of opposed-piston engines differ, to a degree, from those of single-acting engines. Some of the differences occur because power is supplied by two crankshafts in an opposed-piston engine instead of one. Other differences occur in actuating gear because ports are used instead of valves for both intake and exhaust in an opposed-piston engine. Regardless of the differences in mechanisms, the basic types of drives (gear and chain) are found in both single-acting and opposed-piston engines. The two engines (GM 6-71 and GM 16-278A) described previously have only gear drive mechanisms. The opposed-piston engine (Fairbanks-Morse [FM] 38D 1/8) used as an example, has chain assemblies as well as gear trains incorporated in the mechanisms which supply power to engine parts and accessories. The FM 38D 1/8 opposed-piston engine has three separate drive mechanisms. The drive which furnishes power to the camshaft and fuel injection equipment is the chain type. The blower and the accessories are operated by gear-type drives.

CAMSHAFT DRIVE AND FUEL PUMP ACTUATING GEAR

7-17. The camshaft drives of the engines discussed previously supply power to one or more accessories as well as to the valve-actuating gear. This is not true of the drive in an FM 38D 1/8 opposed-piston engine. The FM 38D 1/8 does not have cylinder valves and two other drives are provided to operate the accessories. Therefore, the primary purpose of the camshaft drive is to transmit power for operation of the fuel injection pumps.

Chain Assembly

7-18. The power required to operate the fuel injection pumps at the proper instant during the cycle of operation is transmitted through the camshafts from the crankshaft by a chain drive (frequently referred to as the timing mechanism). The names and arrangement of the components of the drive are shown in Figure 7-6. The drive sprocket is attached to the upper crankshaft at the control end of the engine. A sprocket is attached to the end of each camshaft, and three other sprockets are provided for timing and adjustment purposes. The chain conveys the rotation of the upper crankshaft to the camshaft sprockets by passing over the crankshaft sprocket, under the two timing sprockets, over the two camshaft drive sprockets, and under the tightener sprockets. The timing sprockets are mounted on an adjustable bracket or lever. By moving the lever, the timing of the two camshafts can be adjusted. The adjustable tightener sprocket provides a means of obtaining and maintaining the proper slack in the chain.

Actuating Gear

7-19. The camshafts are located in the upper crankshaft compartment. Since engines of the opposed-piston type use ports for both intake and exhaust, the camshafts function only to actuate the two fuel injection pumps at each cylinder in unison and at the proper time. The shafts turn at the same rate of speed as the crankshaft. The camshafts for the engine shown are of case-hardened alloy steel and are made in sections. The sections are made with matching marked flanges and are joined with fitted bolts. In some opposed-piston engines made by the same manufacturer, the shafts are of construction. The cams are an integral part of the shaft, and one cam is provided on each shaft for each cylinder. The cams transform the rotary motion of the shaft into the up-and-down motion of the fuel pump plunger, through a tappet assembly attached to the top of the fuel pump body. The push rod spring of the tappet assembly holds the push rod and the cam roller against the camshaft cam. As the camshaft rotates, the cam acts against the cam roller to force the push rod down against the spring tension and to actuate the injection pump plunger.

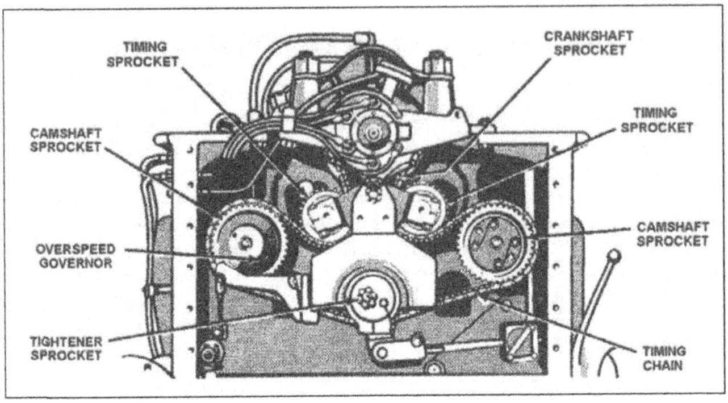

Figure 7-6. Camshaft Drive and Timing Mechanism (FM 38D 1/8)

BLOWER DRIVE MECHANISM

7-20. The power to drive the blower is transmitted from the upper crankshaft through a gear train. The train consists of a drive gear, a pinion gear, and the two timing (impeller) gears of the blower.

Flexible Drive Gear

7-21. The principal parts of the flexible drive gear are a spider drive hub which is keyed to the crankshaft, a gear within which spring spacers are bolted, and springs which absorb torsional oscillations transmitted by the crankshaft. A view of the flexible drive gear, with end plate removed, is shown in Figure 7-7.

Drive Pinion

7-22. The flexible drive gear meshes with the drive pinion. The pinion is keyed to the lower impeller shaft and is held in place by a locknut. The lower impeller driving (timing) gear meshes with the upper impeller driving gear.

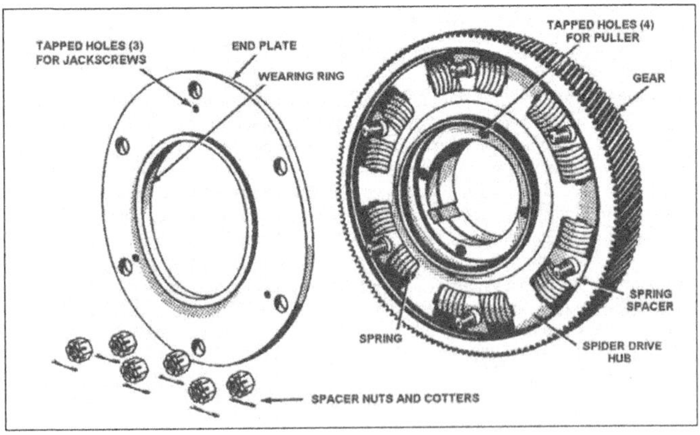

Figure 7-7. Blower Flexible Drive Gear (FM 38D 1/8)

Accessory Drive

7-23. The majority of the accessories for the FM 38D 1/8 are driven by a gear mechanism which receives power from the lower crankshaft at the control end of the engine. Refer to Figure 7-7 as you read. The following description will help you to become familiar with the components of the drive and with the manner in which power is transmitted to the driven units. The accessory drive transmits power to the following:

- Water pumps.
- Fuel oil pump.
- Lubricating oil pump.
- Governor.

Engine shocks transmitted by the crankshaft are absorbed by the drive springs of the gear. The water pump drive gears mesh directly with the flexible drive gear. The fuel pump drive gear (attached to the flexible drive gear) transmits power to the fuel pump driven gear through an idler. The lubricating oil pump drive gear meshes directly with the flexible drive gear. Power is transmitted to the pump through a shaft and an internal gear coupling (the lubricating oil pump drive). The shaft of the lubricating oil pump drive also transmits power to the governor. A gear on the shaft meshes with a mating gear on the governor drive gear shaft. This shaft drives the governor coupling shaft which, in turn, drives the governor through a bevel gear drive.

OPERATING MECHANISMS FOR A 4-STROKE CYCLE, DIESEL ENGINE

7-24. The operating mechanisms discussed so far have applied to 2-stroke cycle diesel engines. We now will take a look at the operating mechanisms of a 4-stroke cycle diesel engine. They are similar except for some minor differences in design and arrangement and the reduced operating speed of the camshaft. The drive has no provision for driving a blower, since 4-stroke cycle diesel engines are either naturally aspirated or are turbocharged. Turbocharging units are exhaust driven and require no mechanical drive. The valve actuating gear operates valves for both intake and exhaust. The arrangement of the operating mechanism is shown in Figure 7-8.

Drive Mechanisms

7-25. The 2-stroke cycle engine drive mechanisms considered are the gear type, the chain type, or a combination of both. Very few chain drives are used on 4-stroke cycle engines by the Navy. Therefore, only the gear drives are discussed. The primary drive, which takes power from the crankshaft, is a gear assembly or gear train. The arrangement of the gear train which forms the drive varies according to the engine.

Valve Actuating Gear

7-26. In addition to furnishing power for the various engine accessories, the drive mechanism provides the power that operates the engine valves. The basic job of the valve actuating gear is to cause and to control the opening and closing of the inlet and exhaust valves. In most engines, this gear consists of rocker arms that actuate the valves, push rods that connect the rocker arms and the cams on the camshaft, and a drive that connects the camshaft to the crankshaft.

Camshaft Drives

7-27. In 4-stroke cycle diesel engines, the camshaft speed must be exactly one-half the crankshaft speed; so that the camshaft makes one complete revolution while the crankshaft makes two. Since these speed relations must be exact, the connecting drive must be positive. The drive arrangement used for any particular engine depends to a large extent on where the camshaft is located. The camshaft may be located low, near the crankshaft, using long push rods, or on the cylinder block using short push rods, or at the cylinder head level without push rods (see Figure 7-9). The gears of the drive must be accurately cut and heat-treated to resist wear. Helical teeth (teeth placed at an angle) are frequently used in place of spur teeth (teeth placed straight) for greater quietness and more transmission of power. Gears and shafts are used in arrangements to drive the vital components and accessories of the engine.

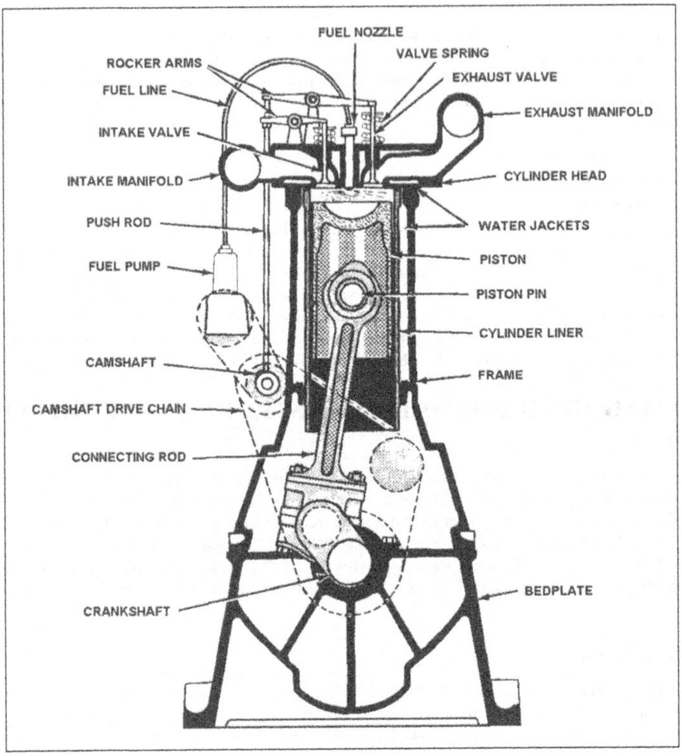

Figure 7-8. Arrangement of 4-stroke Cycle Diesel Engines

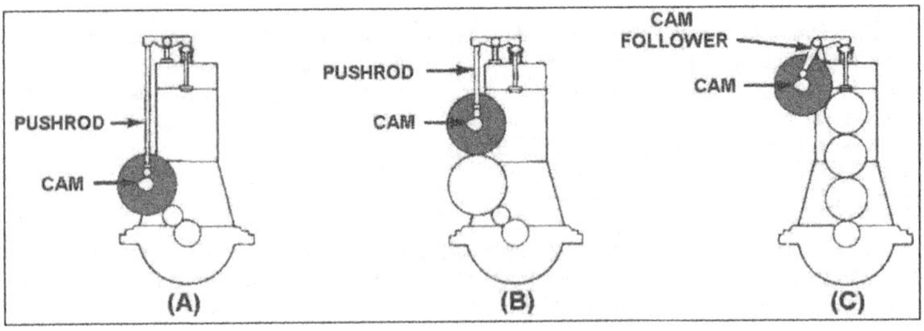

Figure 7-9. Cam Location, 4-stroke Cycle Diesel Engine

CAMSHAFTS

7-28. The camshafts in 4-stroke cycle diesel engines carry the cams for actuating the inlet and exhaust valves. The camshaft may also carry cams for fuel injection pumps, fuel-spray valves, or air-starting valves. Depending on the engine design, some engines have two camshafts, others have only one. The camshaft may be constructed in several ways. It can be forged in one piece, including the cams themselves, as integral cams. It may consist of a steel shaft with separate forged-steel or cast-iron cams keyed onto the shaft. Very few of the cast-iron cams keyed onto the shaft are still in service. Another construction used on large engines is to make up the camshaft in sections. Each section handles one cylinder and enough sections are bolted together to handle the whole engine.

VALVE GEAR IN CYLINDER HEAD

7-29. The remaining parts of the valve actuating gear and the valves are mounted in the cylinder head.

DIESEL ENGINE FAILURE

7-30. Some of the most severe diesel engine failures have occurred because of a broken or chipped gear. A metallic clicking noise in the gear housing usually indicates a broken tooth. Serious trouble can occur if a piece of this broken tooth, or any piece of metal, becomes lodged between meshing gear teeth. A severe shuddering noise can be heard if the gears are about to fail. The damage may be such that the engine suddenly stops during operation. This can mean that the auxiliaries are not being driven correctly by the gear train from the crankshaft, with the result that the engine fails to operate.

GEAR FAILURE (CAUSES AND PREVENTION)

7-31. The following are the most likely causes of gear failure:

- Improper lubrication.
- Corrosion, misalignment.
- Torsional vibration.
- Excessive backlash.
- Wiped gear bearings and bushings.
- Metal obstructions.
- Improper manufacture of gears.

Improper Lubrication

7-32. The method of supplying lubrication to the gears varies in different engines. In some of the smaller engines, a splash system is often used, with the lower gears dipping in oil and carrying it to the other members by gear contact. This method does not always prove satisfactory with the larger engines because of the increased number and size of gears used. In the forced feed system, oil is piped to the gear bearings or bushings, from which point it is conducted to the gears through drilled passages. There may also be a jet that sprays the lubricant directly into the gear teeth as they mesh. Lubricating oil, at the designated pressure and temperature, must be supplied to the entire gear assembly at all times during operation. If the lubricant is not maintained at the specified temperature, its fluidity will be changed. The oil must not be so viscous that it cannot flow through the restricted passages or through the jets. Neither should it be so thin that the oil film will be squeezed from between the teeth. Such practice would damage the gears. The jets must be kept open at all times and should not be altered or rendered inoperative without authority of NAVSEA. It is imperative that the lubricant be kept clean. Any particle of dirt, lint, or metal flake from the wearing-in of new gears must be removed from the system. If not, it will clog up the passages and jets, shutting off the flow of oil. This will result in metal-to-metal gear contact with eventual gear failure. Also, inclusion of metal chips between gear teeth is detrimental. Some lubricating oil strainers have magnets for removing metal particles. The gear train must be inspected periodically. Remove the gear case and inspect the entire train. Pay special attention to the jets and oil passages. They are minute and can become clogged easily. Refer to the instruction manual for the particular engine for the interval between inspections.

Corrosion

7-33. Gear failure, due to corrosion, is not uncommon. Saltwater or acid within the oil is harmful to metal gears. Even a slight amount of either will cause pitting and rusting. Corrosion can also be detrimental to gear bearings and bushings. Bearings that are corroded soon become wiped, causing gear misalignment. This leads to excessive wear that will result in gear failure. To remove water which may contain salt or acid, centrifuge the oil at specified intervals.

Misalignment

7-34. Faulty alignment between meshing gears can cause one gear to wear excessively, resulting in a "feather edge" on the teeth and the probability of a chipped gear. See that auxiliary couplings are aligned accurately before they are bolted together. Gear alignment is particularly important in reverse and reduction gear units. If the pinions and gears are not lined up perfectly, vibration will be induced that will cause gear failure.

Torsional Vibration

7-35. When engines are operated at certain critical speeds, torsional vibrations are established that will disrupt the internal structure of the gears, causing them to break. Each multi-cylinder engine has a critical speed and, in some cases, this lies within the normal operating range. If this is the case, the instruction manual for the engine will state definitely that the engine should not be operated for any length of time within that speed range. If it falls within the normal range, the critical range should be conspicuously marked on the engine tachometer. If necessary to pass through the critical range to reach rated speed, then pass through as quickly as possible.

Excessive Backlash

7-36. Backlash is the play between surfaces of the teeth in mesh. It is measured at the point of least play. It increases with wear and can increase considerably without causing damage. However, excessive backlash can be detrimental to the gears and can cause gear failure. It also will change the timing of the camshaft, fuel pump, and supercharger, causing the engine to operate inefficiently and incorrectly. Check backlash during the regular periodic inspection of the gear train. The clearances taken should be recorded and then compared with the allowable amount as specified in the instruction manual. If the readings are in excess of the specified amount, the gears must be replaced. Frequent inspections for scoring, along with an adequate supply of pure lubricating oil and maintenance of proper clearances, will tend to prevent this trouble. Excessive backlash in governor drives may often cause governor hunting. Wear cannot take place without metallic contact, which will cause scoring of teeth. In the installation of gears, any misalignment will cause improper tooth contact and will result in wear. Fiber gears wear more quickly than metal gears. Water must be kept away from fiber material; if not, the material will soften and fail.

Wiped Gear Bearings and Bushings

7-37. Gear failures have frequently been the result of bearing seizures caused by improper lubrication. If the bearings become wiped, the shaft will drop a slight amount and cause incorrect alignment. This will result in improper tooth contact as well as a scored shaft. The same applies to bushings. In one certain incident, an idler gear bushing wore considerably. This caused misalignment that resulted in the shearing of the studs in the idler gear bracket and failure of the entire gear train when these stripped studs became lodged between meshing teeth. The gear bearings and bushings must be supplied with a lubricant free from metal flakes, dirt, and saltwater.

Metal Obstructions

7-38. Any piece of metal that becomes lodged between meshing teeth will probably cause the teeth to strip or become chipped. The obstruction may be metal flakes from the gear or it may be from an outside source (such as a broken oil line, loose nuts, sheared studs, or loose dowel pins). The gearbox must be kept clean. All wires, cotter pins, and keys must be removed. An inspection should be made for loose nuts and failing lube oil lines.

Improper of Gears

7-39. Gear failures are sometimes the result of the manufacturer's procedure in the manufacture of steel (such as cooling the ingot too rapidly). This will cause a defective gear that will have fractures within its structure.

CHAIN FAILURE (CAUSES AND PREVENTION)

7-40. The outstanding causes of the wearing or breaking of chains are the following:

- Chain too tight.
- Chain too loose.
- Lack of lubrication.
- Sheared cotter pins.
- Misalignment.

Chain Too Tight

7-41. Excessive wear becomes evident if the chain is under too much tension. The chain will become overloaded and subject to breakage. Excessive wear in fuel pump chain drives will become evident by retarded timing. This will cause a combustion knock and generally, inefficient engine operation. Correct tension in the chain should always be maintained. Refer to the engine instruction manual for correct tension and for allowable sag. Most engines employ idler sprockets to adjust the chain tension.

Chain too Loose

7-42. A loose chain will flap and vibrate, causing excessive wear. If a fuel pump drive chain becomes too loose, timing will be incorrect and a combustion knock will develop. Any wear in a camshaft drive chain will retard timing.

Lack of Lubrication

7-43. Chains are lubricated similarly to gears. Quite often there is a jet spray that forces oil between the meshing chain and sprocket. Any clogging of this passage will stop the flow of oil to the chain, resulting in metal-to-metal contact between the sprockets and chains.

Sheared Cotter Pins

7-44. Frequent inspections of the chain link cotter pins is imperative in order to detect any defects that may lead to later failures.

Misalignment

7-45. The camshaft drive chain must be in perfect alignment with the idler sprocket. If it is not, a side thrust will be induced, which in time, will rupture the chain. Misalignment can be readily observed by checking the connecting links for wear as indicated by a polished inside surface of the connecting links of the chain. The sides of the connecting links may also exhibit a polished surface. Correct alignment between the camshaft drive chain and the idler sprocket can usually be obtained by adjusting the crankshaft thrust bearing clearance.

Chain Repair

7-46. Consult the engine instruction manual to obtain the interval between inspections. The tension should be adjusted as required during these inspections. The tension in a new chain should be checked after a few hours of operation (50 hours in some engines). Any needed adjustments should be made. This procedure also applies to a new engine. During operation, the chain will increase slightly in length due to stretch and wear. Most engine manufacturers will insist that the chain be inspected every 1,000 hours, after the initial stretch has been taken up. Spray nozzles should be cleaned at frequent intervals to prevent clogging. In installing a new chain, the connecting link pins should be peened in place, but excessive peening should be avoided. After this procedure, the chain should be checked to see that the links move freely without binding in any position. All cotter pins must be secure. Spare links should be carried in stock. Engine timing should always be checked after new drive chains have been installed. Belts are rarely used in marine diesel engines. A few engines have belt-driven generators and circulating water pumps. Belts, like chains, should be inspected for wear and for sag at frequent intervals. Always consult the engine instruction manual to obtain the allowable tension in a belt. Oil is most injurious to rubber. It attacks the rubber material of the belt in addition to reducing its coefficient of friction.

Chapter 8

FUEL SYSTEMS AND ENGINE SPEED CONTROLS

INTRODUCTION

8-1. This chapter deals with diesel engine fuel systems and engine speed controls. Section I discusses fuel system requirements; Section II, external fuel systems; Section III, fuel injection systems; and Section IV, engine speed control devices.

SECTION I. REQUIREMENTS OF FUEL SYSTEMS

The external fuel system stores and delivers clean fuel to the fuel injection system. In delivering fuel to the cylinders, the fuel injection system must fulfill the following requirements:

- Meter or measure the correct quantity of fuel injected.
- Time the fuel injection.
- Control the rate of fuel injection.
- Atomize or break up the fuel into fine particles according to the type of combustion chamber.
- Pressurize and distribute the fuel to be injected.

The desired condition is to create a homogeneous mixture in the combustion space, in the correct proportions, of the smallest possible fuel particles and air. Although it is not possible to achieve an ideal condition, a good fuel injection system will come close.

METERING

8-2. Accurate metering or measuring of the fuel means that, for the same fuel control setting, the same quantity of fuel must be delivered to each cylinder for each power stroke of the engine. Only with accurate metering can the engine operate at uniform speed with a uniform power output. Smooth engine operation and an even distribution of the load between the cylinders depend on the same volume of fuel being admitted to a particular cylinder each time it fires, and upon equal volumes of fuel being delivered to all cylinders of the engine.

TIMING

8-3. In addition to measuring the amount of fuel injected, the system must properly time the injection to ensure efficient combustion so that maximum energy can be obtained from the fuel. When fuel is injected too early into the injection cycle, it may cause the engine to detonate and lose power and also have low exhaust temperatures. If the fuel is injected late into the injection cycle, it will cause the engine to have high exhaust temperatures, smoky exhaust, and a loss of power. Both situations will have an adverse effect on fuel economy.

CONTROLLING RATE OF FUEL INJECTION

8-4. A fuel system must also control the rate of injection. The rate at which fuel is injected determines the rate of combustion. The rate of injection at the start should be low enough that excessive fuel does not accumulate in the cylinder during the initial ignition delay (before combustion begins). Injection should proceed at such a rate that the rise in combustion pressure is not excessive, yet the rate of injection must be such that fuel is introduced as rapidly as possible to obtain complete combustion. An incorrect rate of injection will affect engine operation in the same way as improper timing. If the rate of injection is too high, the results will be similar to those caused by an excessively early injection. If the rate is too low, the results will be similar to those caused by an excessively late injection.

ATOMIZING FUEL

8-5. As used in connection with fuel injection, atomization means the breaking up of the fuel, as it enters the cylinder, into small particles which form a mist-like spray. Atomization of the fuel must meet the requirements of the type of combustion chamber in use. Some chambers require very fine atomization; others can function with coarser atomization. Proper atomization helps to start the burning process and ensures that each minute particle of fuel will be surrounded by particles of oxygen with which it can combine. Atomization is generally obtained when the liquid fuel, under high pressure, passes through the small opening or openings in the injector or nozzle. The fuel enters the combustion space at high velocity because the pressure in the cylinder is lower than the fuel pressure. The friction, resulting from the fuel passing through the air at high velocity, causes the fuel to break up into small particles.

PRESSURIZING AND DISTRIBUTING FUEL

8-6. Before injection can be effective, the fuel pressure must be sufficiently higher than that of the combustion chamber to overcome the compression pressure. The high pressure also ensures penetration and distribution of the fuel in the combustion chamber. Penetration is the distance through which the fuel particles are carried by the energy imparted to them as they leave the injector or nozzle. Friction between the fuel and the air in the combustion space absorbs this energy. Proper dispersion is essential if the fuel is to mix thoroughly with the air and to burn efficiently. While pressure is a prime contributing factor, the dispersion of the fuel is influenced in part by atomization and penetration of the fuel. If the atomization process reduces the size of the fuel particles too much, they will lack penetration. Lack of sufficient penetration results in ignition of the small particles of fuel before they have been properly distributed or dispersed in the combustion space. Penetration and atomization tend to oppose each other. Therefore, a degree of compromise in each is necessary in the design of fuel injection equipment, particularly if uniform distribution of fuel within the combustion chamber is to be obtained. The pressure required for efficient injection and, in turn, proper dispersion depends on the following factors:

- Compression pressure in the cylinder.
- Size of the opening through which the fuel enters the combustion space.
- Shape of the combustion space.
- Amount of turbulence created in the combustion space.

To control an engine means to keep it running at a desired speed, either in accordance with or regardless of the changes in the load carried by the engine. The degree of control required depends on the engine's performance characteristics and the type of load which it drives. In diesel engines, a varying amount of fuel is mixed with a constant amount of compressed air inside the cylinder. A full charge of air enters the cylinder during each intake event. The amount of fuel injected into the cylinders controls combustion and therefore determines the speed and power output of a diesel engine. A governor is provided to regulate the flow of fuel. Other devices, either integral with the governor or mounted separately on the engine, are used to control overspeed or overload.

SECTION II. EXTERNAL FUEL SYSTEMS

The fuel oil system in a diesel-powered vessel must be installed, operated, and maintained with the same care and supervision as the ship's engines. Inspection, maintenance, and operation of fuel oil tanks and fuel-handling equipment must be carried out according to the applicable technical manuals (TMs) for the vessel.

PUMPING OF FUEL

8-7. Fuel is pumped from the storage tank to the supply or day tank. From there it is delivered to the fuel injection pumps on the engine. It is good practice to clean the fuel of sediment and water before it enters the supply tank. This is usually done with a centrifugal purifier. The fuel is transferred from the supply tank by means of an engine-driven pump (sometimes called a booster, transfer, or primary pump). It travels through a metal-edge strainer and a cartridge-type, replaceable-element filter to a header on the suction side of the fuel injection pumps. Excess fuel pumped to the fuel header and from the fuel injection pumps is returned to the supply tank.

PURIFYING FUEL

8-8. The centrifugal purifier is a machine similar to the separator used for separating cream from milk. Oil enters a revolving bowl which tends to throw any heavy, solid contaminants to the outside of the bowl. Next enters an intermediate layer of water which is heavier than the oil but lighter than the solids, and finally a central core of oil. The discharge holes in the top of the bowl are so located that the water can be drawn off separately from the fuel oil. The solid material collects around the periphery of the bowl and must be cleaned out periodically. This cleaning can be done once a day unless idleness of engine or exceptionally clean fuel indicates that more extended periods are permissible. The supply or day tank is usually vented to the atmosphere and mounted at a high point in the fuel system to allow all air to escape. Extreme care should be taken to keep all fuel lines under a pressure greater than atmospheric. Also, excessive foaming or splashing of fuel oil in the tanks should be avoided to prevent the air from becoming entrained with the fuel because air interferes with proper operation of the injection pump.

MAINTAINING FUEL PRESSURE

8-9. The engine-driven fuel transfer pump is of the positive-displacement type, usually with a built-in relief valve to ensure constant pressure to the injection system. The fuel strainer is located on the suction side of the transfer pump, and the filter is connected into the system on the discharge side of the pump. The pressure drop across these filters and strainers increases with time. For this reason, some systems have a relief valve in the line before the cartridge filter to allow the bypassed oil to be returned directly to the supply tank. The relief valve ensures a more constant fuel supply pressure and also tends to vent entrained air from the suction side of the system.

FUEL STRAINING

8-10. The primary metal-edge fuel strainer used is a duplex type which is actually two complete strainers connected by suitable headers or piping. This arrangement allows either strainer to be completely cut out of the system for cleaning or repair while all of the oil is flowing through the other strainer. Figure 8-1 shows a typical metal-edge strainer (a magnified view of a portion of the element is shown in Figure 8-2). The oil flow is from the outside to the inside. In this type of strainer, the spaces between the leaves or ribbons which act as oil passages are between 0.001 and 0.0025 inches. The pressure drop across these strainers must not exceed 1.5 pounds per square inch (psi), with an oil flow equal to the full capacity of the fuel oil pump. In some engines, the duplex strainer is placed between the fuel supply tank and the transfer pump and, in operation, may be working under a vacuum.

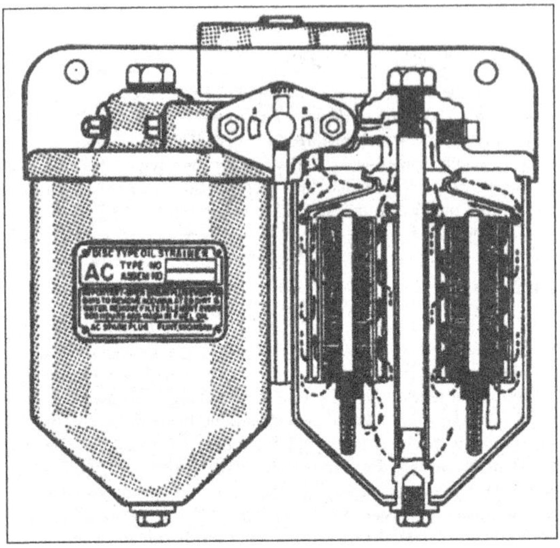

Figure 8-1. Primary Fuel Strainer

Figure 8-2. Enlarged Section of Ribbon in Strainer

SECTION III. FUEL INJECTION SYSTEMS

The primary function of a fuel injection system is to deliver fuel to the cylinders at the proper time and in the proper quantity under various engine loads and speeds. Although there are several types of fuel injection systems in use, their functions are the same. The fuel injection system may be of the air injection type or of the mechanical (solid) type. Since there are no air injection systems now in use, this discussion is limited to mechanical injection systems.

MECHANICAL INJECTION SYSTEMS

8-11. Mechanical injection systems may be divided into the following three main groups:

- Common-rail type.
- Individual pump or jerk type.
- Distributor type.

These systems may be further subdivided as follows:

- The common-rail may be divided into the basic or original system and the modified common-rail system (such as that used in the Cooper-Bessemer engines).
- The individual pump system may be divided into the original system and the modified system. The original system has a separate pump and separate fuel injector nozzles for each cylinder in the engine. In the modified system (such as the General Motors type) the injection pump and fuel nozzle are contained in one unit called the unit injector.
- Examples of the distributor type are the Bosch PSB series system, the Roosa-Master system, and the Cummins PT system.

BASIC COMMON-RAIL INJECTION SYSTEM

8-12. The basic common-rail system consists of a high-pressure pump which discharges fuel into a common-rail, or header, to which each fuel injector is connected by tubing. A spring-loaded bypass valve on the header maintains a constant pressure in the system, returning all excess fuel to the fuel supply tank. The fuel injectors are operated mechanically. The amount of diesel fuel oil injected into the cylinder at each power stroke is controlled by the lift of the needle valve in the injector. The principal parts of a basic common-rail system are shown in Figure 8-3.

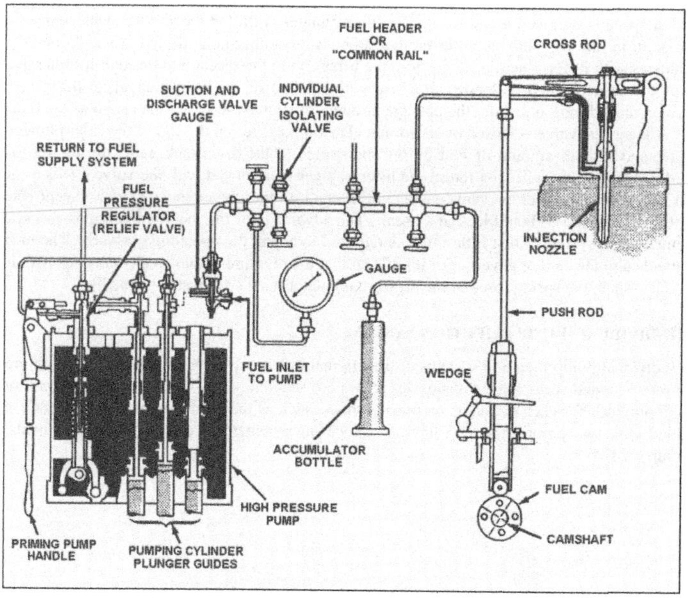

Figure 8-3. Basic Common-rail Injection System

MODIFIED COMMON-RAIL (CONTROLLED-PRESSURE) INJECTION SYSTEM

8-13. The modified common-rail system (sometimes referred to as the controlled-pressure system) differs from the basic system in that mechanically operated fuel injectors are included and the nozzles are operated hydraulically instead of mechanically. The nozzles of the modified system (shown in Figure 8-4) do not meter the fuel. Instead, the fuel is metered by the injectors. In addition, pressure is regulated by a high-pressure pump instead of by a pressure regulator. The pump plunger, on its downward stroke, first closes small holes that connect the pump barrel with the fuel admission line. Additional downward motion increases the oil pressure in the pump until it opens the spring-loaded discharge valve and delivers the oil into the injection system. During the return stroke, the spring moves the plunger upward, creating a vacuum, and when the plunger uncovers the holes on top, oil from the suction side enters into the pump. The oil from the fuel oil pressure tank, on its way to the suction side of the pump, is admitted first to the inner side of a sleeve (sleeve valve). The inner and outer sleeves have two mating holes. By turning the sleeves, one relating to the other and to the housing, the amount of fuel admitted to the pump is adjusted to meet the load and speed requirements. The outer sleeve is set and turned by the governor to admit a sufficient amount of fuel to correspond to the load carried by the engine. The inner sleeve is turned by a mechanism set to maintain a prescribed constant pressure in the system. If the pressure goes up, the sleeve is turned to decrease the effective area of the opening between the two sleeves. The amount of fuel taken by the pump is therefore reduced and, as a result, the pressure in the system is decreased. However, when the pressure begins to drop, the sleeve is turned in the opposite direction to increase the effective area of the opening, resulting in more fuel going into the pump and the pressure going up. The injection nozzle consists of a spring-loaded plunger with a conical end which acts as a valve. When the injection nozzle valve is raised from its seat by the oil pressure, the valve is opened. When the oil pressure drops, the spring-loaded injection nozzle valve is returned to its seat, closing the valve. A quick closing of the injection nozzle and elimination of after dribbling of the fuel into the combustion space is obtained as follows. The lifter plunger is drilled lengthwise at its center from the valve end to a point in line with the recess in the nozzle body. Another hole, drilled at a right angle to the central hole, connects with it, forming a passage from the lifter end to the recess and through it to the drain tank. The bottom of the injector valve is lapped to a seat with the end of the lifter plunger so that when the two are brought in contact during injection, the passage through the plunger is sealed. As soon as the fuel cam releases the lifter plunger, the valve is closed by its spring. The oil pressure on the end of the lifter plunger will move it downward, and a small amount of fuel oil will be spilled to the drain tank, relieving the oil pressure in the nozzle. The lifter spring will then return the lifter plunger to a contact with the valve. This arrangement also acts as a safety feature which prevents passage of the fuel oil into the engine cylinder, except when necessary, even if the injector valve should leak at its seat. The advantage of the modified common-rail system over the basic common-rail system is that little effort is required to adjust the operating pressure. The modified system can be attached to the engine governor or throttle so that the pressure automatically changes with load or speed. In the basic system, the spring force of the pressure regulator must be adjusted manually.

ORIGINAL INDIVIDUAL-PUMP INJECTION SYSTEM

8-14. Individual-pump injection systems of the original jerk pump or basic type include high-pressure pumps and pressure-operated spray valves or nozzles which are separate units. In some engines, only one pump and nozzle are provided for each cylinder. In other engines, such as the Fairbanks Morse (FM 38C), each cylinder is provided with two pumps and two nozzles. Such an arrangement can also be found in the appropriate manufacturer's TM.

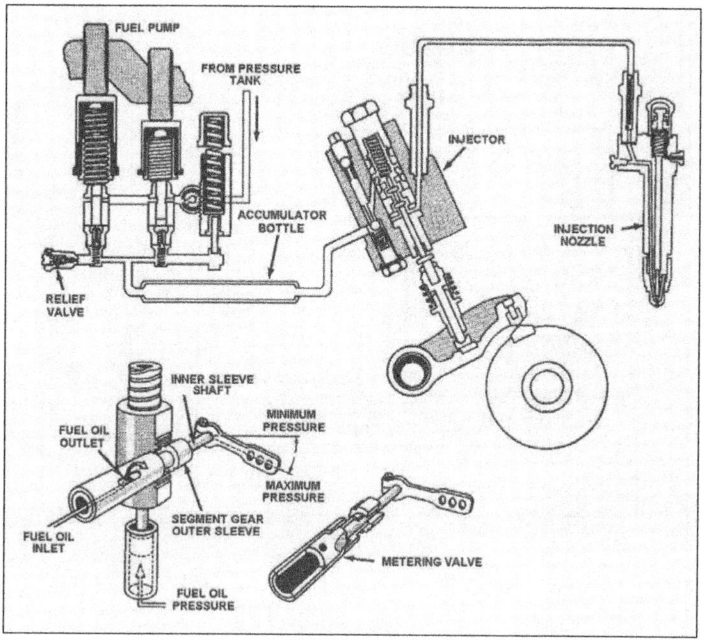

Figure 8-4. Modified Common-rail Injection System

MODIFIED INDIVIDUAL-PUMP INJECTION SYSTEM

8-15. Of all the individual-pump injection systems, the modified system is the most compact. A high-pressure pump and an injection nozzle for each cylinder are combined into one unit. This type of unit (generally used with General Motors engines) is often referred to as a unit injector system. This system is discussed later in this section.

DISTRIBUTOR-TYPE INJECTION SYSTEM

8-16. As previously stated, the Bosch PSB series, the Roosa-Master, and the Cummins PT systems are of the distributor-type mechanical injection system. Before discussing these systems, a general discussion of the Bosch fuel system is given.

Bosch Fuel System

8-17. The Bosch fuel injection system is used on many of the Army's diesel engines. Figure 8-5 shows the fuel piping of a Bosch fuel injection system and governor. A fuel supply pump (on the side of the injection pump housing) draws fuel from the supply tank and pumps it to the injection pumps through a duplex fuel oil strainer. The supply pump furnishes an excess of oil to the injection pumps. The excess oil is returned through a bypass valve and return line to the supply tank. The fuel supply pump is a plunger-type pump. It is mounted directly on the injection pump housing and driven by one of the cams on the injection pump camshaft. It is self-regulating to build up to only a certain maximum pressure. Figure 8-6 is a phantom view of the Bosch fuel supply pump, showing actual positions of the parts.

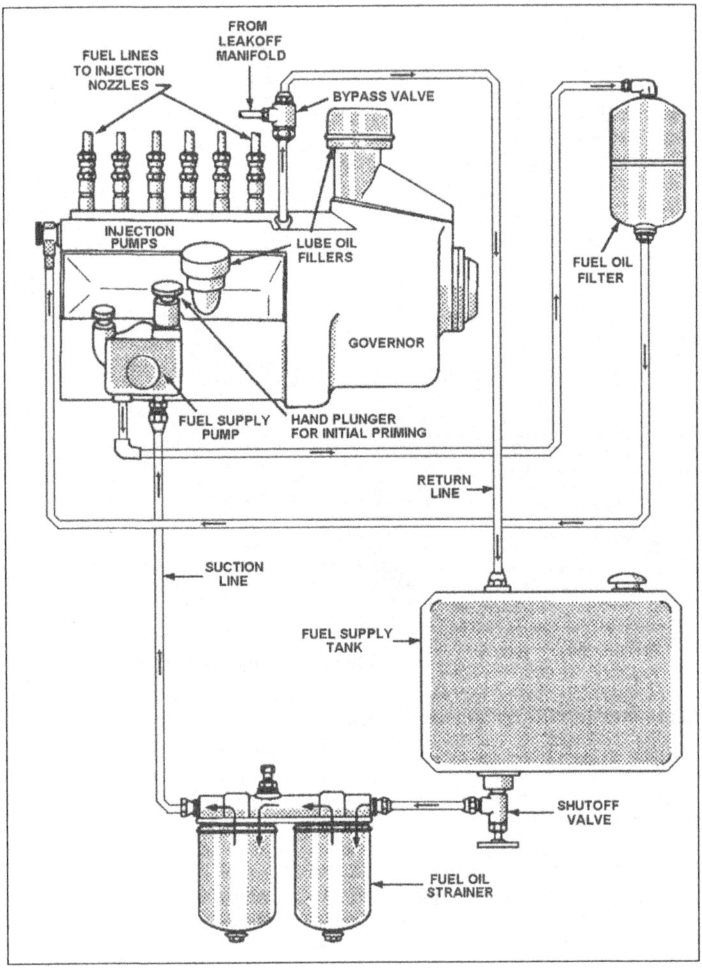

Figure 8-5. Fuel Piping System With Bosch Injection Pump and Governor

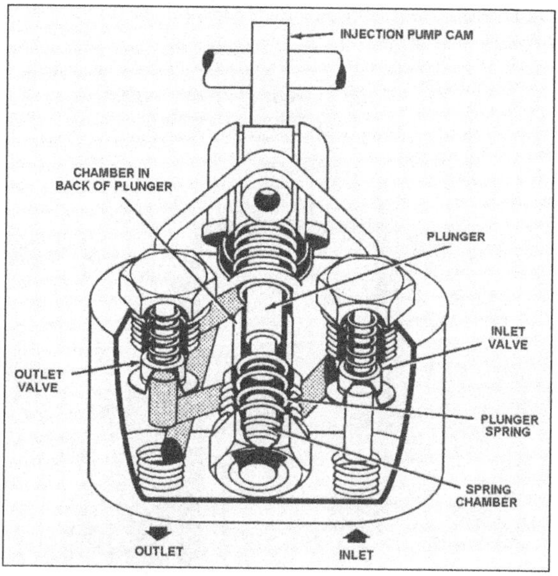

Figure 8-6. Phantom View of the Bosch Fuel Supply Pump

Bosch Fuel Injection Pumps

8-18. Bosch fuel injection pumps are of two basic designs (type APE and type APF). Figure 8-7 shows the two types of injection pumps, along with high-pressure lines and spray nozzles.

APF Pump

8-19. The APF pumps are of single-cylinder design, the plunger pump for each cylinder being in a separate housing. For example, in a 6-cylinder engine, there are six separate APF pumps. Each pump is cam-driven and regulated by an individual control rack.

APE Pump

8-20. The APE pumps are assembled with all the individual cylinder plungers in a single housing. The right side of Figure 8-7 shows the pump assembly for a 6-cylinder engine. The injection pumps are operated from a single camshaft in the bottom part of the housing. The cams dip into lubricating oil and brush against felt cushions at the bottom of each revolution. At the top of each revolution, the cams force the spring-loaded plungers up against the plunger spring resistance. Each plunger moves up and down in a barrel which contains fuel oil at the supply pressure. The plunger traps oil above it during part of the upward stroke and forces it through the delivery valve and high-pressure tubing to the spray nozzle. From there it is injected into the combustion chamber. The action of the plunger, control rack, delivery valve, and spray nozzle are the same in both APE and APF pumps. Study Figure 8-7 to get a better understanding of the fuel injection mechanism and control of the amount of fuel injected. The fuel oil sump is filled with clean oil from the supply pump and fuel oil filter. Oil enters the barrel above the plunger through a pair of ports. The amount of fuel forced out through the spray nozzle of each upward stroke of the plunger depends on how the plunger is rotated. Notice that the control rack has teeth all along the side, meshing with a gear segment on each pump. Lengthwise movement of the control rack rotates all the plungers the same amount and in the same direction. Rotation of the plungers changes the part of the plunger helix that passes over the spill port (on the right side of each barrel) thereby hanging the time at which injection ends.

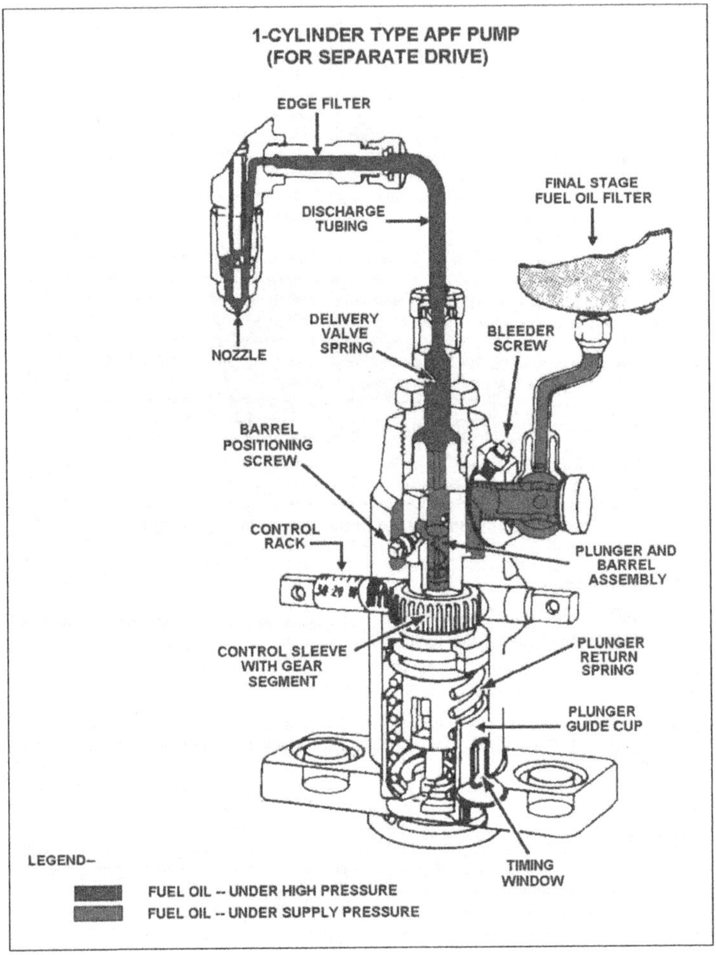

Figure 8-7. Phantom Views of APE and APF Bosch Fuel Injection Pumps

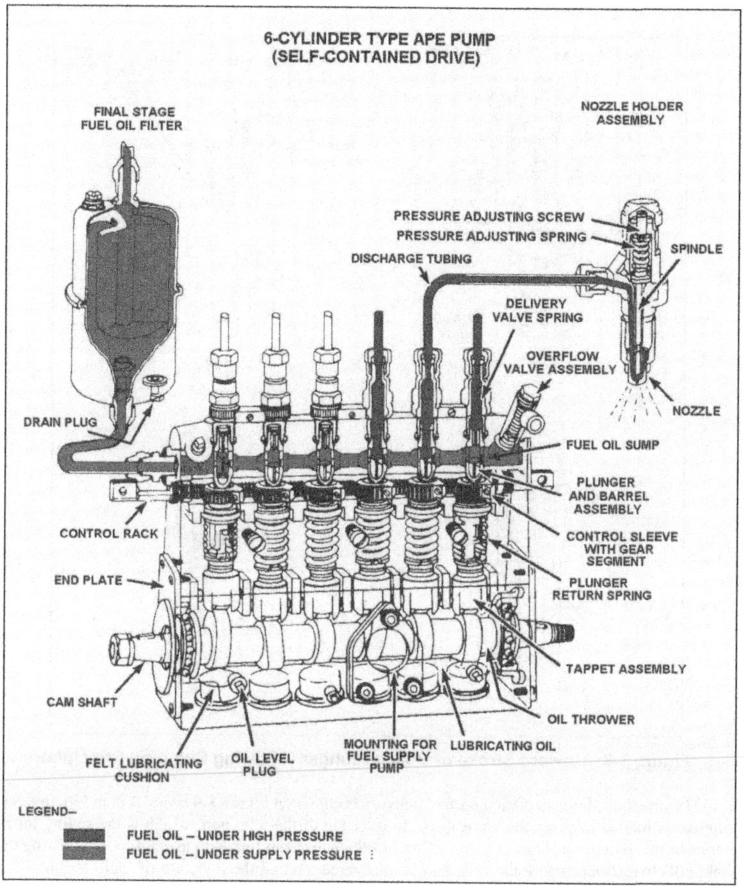

Figure 8-7. Phantom Views of APE and APF Bosch Fuel Injection Pumps (continued)

8-21. The pumping principle of the Bosch pump (four steps of a pumping stroke) is shown in Figure 8-8. In Figure 8-8 (view A), the plunger is below the inlet and spill ports. Fuel oil enters the barrel, as indicated by the arrow, and fills the barrel chamber (between the plunger and the delivery valve). The plunger has a flat top, and the two ports are set at the same level. The two ports are closed by the plunger at exactly the same moment the plunger travels upward. In Figure 8-8 (view B), the ports have just closed. The fuel above the plunger is trapped and placed under high pressure by the rising plunger. The pressure forces the delivery valve up at once, allowing the high-pressure oil to go to the spray nozzle. In Figure 8-8 (view C), the plunger is in the effective part of its stroke with both ports closed. Fuel is passing through the delivery valve to the spray nozzle. The effective stroke will continue as long as both ports remain covered by the plunger. At the moment that the spill port is uncovered by the edge of the helix (see Figure 8-8 (view D)), fuel injection ends. As soon as the port is opened, the fuel oil above the plunger flows out through the vertical slot in the plunger and goes to the low-pressure fuel oil sump. Therefore, the pressure above the plunger is released and the delivery valve is returned to its seat by the valve spring.

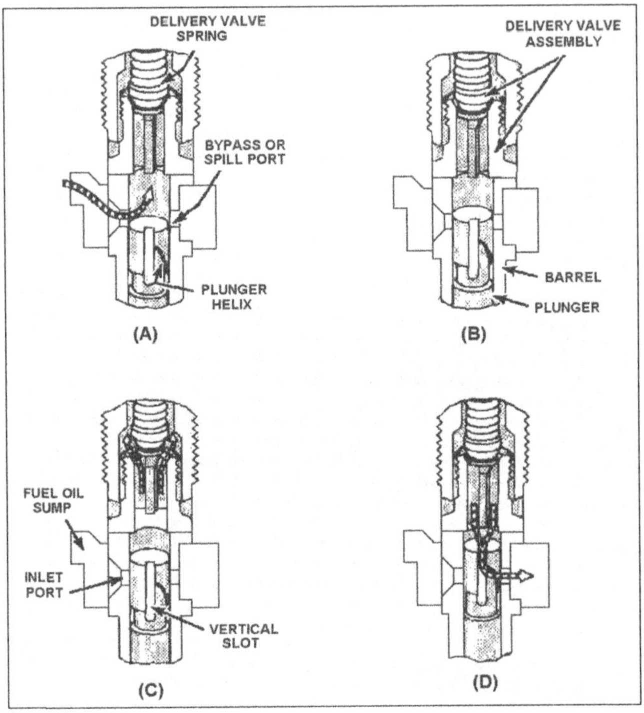

Figure 8-8. Upward Stroke of Bosch Plunger (Showing Pumping Principle)

8-22. The effect of plunger rotation on fuel delivery is shown in Figure 8-9 (view A). In Figure 8-9 (view B), the plunger is rotated to bring the vertical slot to the edge of the inlet port, which is the setting for maximum delivery. In this plunger position, the lowest part of the helix is in line with the spill port, allowing the longest possible effective stroke before the spill port is uncovered, ending the injection of fuel. Figure 8-9 (view C) shows the setting for medium or normal delivery. This brings a higher part of the helix in line with the spill port and leaves a short effective stroke before the spill port is uncovered. The position of no fuel delivery is reached when the plunger has been rotated to bring the vertical slot in line with the spill port (see Figure 8-9 (view A). In this plunger position, the fuel above the plunger will not be under compression during any part of the upward stroke. The amount of fuel injected can be regulated by setting the plunger in any position between no delivery and maximum delivery. The plunger setting is controlled by the position of the control rack, which regulates all the plungers at the same time. Movement of the control rack (either manually or by governor action) rotates the plunger and varies the quantity of fuel delivered by the pump.

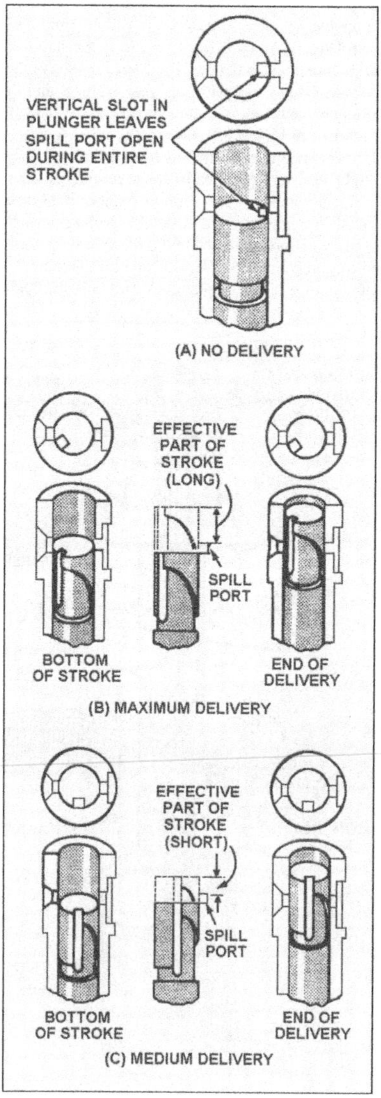

Figure 8-9. Effect of Plunger Rotation on Fuel Delivery

8-23. Figure 8-10 shows a cutaway view of the Bosch injection pump and control rack assembly. The gear segment is secured to the control sleeve which is free to rotate on the stationary barrel. The control sleeve has a slot at the bottom into which fits the plunger flange. The flange moves in the slot as the plunger moves up and down. When the control rack is moved lengthwise, the gear segment and the control sleeve rotate around the outside of the barrel. The plunger flange and the plunger (inside the barrel) follow the rotation of the control sleeve. The Bosch plunger (shown in Figure 8-8, Figure 8-9, and Figure 8-10) has a flat top surface and has only a lower helix. With this type of plunger, fuel injection will always begin at the same point in the piston cycle, whether it is set for light load or heavy load. Injection begins when the ports are closed; the end of injection can be varied by plunger rotation. This type of plunger is used in pumps marked "Timed for Port Closing." Injection has a constant beginning and a variable ending. Another type of Bosch plunger has an upper helix. Rotation of the plunger varies the beginning of the effective stroke, while the ending is constant. This type of plunger is used in pumps marked "Timed for Port Opening." A third type of plunger has both upper and lower helixes. Rotation of this type of plunger varies both the beginning and ending of injection.

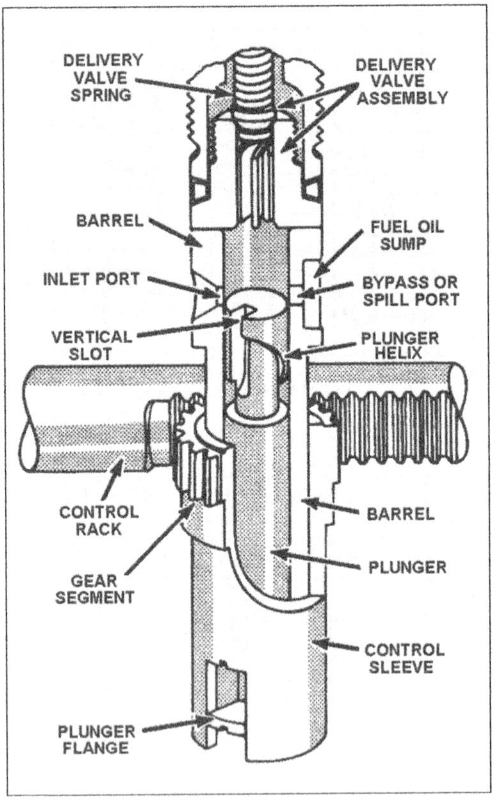

Figure 8-10. Bosch Injection Pump and Control Rack Assembly

Bosch PSB Fuel Injection System

8-24. The Bosch PSB fuel injection system (see Figure 8-11) is used on diesel engines for small boats. It is similar to the Bosch APE fuel system, having a supply pump, a governor, and high-pressure injection pumps built into a single assembly. The PSB injection pump is a single-plunger, constant-stroke, six-outlet unit. It is driven at crankshaft speed and actuated by a cam and tappet arrangement which also carries gearing for the distribution function. Its replaceable hydraulic head contains a single delivery valve and a single plunger which, in addition to being reciprocated by the multi-lobe cam, is rotated continuously to serve as its own fuel distributor. Fuel discharge is varied by axial movement of the plunger control sleeve, similar to that used in most other pumps of this type, which is linked to the governor. The governor is mounted integrally with the injection pump. A positive-displacement, gear-type fuel pump, mounted on the front side of the housing, is driven from the distributor drive gear on the pump camshaft. A hand priming pump for bleeding of fuel lines is also provided. It is mounted either beside the injection pump on some models or integrally with the fuel transfer pump. The Bosch PSB single-plunger fuel pump has the same function as the multi-plunger Bosch APE fuel pump used on earlier models. It meters the fuel accurately and delivers it under high pressure to the spray nozzles. Through these nozzles, the fuel is injected into the engine cylinders at a definite timing in relation to the engine firing cycle and within the required injection period. The PSB pump is designed to simplify servicing. It has substantially fewer parts than the APE pump. The PSB pump requires less mounting space than the APE fuel pump because the PSB is a much shorter, more compact unit. The PSB assembly consists of following three main divisions:

- The housing with drive mechanism.
- The hydraulic head.
- The governor.

The PSB injection pump housing contains a camshaft compartment in the lower half. The camshaft is supported at the rear by a sleeve bearing and at the front by a ball bearing. The cam, just below the plunger assembly in the hydraulic head, has three lobes. The camshaft has a spline at the forward end to receive the drive hub. It also has a spiral gear cut into it next to the cams to mate with and drive the lower gear of the quill shaft assembly of the distributing system. The same spiral gear also drives the fuel transfer pump which is attached to the front of the pump housing at this location. There is also a pointer at the drive end of the pump to align with a mark on the gear hub for an accurate setting of the injection timing. The hydraulic head of the PSB pump is a complete assembly fastened to the housing by four studs. It is easily removed for service or replacement. The assembly consists of the head block, the lapped plunger, the control sleeve, the delivery valve assembly, and the plunger return springs. The PSB governor, whose weight assembly is attached to the fuel pump camshaft, is considered an integral part of the PSB pump assembly. The governor is of the variable-speed, mechanical-centrifugal type. The governor action is accomplished through flyweight action against a movable sleeve which is backed by springs loaded in the opposite direction.

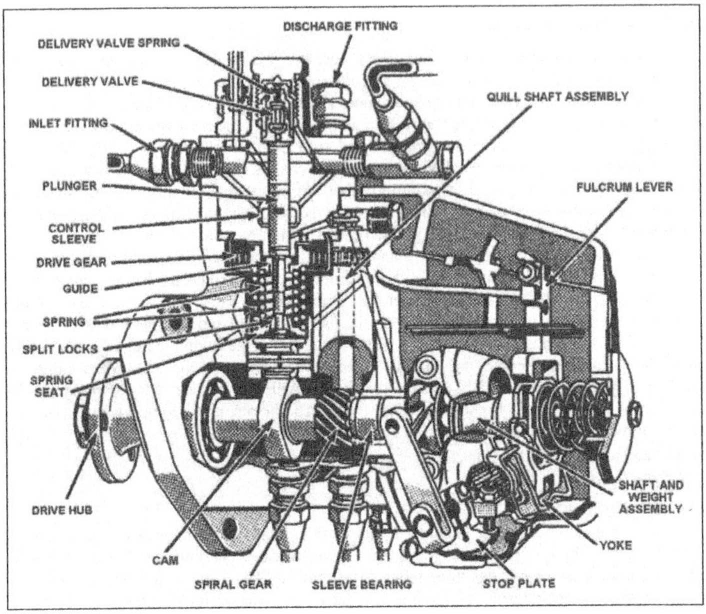

Figure 8-11. Bosch PSB Fuel Injection System

Roosa-Master Fuel System

8-25. The Roosa-Master fuel system is used on some diesel engines such as the Gray Marine. This fuel system is somewhat similar to the Bosch APE and the Bosch PSB in that it also contains a supply pump, a governor, and a high-pressure injection pump built into one assembly. The difference is in the method of pumping and measuring fuel. The function of the Roosa-Master assembly is to draw fuel from the supply tank, force it through a filter, and then meter, distribute, and deliver the fuel at the proper time under high pressure to the injection nozzles. The assembly itself is oil-tight, pressurized, and self-lubricated by the fuel oil it pumps. It is gear-driven from the camshaft. The assembly housing contains the pumping and metering elements plus the drive shaft and a hydraulic governor. Figure 8-12 shows the pumping and metering mechanism. This mechanism consists of five major parts:

- *The transfer pump.* The transfer pump is a four-vane, positive-displacement pump which supplies fuel to the metering device at an even and predetermined pressure. The shape of the pumping cavity has been modified from the usual crescent shape to provide an almost constant flow. A spring-loaded bypass valve controls the pressure.
- *The hydraulic head.* The hydraulic head houses the metering mechanism and most of the oil passages. The metering valve is essentially a closely fitted plunger moving in and out within its cylindrical barrel in the hydraulic head. The metering plunger accurately measures the fuel injected into each cylinder, in response to governor action. The outlet connections to the injection nozzles are mounted on the hydraulic head.

- *The rotor.* The rotor is the major rotating assembly. It contains bored passages for fuel oil distribution. As the rotor turns, it causes different fuel oil passages to align with those in the hydraulic head, so that fuel oil from the injection pump can be distributed at the proper time to the proper injection nozzle. The injection pump plungers are mounted in the drive end of the rotor. The upper end of the rotor is used to drive the transfer pump.
- *The injection pump plungers.* The injection pump contains two opposed plungers closely fitted in a bore perpendicular to the shaft axis. The plungers are actuated simultaneously by shoes and rollers bearing on the cam ring.
- *The cam ring.* The cam ring is stationary and is mounted internally in the housing. As the entire rotor (including the plungers) turns in the hydraulic head, opposing internal lobes in the cam ring simultaneously force the injection pump plungers inward toward each other. This action exerts pressure on the fuel between the plungers and causes the fuel to be delivered at high pressure to the injection nozzles. As the rotor continues to turn, the rollers ride down the lobes of the cam ring, allowing the plungers to move outward away from each other. This allows another charge of fuel to be forced into the pump cavity between the plungers. The fuel can enter the pump cavity and then be discharged to the injection nozzles only at the proper time. The timing relationship is established by the position of the drilled passageways in the rotor and can never be altered by adjustment or improper assembly. Therefore, there is always uniform fuel delivery to each injection nozzle.

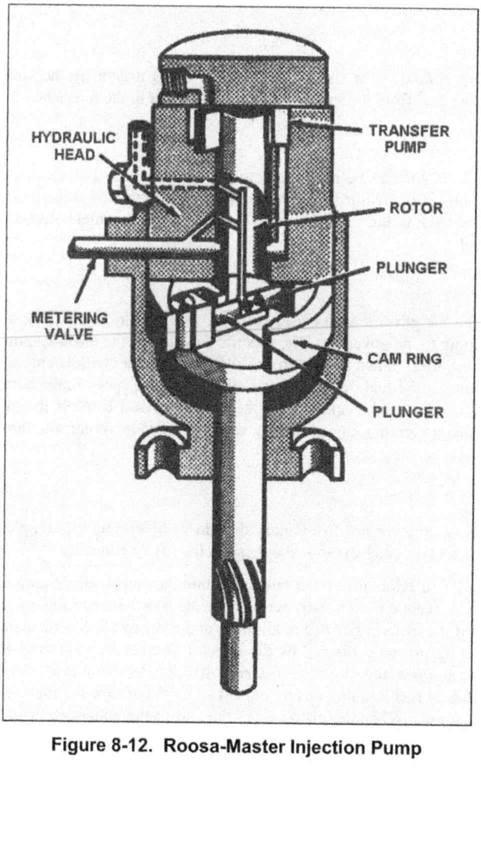

Figure 8-12. Roosa-Master Injection Pump

Cummins PT Fuel System

8-26. The Cummins PT fuel system is designed for use on all diesel engines manufactured by Cummins. It can be adapted, with some modifications, for all Cummins diesel engines now in use. This fuel system is somewhat different from the fuel systems previously discussed. The identifying letters "PT" are an abbreviation for "pressure time." The PT fuel system operates on the principle that a change in the pressure of a liquid flowing through a pipe will change the amount of liquid coming out the open end. Increasing the pressure increases the flow and decreasing the pressure decreases the flow. The PT fuel system consists of a fuel pump assembly, supply and drain lines, and injectors. The system is designed to deliver fuel to each cylinder in equal and predetermined amounts. The fuel pump assembly (see Figure 8-13, view A) consists of the following five main units:

- Gear pump.
- Pressure regulator.
- Throttle.
- Governor.
- Shut-down valve.

Gear Ppump

8-27. The gear pump consists of a single set of gears. It is driven by the camshaft gear and it turns at camshaft speed. It draws fuel from the supply tank and delivers it to the injectors.

Pressure Regulator

8-28. The pressure regulator can be referred to as a bypass valve. Its primary function is to control fuel pressure at the gear pump. The pressure regulator controls and limits fuel pump pressure by the bypass method. Excess fuel is delivered back to the suction side of the pump, limiting fuel delivery to the required amount for any given speed or load.

Throttle and Governor

8-29. The throttle and the governor are closely related. All fuel for engine operation must pass through the throttle shaft before going to the governor. The specific function of the throttle varies somewhat, depending on the type of governor installed. When the governor is of the type that controls only the idling and the maximum speeds, the throttle controls the fuel oil flow to the injectors in the speed range between idling and maximum. When the governor is of the variable-speed type, the governor itself controls the fuel oil flow to the injectors for any speed within the governor's range. With a variable-speed governor, the throttle acts only as a positive shutdown device.

Shutdown Valve

8-30. The shutdown valve gives positive engine shut-down by cutting off completely the flow of fuel to the injectors. The shutdown valve can be operated either electrically or manually.

8-31. The Cummins PT injectors meter and inject fuel into the combustion chambers. Fuel circulates through the injectors at all times except during a short period after the completion of the injection cycle. The circulation of the fuel keeps fuel at the metering orifice at all times and reduces injector temperature. Metering of the fuel occurs when the metering orifice is opened by the injector plunger (see Figure 8-13 (view B)). The injector plunger acts as a valve to open and close the metering orifice. As the injector plunger uncovers the metering orifice, fuel pressure forces fuel into the injector cup. The length of time the fuel meters into the injector cup is determined by engine speed. When the piston nears the end of the compression stroke, injection begins. The falling plunger closes the metering orifice and forces the fuel trapped in the injector cup out through the spray holes and into the combustion chamber. The injector plunger then remains in the injector cup seat throughout the power and exhaust events. The amount of fuel that is injected into the combustion chamber is regulated by both the time that the metering orifice remains open and by the fuel pressure to the orifice.

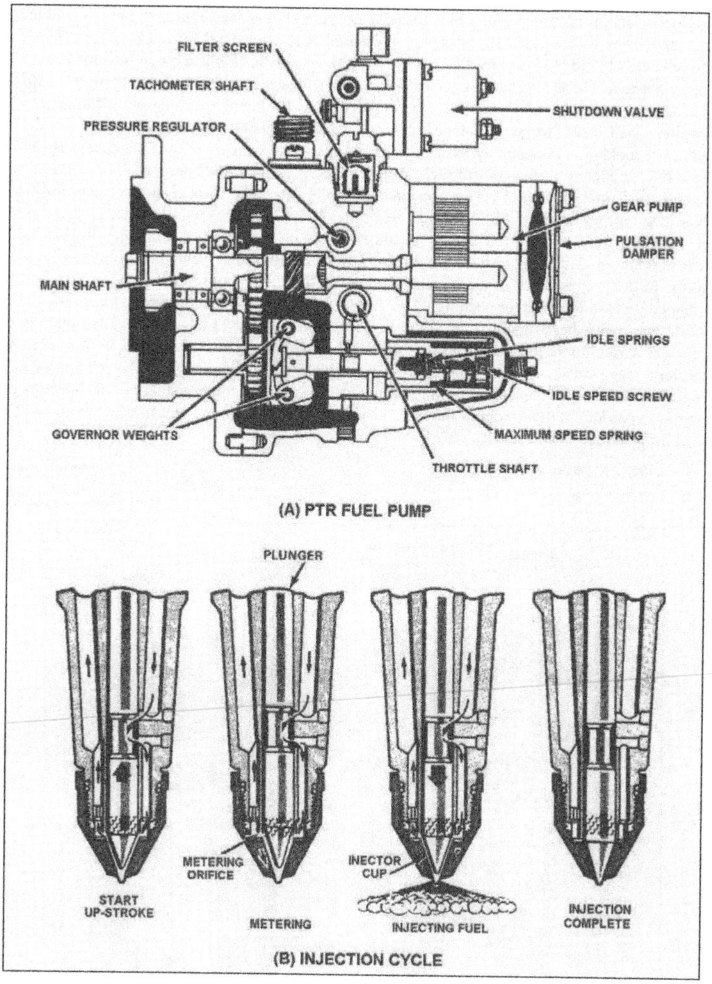

Figure 8-13. Cummins PT Fuel Pump and Injection Cycle

Unit Injector System

8-32. All General Motors marine diesel engines (GM 645, 567, 278, 268, 71, and 53) use the unit injector system which combines a pump and a fuel spray nozzle in one unit (see Figure 8-14). The fuel piping to the unit injector carries oil at the filter discharge pressure, generally about 40 to 50 psi. The unit injector is generally installed in the cylinder head (see Figure 8-15). It is held in place by an injector crab. The cylinder head has a copper tube into which the injector fits snugly with the spray tip projecting slightly into the cylinder clearance space. Water circulates around the copper tube and cools the lower part of the injector. Two fuel lines are connected to each injector. One carries fuel to the injector and the other carries away fuel that is bypassed. The injector is operated by a rocker arm and push rod, which work off the camshaft. The amount of fuel injected is regulated by the control rack, which is operated by a lever secured to the control tube. In the unit injector, fuel under pressure enters the injector at the inlet side through a filter cap and filter element. Injectors of the type shown in Figure 8-14 may be used without a filter. From the filter element, the fuel passes through a drilled passage into the supply chamber (that area between the plunger bushing and the spill deflector) in addition to that area under the injector plunger within the bushing. The plunger operates up and down in the bushing. The bore of the housing is open to the fuel supply in the annular chamber through two funnel-shaped ports in the plunger bushing. The motion of the injector rocker arm is transmitted to the plunger by the follower, which bears against the follower spring. In addition to the reciprocating motion, the plunger can be rotated during operation around its axis by the gear which meshes with the rack. For metering the fuel, an upper helix and a lower helix are machined in the lower part of the plunger. The relation of the helixes to the two ports changes with the rotation of the plunger. Several types and models of unit injectors exist. They are classified according to the design of the injector valve. Three common types in use are the--

- High valve.
- Spherical valve.
- Needle valve.

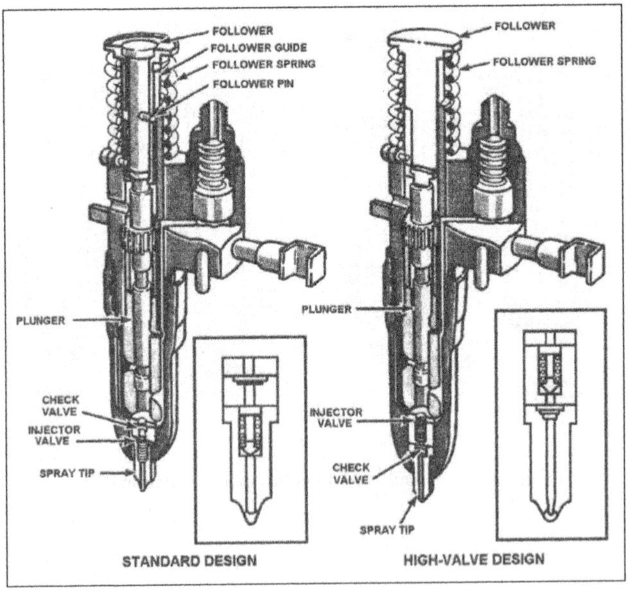

Figure 8-14. Standard and High-valve Unit Injectors

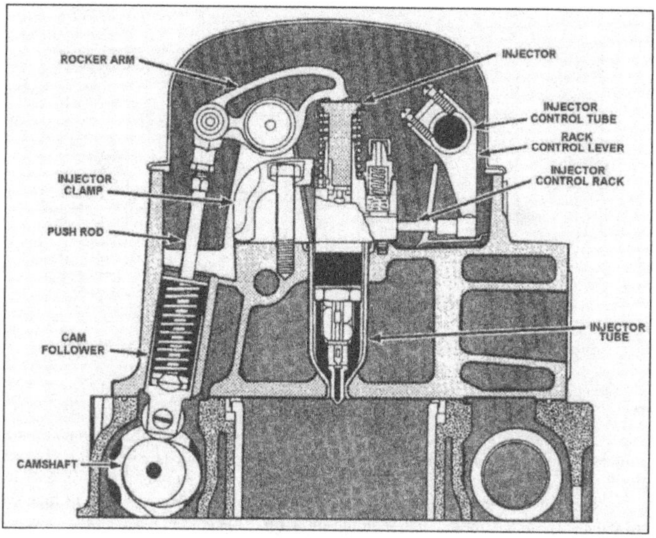

Figure 8-15. Typical Unit Injector Installation

High Valve

8-33. In the high valve injector, the plunger moves downward under pressure of the injector rocker arm. As it does, a portion of fuel trapped under the plunger is displaced into the supply chamber through the lower port until the port is closed off by the lower end of the plunger. Then, a portion of fuel trapped below the plunger is forced upward through a central passage of the plunger into the recess and into the supply chamber through the upper port until that port is closed off by the upper helix of the plunger. With the upper and lower ports both closed off, the remaining fuel under the plunger is subjected to increased pressure by the continued downward motion of the plunger. When sufficient pressure is built up, the injector valve is forced off its seat and fuel is forced through small orifices in the spray tip and atomized into the combustion chamber. A check valve mounted in the spray tip prevents air leakage from the combustion chamber into the fuel injector if the valve is accidentally held open by a small particle of dirt. The injector plunger is then returned to its original position by the injector follower (plunger) spring. On the return upward movement of the plunger, the high-pressure cylinder within the bushing is again filled with fuel oil through the ports. The constant circulation of fresh, cool fuel through the injector; renews the fuel supply in the chamber and helps cool the injector. It also effectively removes all traces of air which might otherwise accumulate in the system and interfere with accurate metering of the fuel. The fuel injector outlet opening returns the excess fuel oil to the fuel return manifold and from there back to the fuel tank. Changing the position of the helixes, by rotating the plunger, retards or advances the closing of the ports and the beginning and ending of the injection period. At the same time, it increases or decreases the amount of fuel injected into the cylinder. Figure 8-16 shows the various plunger positions from light load to heavy load. With the control rack pulled out all the way (no injection position), the upper port is not closed by the helix until after the lower port is uncovered. Consequently, with the rack in this position, all of the fuel is forced back into the supply chamber and no injection of fuel takes place. With the control rack pushed in (full injection position), the upper port is closed shortly after the lower port has been covered. This will produce maximum effective stroke and maximum injection. From the "no injection" position to the "full injection" position (full rack movement), the contour of the upper helix advances the closing of the ports and the beginning of injection.

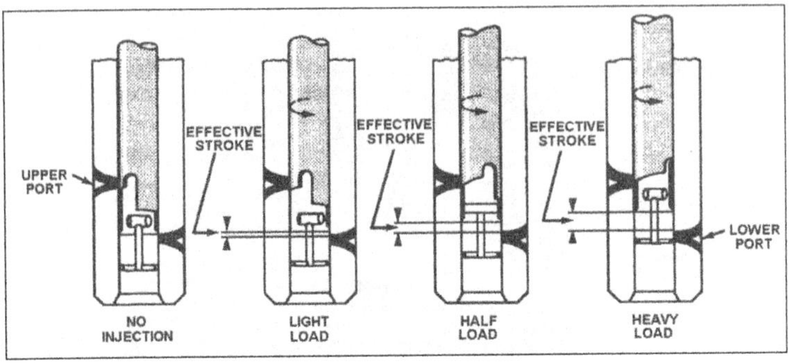

Figure 8-16. Injection and Metering Principles of a Unit Injector

Spherical Valve

8-34. In spherical valve fuel injector assemblies. the injector valve assembly is located in the spray tip. The high valve injectors incorporate the valve assembly in a valve cage located between the valve seat and the spray tip. The check valve in spherical valve injectors rests in a recess on the top of the valve seat. It is held in place by a spacer located between the valve seat and injector bushing. The check valve in high valve injectors is located below the valve assembly in a recess in the spray tip. It is held in place by the valve cage. Due to the difference in location of the injector valve assembly, spherical valve injectors require the following:

- A longer plunger and bushing assembly.
- A longer spill deflector.
- A different spray tip and injector nut.

Spherical valve injector assemblies incorporate a lighter plunger spring and a three-piece follower assembly consisting of the follower, follower guide, and follower pin. High valve injectors use a one-piece follower. Operation and servicing procedures are essentially the same for both spherical valve and high valve injector assemblies.

Needle Valve

8-35. In later type GM unit injector is the needle valve injector (N-type high valve design). The needle valve injector has a higher opening pressure (2,800 psi at the needle valve as compared to 600 psi) than was used in earlier injector models. A higher opening pressure provides better atomization of the first part of the fuel charge and better atomization at low fuel input. The needle valve injector differs from earlier designs primarily in the plunger, bushing, and spray tip. Do not intermix needle valve injectors with other type injectors in the same engine because injection characteristics are not the same in different types of injectors. The principles by which injecting and metering are accomplished are almost identical in all GM unit injectors.

FUEL INJECTION EQUIPMENT (NOZZLES)

8-36. Injection equipment (such as nozzles) plays an important part in the efficient operation of diesel engines. It is very important to keep this equipment clean and in good working order. Two general types of injection nozzles are the open and the closed types. Each is described below.

OPEN NOZZLE

8-37. The open nozzle is usually a simple spray nozzle with a check valve which prevents passage of the high-pressure gases from the engine cylinder to the pump. Although the open nozzle is simple, it does not give proper atomization. Therefore, it is not generally used.

Closed Nozzle

8-38. The closed nozzle is more commonly used. Two main classifications of the closed type nozzles are the hole nozzle and pintle nozzle.

Hole Nozzle

8-39. Figure 8-17 shows a cutaway sectional view of a Bosch injector, showing details of the nozzle holder and a hole nozzle. The high-pressure oil from the injection pump enters the nozzle holder body through a metal edge strainer. From the strainer, the oil goes through a drilled fuel passage which extends to the bottom of the nozzle holder body. The nozzle, with its spray tip, is held against the bottom of the nozzle holder by the cap nut. A groove in the top of the nozzle forms a circular passage for the fuel oil between the nozzle and holder. Several vertical ducts carry the oil from the circular passage to the oil cavity, near the bottom of the nozzle. The nozzle valve cuts in sharply to a narrower diameter in the oil cavity, providing a surface against which the high-pressure oil in the oil cavity can act to raise the valve from its seat in the spray tip. When the valve is raised from its seat, the oil sprays out to the combustion chamber through a ring of small holes. The valve has a narrow stem which projects into the central bore of the nozzle holder where it bears against the bottom of the spindle. The spindle is held down by the pressure-adjusting spring. Whenever the upward force of the high-pressure oil acting on the needle valve exceeds the downward force of the spring, the valve can rise. The moment the spring force is greater, the valve will snap back to its seat. The spring tension is regulated by a pressure-adjusting screw which is held by a locknut. Regardless of the close lapped fit of the valve, some oil leaks past the valve and rises through the central bore of the nozzle holder. This oil lubricates the moving parts of the injector and then drains off through the oil drain connection to a drip tank. A bleeder screw is installed which can be used to bypass fuel oil to the nozzle, sending the fuel directly to the oil drain. Bleeder screws can be used to determine which injector is at fault when a cylinder is misfiring. The pintle nozzle and the hole nozzle are shown in Figure 8-18.

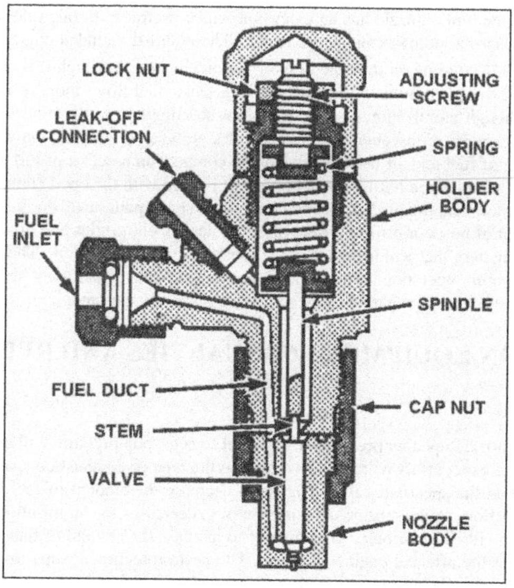

Figure 8-17. Sectional View of Bosch Injection Nozzle Assembly

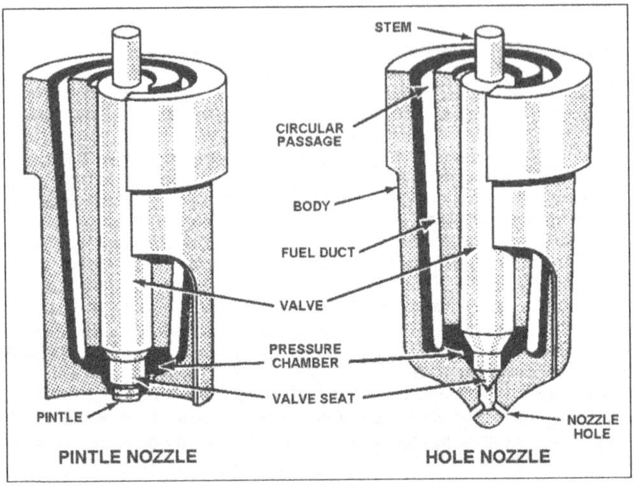

Figure 8-18. Pintle Nozzle and Hole Nozzle

Pintle Nozzle

8-40. The valve of the pintle nozzle has an extension which protrudes through the hole in the bottom of the nozzle body and produces a hollow, cone-shape spray. The nominal included angle of the spray cone may be between zero and 60°, depending on the type of combustion chamber in which it is employed. A pintle nozzle generally opens at a lower pressure than the hole nozzle because fuel flows more readily from the large hole of the pintle type. Although atomization of the fuel is not so complete in the pintle type, penetration into the combustion space is greater. Consequently, pintle nozzles are used in conjunction with the auxiliary chamber system, where mixing of fuel and air largely depends on combustion reaction or turbulence. The multiple hole nozzle provides good atomization but less penetration. It is used with the open combustion chamber, in which high atomization is more important than penetration. The spray pattern of the hole nozzle depends on the number and placement of holes or orifices. Regardless of design, all nozzles and tips function to direct the fuel into the cylinder in a pattern that will bring about the most efficient combustion. The slightest defect in nozzles and tips will affect engine operation. The troubles and their causes which may be encountered in either the spray tip of a unit injector or in a separate spray nozzle are relatively the same.

FUEL INJECTION EQUIPMENT CASUALTIES AND REPAIRS

INCORRECT PRESSURE

8-41. An incorrect nozzle opening pressure or incorrect injector pop pressure will influence engine efficiency and performance. The exact effect will vary according to the type of combustion space served. When opening pressure is greater than the specified value, it tends to decrease the amount of fuel injected and also tends to retard the start of injection. A low nozzle opening pressure, decreases the atomization of the fuel at low speeds and in extreme cases, will cause dribble. It also tends to increase the amount of fuel injected, which will cause a smoky exhaust from the affected engine cylinder. The best protection against trouble from this source is a periodic check of the nozzle or injector with an appropriate tester. Test stands vary, depending on the manufacturer, but all operate on the same principle. The following are the tests that may be performed on this type of fixture:

- Spray tip orifice test.
- Valve opening pressure test.
- Holding pressure test.

Complete details on test procedures should be obtained from the appropriate fuel injection equipment maintenance manual.

WARNING

When personnel are using a tester, they must be aware that the penetrating power of the fuel oil spray is great enough to drive oil through their skin. Since the fuel oil can cause blood poisoning, it is essential that all parts of the body be kept out of the line of the fuel spray.

When the opening pressure of a nozzle is too high, the cause depends on the fuel system involved. For example, if a Bosch nozzle opening pressure is too high or if the nozzle fails to open, the following may be the problem:

- The pressure spring may be improperly adjusted.
- The nozzle valve may be stuck in the nozzle body.
- The nozzle orifices may be clogged.

If the pressure is too high because of a maladjusted pressure spring, the valve must be disassembled and a new spring installed. Since the cause of excessive opening pressure depends on the fuel system involved, the mechanic must be familiar with the equipment with which he/she is working and must follow the applicable injection system TM. Unit injectors should be tested if improper opening pressure is suspected. Injectors equipped with needle valve assemblies cannot be tested for opening pressure without special equipment. Complete instructions for testing this type of injector are given in the fuel injector maintenance manual. When a pressure can be built up considerably higher than the pop pressure prescribed for a unit injector, it is usually due to improper assembly of the injector. In certain models of some unit injectors, it is possible to reverse one of the check valves. When this is done, the check valve will seat when the fuel tends to flow from the injector to the engine. A pop pressure that is too low for a unit injector may be due to a weak valve spring or dirty sealing surfaces. In nozzles where adjustments are possible, a low opening pressure may be due to a broken or an improperly adjusted pressure spring.

Distorted Spray Pattern

8-42. Distortion of the spray pattern of a nozzle or injector may be indicated by such symptoms as the following:

- Low firing pressure.
- Loss of power.
- Smoky exhaust.
- Local deposits of carbon within the combustion space.

Nozzles and injector spray tips are so designed that combustion should start before any appreciable quantity of fuel has struck the relatively cold surfaces of the combustion space. Orifices are drilled to take advantage of air currents in creating turbulence. For efficient combustion, the spray pattern must not become distorted. Spray pattern can be checked with a tester similar to the one shown in Figure 8-19. However, throttling-type pintle nozzles are difficult to test with a hand tester because of the high number of strokes per minute required to produce a representative spray pattern. A motor-driven test stand is highly desirable when testing a nozzle of this type. Depending on the type of nozzle or spray tip, distortion of a spray pattern may be caused by the following:

- Eroded valves.
- Eroded or clogged orifices.
- Broken pintle.

Erosion of orifices or valves generally results when filtration or centrifuging of the fuel is inadequate. A clogged nozzle orifice or injector spray hole will prevent complete mixing of the fuel charge with the available air in the cylinder. This will result in a drop in the power output of that cylinder, causing other cylinders to be overloaded. Nozzles and injectors must be inspected carefully when removed from the engine to determine whether any of the orifices are clogged. When pintle nozzles are handled, they should receive extra care to avoid damaging that portion of the nozzle valve which protrudes below the bottom of the nozzle body. When carbon is being removed from the nozzle body, it is possible to inadvertently remove the pintle. Pintles may also be broken off by striking or dropping the nozzle and holder on a hard surface. Distortion of the spray pattern is usually caused by dirt in the nozzle. Nozzle orifices should be cleaned as shown in Figure 8-20. The cleaning wire should be held at the same angle as the drilled orifice. Otherwise, the orifice may be damaged or the cleaning wire may break as it passes through the opening. If the spray is still distorted after the nozzle has been thoroughly cleaned, the cause is probably eroded parts. Erosion will require replacement of the nozzle.

Figure 8-19. Testing a Spray Nozzle

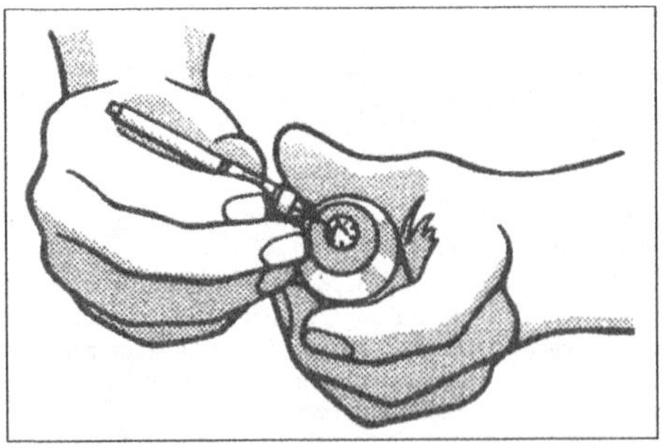

Figure 8-20. Cleaning Nozzle Orifices

LEAKY NOZZLES AND INJECTOR SPRAY TIPS

8-43. Dribbling from a nozzle or an injector may result in smoky operation, detonation, loss of power, and crankcase dilution. It can also result in excessive carbon formation on cylinder injection equipment and other surfaces of the combustion chamber. When a nozzle or injector spray tip is suspected of leakage, it should be checked using the appropriate tester. Dribbling of nozzles may be due to the following:

- Damaged valve or valve body seat.
- Dirty nozzle.
- Broken pressure-adjusting spring or screw.
- A nozzle valve stuck in the nozzle body.

Leakage from a spray tip or an injector may be due to damaged sealing surfaces or broken valve springs. Many leaking or dribbling nozzles may be repaired and placed back in service. When a nozzle is found to be leaking, it should first be soaked overnight in a suitable solvent. Then the accumulated deposits should be removed with brass tools or tools of softer metal. The nozzle body should be secured in an appropriate holding device. The nozzle valve should be coated with clean mutton tallow and rotated into the valve body. This will usually remove the surface deposits responsible for leakage. If the leakage does not stop, lapping (explained in detail in the fuel injection equipment maintenance manual) may be necessary. Lapping is a precision operation and unless all the precautions listed in the manual are followed, the nozzle may be ruined. Any nozzle must be replaced if it is cracked extensively, eroded, or so badly stuck that valve removal is not possible. Defective nozzles should be shipped to a reconditioning center. When a nozzle or injector is being assembled, the parts must be carefully inserted in their proper positions. After assembly, the unit must be tested to determine whether it performs as specified in the applicable maintenance manual. The unit must not be used if it does not meet the specifications.

PURGING THE DIESEL ENGINE FUEL INJECTION SYSTEM

8-44. When an engine fails to operate, stalls, misfires, or knocks, there may be air in the high-pressure pumps and fuel lines. If air is present in the system, compression and such air may take place without the injection valves opening. The presence of air in a fuel system can be determined by bleeding a small amount of fuel from the top of the fuel filter or by slightly loosening the bleeder screw or plug. If the fuel appears quite cloudy, it is likely that there are small bubbles of air in the fuel. In working with fuel systems, remember that if air is entering a fuel line, the pressure within the fuel line must be lower than atmospheric pressure. The smallest of holes in the transfer pump suction piping will permit air to flow into the system in quantities sufficient to air bind the high-pressure pumps. Carefully inspect all fittings in the suction piping. A loose fitting or a damaged thread will allow air to enter the system. On installations where flanged connections are used, the gaskets should be checked. Tubing, especially copper, should be inspected carefully for cracks which may result from constant vibration. If an engine is permitted to run out of fuel, trouble can be expected from air which enters the fuel system. If a considerable quantity of air exists in the filter, a quick method of purging the system of air is to remove the filling plugs on top of the filter and pour in clean fuel oil until all air is displaced. Any air remaining in the system can then be removed by using the hand-priming pump. For example, in the fuel system shown in Figure 8-21, the line is opened between the pump and strainers. The pump is operated until all air is removed and only clear fuel flows from the line. The line is then closed. The same procedure is repeated at the following different points in the system.

- Between the strainers and the filters.
- Between the filters and the high-pressure pumps.
- At the overflow line connection on the high-pressure pump housing.

In small, high-speed diesel engines, priming at the overflow connection may be all that is necessary. Since priming high-pressure lines is time-consuming, attempts to start the engine should be made before purging these lines. However, the engine must not be cranked for more than the specified time interval. If the engine still fails to start, priming the high-pressure lines will be necessary. Since the procedure to prime high-pressure lines varies considerably with different installations, follow the manufacturer's instructions. Table 8-1 gives information on troubleshooting a fuel system when there is no fuel or insufficient fuel.

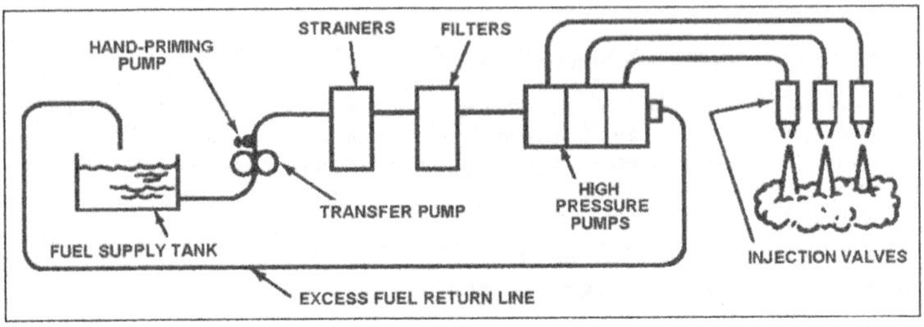

Figure 8-21. Schematic Drawing of a Fuel System

Table 8-1. Troubleshooting (No Fuel or Insufficient Fuel)

NO FUEL OR INSUFFICIENT FUEL *Probable Causes*	
AIR LEAKS	**FAULTY INSTALLATION**
• LOW FUEL SUPPLY • LOOSE CONNECTIONS OR CRACKED LINES BETWEEN FUEL PUMP AND TANK OR SUCTION LINE IN TANK • DAMAGED FUEL OIL STRAINER GASKET • FAULTY AIR HEATER • FAULTY INJECTOR TIP ASSEMBLY	• DIAMETER OF FUEL SUCTION LINES TOO SMALL • RESTRICTED FITTING MISSING FROM RETURN LINE • INOPERATIVE FUEL INTAKE LINE CHECK VALVE • HIGH FUEL RETURN TEMPERATURE
FLOW OBSTRUCTION	**FAULTY FUEL PIPE**
• FUEL STRAINER OR LINES RESTRICTED • TEMPERATURE LESS THAN 10° F ABOVE POUR POINT OF FUEL • FUEL SHUTOFF VALVES CLOSED	• RELIEF VALVE NOT SEATING • WORN GEARS OR PUMP BODY • FUEL PUMP NOT ROTATING

SECTION IV. ENGINE SPEED CONTROL DEVICES

Generally, diesel engines have devices installed for controlling speed. These are governors and overspeed safety devices such as air shutoff mechanisms.

GOVERNORS

8-45. A speed-sensitive device, called a governor, is designed to maintain a reasonably constant engine speed regardless of load variation. All governors used on diesel engines control engine speed by regulating the quantity of fuel delivered to the cylinders. Therefore, these governors may be classified as regulating governors. Governors may also be classed according to the following:

- Function or functions performed.
- The forces used in operation.
- The means by which the governor operates the fuel control mechanism.

The function which a governor must perform on a given engine is determined by the load on the engine and the degree of control required. Governors are classed according to the following:

- Function as speed.
- Variable-speed.
- Speed-limiting.
- Limiting.

Some installations require that engine speed remain constant from a no-load condition to a full-load condition. Governors which function to maintain a constant speed, regardless of load, are called constant-speed governors. Governors which maintain any desired engine speed between idle and maximum speed are classed as variable-speed governors. Speed-control devices which are designed to prevent an engine's exceeding a specified maximum speed and dropping below a specified minimum speed are classed as speed-limiting governors. However, some speed-limiting governors function only to limit maximum speed. Some engine installations require a control device that limits the load that the engine will handle at various speeds. Such devices are called load-limiting governors. Some governors are designed to perform two or more of these functions.

OPERATING PRINCIPLES

8-46. In most of the governors installed on diesel engines, the centrifugal force of rotating weights (flyballs) and the tension of a helical coil spring (or springs) are used in governor operation. On this basis, these governors are generally referred to as spring-loaded centrifugal governors. In spring-loaded centrifugal governors, two forces oppose each other. One of these forces is the pressure of a spring (or springs), which may be varied either by means of an adjusting device or by movement of the manual throttle. The other force is produced by the engine. Weights attached to the governor drive shaft are rotated, creating a centrifugal force when the shaft is driven by the engine. The magnitude of the centrifugal force varies directly with the speed of the engine. Transmitted to the injectors through a connecting linkage, the pressure of the spring (or springs) tends to increase the amount of fuel delivered to the cylinders. However, the centrifugal force of the rotating weights, through connecting linkage, tends to reduce the quantity of fuel injected. When the two opposing forces are equal or balanced, the speed of the engine remains constant. To illustrate how the centrifugal governor works, assume that an engine is operating under load and that the opposing forces in the governor are balanced so that the engine speed is constant. If the load is increased, the engine will decrease and a resultant reduction will occur in the centrifugal force of the flyballs. The spring pressure then becomes the greater force, and it acts on the fuel control mechanism to increase the quantity of fuel delivered to the engine. The increase in fuel results in an increase in engine speed until the forces are again balanced. When the load on an engine is reduced or removed, the engine speed increases and the centrifugal force within the governor increases. The centrifugal force then becomes greater than the spring pressure and acts on the fuel control linkage to reduce the amount of fuel delivered to the cylinders. This causes the engine speed to decrease until a balance between the opposing forces is again reached and engine speed becomes constant.

REGULATION OF FUEL CONTROL MECHANISMS

8-47. Governors may be classified according to the method by which the fuel control mechanisms are regulated. In some governors, the centrifugal force of the rotating weights regulates the fuel supply directly through a mechanical linkage which operates the fuel control mechanism. Other governors are designed so that the centrifugal force of the rotating weights regulates the fuel supply indirectly by moving a hydraulic pilot valve which controls oil pressure. Oil pressure is then exerted on either side of a power piston which operates the fuel control mechanism. Governors which regulate the fuel supply directly (through mechanical linkage) are called mechanical governors. Those which control the fuel indirectly (through oil pressure) are called hydraulic governors. A simple mechanical governor is shown in Figure 8-22. Note that the weights (or flyballs) are in an upright position. This indicates that the centrifugal force of the weights and the pressure of the spring are balanced. In other words, the engine is operating at constant load and speed. The positions of the parts in the hydraulic governor shown in Figure 8-23 indicate that the engine is responding to an increase in load with a resultant decrease in engine speed. Note that the weights tilt inward at the top. As engine speed decreases, the spring pressure overcomes the centrifugal force of these rotating weights. When the spring pressure is greater than the centrifugal force of the flyballs, the governor mechanism acts to permit oil under pressure to force the piston to increase the fuel valve opening. The increased fuel supply causes an increase in engine power output and speed. The governor regulates the fuel supply so that sufficient power is developed to handle the increase in load. There is always a lag between the moment a change is made in fuel setting and the time the engine reaches the new desired speed. Even though the fuel controls are set as required during a speed change, hunting caused by overshooting will occur. As long as engine speed is above or below the desired new speed, the simple hydraulic governor will continue to adjust (over-correct) the fuel setting, to decrease or increase the delivery of fuel. For this reason, a hydraulic governor must have a mechanism that will .discontinue changing the fuel control setting slightly before the new setting required for sustaining the desired speed has actually been reached. The mechanism which accomplishes this process (compensating) in all modern hydraulic governors is called a compensating device (see Figure 8-24). In the device illustrated, the pilot valve plunger operates in a movable pilot valve bushing in which are located the parts that control the oil flow. The movement of the valve bushing during a speed change is controlled by the receiving compensating plunger. The compensating action of the valve bushing is controlled hydraulically by transfer and by leakage of oil between the compensating receiving plunger and the compensating actuating piston. The rate of compensation is adjusted to fit the engine characteristics by regulating the oil leakage through the compensating needle valve. If the compensating needle valves are adjusted correctly, only a slight amount of hunting will occur following a load change. This hunting will quickly be dampened, resulting in stable operation through the operating range of the governor. Hydraulic governors are more sensitive than mechanical ones. The mechanical governor is more commonly used on small engines which do not require extremely close regulation of fuel. Hydraulic governors are more suitable for larger engines which require more accurate regulation of fuel.

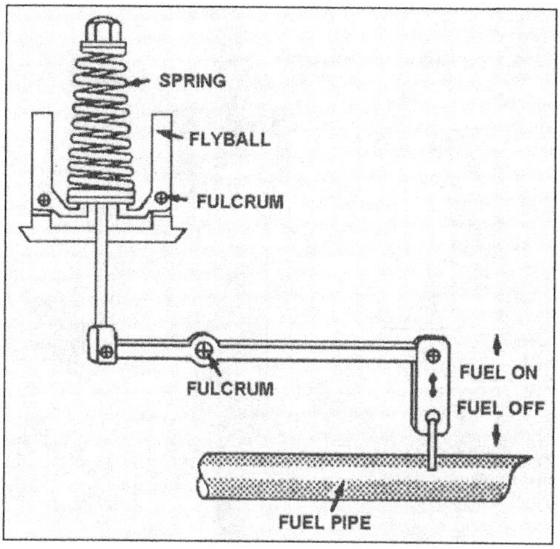

Figure 8-22. Simple Mechanical Governor

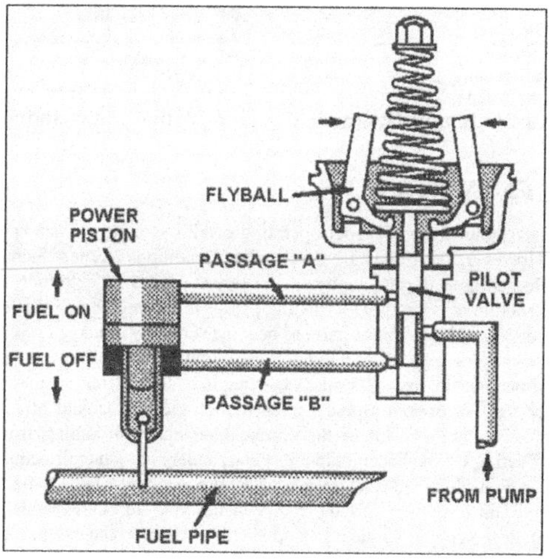

Figure 8-23. Simple Hydraulic Governor

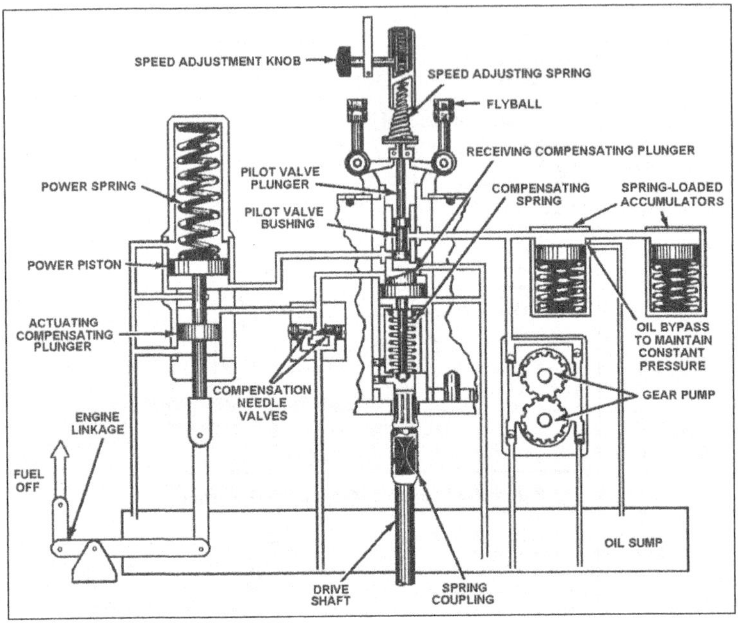

Figure 8-24. Simple Hydraulic Governor With Compensating Device

OVERSPEED SAFETY DEVICES

8-48. Engines that are maintained in proper operating condition seldom reach speeds above those for which they are designed. However, there may be times when conditions occur which result in speeds becoming excessively high. Operating a diesel engine at excessive speeds is extremely dangerous because of the relatively heavy construction of the engine's rotating parts. If the engine speed is sufficiently high, the inertia and centrifugal force developed may cause parts to become seriously damaged or even to fly apart. Therefore, you must know why an engine may reach a dangerously high speed and how it can be brought under control when excessive speed occurs. In some 2-stroke cycle engines, lubricating oil may leak into the cylinders as a result of leaky blower seals or broken piping. Even though the fuel is shut off, the engine may continue to operate or even "run away" as a result of the combustible material coming from the uncontrolled source. Engines in which lubricating oil may accumulate in the cylinders are generally equipped with an automatically operated mechanism which shuts off the intake air at the inlet passage to the blower. If no air shutoff mechanism is provided and if shutting off the fuel will not stop an engine which is overspeeding, anything which can be placed over the engine's intake to stop air flow will stop the engine. Excessive engine speeds are more commonly associated with an improperly functioning regulating governor than with lubricating oil accumulations in the cylinders. Stopping the flow of intake air is used as one means of stopping an engine which is overspeeding because of lubricating oil in the cylinders. However, a means of shutting off or decreasing the fuel supply to the cylinders is more commonly used to accomplish an emergency shutdown or reduction of engine speed when the regulating governor fails to function properly. Shutting off the fuel supply to the cylinders of an engine may be accomplished in various ways. The fuel control mechanism may be forced to the "no fuel" position or the fuel line may be blocked by closing a valve. The pressure in the fuel injection line may also be relieved by opening a valve or the mechanical movement of the injection pump may be prevented. These methods of shutting off the fuel supply may be either manual or automatic. Fuel and air control mechanisms are operated automatically by overspeed safety devices. As emergency controls, these safety devices operate only if the regular speed governor fails to maintain engine speed within the maximum design limit. Devices which function to bring an overspeeding engine to a full stop by completely shutting off the fuel or air supply are generally called overspeed trips. Devices which function to reduce the excessive speed of an engine but allow the engine to operate at safe speeds are more commonly called overspeed governors. All overspeed governors and trips depend on a spring-loaded centrifugal governor element for their operation. In overspeed devices, the spring tension is sufficiently great to over-balance the centrifugal force of the weights until the engine speed rises above the desired maximum. When an excessive speed is reached, the centrifugal force overcomes the spring tension and operates the mechanism which stops or limits the fuel or air supply. When a governor serves as the safety device, the fuel or air control mechanism may be actually operated by centrifugal force directly, as in a mechanical governor, or indirectly, as in a hydraulic governor. In an overspeed trip, the shutoff control is operated by a power spring. The spring is placed under tension when the trip is manually set and is held in place by a latch. If the maximum speed limit is exceeded, a spring-loaded centrifugal weight will move out and trip the latch, allowing the power spring to operate the shutoff mechanism. Overspeed safety devices must always be operative and must never be disconnected for any reason while the engine is operating. They are to be tested either weekly or quarterly depending on the application of the engine.

Chapter 9

AIR INTAKE AND EXHAUST SYSTEMS

INTRODUCTION

9-1. Combustion requires air, fuel, and heat. Certain amounts of all three are necessary for engine operation. This chapter deals only with air as it relates to combustion in the cylinder of an engine. The processes of scavenging and supercharging are also covered. The group of parts involved in supplying the cylinders of an engine with air and removing the waste gases after combustion and after the power event is finished are discussed. The engine parts which accomplish these functions are commonly referred to as the intake and exhaust systems.

SECTION I. INTAKE SYSTEMS

9-2. The primary function of a diesel engine intake system is to supply air required for combustion. The system also cleans the air and reduces the noise created by the air as it enters the engine. An intake system may include the following:

- Air silencer.
- Air cleaner and screen.
- Air box or header.
- Air heater.
- Air cooler.
- A blower.
- Intake valves or ports.

Not all of these parts are common to every intake system. A cross-sectional view of the air systems of one type of high-speed diesel engine is shown in Figure 9-1. Shown in the lower portion of the illustration is an intake system with a silencer, a screen, a blower, an air box, and intake ports. This system provides a clean supply of air with minimum noise to the combustion spaces.

SCAVENGING AND SUPERCHARGING

9-3. Before discussing the parts which may be included in an air-inlet system, the terms "scavenging" and "supercharging" are explained. You should understand the meaning and significance of these two terms because they are used frequently in discussing intake systems of diesel engines. Scavenging and supercharging and the processes they identify are not common to all diesel engines. In a few 4-stroke cycle engines, air enters the cylinder as a result of a pressure differential created by the piston as it moves away from the combustion space during the intake event. This is sometimes referred to as the naturally aspirated intake. However, the air is actually forced into the cylinder because of the atmospheric pressure outside the cylinder.

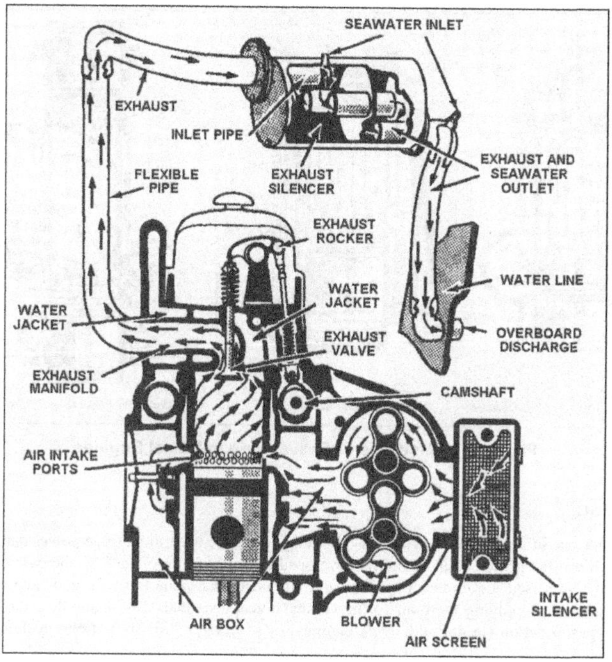

Figure 9-1. Air Systems of a 2-stroke Cycle Engine

SCAVENGING

9-4. In the intake systems of all modern 2-stroke cycle engines and of some 4-stroke cycle engines, a device, usually called a blower, is installed to increase the flow of air into the cylinders. The blower compresses the air and forces it into an air box or manifold which surrounds, or is to, the cylinders of an engine. Therefore, an increased amount of air under constant pressure is available as required during the cycle of operation. This increased amount of air is used to fill the cylinder with a fresh charge of air. During the process, it aids in clearing the cylinder of the gases of combustion. The process is called scavenging. Therefore, the intake system of some engines, especially those operating on the 2-stroke cycle, is sometimes called the scavenging system. The air forced into the cylinder is called scavenge air and the ports through which it enters are called scavenge ports. The process of scavenging must be done in a relatively short portion of the operating cycle. The duration of the process differs in 2-and 4-stroke cycle engines. In a 2-stroke cycle engine, the process takes place during the latter part of the downstroke (expansion) and the early part of the upstroke (compression). In a 4-stroke cycle engine, scavenging takes place when the piston is nearing and passing TDC during the latter part of an upstroke (exhaust) and the early part of a downstroke (intake). The intake and exhaust openings are both open during this interval. The overlap of intake and exhaust permits the air from the blower to pass through the cylinder into the exhaust manifold. Therefore, it cleans out the exhaust manifold and the exhaust gases from the cylinder and, at the same time, cools the hot engine parts. Scavenging air must be so directed, when entering the cylinder of an engine, that the waste gases are removed from the remote parts of the cylinder. The two principal methods by which this is accomplished are sometimes referred to as port scavenging and valve scavenging (see Figure 9-2). Port scavenging may be of the direct (cross-flow or loop (return)), or the uniflow type. Valve scavenging is of the uniflow type.

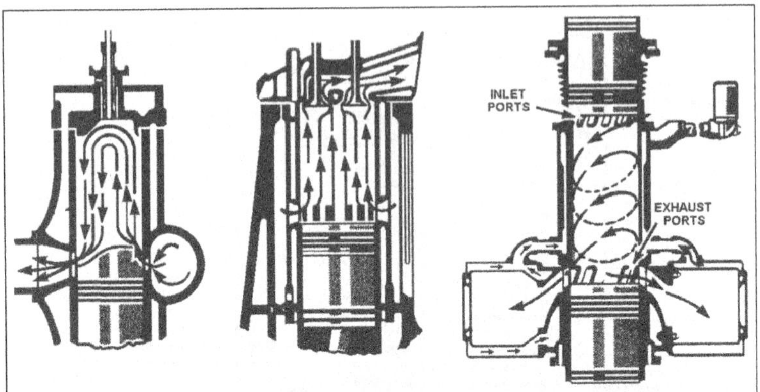

Figure 9-2. Methods of Scavenging in Diesel Engines

SUPERCHARGING

9-5. An increase in air flow into cylinders of an engine can be used to increase power output as well as for scavenging. Since the power of an engine is developed by the burning of fuel, an increase of power requires more fuel. The increased fuel in turn requires more air, since each pound of fuel requires a certain amount of air for combustion. Supplying more air to the combustion spaces than can be supplied through atmospheric pressure and piston action (in 4-stroke cycle engines) or scavenging air (in 2-stroke cycle engines) is called supercharging.

Supercharging of 2-stroke Cycle Diesel Engines

9-6. In some 2-stroke cycle diesel engines, the cylinders are supercharged during the air intake simply by increasing the amount and pressure of scavenge air. The same blower is used for both supercharging and scavenging. Scavenging is accomplished by admitting air under low pressure into the cylinder while the exhaust valves or ports are open. Supercharging is done after the exhaust ports or valves close. This enables the blower to force more air into the cylinder and thereby increase the amount of air available for combustion. The increase in pressure, resulting from the compressing action of the blower, depends on the installation involved. However, it is usually low, ranging from 1 psi to 5 psi. The increase in pressure and in the amount of air available for combustion results in a corresponding increase in combustion efficiency within the cylinder. In other words, a given size engine which is supercharged can develop more power than the same size engine which is not supercharged.

Supercharging of 4-stroke Cycle Diesel Engine

9-7. Supercharging a 4-stroke cycle diesel engine requires the addition of a blower to the intake system. This is because the operations of exhaust and intake in an unsupercharged engine are performed by the action of the piston. The timing of the valves in a supercharged 4-stroke cycle engine is also different from that in a similar engine which is not supercharged. In the supercharged engine the intake-valve opening is advanced and the exhaust-valve closing is retarded so that there is considerable overlap of the intake and exhaust events. The overlap increases power (the amount of the increase depends on the supercharging pressure). The increased overlap of the valve openings in a supercharged 4-stroke cycle engine also permits the air pressure created by the blower to be used in removing gases from the cylinder during the exhaust event. Study Figure 9-3 and Figure 9-4 to see how the opening and closing of the intake and exhaust valves or ports affect both scavenging and supercharging and the differences in those processes as they occur in supercharged 2- and 4-stroke cycle engines.

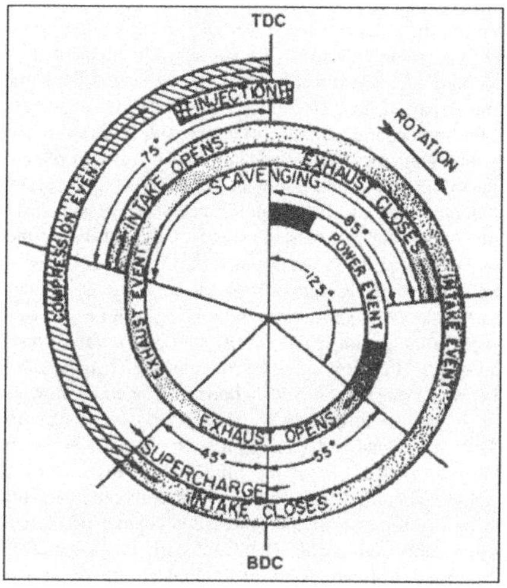

Figure 9-3. Scavenging and Supercharging in a 4-stroke Cycle Diesel Engine

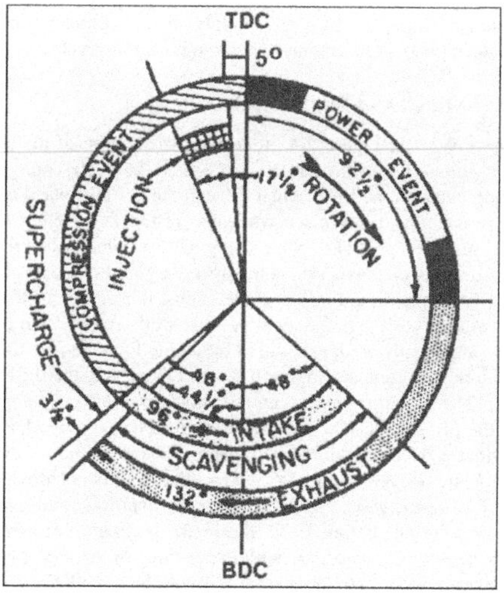

Figure 9-4. Scavenging and Supercharging in a 2-stroke Cycle Diesel Engine

FOUR-STROKE CYCLE SCAVENGING AND SUPERCHARGING

9-8. Figure 9-3 is based on the operating cycle of a 4-stroke cycle engine. It uses a centrifugal blower (turbocharger) to supply the cylinders with air under pressure. The mechanical details of the turbocharger are discussed later in this chapter. In a supercharged 4-stroke cycle engine, the duration of each event differs somewhat from the length of comparable events in a 4-stroke cycle engine that is not supercharged. The intake and exhaust valves are open much longer in a supercharged engine and the compression and power events are shorter. This permits a longer period for scavenging. When the exhaust event is completed and before the compression event begins, the turbocharger fills the cylinder with fresh air under pressure. In other words, the turbocharger supercharges the cylinders. To understand the relationship of scavenging and supercharging to the events of the cycle, refer to Figure 9-3 and follow through the complete cycle. Start your study of the cycle at TDC, the beginning of the power event. At this point, peak compression has been reached, fuel injection is nearly completed, and combustion is in progress. Power is delivered during the downstroke of the piston for 125° of crankshaft rotation. At this point in the downstroke) 55° before BDC) the power event ends and the exhaust valves open. The exhaust valves remain open throughout the remainder of the downstroke (55°) throughout all of the-next upstroke (180°) and throughout 85° of the next downstroke-(a total of 320° of shaft rotation). At a point 75° before the piston reaches TDC, the intake valves open and the turbocharger begins forcing fresh air into the cylinder. For 160° of shaft rotation, the air passes through the cylinder and out the exhaust valve, clearing the waste gases from the cylinder. The rapid flow of gases escaping through the exhaust manifold is used to drive the turbocharger. The process of scavenging continues until the exhaust valves close at 85° past TDC. The intake valves remain open, after the exhaust valves close, for an additional 140° of shaft rotation (45° past BDC). From the time the exhaust valves close until the piston reaches approximately BDC, the cylinder is being filled with air made available by the blower. During this interval, the increase in pressure is negligible because of the increasing volume of the cylinder space (the piston is in downstroke). However, when the piston reaches BDC and starts the upstroke, the volume of the space begins to decrease as the blower continues to force air into the cylinder. The result is a supercharging effect with the pressure reaching 3 to 5 psi by the time the intake valves close. During the remainder of the upstroke (after the intake valves close), the supercharged air is compressed. Fuel injection begins several degrees before TDC and ends shortly after TDC. The actual length of the injection period in a specific engine depends on engine speed and load. When the piston reaches TDC, a cycle, involving two complete crankshaft revolutions and four strokes of the piston, has taken place. The engine is ready to repeat the cycle.

TWO-STROKE CYCLE SCAVENGING AND SUPERCHARGING

9-9. In comparing Figure 9-3 and Figure 9-4, note that the duration of the supercharging and scavenging period in a 2-stroke cycle engine is not the same as in a 4-stroke cycle engine. There is also considerable difference in piston location between the times when these processes take place in the two types of engines. In a 4-stroke cycle engine, scavenging takes place while the piston is traveling through the latter part of the upstroke and the early part of the downstroke. Supercharging takes place when the piston is near BDC. In a 2-stroke cycle engine, the processes of scavenging and supercharging both take place while the piston is in the lower part of the cylinder. A piston in a 4-stroke cycle engine does much of the work of intake and exhaust. However, a piston in a 2-stroke cycle engine does very little work in these two processes. Because of this, 2-stroke cycle diesel engines are usually equipped with a blower to force air into the cylinder and to clear out the exhaust gases. If you continue to compare Figure 9-3 and Figure 9-4, the differences in the scavenging and supercharging processes in 2- and 4-stroke cycle engines should become more apparent. Start your study of the cycle in Figure 9-4 with the piston at TDC. Fuel has been injected, ignition has occurred, and combustion is taking place. The power that is developed forces the piston through the power event until the piston is 92 1/2° (compared to 125° for the 4-stroke cycle) past TDC, just a little more than halfway through the downstroke. At this point, the exhaust valves open, gases escape through the manifold, and cylinder pressure drops rapidly. When the piston reaches a point 48° before BDC, the intake ports are uncovered by the piston as it moves downward and scavenging begins. Compare this with the opening of the intake valves in a 4-stroke cycle. The scavenging air, under blower pressure, swirls upward through the cylinder and clears the cylinder of exhaust gases. The situation in the cylinder when scavenging starts is approximately the same as that shown in Figure 9-1 and in valve uniflow scavenging in Figure 9-2. Note the position of the piston, the open scavenging ports, the open exhaust valves, and the flow of air through the cylinder. The flow of scavenge air through the cylinder aids in cooling the parts which are heated by combustion. Scavenging continues until the piston is

44 1/2° past BDC (a total of 92 1/2° as compared with 160° in the 4-stroke cycle), at which point the exhaust valves close. The exhaust valves remain open during only 132°, as compared with the 320° in the example of the 4-stroke cycle. The scavenge ports remain open for another 3 1/2° of shaft rotation (45° in the 4-stroke cycle) and the blower continues to force air into the cylinder. Even though the ports are open only for a short interval after the exhaust valves close, there is sufficient time for the blower to create a supercharging effect before the compression event starts. The piston closes the intake ports at 48° past BDC. The compression event takes place during the remainder of the upstroke with injection and ignition occurring near TDC. At this point, one cycle is ended and another is ready to start.

ENGINE AIR-INTAKE SYSTEMS PARTS

9-10. Many variations exist in the design of engine parts which function as a group to conduct clean air to intake valves or ports under proper conditions. Regardless of design differences, the purpose of each kind of part remains the same. It is beyond the scope of this text to cover every type of model of each part of engine air-intake systems. Therefore, only a few of the common types of each of the principal parts of these systems are discussed here.

SILENCERS AND SCREENS

9-11. The air that enters the intake system must do so relatively quietly and be as clean as possible. A diesel engine uses great quantities of air and, unless a silencer is installed, the rush of air through the air-cleaning devices creates an extremely high-pitched-whistle. Devices used to reduce the noise of intake air are generally constructed as parts of air-cleaning components. One type of air-intake silencer assembly (shown in Figure 9-5) is used on some models of Gray Marine and GM 71 engines. The silencer assembly is bolted to the intake side of the blower. A perforated steel partition divides the silencer lengthwise into two sections. Air enters the end of the silencer and passes through the inner section into the blower. The noise created by the air as it passes through the silencer is reduced by a sound-absorbent, flameproof, felted cotton waste which fills the outer section of the silencer. Upon leaving the silencer, air enters the blower through an air-intake screen. The purpose of the air-intake screen is to prevent particles of foreign material from entering the engine. Unless filtered from the intake air, foreign material might seriously damage the blower assembly and internal engine parts such as pistons, piston rings, and liners. The silencer and screen assembly just described is sometimes referred to as a dry-type cleaner and silencer. Another kind of air cleaner and silencer is the viscous type. In both the dry and the viscous types, intake air is drawn through a fine mesh or screen which filters the air. The mesh of such cleaners may consist of cotton fabric, wire screens, specially wound copper crimp, or metal wool. The principal difference between cleaners of the dry and viscous types is that the mesh of the viscous cleaner is wet, usually with medium-weight oil. The cleaning element (metal wool in the example shown) is oil-stroked to collect the dust and dirt from air passing through the assembly. The hollow housing which supports the elements also serves as a silencing chamber. Assemblies of the type shown in Figure 9-6 are also used on some models of Gray Marine and GM 71 engines. The silencer and cleaner assemblies described here are representative of the devices used to clean intake air and to reduce the noise it makes when entering the engine. To clean the intake air sufficiently, air filters, regardless of type, should be cleaned as specified by the preventive maintenance schedule.

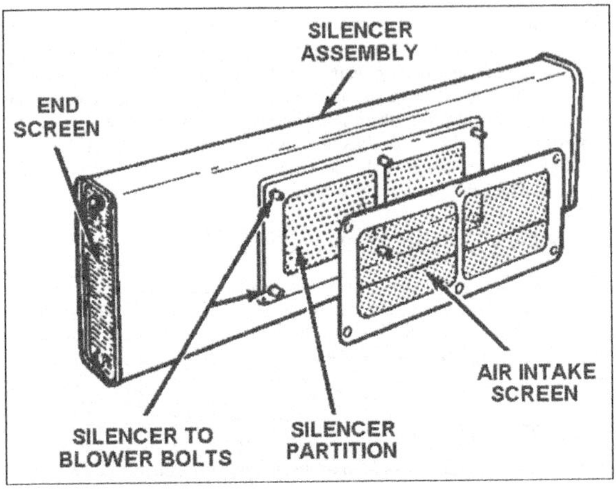

Figure 9-5. Air-intake Silencer Assembly

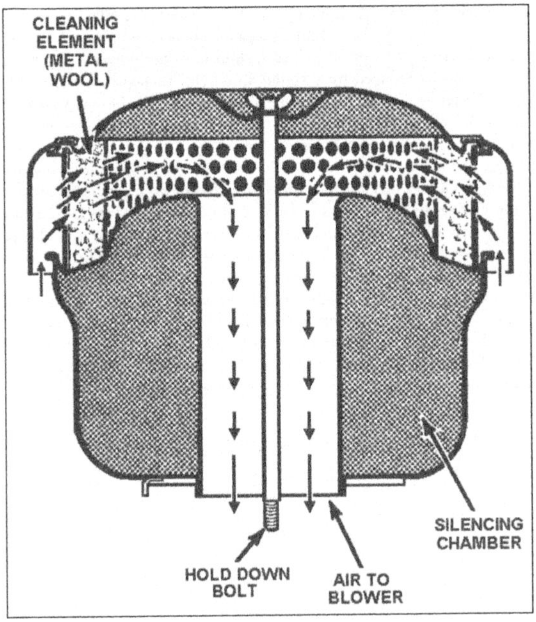

Figure 9-6. Viscous-type Air Silence and Cleaner

AIR CLEANERS

9-12. The air cleaners used on GM 71 engines are designed to remove foreign matter from the air and pass the required volume of air for proper combustion and scavenging. They are also designed to maintain their efficiency for a reasonable period before requiring service. The importance of keeping dust and grit-laden air out of an engine must be emphasized since clean air is so essential to satisfactory engine operation and long engine life. The air cleaner must be able to remove fine materials (such as dust and blown sand) as well as coarse materials (such as chaff, sawdust, or lint) from the air. It must also have a reservoir capacity large enough to retain the material separated from the air to permit operation for a reasonable period before cleaning and servicing are required. Dust and dirt entering an engine will cause rapid wear of piston rings, cylinder liners, pistons, and the exhaust valve mechanism. This will result in a loss of power and a high consumption of lubricating oil. Dust and dirt, which is allowed to build up in the air cleaner passages, will also eventually restrict the air supply to the engine and result in heavy carbon deposits on pistons and valves due to incomplete combustion.

Light-duty, Oil-wetted Air Cleaner

9-13. The oil-wetted air cleaner consists essentially of a metal screen element saturated with oil and supported inside a cylindrical housing. Air drawn into the cleaner passes through the element where dust and other foreign matter are removed. Then it travels down the central tube in the cleaner to the blower inlet. This lower portion serves as a chamber to reduce air intake noise.

Light-duty, Oil-bath Air Cleaner

9-14. The light-duty, oil-bath air cleaner (see Figure 9-7) consists essentially of a wire screen element supported inside a cylindrical housing which contains an oil bath directly below the element. Air drawn through the cleaner passes over the top of the oil bath. The air stream direction reverses when the air impinges on the oil in the sump and is then directed upwards by baffles. During this change in the direction of air flow, much of the foreign matter is trapped by the oil and is carried to the sump where it settles out. The air passes upward through the metal-wool elements where more dust and the entrained oil are removed. A second change of air direction, at the top of the cleaner, directs the air downward through the center tube into the blower inlet housing.

Service

9-15. To service the light-duty air cleaner for GM 71 engines, loosen the wing bolt and remove the cleaner from the air inlet housing. The cleaner may then be separated into two sections. The upper section contains the metal-wool elements and the lower section consists of the oil sump, removable baffle, and center tube. The upper shell and metal-wool elements may be cleaned by soaking the entire section in kerosene or fuel oil. This will loosen the oil and dust in the elements and help to flush out the dirt. The oil should be emptied from the sump, the baffle removed, and the sump and baffle cleaned in kerosene or fuel oil to remove all sediment. A lintless cloth should be pushed through the center tube of the cleaner before the baffle is installed and the sump refilled to the oil-level mark with clean engine oil. Never use cotton waste to wipe the center tube. Use the same viscosity and grade of oil that is used in the engine crankcase. All gaskets and sealing surfaces should be checked and cleaned to ensure airtight seals.

Installation

9-16. After the filter element has been thoroughly drained of the flushing fluid, the cleaner should be assembled. However, before installing the cleaner on the engine, the air inlet housing and blower inlet screen should be checked for dirt accumulations. If the service period has been too long or dust-laden air has been leaking past the seals, the inlet housing and screen will be dirty. This will serve as a good check on the maintenance of the air cleaner installation. When installing the cleaner (and its seal) on the inlet housing, be sure that the cleaner seats properly and then tighten the wing bolt securely until the cleaner is rigidly mounted.

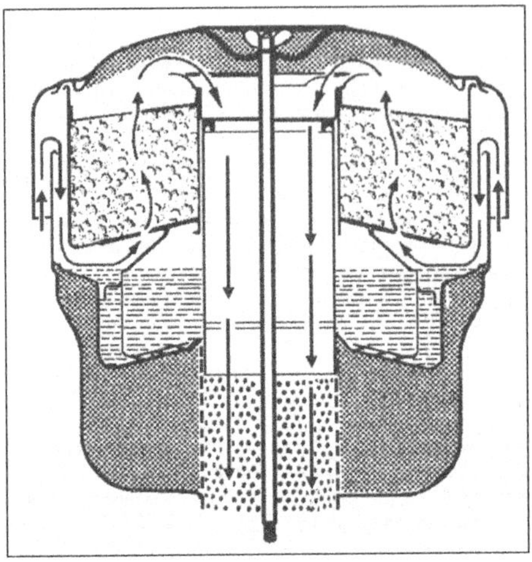

Figure 9-7. Light-duty, Oil-bath Air Cleaner and Silence Assembly

Heavy-duty Air Cleaner

9-17. In all heavy duty air cleaners, air is drawn through the air inlet hood, which acts as a cleaner, down through the center tube. At the bottom of the tube, the direction of air flow is reversed and oil is picked up from the oil reservoir cup. The oil-laden air is carried up into the separator screen where the oil, which contains the dirt particles, is separated from the air by collecting on the separator screen. Figure 9-8 shows a heavy-duty, oil-bath air cleaner and silencer assembly. A low-pressure area is created toward the center of the air cleaner as the air passes a cylindrical opening formed by the outer perimeter of the central tube and the inner diameter of the separator screen. This low pressure is caused by the difference in air current velocity across the opening. The low-pressure area, plus the effect of gravity and the inverted cone shape of the separator screen, causes the oil and dirt mixture to drain to the center of the cleaner cup. This oil is again picked up by the incoming air causing a looping cycle of the oil. However, as the oil is carried toward another cycle, some of the oil will overflow the edge of the cup carrying the dirt with it. The dirt will be deposited in the outer area surrounding the cup. Oil will then flow back into the cup through a small hole located in the side of the cup. Above the separator screen, the cleaner is filled with a wire screen element which will remove any oil passing through the separator screen. This oil will also drain to the center and back into the pan. The clean air then leaves the cleaner through a tube at the side and enters the blower through the air inlet housing. Depending on operating conditions, an air inlet hood or prescreener must be used with heavy-duty air cleaners. This equipment normally requires cleaning more frequently than the main air cleaner. The usual installation uses an air inlet hood which serves only to prevent rain and such items as rags, paper, and leaves from entering the air cleaner. The smaller cleaners use a spherical-shape hood. Air enters the hood through a heavy screen which forms the lower portion of the hood. The air is reversed in the hood and pulled downward into the air cleaner. The hood is mounted on the air cleaner inlet tube and is held in place by the fit of the hood in the inlet tube. The larger cleaners use a dome-shape hood. A heavy screen inside the dome guards against large pieces of foreign material entering the cleaner. The hood is mounted on the air inlet tube of the cleaner and is secured to it by a screw clamp. As previously mentioned, the hoods serve only to prevent rain and large pieces of foreign material from entering the cleaner. The openings in the hoods should be kept clear to prevent excessive restriction of the air flow. A prescreener should be used on the inlet tube of the air cleaner instead of the inlet hood for those operations in which the air cleaner elements load up with lint or chaff. The purpose of the prescreener is to remove as much lint or chaff as possible before the air enters the cleaner.

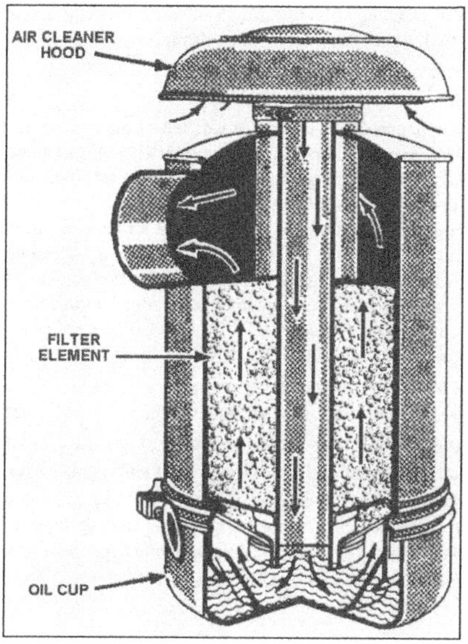

Figure 9-8. Heavy-duty, Oil-bath Air Cleaner and Silencer Assembly

Service

9-18. The air inlet hoods used on heavy-duty air cleaners are not intended to do any cleaning. However, some dirt will collect on the heavy screens and in the hood itself. Therefore, the hood must be removed occasionally for cleaning. It can be cleaned by brushing or by using compressed air. Some applications may be equipped with a prescreener. This device catches the lint and chaff on the screen surrounding the shell. This screen can be removed by unhooking the retaining springs. It can be cleaned by brushing or by using compressed air. The shell can be cleaned by wiping it with a lintless cloth. The prescreener may then be assembled and installed on the cleaner inlet tube. Although the prescreener will remove most of the lint and chaff from the air, some may still find its way into the main cleaner. Therefore, the fixed element of the main cleaner must be checked each time the cleaner is serviced to prevent excessive lint deposits. When the oil sump is removed on some heavy-duty air cleaners, a tray-type screen attached to the tube will be visible. It may be removed by loosening the wing nuts and rotating the tray so that it unlocks from the tube. On other heavy-duty models, the tray rests on the lip of the inner oil cup of the sump and is not retained by wing nuts. The efficiency of the tray-type, oil-bath air cleaner can be greatly reduced if the fibrous material caught in the tray is removed. It is extremely important that the tray be cleaned regularly and properly. If a tray is plugged with lint or dirt, wash the tray in a solvent or similar washing solution and blow out with high-velocity compressed air or steam. An even pattern of light should be visible through the screens when a clean tray is held up to the light. As a last resort, you may have to burn off the lint, but be extremely careful not to melt the galvanized coating in the tray screens. Some trays have equally spaced holes in the retaining baffle. Check to make sure that they are clean and open. It is advisable to have a spare tray on hand to replace the tray that requires cleaning. Having an extra tray available provides for better servicing and the plugged tray can be cleaned thoroughly at leisure. Check for dirt accumulation in the air cleaner center tube. Remove dirt by passing a lintless cloth through the center tube. Some tubes have a restricted portion at the lower end. Be careful not to damage this end. Check oil sump for any dirt accumulation in the inner and outer cups and clean, if necessary. At some regular period of engine service, remove the entire air cleaner from the engine and clean the fixed element. This can be done by passing a large quantity of cleaning solvent through the air outlet and down into the fixed element. When clean, allow element to dry thoroughly before installing cleaner.

Assembly and Installation

9-19. When all of the components have been cleaned, the cleaner is ready for assembly. The removable screen should be installed and the oil sump should be filled with clean engine oil to the indicated level and installed on the cleaner. Make sure that all gaskets and joints are tight. All connections from the cleaner to the engine should be checked for air leaks to prevent air bypassing the air cleaners. If it is found that unfiltered air is being admitted into the engine through the duct work of the air cleaner installation, the following procedure may be used for finding air leaks in the air duct system. The air cleaning system does not have to be dismantled. This will save valuable time. Suitable plugs must be provided to block the inlet and outlet ends of the air cleaner system. The air cleaner inlet plug should contain a suitable air connection and shutoff valve to maintain 2 pounds pressure in the air duct system. The outlet plug need only be of sufficient size to form a completely airtight seal at the outlet end of the system. Check the system as follows:

- Remove air inlet hood or prescreener.
- Insert the plug (with the fitting for the air hose) in the air cleaner inlet to form an airtight seal.
- Insert the other plug in the outlet end of the system to form an airtight seal.
- Attach an air hose to the plug in the air cleaner inlet and regulate pressure not to exceed 2 psi.
- Brush a soapsuds solution on all air duct connections. Any opening which would allow dust to enter the engine can then be detected by the escaping air causing bubbles in the soapsuds solution. All leaks discovered should be remedied until the system checks airtight.
- Remove plugs and install air inlet hood or prescreener.

Dry-type Air Cleaner

9-20. The Donaldson dry-type air cleaner is designed to provide highly efficient air filtration under all operating conditions. The cleaners have a replaceable impregnated paper filter element that can be cleaned. Fins on the element give high-speed rotation to the intake air, which separates a large portion of dust from the air by centrifugal action. The plastic fins, the element, and the gasket make up a single replaceable element assembly. Dust is swept through a space in the side of the baffle and collects in the lower portion of the body or dust cup. Dust remaining in the precleaned air is removed by the element. The dry-type cleaner cannot be used where the atmosphere contains oil vapors or where fumes from the breather can be picked up by the air cleaners.

Servicing

9-21. The air cleaner should be serviced as operating conditions warrant. Disassemble the dry-type air cleaner as follows:

- Loosen the bolted cover and remove the cover and bolt as an assembly.
- Remove the element assembly and baffle from cleaner body.
- Remove the dust and clean the cleaner body thoroughly.

Disassemble the heavy-duty, dry-type air cleaner as follows:

- Loosen dust cup clamp and remove dust cup.
- Loosen wing bolt in the dust cup and remove the baffle from dust cup.
- Remove wing bolt from cleaner body and remove the element assembly.
- Remove the dust and thoroughly clean the cleaner body, dust cup, and baffle.

The paper pleated element assembly can be cleaned as follows:

Note: The precleaning fins are not removable.

- The element can be dry-cleaned by directing clean air up and down the pleats on the clean-air side of the equipment.

CAUTION

Air pressure at the nozzle must not exceed 100 psi. Maintain a reasonable distance between the nozzle and the element.

- To wash the element, use the Donaldson filter cleaner or a nonsudsing equivalent. Proportions are 2 ounces of cleaner to 1 gallon of water. For best mixing results, use a small amount of cool tap water, then add it to warm (100^{0} F) water to give the proper proportion. Soak the element for 15 minutes; then rinse it thoroughly with clean water from a hose (maximum pressure 40 psi). Air-dry the element completely before reusing (a fan or air draft may be used, but do not heat the element to hasten drying).
- Inspect the cleaned element with a light bulb after each cleaning for thin spots, pin holes, or the slightest rupture. These could admit sufficient airborne dirt to render the element unfit for further use and cause rapid failure of the piston rings. Replace the element assembly if necessary.
- Inspect the gasket on the end of the element. If the gasket is damaged or missing, replace the element.
- Reassemble the air cleaner in reverse order of disassembly. Replace the air cleaner body gasket if necessary.

Note: The element assembly should be replaced after six cleanings, or annually.

CAUTION

Do not use oil in the bottom of the cleaner body.

Air Cleaner Mounting

9-22. Air cleaner mountings vary according to the air cleaner installation and the engine units on which they are used. The light-duty, oil-wetted type, the oil-bath type, and the dry-type air cleaners are mounted on the air inlet housing. Heavy-duty air cleaners are remotely mounted from the air inlet housing and are connected to it by airtight ducts.

Air Cleaner Maintenance

9-23. Although the air cleaner is highly efficient, this efficiency depends on proper maintenance and periodic servicing. No set rule for servicing an air cleaner is given since servicing depends on the type of air cleaner, the condition of the air supply, and the type of application. An air cleaner operating in severe dust will require more frequent service than one operating in comparatively clean air. The most satisfactory service period should be determined by frequently inspecting the air cleaner under normal operating conditions, then setting the service period to best suit the requirements of the particular engine application. The following maintenance procedures will help to assure efficient air cleaner operation:

- Keep the air cleaner tight on the air intake pipe to the engine.
- Keep the air cleaner properly assembled so the joints are strictly oiltight and airtight.
- Immediately repair any damage to the air cleaner or related parts.
- Inspect and clean or replace the air cleaner element as operating conditions warrant.

Note: In the dry-type cleaner, it is possible to clean and reuse the element several times as long as the paper is not ruptured in the process. In an oil-bath cleaner, keep the oil at the level indicated on the air cleaner sump. Overfilling may result in oil being drawn through the element and into the engine and this can carry dirt into the cylinder and also result in excessive engine speed.

- After servicing the air cleaner, remove the air inlet housing and clean accumulated dirt deposits from the blower screen and the inlet housing. Keep all air intake passages and the air box clean.
- Where rubber hose is used, cement it in place. If necessary, use new hose and hose clamps to get an airtight connection.
- Carefully inspect the entire air system periodically.

BLOWERS

9-24. As discussed earlier, blowers are necessary on most 2-stroke cycle engines to force scavenging air through the cylinders. In addition, a supercharged engine of either the 2- or 4-stroke cycle must have a blower to fill the cylinder with fresh air at a pressure above atmospheric pressure before the compression event starts. Basically, the primary function of an engine blower is to deliver a large volume of air at a low pressure (1 psi to 5 psi). The two principal types of blowers that exist are the positive-displacement blower and the centrifugal blower. A positive-displacement blower is usually gear-driven directly by the engine. A centrifugal blower is usually driven by an exhaust-gas turbine. Positive-displacement blowers may be divided into the following two groups:

- The multiple-lobe type, commonly called the lobe or roots blower.
- The axial-flow blower.

There are many variations in blower design. However, examples of only the most common designs are described here. The first example, a lobe blower, is commonly used on many 2-stroke cycle engines. The others are exhaust-driven, centrifugal blowers found on some 4-stroke cycle diesel engines.

Positive-displacement Blowers

9-25. The drive mechanism which transmits power from the crankshaft to the blower is attached to the end of the blower. Drive mechanisms are discussed in Chapter 13 of this manual. Blowers of this general type are commonly found on General Motors, Gray Marine, and many other engines. This type blower has two, three-lobe rotors which rotate in opposite directions within the closely fitted inner wall of the housing. The rotors do not touch each other or the housing wall which surrounds them. The rotors are mounted on tubular, serrated shafts to which the helical rotor gears are attached. The closely fitted rotor gears are rigidly attached to the shafts, thereby assuring that the meshing lobes of the rotors will not touch each other. The radial position of each rotor and the clearance between the rotor lobes and the housing are maintained by babbitted bearings located in the blower end plates. The rotor bearings at the gear end have thrust surfaces which maintain the correct axial position of the rotors and prevent contact between the rotors and the end plates. Oil passages in the end plates conduct lubricating oil under pressure to the bearings. Oil seals are provided at each bearing to prevent the oil from entering the housing. An airtight seal is maintained between the end plates and the housing by a fine silk thread and a very thin coat of nonhardening gasket compound. This material is placed around the housing end plate openings inside the stud line. Power to drive the blower is transmitted from the blower-and-accessory drive mechanism (attached to the front of the blower) to the rotor gear train by a shaft which extends through a passage in the blower housing. Air from the silencer enters the top of the blower housing through the scavenging air inlet. The rotor lobes produce a continuous and uniform displacement of air. The lobes carry the air around the cylindrical sides of the housing (in the spaces between the lobes and the housing) and force the air, under pressure, to the bottom of the housing. An opening in the inner wall of the housing permits the air to pass into the space between the inner and outer walls of the housing and then out the discharge opening.

Centrifugal Blowers

9-26. Unlike positive-displacement blowers, which are driven through a gear train by the engine crankshaft, the centrifugal blower (or turbocharger) takes its power from the exhaust gases. It thereby makes use of some of the energy which would otherwise be wasted. The basic principle of operation of a turbocharger is that the gases from the exhaust manifold drive a turbine and the turbine drives an impeller (on the same shaft). This supplies air to the cylinders for scavenging and supercharging. The several types of centrifugal blowers in naval service all operate on this basic principle.

CRANKCASE VENTILATION DEVICES

9-27. During the power event in all engines, some of the products of combustion enter the crankcase. These products contain a great deal of water vapor. The water vapor condenses when it contacts a relatively cool surface. This often causes corrosion and contamination of the oil by forming an emulsion. The lubricating oil system of an engine must be properly ventilated. If so, accumulation of vapors which might be ignited by a local hot spot within the engine can be prevented. Fuel dilution and water contamination will also be reduced considerably. If the crankcase of an engine is kept free of moisture and fuel, the oil will maintain an adequate viscosity for a longer period. Corrosion will therefore be minimized. The method by which crankcases are ventilated varies from one engine to another. The functions of crankcase ventilation devices in all marine engine installations are to prevent contamination of the engine room atmosphere by heated or fume-laden air. These devices also help reduce or eliminate vapors and liquids which might cause dilution of the oil, corrosion of engine parts, or the formation of sludge. They also serve to prevent the accumulation of combustible gases within the crankcase. Crankcases are ventilated by devices and passages called the breather system. Design and arrangement of breather systems vary with types of engines. For information regarding the design of the breather system on each type of engine, consult the manufacturer's technical manual (TM). Air must pass through a number of passages to reach the combustion spaces within an engine. So far, the passage of air through components which clean, silence, and compress the intake air has been covered. However, from the blower, the air is discharged into a unit or passage to conduct the air to the intake valves or ports of the cylinders. The design of such a unit and the terminology used to identify it differ depending on the type of engine. In 2-stroke cycle engines, the passage which conducts intake air to the cylinders is generally called an air box. The air box surrounds the cylinders and, in many engines, is built into the block. In 2-stroke cycle, V-type diesel engines, the air box consists of the space (within the block) included between the two banks of the V-construction and the open space between the upper and lower deck plates of each bank. The scavenging air passages in an opposed-piston engine are referred to as the air receiver. This compartment is at the upper part of the block and surrounds cylinder liners. In some 2-stroke cycle engines, the passage which serves as a reservoir for intake air from the blower is called an air header. Drains are generally provided in air boxes, receivers, and headers to drain off any liquids that may accumulate. Some vapors from the air charge may condense and settle in the air box. However, a small amount of lubricating oil may be blown into the air box as the piston passes the ports on the downstroke following the power event. On some engines, the drains are vented to the atmosphere. In others, a special drain tank is provided to collect the drainage from the air box. Figure 9-9 shows the air box drain tank assembly. The purpose of the tank is to prevent drainage of oil to the engine room. A small connection from the fuel pump also carries to the drain tank any fuel oil that may leak past the seals in the pump. Drains are of prime importance when the air cooler is installed between the blower or turbocharger discharge and the air intake manifold or receiver. The drains are usually left open during engine operation to prevent condensation or water from leaky coolers being carried into the engine cylinders with the combustion air. In 4-stroke cycle diesel engines, the intake air from the blower to the cylinders differs, in general, from the air passage in 2-stroke cycle engines in that the passage is not an integral part of the block. Instead, a separate unit is attached to the block for conducting intake air to the engine cylinders. The attached unit is generally called the air intake manifold. Figure 9-10 shows the manifolds (both intake and exhaust). Note the arrows which indicate the flow of air from the turbocharger, through the intake manifold, to the cylinders, and the flow of the gases from the cylinders back to the turbocharger and the exhaust. The intake manifold shown is a one-piece fabricated steel unit. It is heavily insulated with felt lagging and covered with heavy canvas. The insulation serves to dampen the noise created by the turbocharger.

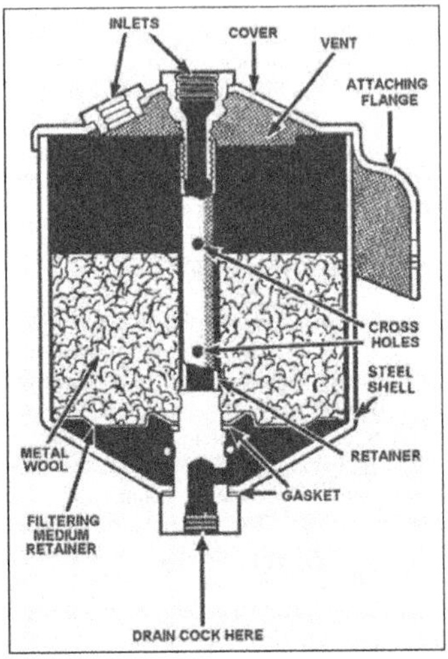

Figure 9-9. Air Box Drain Tank Assembly

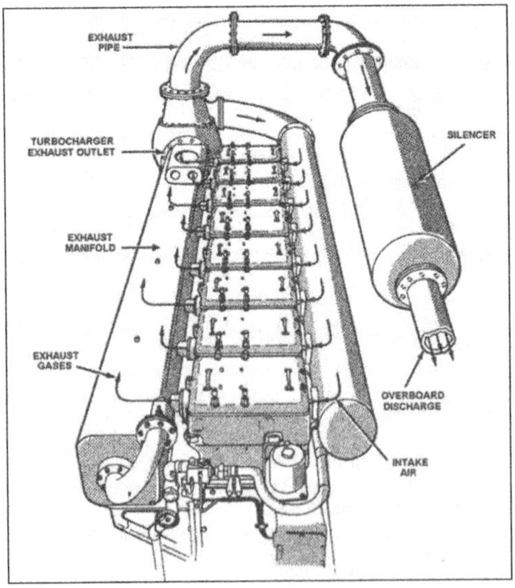

Figure 9-10. Manifolds

SECTION II. EXHAUST SYSTEMS

The parts of engine air systems discussed so far provide a passage for air into the cylinders after combustion. The relationship of blowers or turbochargers to both the intake and exhaust systems has also been pointed out. The system which primarily conveys gases away from the cylinders of an engine is called the exhaust system. In addition to this principal function, an exhaust system may be designed to perform other functions. These may include the following:

- Muffling exhaust noise.
- Quenching sparks.
- Removing solid material from exhaust gases.
- Furnishing energy to a turbine-driven turbocharger.

The following discussion covers the principal parts of an engine exhaust system which may be used in combination to accomplish these functions.

EXHAUST MANIFOLDS

9-28. When the gases of combustion are forced from the cylinders of an engine, the gases enter a unit called the manifold, also sometimes called a header. The manifold shown in Figure 9-11 is a one-piece unit, fabricated of steel. Some manifolds are made of steel plate with welded joints and branch elbows of steel castings. Others are made of aluminum castings. On some engines, all exposed surfaces of exhaust manifolds and related parts are insulated with layers of spun glass held in place by laced-on woven asbestos covers. The insulation aids in reducing heat radiation. Expansion joints are generally provided between manifold sections and turbochargers or other outlet connections. The exhaust manifold serves as the passage for the gases from the combustion spaces to the exhaust inlet of the turbocharger. Therefore, the turbined end of the turbocharger may be considered as part of the exhaust system. This is because it forms a part of the passageway for the escape of gases to the exhaust outlet. A similar arrangement is characteristic of other 4-stroke cycle, turbocharged diesel engines.

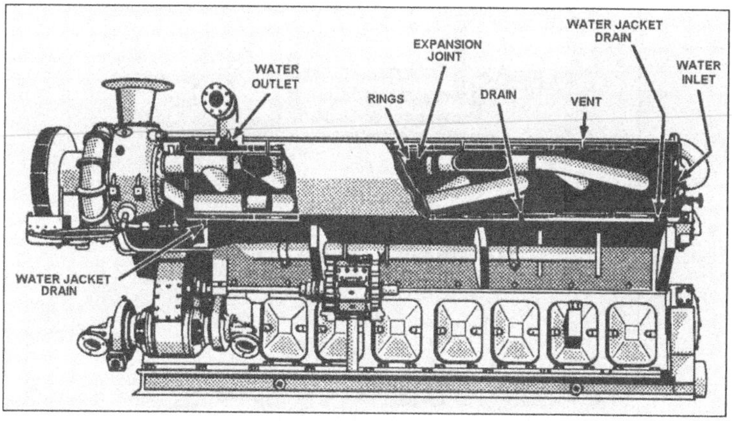

Figure 9-11. Water-jacketed Exhaust Manifold

PYROMETERS

9-29. On some diesel engines the exhaust system is provided with a device for measuring the exhaust temperature. The one most commonly used is a pyrometer (see Figure 9-12). By comparing the exhaust gas temperature of each cylinder, the operator can determine if the load is balanced throughout the engine. The indicating unit of the pyrometer is calibrated to give a direct reading of temperature. However, the pyrometer actually measures the difference between the electric current produced by heat acting on dissimilar metals in the thermocouple (see Figure 9-13) hot junction and the cold junction. The metals most commonly used in the thermocouple are iron and constantan. These are covered by an insulator. The thermocouple is located so that the hot junction is in contact with the exhaust gases. The pyrometer measures the difference between the hot junction and the cold junction. Therefore, the temperature at the cold junction must either be constant or compensated for, if the temperature reading on the indicating unit is to have any meaning. Two types of pyrometers used to take exhaust temperature readings are the fixed installation and the portable, hand-recording instrument. Both types have a thermocouple unit installed in the exhaust manifold. Fixed installation pyrometers (see Figure 9-12) are equipped with a thermostatically operated control spring which makes the required temperature corrections automatically. The portable hand pyrometer (see Figure 9-12) has a zero adjuster which must be set by hand.

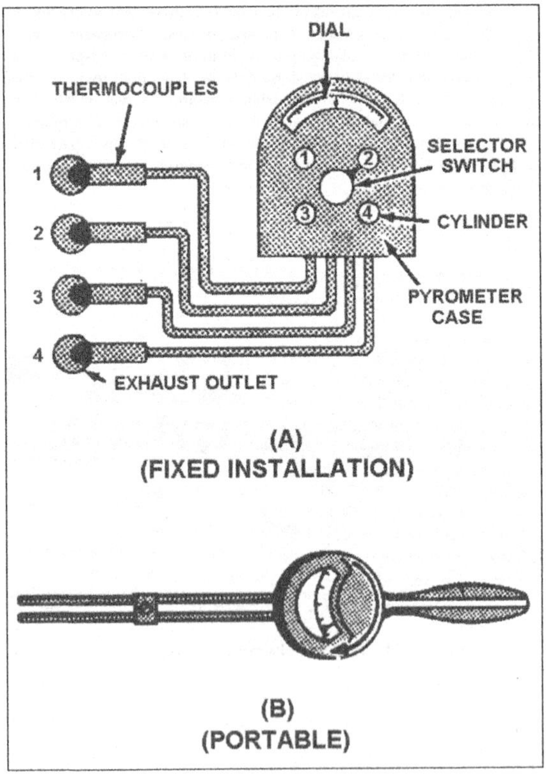

Figure 9-12. Pyrometers Used in Diesel Engine Exhaust Systems

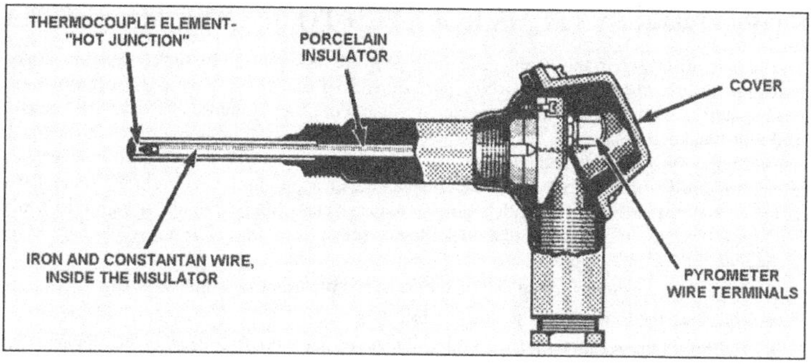

Figure 9-13. Sectional View of a Thermocouple

PIPING AND SILENCERS (MUFFLERS)

9-30. After passing through the turbine end of the supercharger of a 4-stroke cycle engine or after being discharged from the cylinders of a 2-stroke cycle engine, the gases-pass through the exhaust pipe, which may be either flexible or rigid to the silencer or muffler. The gases are discharged from the silencer to the atmosphere through the tail pipe or overboard discharge pipe. As the name implies, silencers or mufflers are provided on internal-combustion engines mainly to reduce noise created by the exhaust as the exhaust valves or ports open. This noise could be reduced by placing sound-absorbent, flameproof material in the exhaust passages, but the resulting back pressure might prevent the engine from operating. Because of the serious effects of back pressure upon the operation of the engine, the noise-reducing device must be so designed that a minimum exhaust back pressure is created. Since seawater is readily available, most modern marine engines use an exhaust silencer of the wet type. The muffler shown in Figure 9-10 is also of the wet type. Wet-type mufflers are usually of cast or sheet iron construction, with a system of internal baffles which break up the exhaust gas pulsation. Therefore, a silencing effect is obtained without producing a back pressure in the system. The water used in wet-type silencers also aids in reducing noise. The water cools the exhaust gases, causing the gases to contract. The decrease in volume reduces the velocity of the exhaust gases and thereby reduces the exhaust noise. The water itself absorbs some of the sound. The silencer consists of a steel drum which is divided into two compartments by a transverse baffle plate. The exhaust inlet pipe extends part way through the inlet compartment so that the exhaust gases must circle back, passing through a stream of water, before entering the pipes which extend through the baffle plate. In the outlet compartment, the exhaust gases are again deflected before they enter the water outlet pipe. A second stream of water enters the tail pipe and helps carry the gases to the overboard discharge. Some marine engines use a dry-type exhaust silencer. In both the wet-and dry-type silencers of most marine installations, circulating water is used to reduce the temperature of the exhaust gases. The principal difference between the two is that in the dry-type, the exhaust gases do not come in contact with the cooling water. In other words, the water does not flow through the silencer compartment but flows instead through a jacket around the silencer. In wet-type silencers, the gases are expanded into the silencer and come in direct contact with the water. In passing through the baffles and through the water, the gases are cooled, condensed, decreased in volume, and effectively silenced. In addition to acting as silencers, most mufflers also function as spark arresters. The water in wet-type silencers serves as a spark arrester. In some dry-type silencers, a device is incorporated to trap burning carbon particles and soot. The interiors of all mufflers are subjected to moisture-condensation in the dry-type mufflers and the supplied water in the wet-type mufflers. Because of this, most silencers are coated inside and out with a resistant material.

PARTS OF OTHER SYSTEMS RELATED TO ENGINE AIR SYSTEMS

9-31. The parts discussed in this chapter are those generally associated with the air systems of most engines. Some engines are equipped with parts which perform special functions. Such parts, even though related to engine air systems, are more frequently considered as parts of other systems. For example, some engines are equipped with intake air heaters or use other methods to overcome the influence of low temperature in cold-weather starting. Devices and methods used for this purpose are discussed in subsequent chapters. Some engines are equipped with devices to cool the compressed air being discharged from the supercharger. These devices are considered in connection with engine cooling systems. Many engines are equipped with devices of systems which provide crankcase ventilation. Blower action is necessary in many of these systems. The systems operate for the following reasons:

- To prevent contamination of engine room space with heated or fume-laden air.
- To reduce the formation of sludge in lubricating oil.
- To prevent the accumulation of combustible gases in the crankcase and in the oil pan or sump.

The devices used to ventilate engine crankcases are discussed in connection with lubricating systems.

Chapter 10

LUBRICATING OIL SYSTEMS

INTRODUCTION

10-1. The reliability and performance of modem diesel engines directly depend on the effectiveness of their lubricating systems. The primary functions of lubrication are to minimize friction between the bearing surfaces of moving parts and to dissipate heat. Another function is to keep internal engine parts clean by removing carbon and other foreign matter. The engine system that supplies the oil to perform these functions is of the pressure type in practically all modem internal-combustion engines. Even though many variations exist in the details of engine lubricating systems, the parts of such systems and their operation are basically the same. This chapter provides information on the following:

- Parts of the lubricating oil system.
- Filtering devices.
- Types of filtering systems.
- Ventilation of the crankcase.
- Purification of lubrication oil.

PARTS OF A LUBRICATING OIL SYSTEM

10-2. The lubricating system of an engine consists of two main divisions (that external to the engine and that within the engine). The internal division, or engine part, of the system consists principally of passages and piping. The external part of the system includes several components which aid in supplying oil in the proper quantity at the proper temperature and free of impurities. To meet these requirements, the lubricating systems of many engines include, external to the engine, such parts as tanks and sumps, pumps, coolers, strainers and filters, and purifiers.

TANKS AND SUMPS

10-3. The lubricating systems of propulsion installations include tanks which collect oil that has been used for lubricating and cooling so that this oil can be recirculated through the engine. The oil system of some installations also includes a sump or drain tank under the engine. These tanks collect the oil as it drains flushing from the engine crankcase. Storage and sump tanks are not common in auxiliary engines. In these engines the oil supply is generally contained directly within the engine oil pan.

PUMPS

10-4. Positive-displacement, rotary-gear pumps are used to deliver oil under pressure to the various parts of the engine. These pumps are engine-driven. Lubricating oil pumps supply oil at pressures that are closely adjusted to the engine requirements, since the pump is driven from the engine camshaft or, in some engines, directly from the crankshaft. Changes in engine speed cause corresponding changes in pump output. Detached lubricating pumps on large diesel engines are used to supply the purifier, to fill the sump tanks from the storage tanks, and to flush and prime the lubricating oil system. Operators should be thoroughly familiar with the lubricating oil system before attempting to transfer oil or to flush and prime the engine. When priming or flushing the engine, operators should be aware that prolonged flushing or priming of the lubricating oil system on any engine may result in an accumulation of oil in the air intake passages and cause overspeeding upon starting. Observance of the following precautions will prevent this:

- The engine lubricating oil system should be primed by hand or by motor-driven jacking gear before the engine is turned over.
- Priming of the engine should be continued only until a slight pressure is registered on the engine lubricating oil pressure gauge or until oil is observed at each main bearing.
- Before starting the engine after a prolonged shutdown, the air receiver and blower discharge passages should be inspected and any excessive accumulation of lubricating oil removed.

The pressure maintained by most lubricating oil pumps is controlled by pressure-regulating valves or pressure-relief valves built directly into the pump. In some pumps, pressure-relief valves separate from the pump are used for pressure control purposes. Most oil pressure regulating devices recirculate excess oil back to the pump. However, some pumps are designed to discharge excess oil directly into the engine sump or the crankcase.

COOLERS

10-5. The lubricating oil systems of most engines, especially large ones, must include coolers (heat exchangers) to maintain the oil temperature the most efficient operating temperature range. The oil, as it passes through the operating engine, will absorb heat from the metal parts with which it comes in contact. Since engine oil is recirculated (used over and over), the oil is continually absorbing additional heat. Unless provision is made to remove the heat, the oil temperature will rise to excessive values. At extremely high temperatures, oil tends to oxidize rapidly and form carbon deposits. Excessively high temperatures also cause an increase in the rate of oil consumption. Consequently, to retain the lubricating qualities of the oil, oil coolers are required to remove excess heat from the oil. Coolers used to remove heat from lubricating oil are of the same types as those used to remove heat from other fluids common to internal-combustion engines. Additional information on heat transfer and coolers is contained in the next chapter, since the liquid to which heat is transferred and the cooler are both commonly associated with the engine cooling system.

FILTERING DEVICES

10-6. Strainers and filters are incorporated in the lubricating oil system of an engine. Their purpose is to remove abrasives and foreign materials which tend to increase wear of engine parts and cause deterioration of the lubricating oil. A variety of strainers and filters are used in marine engines. In marine terminology, all metal-edge and wire-mesh devices are classed as strainers; devices that have replaceable, absorbent cartridges are called filters. Filters function to remove small particles. Depending on the type of installation, the location and number of strainers and filters vary.

STRAINERS

10-7. Lubricating oil strainers are made in both simplex and duplex types. A duplex strainer is essentially two strainer elements incorporated in one assembly. A manual valve is provided for directing the flow of oil through either of the elements. When duplex strainers are included in a lubricating system, one element can be bypassed and the element removed and cleaned without disturbing the flow of oil through the other element to the engine. Every approved lubricating oil strainer contains a built-in, spring-loaded or differential area pressure-relief valve. The valve must be sufficiently large to bypass all of the oil around a clogged strainer, so that an uninterrupted flow of oil may be maintained to the engine. Metal-edge strainers consist principally of a strainer element surrounded by a case which serves as a sump to collect foreign material and water. The element consists of an edge-wound metal ribbon or a series (stack) of edge-type disks. Most strainers have devices for manually rotating the strainer element against metallic scrapers which remove material caught by the element. Strainers are usually provided with vents for releasing air from the system.

Edge-wound Metal-ribbon Strainer

10-8. A strainer of the edge-wound, metal-ribbon type is shown in Figure 10-1. This strainer is so constructed that the oil required by the engine is continuously filtered, except when the element is removed for cleaning or servicing. When an element is to be removed, the control valve handle is turned to the BYPASS position. Shunting the oil flow through the strainer head, the element may then be removed without interruption of the oil flow to the engine. Under normal operating conditions, the oil comes into the strainer at the top and descends to surround the ribbon element. The oil then passes through the element, into the center, and then upward to the outlet passage. Another type of element consists of a closely compressed coil of stainless steel wire. The wire has been passed between rollers so that it is a wedge-shape wire or ribbon with one edge thicker than the other. On one side of the wire, projections are spaced at definite intervals; the other side is smooth. The projections on one side of the wire touch the smooth side of the wire on the adjacent coil to provide appropriate spacing between adjacent coils. The thick edge of the wire is on the outside of the coil. A tapered slot is therefore formed from the outside to the inside of the coil, with the narrowest part of the slot on the outside. This arrangement ensures that dirt particles which are small enough to pass through the outside or narrowest portion of the slot will not become stuck halfway through the slot and clog the oil flow. The dirt removed from the oil remains on the outside of the element and can readily be removed by rotation of the element with the cleaning handle. As the element rotates, foreign material is removed by the cleaning blade. The control-valve handle on the strainer operates the bypass valve. When the handle is in the ON position, the lubricating oil is flowing through the strainer. When the handle is in the BYPASS position, the oil is flowing directly through the head of the unit, and the strainer case and element can be removed and cleaned. The ON and BYPASS positions are indicated on the strainer head.

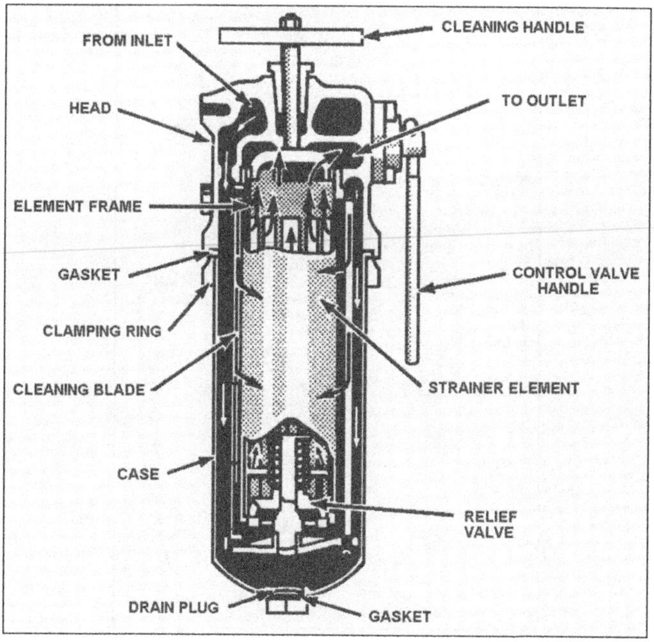

Figure 10-1. Edge-wound, Metal-ribbon Lubricating Oil Strainer

Edge-disk Strainer

10-9. The duplex strainer of the edge-disk type is shown in Figure 10-2. This strainer consists of, two sections, each of which includes three strainer elements. A control valve between the two sections is used to secure one section while the other remains in operation. The secured section acts as a stand-by unit. It may be opened for cleaning and inspection without interrupting the straining operation. An edge-disk strainer element consists of an assembly of thin disks separated slightly by spacer disks. The assembly of one type of strainer element is shown in Figure 10-3. The lower end of a disk assembly is closed; the upper end is open to the strainer discharge. Oil entering the strainer assembly is forced down between the casing and disk assembly and then through the disks into the center of the disk assembly. The oil then passes up through the assembly and out the discharge outlet. In passing through the strainer, the oil must pass through the slots between the strainer disks. At the bottom of the strainer element a relief valve is provided to relieve pressure which will build up on the strainer if the slots become filled with foreign matter. The relief valve bypasses the oil up through the center of the strainer element and out the strainer discharge when a predetermined pressure is reached. When the assembly is turned by the external handle, the solids which have lodged against or between the disks are carried around until they meet the stationary cleaner blades. These blades clean the solids from the strainer surface. The solids are compacted by the action of the cleaner blades and fall into the sump of the strainer. To keep the strainer in a clean, free-filtering condition, the external handle is given one or more complete turns in a clockwise direction at frequent intervals. It is therefore not necessary to break any connections or to interrupt the flow of oil through the strainer to clean the strainer unit. If it is difficult to turn the handle, the strainer surfaces have deposits of solids on them. The handle should be turned frequently. There is no danger of turning the handle too often since there are not parts to wear out. If the strainer cannot be cleaned by turning, the head and disk assembly must be removed and soaked in a solvent until the solids have been removed.

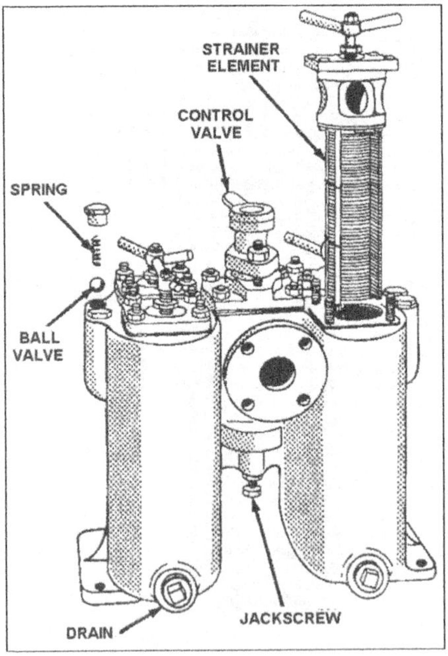

Figure 10-2. Edge-disk Oil Strainer

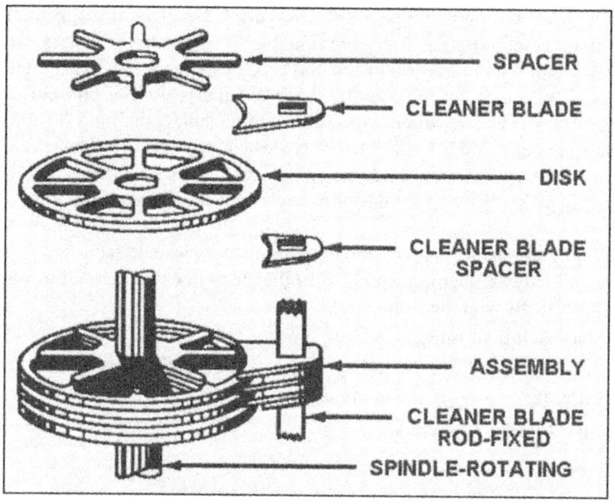

Figure 10-3. Edge-disk Strainer Element Assembly

Wire-mesh (Screen) Strainer

10-10. Strainers installed on the suction or intake side of the pressure pumps are generally of the wire-mesh (screen) type. These units are generally referred to as coarse strainers. Some screen strainers are located in the oil pan or sump. One type of screen strainer is shown in Figure 10-4.

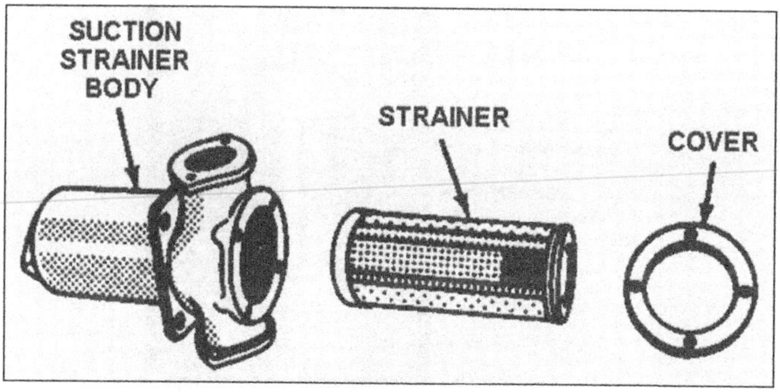

Figure 10-4. Screen Lubricating Oil Strainer

Filters

10-11. In approved filters, the absorbent material is composed of such substances as cellulose, cotton yarn, cotton waste, and paper disks. Materials such as mineral wool, fuller's earth, and activated charcoal remove the compounds from additive oils. Such materials are not permitted in filters used on engines in naval service. Lubricating oil filters are of the simplex type only. Filters may be located directly in the pressure-lubricating oil system or they may be installed as bypass filters. When installed in the pressure system, a filter must contain a built-in, spring-loaded, pressure-relief valve. The valve must be sufficiently large to bypass all oil to the engine in the event the filter element becomes restricted. Bypass filters require an orifice plate in the line to the filter. This plate controls the amount of oil which is removed from the lubricating oil pressure system. The

amount of oil which flows through a bypass filter is only a small percentage of that flowing through the pressure system. The oil from a bypass filter is returned to the sump tank. Filters vary as much in design and construction as do strainers. Filters may be of the unit type or the tank type. The unit filter may be the single, double, or triple unit. In double or triple units, the individual units are connected by manifolds for inlet, discharge, and sludge drain. The tank filter consists of a single tank containing several filter elements. In some tank filters, each filter-element holder has a relief valve which protects the element against excessive pressure. Other tank filters are constructed to withstand pressure greater than that of the relief valve setting on any of the pumps in the lubricating oil system. Examples of a single-unit filter and a tank filter are shown in Figure 10-5 and
Figure 10-6. So far, only the main parts of an engine lubricating system have been discussed. You should be familiar with piping and gauges, thermometers, and other instruments essential to complete the system. The remainder of this chapter deals with the following:

- Types of lubricating oil filtering systems.
- Path of oil through a system (including the internal or engine part of the system).
- How and why these systems are ventilated.
- Methods of purifying lubricating oil.

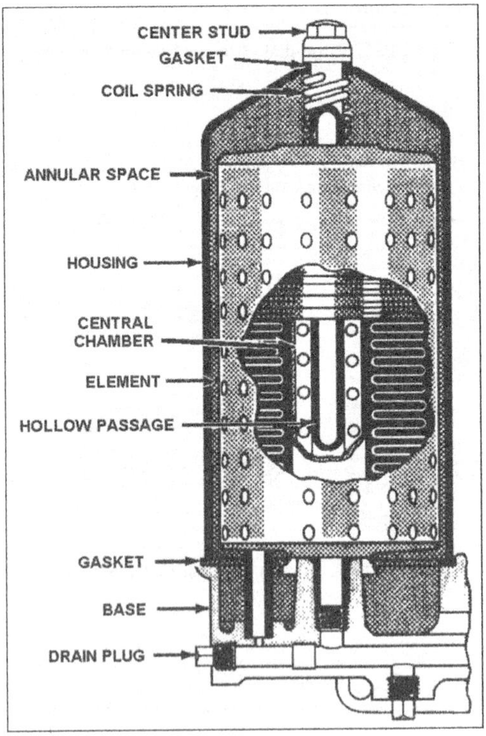

Figure 10-5. A Single-unit Lubricating Oil Filter

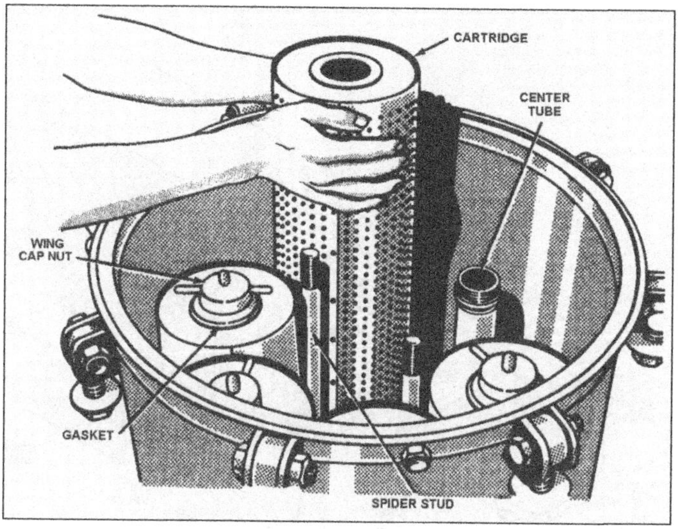

Figure 10-6. Tank Lubricating Oil Filter

LUBRICATING OIL FILTERING SYSTEMS

10-12. The strainers and filters of marine diesel engine installations are incorporated in the lubricating oil filtering system. An engine may use one of four of the following types of filtering systems:

- Shunt.
- Full-flow.
- Sump.
- Bypass.

The type used depends on the engine installation and its applications.

Shunt Filtering System

10-13. In a shunt filtering system, oil is taken from the sump by the pressure pump and discharged first through a strainer, then through a filter, and finally through a cooler, to the engine. The pump delivers a constant amount of oil per revolution. However, the resistance in the strainer and the filter varies, depending on the condition of these units and the temperature of the oil. In order that an adequate flow of oil will be delivered to the engine at any particular engine speed, the filter and the strainer are each provided with a bypass. This is fitted with a spring-loaded bypass valve through which a portion of the oil flows. If the filter becomes clogged or if the oil is cold, a relatively large portion of the lubricating oil is shunted through the bypass. Strainers and filters (in a shunt filtering system) may also be manually bypassed. Three-way valves are provided for bypassing each unit so that strainers may be cleaned or filter elements may be replaced while the engine is operating. A schematic diagram of a shunt filtering system is shown in Figure 10-7.

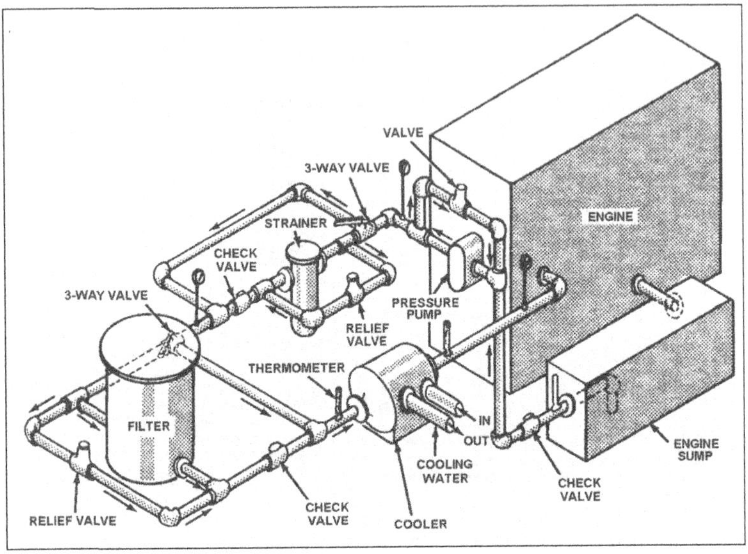

Figure 10-7. Shunt Lubricating Oil Filtering System

FULL-FLOW FILTERING SYSTEM

10-14. The full-flow filtering system is found on most new diesel engines. It is similar to the shunt system. The exception is that the filter elements are designed for high flow rates permitting the entire pump delivery to pass through the elements. The bypass valve serves the same function as in the shunt system. However, the valve will remain closed during normal engine operation. A schematic arrangement of a full-flow filtering system is shown in Figure 10-8. In some newer engines, the lube oil strainer is located after the lube oil cooler, and the relief valve for the lube oil filter is external. The lube oil strainer was moved because, on some occasions, it was not cleaned properly when being worked on. The dirt was therefore carried directly into the engine, damaging the bearings. With the strainer located after the cooler, dirt from the strainer cannot enter the engine.

SUMP FILTERING SYSTEM

10-15. The sump filtering system is similar to the shunt system. The exception is that the filter is placed in a separate recirculating system which includes a separate motor-driven pump. The sump system permits the lubricating oil to circulate through the filter even when the engine is secured. In the sump filtering system, the oil to be filtered is taken from the sump by the motor-driven pump, forced through the filter, and then discharged back into the sump. Oil to the engine is taken from the sump by the engine-driven pump and forced through the cooler and strainer to the engine. The path of the oil through a sump filtering system may be seen in

Figure 10-9.

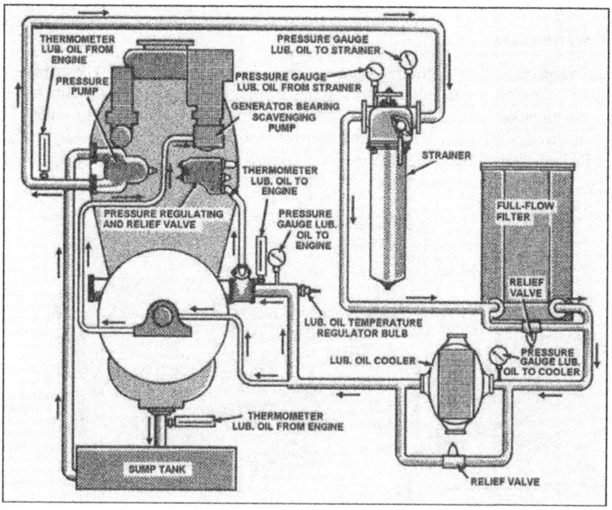

Figure 10-8. Full-flow Lubricating Oil Filtering System

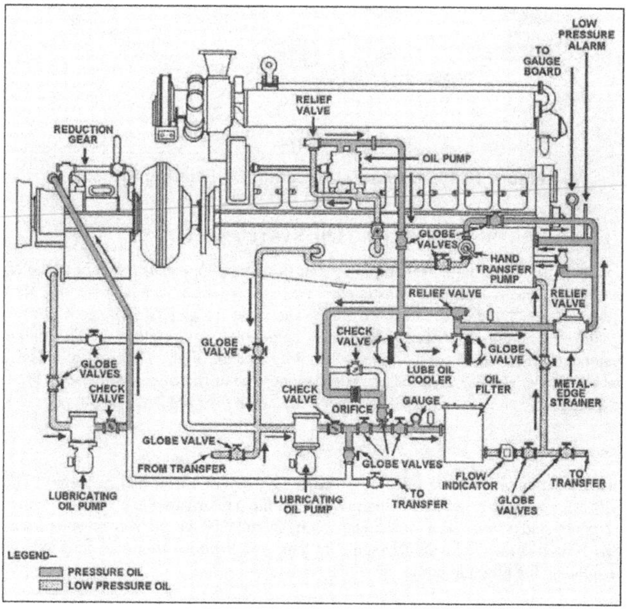

Figure 10-9. Sump Lubricating Oil Filtering System

Bypass Filtering System

10-16. In many respects, a bypass filtering system is similar to a shunt system. The primary exception is that a portion of the oil discharged by the pressure pump in the bypass system is continuously bypassed back into the sump through the filter or filters. To ensure that sufficient oil is supplied to the engine, the amount of oil permitted to flow through the filter is limited by the size of the piping and, if necessary, by an orifice. A valve is provided so that the flow of oil to the filter may be stopped. The arrangement of a bypass filtering system for one type of engine is shown in Figure 10-10.

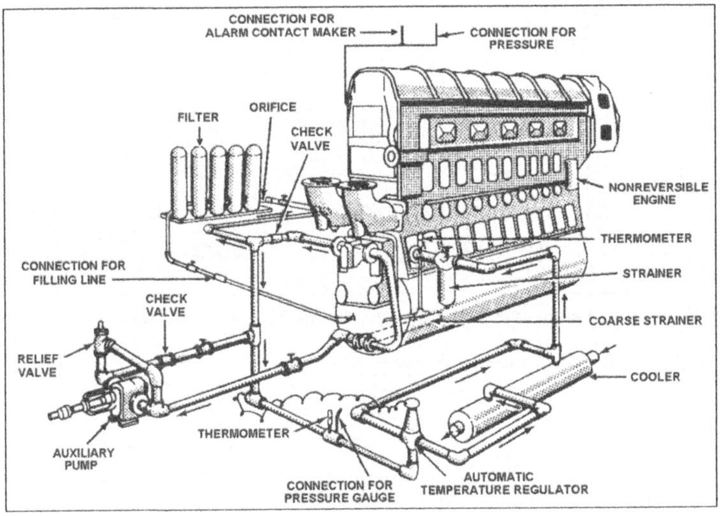

Figure 10-10. Bypass Lubricating Oil Filtering System

Path of Oil Through the Lubricating Oil System

10-17. The path of the oil through system piping and through components external to the engine for the four types of diesel installations has been illustrated. These illustrations demonstrate that the sequence of the components through which the oil flows is not always the same. In general, arrangement of the parts of the oil system external to the engine depends on the type of engine and the installation. The path of oil through the external components is governed by the type of filtration system used. That part of a lubricating oil system which is external to an engine is classified according to the type of filtering system used. The engine or internal part of a lubricating oil system is classified as either a dry-sump type or a wet-sump type.

Dry-sump System

10-18. In the dry-sump system, the oil is returned by gravity to an oil pan or sump. The oil is delivered continuously by an engine-driven scavenging pump from the pan or sump to a separate sump tank, which may include the strainer and filter. The scavenging pump maintains the oil pan or sump nearly empty of oil. Therefore, this system is known as the dry-sump system. Oil is drawn from the sump tank and is recirculated through the engine by the pressure pump.

Wet-sump System

10-19. In the wet-sump system, the oil is returned to the oil pan or sump by gravity flow after lubricating the various parts of the engine. In wet-sump systems, the pressure pump draws oil directly from the oil pan or sump and recirculates the oil through the filtering equipment and the engine. The systems shown in Figures 10-7 through 10-10 are of the wet-sump type. Note that these systems do not include a scavenging pump in the lubricating oil system.

10-20. Differences exist in the design and arrangement of various diesel engine lubricating systems. However, the systems of most engines are similar in many respects. Some of the similarities and differences are pointed out in preceding descriptions. Learning to trace the path of oil through the lubricating oil system of an engine will be easier if you refer to the illustrations shown in this chapter or in the applicable manufacturer's instruction manual.

VENTILATION OF INTERNAL SPACES

10-21. Most engines have some means of ventilating the internal cavities or spaces which are related to the lubricating oil system. Systems may be vented directly to the atmosphere or through the engine intake-air system. The latter method is preferred in marine installations that have the engine located in an engine room or other compartment. Venting the heated, fume-laden air directly to the atmosphere in a compartment will seriously contaminate the compartment air and may create a fire hazard. However, if the lubricating oil system is not vented in some manner, combustible gases may accumulate in the crankcase and oil pan. Under certain conditions, these gases may explode. Under normal operating conditions, the mixture of oil vapor and air within an engine crankcase is not readily explosive. However, if such a working part (as a bearing or a piston) becomes overheated as a result of inadequate lubrication or clearances, additional oil will vaporize and an explosive mixture will be created. If the temperature of the overheated part is sufficiently high to cause ignition or if a damaged part strikes another part and causes a spark, an explosion may occur. In addition to the vapor created when lubricating oil contacts extremely hot surfaces, vapor may accumulate in the crankcase as a result of blowby past the pistons. Blowby occurs when the piston is compressing the air and during the power event. There is little danger of a crankcase explosion or of other troubles caused by vapors within an engine, if the engine is in condition according to the prescribed program. The ventilation system of an engine greatly reduces the possibility of troubles which might occur because of an accumulation of vapors in the crankcase. Nevertheless, even when an engine is maintained according to prescribed procedures, casualties may occur or conditions may be created which will lead to an explosion in the crankcase of the engine. When such casualties or conditions occur, they are generally due to abnormal operating circumstances or to the failure of a part. Crankcase explosion can result from overheated lubricating oil or diluted lubricating oil. These conditions may also lead to crankcase-bearing failure, damaged or excessively worn liners or piston rings, and cracked or seized pistons. You should be familiar with the possible causes of crankcase explosions to prevent their occurrence. The importance of knowing the causes and the precautionary or preventive measures required are apparent when the after effects of an explosion are considered. A crankcase explosion may cause serious injury to personnel, extensive damage to the engine, and fires.

OVERHEATED LUBRICATING OIL

10-22. A rise in the temperature of the lubricating oil may cause the formation of explosive vapor. A rise in temperature may be due to the following factors:

- Insufficient circulation of the oil.
- Inadequate cooling of the oil.
- Faulty temperature-regulating valve.
- Overloading of the engine.
- Damaged or excessively worn parts.

The causes of overheated lubricating oil should be corrected immediately. In addition to creating explosive vapors, overheated lubricating oil has other serious effects. The viscosity of the oil will be greatly reduced and the tendency to form gum will be increased. Lubricating oil temperatures should be maintained at the value specified in the manufacturer's technical manual (TM).

DILUTED LUBRICATING OIL

10-23. Dilution of engine lubricating oil with diesel fuel oil increases the tendency toward vapor formation in the crankcase. This is because diesel oil has a lower flashpoint than lubricating oil. Petroleum products vary greatly in their flashpoints. In general terms, gasolines give off sufficient vapors to ignite at temperatures well below freezing. Diesel fuel oil gives off vapor in sufficient quantities to be ignited when the oil is heated to approximately 140° F. However, a lubricating oil must be heated to a higher temperature (325° F to 580° F, depending on the series and symbol of the oil) before it reaches its flashpoint. Dilution of engine lubricating oil in diesel engines may be caused by a variety of problems. In general, it may result from worn or stuck rings, worn liners or pistons, fuel leaks, or leaky nozzles or injectors. Remember that, even though an engine is in good condition, dilution will occur during continuous engine operation at low speeds and under idling conditions. Under these conditions it occurs as a result of the blowby of unburned fuel particles which accumulate in the combustion spaces.

Note: Flashpoint is the minimum temperature at which a product of petroleum gives off sufficient flammable vapors to ignite, or momentarily flash.

WARNING

Dilution alone cannot cause a crankcase explosion. However, it may contribute to making one possible.

PURIFICATION OF LUBRICATING OIL

10-24. Oil must be clean before it goes into the lubricating system of an engine. It must be cleaned or purified regularly while it is being recirculated through the engine. Dust and dirt particles from the intake air get into the oil system. Bits of metal from the engine parts are picked up and carried in the oil. Carbon particles from combustion in the cylinders work into the oil, even in the best engine. The oil itself deteriorates and leaves some sludge and gummy material which circulates through the oil system. Some water will get into the oil, even when precautions are taken. Contamination must be removed or the oil will not meet the requirements of lubrication. Dirt and other hard particles score and scratch the rubbing metal surfaces within the engine. This abrasive action greatly increases friction and heating of the moving parts and causes them to wear faster. Sludge and water interfere with the oil's ability to hold a good lubricating film between rubbing surfaces within the engine. Several devices, such as strainers, filters, settling tanks, and centrifugal purifiers, are used to keep the oil as pure as possible. Each device is designed to remove certain kinds of contamination. Therefore, several types are required in the lubricating oil system to maintain the oil in the best possible condition. Settling tanks and purifiers, with emphasis on purifiers, are discussed here.

Settling Tanks and Purifiers

10-25. The lubricating system of many shipboard diesel engine installations includes settling tanks (used-oil tanks). These tanks allow the oil to stand while accumulated water and other impurities settle. Settling is due to the force of gravity. A number of layers of contamination may form. The number depends on the different specific gravities of the various substances which contaminate the oil. Settling takes place more rapidly and efficiently when the oil is heated. Although settling tanks do remove much contamination from lubricating oil, most ships have additional equipment to purify the oil by removing water and impurities that are not removed by other devices. The machines which perform this purification process are usually called purifiers, but are also referred to as centrifuges. Detailed instructions on construction, operation, and maintenance of purifiers are furnished in the applicable TM. In maintaining purifier, follow the instructions carefully. The following provides general information on methods of purification and principles of operation of purifiers.

Methods of Purification

10-26. On diesel-propelled ships, the piping system is generally arranged to permit two methods of purifying (batch purification and continuous purification).

Batch Purification Process

10-27. The batch purification process is used when an engine is secured. Lubricating oil is transferred from the sump to a settling tank by means of a transfer pump. The oil is heated by steam heating coils in the settling tank and maintained at a temperature of approximately 175° F for several hours. Water and other settled impurities are drained from the settling tank through a valve. The oil is then centrifuged and discharged back into the sump from which it was taken.

Continuous Purification Process

10-28. The continuous purification process is used when an engine is operating. The centrifugal purifier takes suction from a sump tank and, after purifying the oil, discharges the purified oil back to the same sump.

Principles of Purifier Operation

Centrifugal force is the principle used in purifying oil. Centrifugal force is that force which is exerted upon a body or substance by rotation. Centrifugal force impels the body or substance outward from the axis of rotation. A centrifugal purifier is essentially a container which is rotated at high speed while contaminated oil is forced through and rotates with the container. The centrifugal force imposed on the oil by the high rotating speed of the container acts to separate the suspended foreign matter from the oil. However, only materials that are insoluble in one another can be separated by centrifugal force. For example, JP-5 or diesel fuel cannot be separated from lubricating oil nor can salt be removed from seawater by centrifugal force. However, water can be separated from oil because water and oil do not form a true solution when mixed. There must also be a difference in the specific gravities of the materials to be separated by centrifugal force. When a mixture of oil, water, and sediment stands undisturbed, the action of gravity tends to form an upper layer of oil, an intermediate layer of water, and a lower layer of sediment. The layers form because of the different specific gravities of the materials in the mixture. If the oil, water, and sediment are placed in a container which is revolving rapidly around a vertical axis, the effect of gravity is negligible in comparison with that of the centrifugal force. Since centrifugal force acts at right angles to the axis of rotation of the container, the sediment with its greater specific gravity assumes the outermost position, forming a layer on the inner surface of the container. Water, being heavier than oil, forms an intermediate layer between the layer of sediment and the oil, which forms the innermost layer. Centrifugal purifiers are so designed that the separated water is discharged as waste and the oil is discharged for reuse. The solids remain in the rotating unit. The units are cleaned when necessary. The effectiveness of separation by centrifugal force is further affected by the following:

- Size of the particles.
- Viscosity of the fluids.
- The time which the materials are subjected to the centrifugal force.

In general, the greater the difference in specific gravity between the substances to be separated and the lower the viscosity of the oil, the greater will be the rate of separation.

USE OF PURIFIERS

10-29. Centrifugal purifiers are used to purify both fuel oil and lubricating oil. A purifier may be used either to remove water and sediment from oil or to remove sediment only. When water is involved in the purification process, the purifier is usually called a separator. When the principal item of contamination is sediment, the purifier is used as a clarifier. Purifiers are generally used as separators when purifying fuel oils. When used to purify a lubricating oil, a purifier may be used as either a separator or a clarifier. Whether it is used as a separator or a clarifier depends on the moisture content of the oil being purified. An oil which contains no moisture needs only to be clarified. This is because the oil will be discharged in the purified state after the solids deposit in the bowl of the purifier. However, if the oil contains some moisture, the continued feeding of "wet" oil into the bowl results eventually in a bowl that is filled with water. From that time on, the centrifuge is not accomplishing any separation of water from the oil. Even before the bowl is completely filled with water, the presence of a layer of water in the bowl reduces the depth of the oil layer. As a result, the incoming oil passes through the bowl at an increased velocity. Because of this, the liquid is subject to centrifugal force for a shorter time. The separation of water from the oil is therefore not as complete as it would be if the bowl were without the water layer, or if the water layer were a shallow one. For this reason, the centrifuge should not be operated as a clarifier unless the oil contains little or no water. A small amount of water can be accumulated, along with the solids, and drained when the bowl is stopped for cleaning. However, if there is any appreciable amount of water in the oil, the purifier should be operated as a separator.

TYPES OF CENTRIFUGAL PURIFIERS

10-30. Two types of purifiers used in marine engineering are the disk type and the tubular type. Both operate on the same general principle. The major difference between the two types is in the design of the rotating units. In the disk purifier, the rotating element is a bowl-like container which encases a stack of disks. In the tubular purifier, the rotating element is a hollow tubular rotor.

Disk-type Centrifugal Purifier

10-31. A sectional view of a disk-type centrifugal purifier is shown in Figure 10-11. The bowl is mounted on the upper end of the vertical bowl spindle, which is driven by a worm wheel and a friction clutch assembly. A radial thrust bearing at the lower end of the bowl spindle carries the weight of the bowl spindle and absorbs any thrust created by the driving action. The flexible mount of the top bearing allows the bowl to come to the center of rotation. The parts of a disk-type bowl are shown in Figure 10-12. The flow of oil, through the bowl and additional parts, is shown in Figure 10-13. Contaminated oil enters the top of the revolving bowl through the regulating tube. The oil then passes down the inside of the tubular shaft and out at the bottom into the stack of disks. As the dirty oil flows up through the distribution holes in the disks, the high centrifugal force exerted by the revolving bowl causes dirt, sludge, and water to move outward. The purified oil moves inward toward the tubular shaft. The disks divide the space within the bowl into many separate, narrow passages or spaces. The liquid confined within each passage is restricted so that it can flow only along the passage. This arrangement prevents excessive agitation of the liquid as it passes through the bowl and creates shallow settling distances between the disks. Most of the dirt and sludge remains in the bowl, collecting in a layer on the inside vertical surface of the bowl shell. Water, along with some dirt and sludge separated from the oil, is through the discharge ring at the top of the bowl. The purified oil flows inward and upward through the disks, discharging from the neck of the top disk.

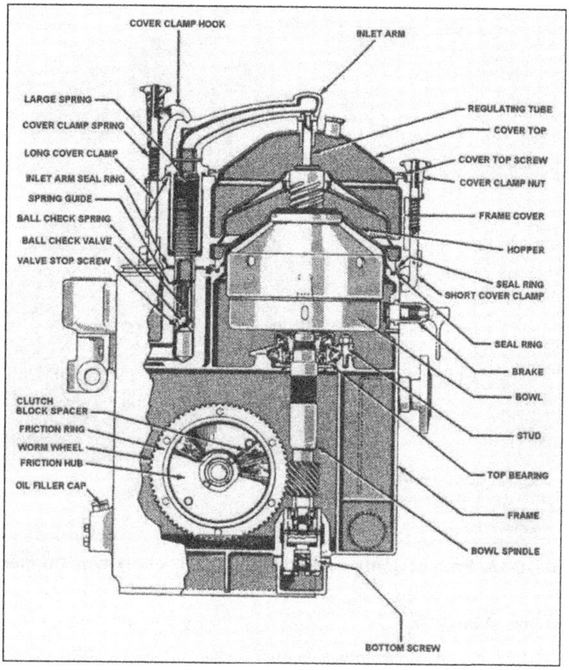

Figure 10-11. Disk-type Centrifugal Purifier

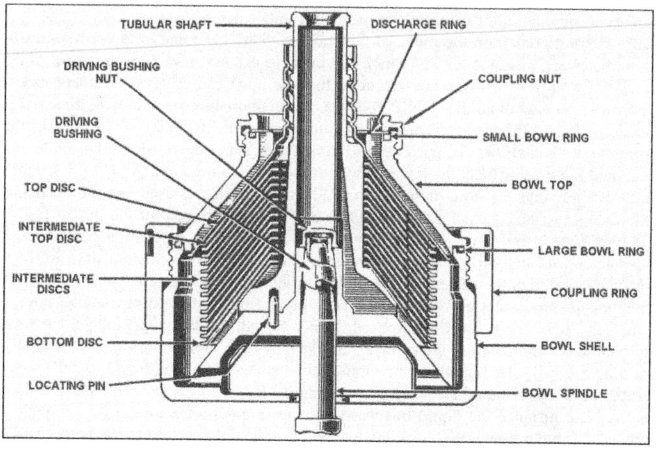

Figure 10-12. Parts of a Disk-type Purifier Bowl

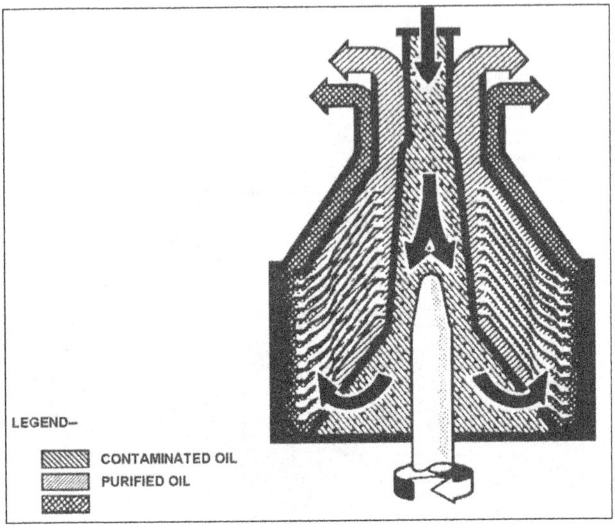

Figure 10-13. Path of Contaminated Oil Through a Disk-type Purifier Bow

Tubular-type Centrifugal Purifier

10-33. A cross section of a tubular centrifugal purifier is shown in Figure 10-14. This type purifier consists essentially of a rotor or bowl which rotates at high speeds. The rotor has an opening in the bottom through which the lubricating oil enters. It also has two sets of openings in the bowl top through which oil and water (in a separator action) or the oil alone (in a clarifier action) discharge. The bowl or hollow rotor of the purifier is connected by a coupling unit to a spindle suspended from a small bearing assembly. The bowl is belt-driven by an electric motor mounted on the frame of the purifier. The lower end of the bowl extends into a flexibly mounted drag bushing. The assembly of which this bushing is a part restrains movement of the bottom of the bowl. However, it allows sufficient movement so that the bowl can center itself about its center of rotation when the purifier is in operation. Inside the bowl is a device consisting essentially of three flat plates which are equally spaced radially. This device is commonly referred to as the wing device (or just the three-wing). The three-wing rotates with the bowl. Its purpose is to force the liquid in the bowl to rotate at the same speed as the bowl. The liquid to be centrifuged is fed into the bottom of the bowl, through the feed nozzle, under pressure so that the liquid jets into the bowl in a stream. When the purifier is used as a lubricating oil separator or clarifier, the feed jet strikes against a cone which is placed on the bottom of the three-wing. This brings the liquid up to bowl speed smoothly, without making an emulsion. The cone is not necessary when the purifier is used as a separator with fuel oil, because fuel does not have the tendency to emulsify. In the tubular purifier, the process of separation is the same as in the disk purifier. In both, the separated oil assumes the innermost position and the separated water moves outward. Both liquids are discharged separately from the bowls. Solids separated from the liquid are retained in the bowls. Even though similar in operation, the two types of purifiers differ somewhat in design features. By comparison, the bowl of a tubular purifier is of a small diameter and is operated at a high speed. The length of the tubular bowl (the distance the liquid travels through the bowl) is many times the depth of the liquid layer (settling distance). The disk-type bowl has a larger diameter and much shorter length. The distance the liquid travels in passing through such a bowl is not much greater than the settling distance. Tubular bowls are fed through a feed nozzle at the bottom of the bowl. Disk-type bowls are ordinarily fed from the top through a center tube which directs the liquid toward the distribution holes in the disk stack. Disks are provided to set up layers, thereby reducing settling distance.

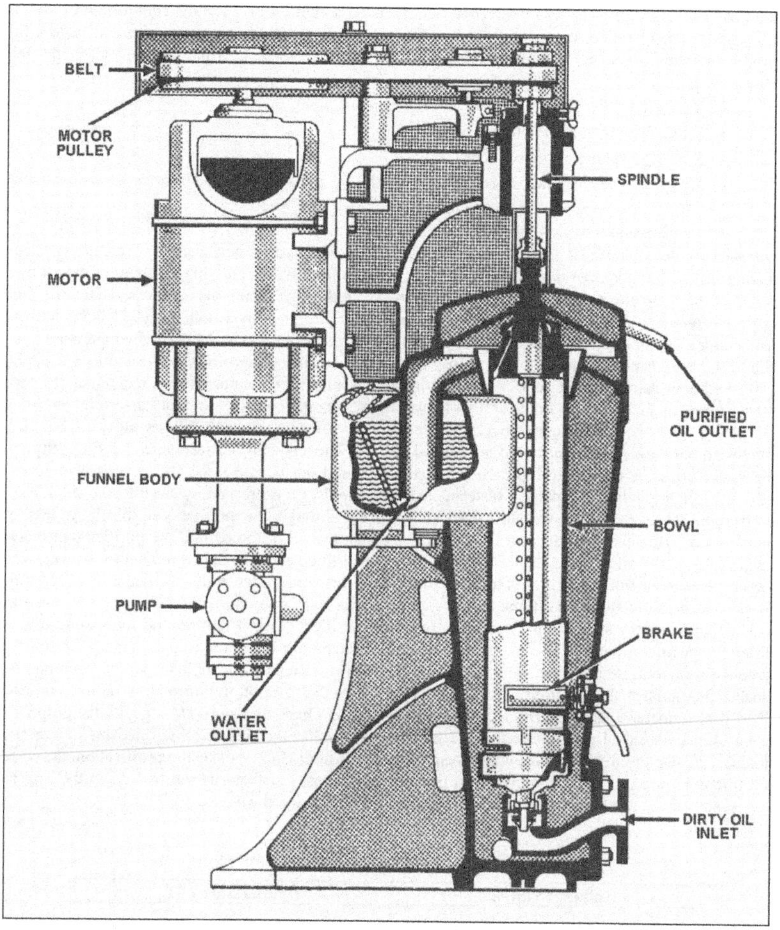

Figure 10-14. Tubular-type Centrifugal Purifier

OPERATION OF PURIFIERS

10-34. Specific directions for the operation of a purifier should be obtained from the manufacturer's instructions provided with the unit. The following general information applies largely to both types of purifiers. For maximum efficiency, purifiers are to be operated at their maximum designed speed and rated capacity. An exception to operating at designed rated capacity is made when the unit is used as a separator with 9000-series detergent oil. Some engine installations using this series oil are exposed to large quantities of water. When the oil becomes contaminated with water, it has a tendency to emulsify. This tendency is most pronounced when the oil is new and gradually decreases during the first 50 to 75 hours of engine operation. During this period, the purifier capacity should be reduced to approximately 80 percent of the rated capacity.

When a purifier is operated as a separator, priming of the bowl with freshwater is essential before any oil is admitted to the purifier. The water serves to seal the bowl and to create an initial equilibrium of liquid layers. If the bowl is not primed, the oil will be lost through the water discharge ports. Several factors influence purifier operation. The time required for purification and the output of a purifier depend on the following factors:

- Viscosity of the oil.
- Pressure applied to the oil.
- Size of the particles of sediment.
- Difference between the specific gravity of the oil and that of the substances which contaminate the oil.
- Tendency of the oil to emulsify.

Oil viscosity determines to a great extent the length of time required for purification. The more viscous the oil, the longer the time required to purify it to a given degree of purity. Decreasing viscosity of the oil by heating is one of the most effective methods of purification. Even though certain oils may be satisfactorily purified at operating temperatures, a greater degree of purification will generally result if the oil is heated to a high temperature. To accomplish this, the oil is passed through a heater where the desired temperature is obtained before the oil enters the purifier bowl. Most oils used in marine engines may be heated to a temperature of 180° F without damage to the oils. Prolonged heating at higher temperatures is not recommended because of the tendency of such oils to oxidize at high temperatures. Oxidation results in rapid deterioration. In general, oil should be heated sufficiently to produce a viscosity of approximately 90 seconds, Saybolt Universal (90 SSU), but the temperature should not exceed 180° F. The temperatures recommended for purifying oils in the 9000 series are shown in Table 10-1. Pressure should not be increased above normal to force a high viscosity oil through the purifier. Instead, viscosity should be decreased by heating the oil. The use of pressure in excess of that normally used to force oil through the purifier will result in less efficient purification. However, a reduction in pressure at which the oil is forced into the purifier will increase the length of time the oil is under the influence of centrifugal force and will therefore tend to improve results. If the oil discharged from a purifier is to be free of water, dirt, and sludge and if the water discharged from the bowl is not to be mixed with oil, the proper size discharge ring (ring dam) must be used. The size ring to be used depends on the specific gravity of the oil being purified. Diesel fuel oil and lubricating oils all have different specific gravities. Therefore, they require different size discharge rings. While all discharge rings have the same outside diameter, their inside diameters vary. Rings sizes are indicated by even numbers (the smaller the number, the smaller the ring size). The inside diameter, in millimeters, is stamped on each ring. Sizes vary in increments of 2 millimeters. Charts provided in manufacturer's TMs specify the proper ring size to be used with an oil of a given specific gravity. Generally, the ring size indicated on such a chart will produce satisfactory results. If the recommended ring fails to produce satisfactory purification, the correct size must be determined by trial and error. In general, oil is most satisfactorily purified when the ring is of the largest size which can be used without loss of oil with the discharged water.

Table 10-1. 9000 Series Temperatures

Military Symbol	Temperature (° F)
9110	140°
9170	160°
9250	175°
9500	180°

Maintenance of Purifiers

10-35. The bowl of a purifier must be cleaned daily, in accordance with the TMs, and all sediment must be carefully removed. The amount of dirt, grit, sludge, and other foreign matter in the oil may warrant more frequent cleaning. If the amount of foreign matter in an oil is not known, the machine should be shut down for examination and cleaning once during each watch, or more often if necessary. The amount of sediment found in the bowl indicates how long the purifier may be operated between cleanings. Periodic tests should be made to ensure that the purifier is working properly. When the oil in the system is being purified by the batch process, it should be tested at approximately 30-minute intervals. When the continuous process of purification is used, tests should be made once during each watch. Analysis of oil drawn from the purifier is the best method of determining the efficiency of the purifier. However, the clarity of the purified oil and the amount of oil discharged with the separated water will also indicate whether the unit is operating satisfactorily.

Chapter 11
COOLING SYSTEMS

INTRODUCTION

11-1. A great amount of heat is generated within an engine during operation. Combustion produces the greater portion of this heat. Compression of gases within the cylinders and friction between moving parts add to the total amount of heat developed. Since the temperature of combustion alone is approximately twice that at which iron melts, it is apparent that without some means of dissipating heat, an engine will operate for only a very limited time. Of the total heat supplied by the burning fuel to the cylinder of an engine, only approximately one-third is transformed into useful work. An equal amount is lost to the exhaust gases, leaving approximately 30 to 35 percent of the heat of combustion to be removed to prevent damage to engine parts. Heat, which may produce harmful results, is transferred from the engine through water, lubricating oil, air, and fuel. This chapter deals with heat transfer and cooling systems of marine diesel engines and with maintaining these cooling systems.

HEAT TRANSFER AND COOLING

11-2. Heat may be transferred by conduction, convection, or radiation. Heat within an engine is transferred by all three methods. However, this heat transfer occurs chiefly by convection of heat in the exhaust gases and in the cooling water flowing from the engine, and by radiation from the hot external engine surfaces to the cooler surfaces of the engine room. The heat lost to the exhaust and by radiation depends mainly on the engine load. It cannot be controlled directly. The operating temperature of the engine must be regulated controlling the cooling system. Cooling must be provided for internal-combustion engines principally for the following reasons:

- To prevent breakdown due to overheating of the lubricating oil film that separates the engine bearing surfaces.
- To prevent loss of strength due to overheating the metal in the engine.
- To prevent excessive stresses in or between the engine parts due to unequal temperatures within the engine.

The cooling medium or coolant used in internal-combustion engines is either water or air. However, marine diesel engines use water. The water used as coolant should not contain any impurities which might form deposits on the inside of the engine water passages and thereby impairing the heat transfer. In a marine engine, the system which functions to keep engine parts and fluids at safe operating temperatures may be either the open or the closed type. In the open system, the engine is cooled directly by saltwater. Since the Army has few, if any, remaining diesel engines in service using an open cooling system, only the closed type is discussed in this chapter. In the closed system, freshwater (or antifreeze) is circulated through the engine. The freshwater is then cooled by saltwater.

CLOSED COOLING SYSTEM

11-3. The closed cooling system of a marine engine actually consists of two entirely separate circuits, sometimes called systems: a freshwater (distilled) circuit and a saltwater circuit. The freshwater circuit is a self-contained system similar to the cooling system in the engine of an automobile. One of the primary differences between the freshwater circuit of a marine installation and that of an automobile is the incorporation of a cooler, rather than a radiator, in the marine installation. The heat is carried away by saltwater instead of by air. The flow of liquid through the two circuits of one type of closed cooling system is shown in Figure 11-1. In some marine installations a separate saltwater circuit is not included in the closed cooling system. In such installations, the freshwater cooler is located on the outside of the hull, well below the waterline, in direct contact with the seawater. In the freshwater cooling circuit, freshwater is reused continuously for cooling the engine. The order of water flow through the parts in the freshwater circuit of a closed cooling system is not always the same. In a majority of installations, water is circulated throughout the engine cooling spaces by an attached circulating freshwater pump. The water then flows to a freshwater cooler where it is cooled by the saltwater of the saltwater cooling circuit. After it leaves the cooler, the freshwater may or may not, depending on the installation, go through the lubricating oil cooler to act as a cooling agent for the lubricating oil. The water finally returns to the suction side of the freshwater pump, completing the circuit. The saltwater circuit of the closed cooling system consists of an attached saltwater pump (usually similar to the freshwater pump). This pump draws saltwater from the sea, through a sea chest with strainer, and through sea valves, and discharges it through the freshwater cooler. In some installations an additional strainer is located in the pump discharge. From the freshwater cooler, seawater may or may not, depending on the installation, pass through the lubricating oil cooler before it is discharged overboard. The overboard discharge system performs various functions, depending on the individual installation. Normally, it is used to cool the exhaust piping and the silencer. On some engine-generator units, the attached saltwater pump furnishes saltwater to the overboard discharge. Throttling valves are frequently placed in lines to the freshwater cooler and the generator air coolers to control the flow of water through these heat exchangers.

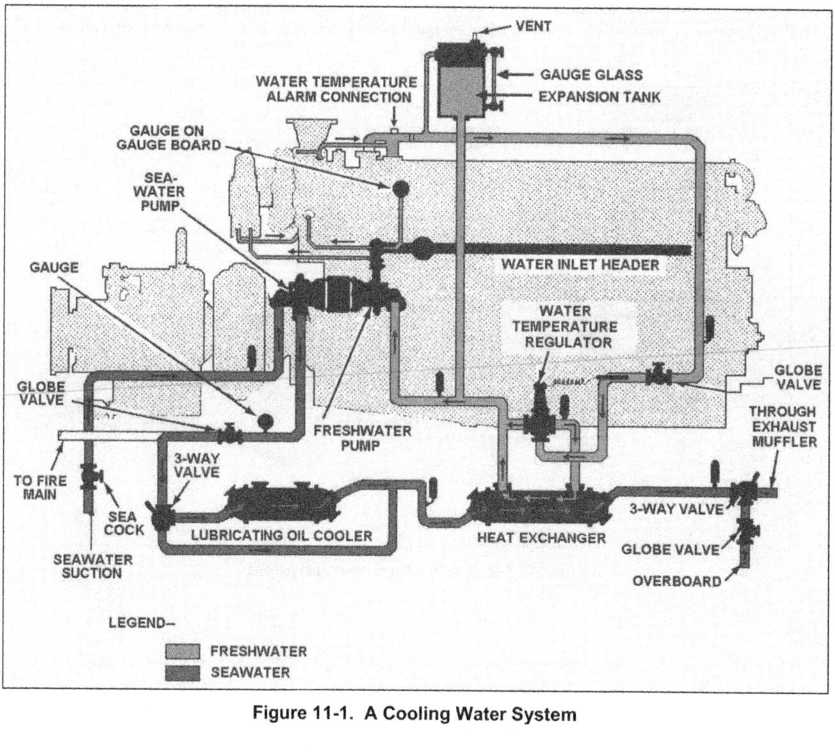

Figure 11-1. A Cooling Water System

PRINCIPAL PARTS OF A CLOSED COOLING SYSTEM

11-4. The closed cooling system of an engine may include some of the following parts:

- Pumps.
- Coolers.
- Engine passages.
- Water manifolds.
- Valves.
- Expansion tank.
- Piping.
- Strainers.
- Connections.
- Instruments.

Some of these parts and their locations on one type of engine are shown in Figure 11-2. Even though many types and models of engines are used, the cooling systems of most of these engines consist of the same type of basic parts. However, design and location of parts may differ considerably from one engine to another. The following discussion points out some similarities and differences found in the various parts of a closed cooling system.

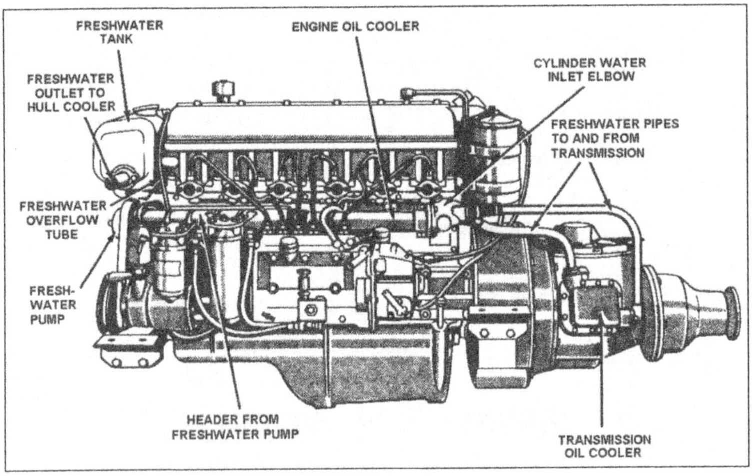

Figure 11-2. Parts of a Cooling System

PUMPS

11-5. All engine cooling systems are provided with an attached freshwater pump. In some installations, a detached auxiliary pump is also provided to keep water circulating through the cooling system. Since attached pumps are engine-driven, it is impossible for cooling water to be circulated in the engine after the engine has stopped, or in the event the attached pump fails. For this reason, some engines are equipped with an electric-driven (detached) auxiliary pump, which may be used if either the freshwater pump or the saltwater pump fails. An auxiliary pump may also be used as an after-cooling pump, when an engine has been secured.

Types of Pumps

11-6. The pumps used in the freshwater and saltwater circuits of an engine cooling system may or may not be of the same type. In some systems, the pumps in both circuits are identical. In other systems, where pumps are of the same type but where variations exist, the principal differences between the pumps of the two circuits are in size and capacity. In the cooling system of some engines, the saltwater pump has a capacity almost double that of the freshwater pump. Centrifugal pumps and gear pumps are the principal types used in engine cooling systems. On some engines, a rotary pump in which the impeller has flexible vanes is used in the saltwater circuit. Centrifugal pumps are more common in engine cooling systems than other types, particularly in large diesel engines. The following types of centrifugal pumps exist. They may be:

- Separately driven or attached to the engine.
- Single or double suction.
- Have open or closed impeller.
- Be reversible or nonreversible.

However, in all centrifugal pumps, water is drawn into the center of the impeller and thrown at a high velocity into the casing surrounding the impeller. This is where velocity decreases and pressure increases correspondingly. In all such pumps, sealing devices, usually of the stuffing box type, are provided to prevent leakage of water, oil, grease, or air around the impeller shaft or the impeller shaft sleeve (when one is used to protect the shaft). Generally, clearances between the impeller and the casing must be small in order to reduce internal leakage. Wear rings are frequently used between the impeller and the casing so that the desired small clearances, when lost, may be regained readily by replacing these rings. The rings are designed to take most of the wear.

Location of Pumps and Method of Drive

11-7. Depending on the engine and the type of installation, location of the pumps and method of drive will vary. In Figure 11-1, notice that the freshwater pump is mounted on the bottom of the lower crankcase at the supercharger end of the engine. This pump is gear-driven at 1 1/2 times engine speed. The saltwater pump of the same engine is mounted on the supercharger housing. The saltwater pump is driven from the supercharger drive shaft through a coupling shaft. It operates at crankshaft speed. The saltwater pump is used only for exhaust cooling. It is mounted on the front of the engine and is belt-driven by the crankshaft. In some models of the Fairbanks-Morse (FM) opposed-piston engine, the freshwater pump and the saltwater pump are located on opposite sides of the engine at the control end. The pumps of these FM engines driven by the lower crankshaft, through the flexible-drive coupling, which also drives the fuel and lubricating pumps and the governor. In the General Motors (GM) 16-278A, the cooling-system pumps are mounted on opposite sides of the blower housing and are driven by the crankshaft through the accessory-drive gear train. The freshwater pump on the GM 6-71 is driven by the lower blower rotor shaft, through a coupling. The examples here are only a few of many which could be given to illustrate variations in pump locations and methods of drive. Regardless of location and drive, the pumps function to keep water circulating through the system so that heat may be dissipated and operating temperatures kept within safe limits. Most of the excess heat developed by an engine is transferred from the freshwater circuit and the lubricating oil to the saltwater circuit through coolers.

COOLERS (HEAT EXCHANGERS)

11-8. Devices that transfer heat from one fluid to another are called heat exchangers. These devices may also be used as either heaters or coolers. The same device can be also be used for both purposes. In internal-combustion engines, heat exchangers are used primarily for cooling. Therefore, heat exchangers used in engines for cooling a hot fluid (liquid or gas) by transferring heat to a cooler fluid are commonly referred to as coolers. Coolers are used principally to cool freshwater and lubricating oil. Sometimes they are used to reduce the temperatures of engine intake air and generator cooling air. In most marine engine installations, the freshwater of the engine cooling systems is cooled by saltwater. Depending on the installation, lubricating oil and air may be cooled by saltwater or by freshwater. Therefore, on the basis of the fluids cooled, you will encounter freshwater coolers, lubricating oil coolers, and air coolers. All coolers operate on the same principle. However, coolers used on various installations and for cooling of various fluids may differ in appearance and design. On the basis of classification by construction features, a cooler is either the shell-and-tube type or the jet type. In jet coolers, hot and cold fluids enter the unit, are mixed, and are then discharged as a single fluid. Since this feature is not desirable in engine cooling, the coolers used with engines are the shell-and-tube type. In a shell-and-tube cooler, mixing of the hot and cold fluids is prevented by the thin walls of the tubes of the element. Shell-and-tube type is a general classification which includes all coolers in which mixing of the two liquids is prevented. The original shell-and-tube cooler is also referred to by that name. Modifications of this cooler have resulted in two other types: the strut-tube cooler and the plate-tube cooler. These modifications are also of the shell-and-tube type in that the fluids are not permitted to mix. The three coolers used in engines are commonly identified as:

- Shell-and-tube type cooler.
- Strut-tube type cooler.
- Plate-tube cooler.

Shell-and-Tube Cooler

11-9. The cooling systems of many engines are equipped with coolers of the shell-and-tube type (frequently referred to as Ross-type coolers). Shell-and-tube coolers are used for cooling lubricating oil and freshwater. Coolers used for cooling lubricating oil are somewhat smaller than those used to cool water. One model of a tube cooler is shown in Figure 11-3. The shell-and-tube cooler consists principally of a bundle (bank or nest) of tubes encased in a shell. Cooling liquid generally flows through the tubes. The cooled liquid enters the shell at one end, circulates around the tubes, and is discharged at the opposite end of the shell. In some coolers the pattern is reversed, the cooling liquid flows through the shell and around the tubes, and the cooled liquid passes through the tubes. The tubes of the cooler are attached to the tube sheets at each end of the shell, forming a tube bundle that can be removed from the shell as a unit. The ends of the tubes are expanded to fit tightly into the holes in the tube sheets and are flared at their outer edges to prevent leakage. One tube sheet and a bonnet, bolted to the flange of the shell, comprise the stationary-end tube sheet. The tube sheet at the opposite end "floats" in the shell, allowing for expansion of the tube bundle. Packing rings, which prevent leakage past the floating-end tube sheet, are fitted between the shell flange and the bonnet at the floating end. The packing joint allows for expansion and prevents mixing of the cooling liquid with the liquid to be cooled inside the shell. This is done by means of a leak-off or lantern gland which is vented to the atmosphere. Transverse baffles are so arranged around the tube bundle that the liquid is directed from side to side as it flows around the tubes and through the shell. Deflection of the liquid ensures maximum cooling. Several of the baffles serve as supports for the bank of tubes. These baffles are of heavier construction than those which only deflect the liquid. The flow of liquid in the tubes is opposite that of the flow of liquid in the shell. On this basis, the cooler can be classified as the counter-flow type. However, since heat transfer is through the walls of the tubes and since cooling liquid enters one end of the cooler and flows directly through the tubes and leaves at the opposite end, the cooler can be more precisely classified as a single-pass, indirect-type cooler.

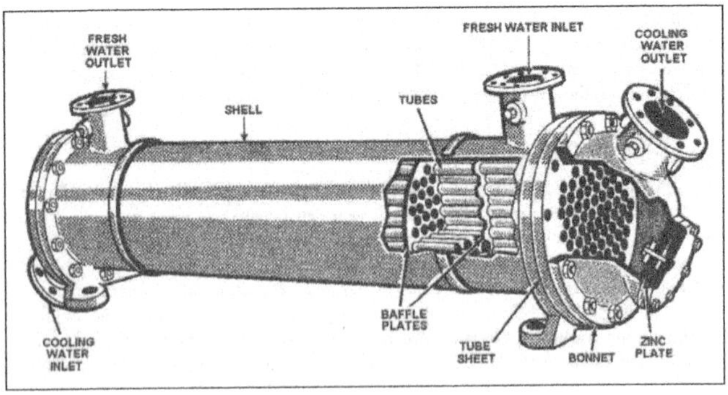

Figure 11-3. Shell-and-tube Cooler

Strut-tube Cooler

11-10. The majority of coolers used in the cooling systems of marine engines are the strut-tube type. The strut-tube cooler has an advantage over the shell-and-tube cooler in that it provides considerable heat transfer in a smaller and more compact unit. However, the shell-and-tube cooler, while larger for an equivalent amount of heat transfer, has an advantage over the strut-tube cooler. This advantage is that it is able to withstand a higher degree of scaling and larger foreign particles without clogging the cooling system. Strut-tube coolers are commonly referred to as Harrison coolers. However, manufacturers other than Harrison produce strut-tube coolers. The term radiator is also used to identify coolers with strut-tube construction. Many different designs of strut-tube coolers exist. The tube assemblies of two of these coolers and the type of tube construction in each are shown in Figure 11-4 and Figure 11-5. Strut-tube coolers are used for cooling water and lubricating oil. Water coolers (Figure 11-4) and oil coolers (Figure 11-5) differ principally in design and in size of the tubes. Each of the tubes in both oil coolers and water coolers is composed of two sections, or strips. In the strut-tube water cooler, both sections of each tube contain either a series of formed dimples or cross-tubes brazed into the tubes. These "struts" (sometimes referred to as baffles) increase the inside and outside contact surfaces of each tube and create turbulence in the liquid flowing through the tube. Therefore, heat transfer from the cooled liquid to the cooling liquid is increased. The struts also increase the structural strength of the tube. In the oil cooler, the tubes are from one-half to one-third as large as the tubes of the water cooler, and the sections of the tubes do not contain either dimples or cross-tubes. Instead, a distributor strip, which serves the same purpose as the struts in the tubes of the water cooler, is enclosed in each tube. The tubes of a strut-tube cooler are fastened in place with a header plate at each end and with an intermediate reinforcement plate. These plates are electroplated with tin to protect the iron parts of the cooler. The tube-and-plate assembly (sometimes called the tube bundle or the core assembly) is mounted in a casing or housing and held in place by the two end covers. The casing, core assembly, frame, covers, and other parts of one model of a strut-tube cooler are shown in Figure 11-6. The header plates, at the ends of the tubes, separate the cooling-liquid space in the casing from the cooled-liquid ports in the end covers. The cooled liquid flows through the tubes in a straight path from the cover inlet port to the cover discharge port at the opposite end of the cooler. The intermediate tube-plate acts as a baffle to create a U-shape path for the cooling liquid, which flows around the outside of the tubes, from the inlet opening of the casing to the discharge opening.

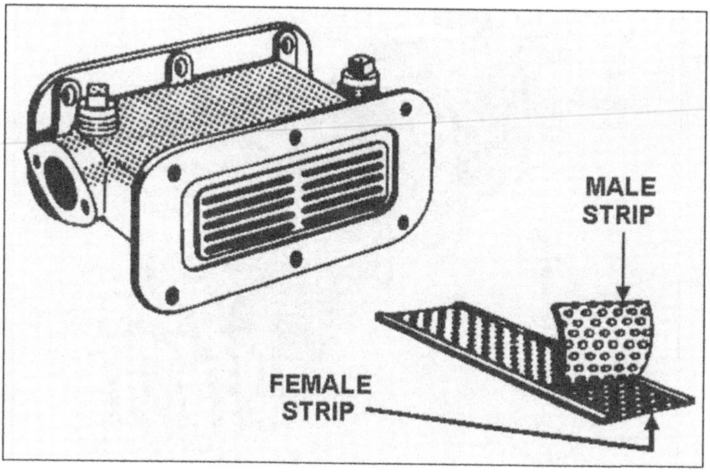

Figure 11-4. Tube Assembly of a Strut-tube Water Cooler (Harrison)

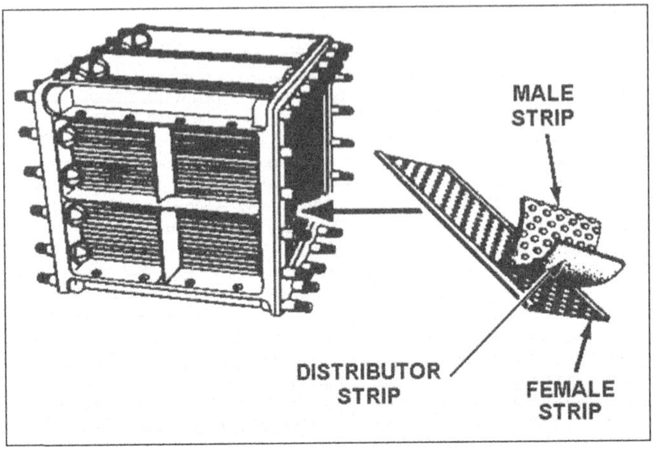

Figure 11-5. Tube Assembly of a Strut-tube Lubricating Oil Cooler (Harrison)

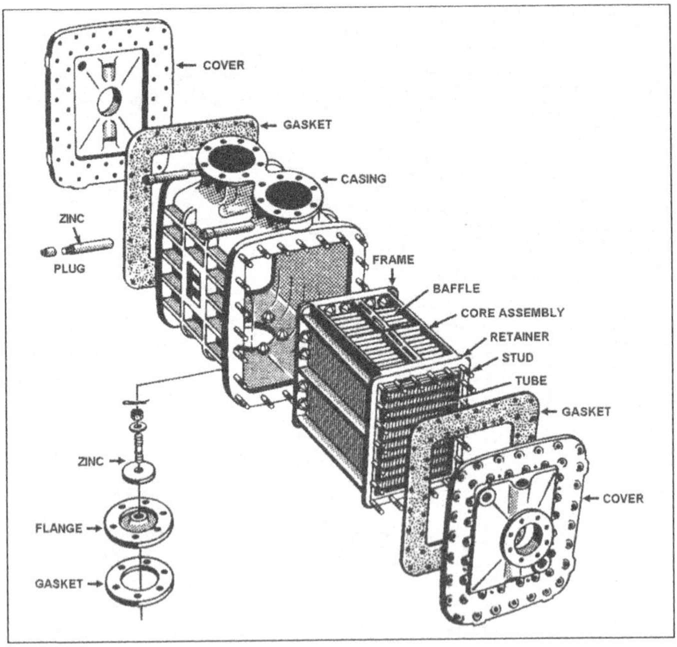

Figure 11-6. Parts of a Strut-tube Cooler

Plate-tube Cooler

11-11. Shell-and-tube coolers and strut-tube coolers are used for cooling both oil and water, usually with seawater as a coolant. However, plate-tube coolers are used only for cooling oil. Depending on the installation, seawater or freshwater may be used as the cooling liquid in plate-tube coolers. An exposed view of one model of a plate-tube cooler is shown in Figure 11-7. A plate-tube cooler consists of a stack of flat, oblong, plate-type tubes. These are connected in parallel with the oil supply and are enclosed in a cast metal housing. Each tube of a plate-tube cooler consists of two sections or stampings of copper-nickel. A distributor strip is enclosed in each tube. Several tubes are assembled to form the cooling element or core of the cooler. Figure 11-8 shows a plate-type core and tube construction. In a plate-tube cooler, the cooling liquid flows through the casing and over the tubes. The heated oil flows through the tubes. The tubes in the core assembly are so spaced that the cooling water circulates freely over their external surfaces.

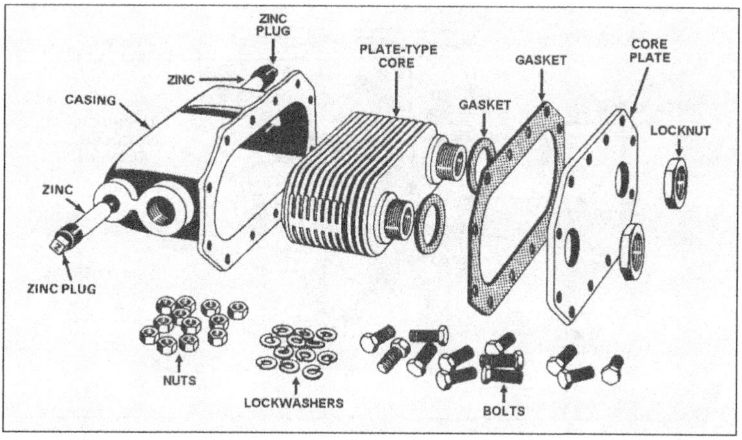

Figure 11-7. Plate-tube Cooler

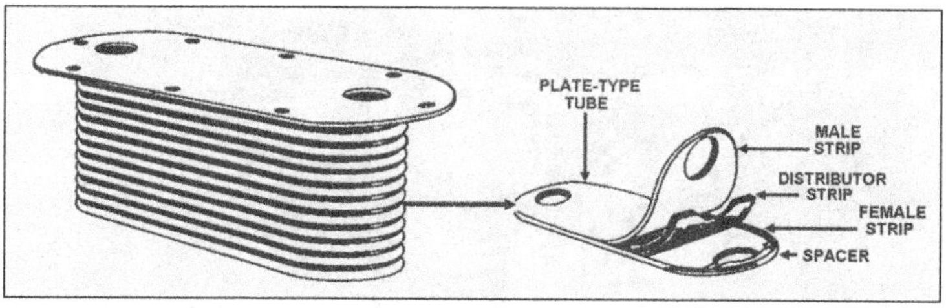

Figure 11-8. Plate-type Core and Tube Construction (Harrison)

Location of Coolers

11-12. Depending on the engine and the fluid cooled, the location of coolers will vary. Some coolers are attached, others are detached. Some freshwater coolers are located on the outside of the hull, well below the waterline. When so located, coolers are frequently referred to as outboard, or keel, or hull coolers (see Figure 11-4 and Figure 11-5). The lubricating oil cooler in Figure 11-1 is attached to the engine block and is located between the saltwater intake and the exhaust-manifold water jacket (heat exchanger). The engine cooling system shown in Figure 11-2 uses two lubricating coolers (one for engine oil and the other for transmission oil). Freshwater serves as the cooling liquid in both coolers. The location of freshwater coolers and lubricating oil coolers of the GM 71 engines (Figure 11-9) is representative of cooler location in many small diesel engines. Compare this location with the location coolers in **a** medium-size diesel, as shown in Figure 11-10.

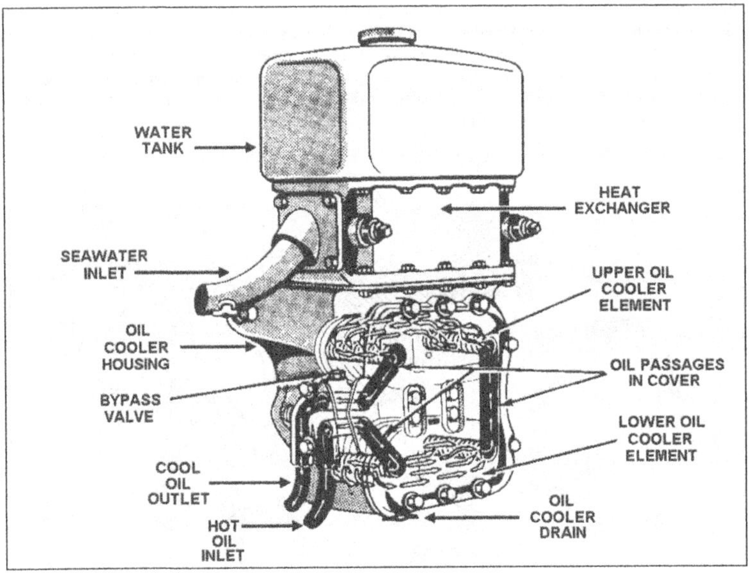

Figure 11-9. Location of Coolers and Other Cooling System Components of GM 71 Engines

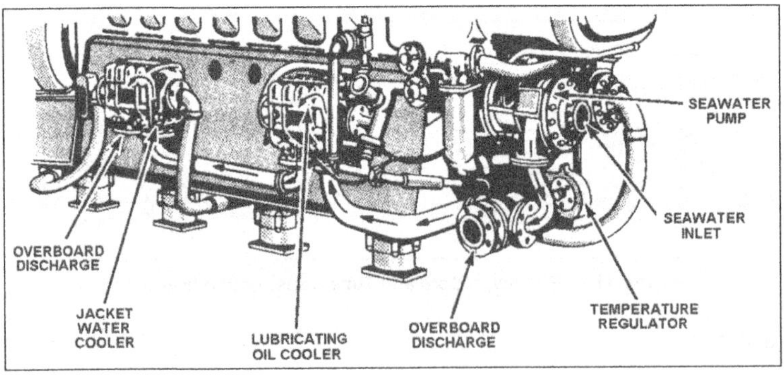

Figure 11-10. Location of Coolers and Other Cooling System Components of GM 8-268A Engines

Air Coolers

11-13. The coolers discussed to this point are those used in lowering the temperature of freshwater and lubricating oil. In some engines, coolers are also used to reduce the temperature of the supercharged intake air. This air is heated by compression within the supercharger. If the temperature of this air is reduced, the size of the air charge entering the cylinder during each intake event as well as the power output of the engine will be increased. Coolers used for lowering the temperature of the intake air are of the shell-and-tube type and the strut-tube type. They operate on the same principle as coolers used for cooling freshwater and lubricating oil. Air coolers used to cool intake air (sometimes called after-coolers) are located between the supercharger and the intake manifold. The heated air from the supercharger passes through the tubes, where the heat is transferred to the cooling water following around the outside of the tubes. Cooling water is generally from the freshwater circuit. The engine cooling system is also used to cool the air around some engine generator sets. Generators, unlike internal-combustion engines, cannot be cooled by liquids. If a generator develops more heat than can be removed by the surrounding air, a supply of cool air must be provided to remove the excess heat. When generator air cooling is necessary, an air cooler is provided in a closed air circuit. The heated air from the generator is forced through the cooler, where the temperature is reduced. The air is then recirculated to the generator. Depending on the installation, either freshwater from the engine cooling system or seawater may serve as the cooling medium.

ENGINE WATER PASSAGES

11-14. The form, location, and number of cooling passages within an engine vary considerably in different engines. The form which a cooling water passage must have and its location are controlled by many factors, some of which are size of engine, cycle of operation, and cylinder arrangement. Many of the water passages in various engines are illustrated and mentioned earlier in this manual in connection with engine parts. A review of these figures will reveal some of the differences in the form and location of the cooling water passages in different engines. Additional information on engine cooling passages is given in the following paragraphs. The examples used for illustration purposes are not all-inclusive. However, those described are representative of the passages to be found in in-line, V-belt, and opposed-piston engines.

In Cylinder Heads

11-15. Most engines have cooling water passages within the cylinder heads. These passages generally surround the valves and, in diesels, the injectors. Usually the passages are cast or drilled as an integral part of the head. In some cylinder heads, such as those of the GM 6-71 and the Gray Marine 64-HN9, the injectors are sealed in a water-cooled tube by means of a neoprene seal. The passages in the cylinder heads of two in-line engines which differ considerably in size are shown in Figure 11-11and Figure 11-12.

In Water Manifolds and Jackets

11-16. The location and form of the water passages in a V-type engine are basically the same as those found in an in-line engine. Differences which exist are generally due to cylinder arrangement. The manifolds receive water from the freshwater pump. From the manifolds, the water flows into the cylinder liner passages, and then through the cylinder head passages. From the cylinder head passages, the water flows through the water jackets of both the exhaust elbows and the exhaust manifold. The water is forced from the water jacket of the exhaust manifold, through a cooler, before the water is recirculated through the system by the pump. Because of differences in engine design, the location and form of the cooling passages in an opposed-piston engine will differ, to a degree, from those in other types of engines. The lack of cylinder heads eliminates some of the passages common to engines of the in-line and V-types. While differences of a minor nature exist in the passages of different types of engines, the cooling passages of all engines are similar in many respects. Some of the ways in which the passages of an opposed-piston engine are similar to those of other types of engines are shown in Figure 11-13. Note that in the FM engine the exhaust manifolds are encased in water jackets similar to those of the GM 16-278A. The liner passages of the FM 38 are similar to those found in other types of engines. The location of the water header (manifold) differs in various engines. In such engines as the GM 16-289A, the water manifold receives water from the pump. In these engines, water from the manifold flows through the liner and head passages and then on to cool the exhaust manifold before it flows through the cooler and back to the pump. In the FM 38D, the water from the pump usually enters the engine through the water jackets of both the exhaust elbows and the exhaust manifolds. In some engines, water enters the cylinder liner through a nozzle adapter. In the usual arrangement, the water header or manifold in the FM 38D receives water from the cylinder-liner water passages (see Figure 11-13). In other words, the water header of an FM 38D is the last passage in the engine through which water flows before it goes through the cooler and back to the pump. In the GM 16-278A, the water header (manifold) is the first part to receive water from the pump.

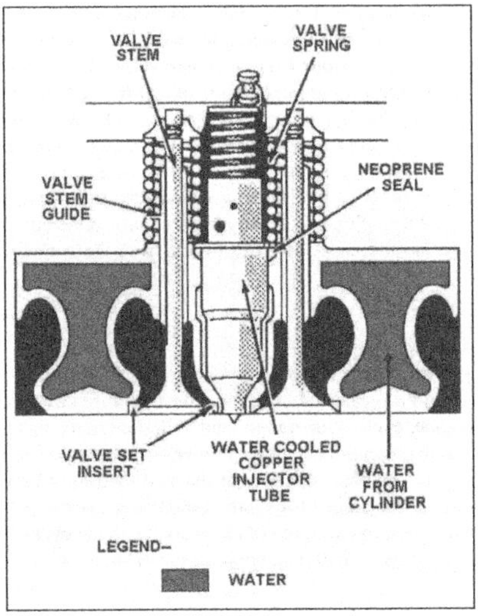

Figure 11-11. Water Passages in Cylinder Head

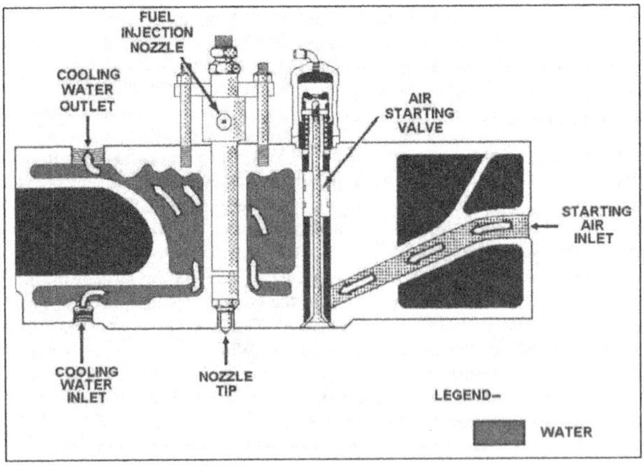

Figure 11-12. Location of Coolers and Other Cooling System Components of GM 8-268A Engines

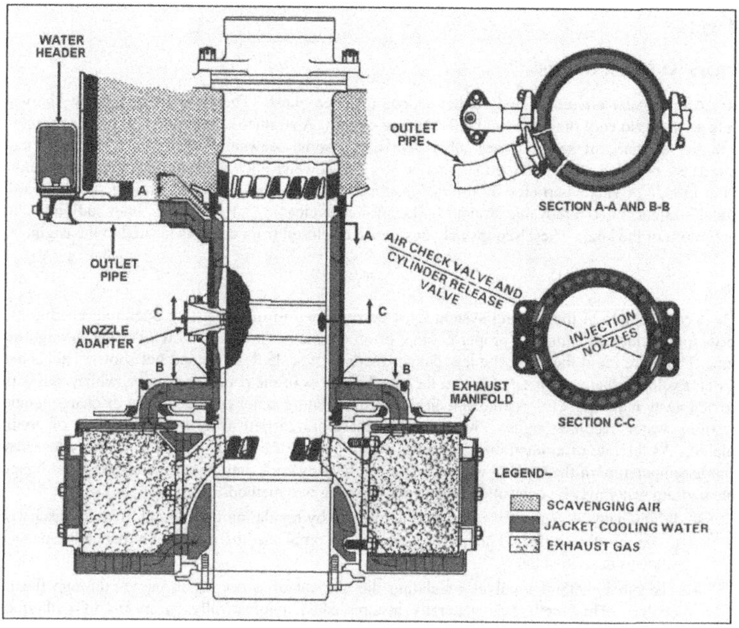

Figure 11-13. Cooling Water Passages in an FM Opposed-piston Engine

FRESHWATER (EXPANSION) TANKS

11-17. The freshwater circuit of an engine cooling system includes a tank commonly referred to as the expansion tank. Some expansion tanks are identified as the surge tank or supply tank. The freshwater tank provides a place for adding water to the system. It also provides a space to accommodate variations in water volume which result from expansion and contraction caused by heating and cooling of the water. The piping arrangement of a cooling system is such that excess water in the system passes back to the tank as the water expands upon becoming warm. Also, water from the tank flows into the system when the water contracts as it cools, and when the water becomes low because of leakage in the system.

Location of Expansion Tanks

11-18. Even though the exact locations of expansion tanks vary in different engines, the tank is always located at or near the highest point in the circuit.

Venting of Expansion Tanks

11-19. The manner in which the freshwater circuit of a cooling system is vented varies depending on the engine. However, it generally involves venting the expansion tank to the atmosphere. In some expansion tanks, particularly in the systems of large engines, a vent pipe at the high point of the circuit carries to the tank steam or air bubbles which may form in the system. When steam comes in contact with the cooler water in the tank, the steam condenses back into water. The condensation keeps the system free from steam or air pockets. A gauge glass, located on the side of the tank, indicates the water level. In many small engines, the freshwater circuit has no vent. It operates under a slight pressure which confines the water vapor, thereby preventing the loss of water. The only escape for water vapor from a circuit which operates under pressure is through a small overflow pipe.

Sea Scoops and Strainers

11-20. All seawater circuits include either scoops or a sea chest. These are located below the waterline to provide seawater to cool the water in the freshwater circuit. A strainer is incorporated in the seawater circuit to prevent the entrance of seaweed and other debris. In some seawater circuits, two strainers (inboard and outboard) are installed. The outboard strainer covers the seawater inlet. The inboard strainer prevents entrance into the circuit of small particles of foreign matter which may have passed through the outboard strainer. Seawater strainers have removable strainer baskets for easy cleaning. Two scoops, inlet and outlet, are located in the bottom of the boat. These scoops may be opened or closed from controls located in the engine room.

Valves

11-21. Temperatures in the cooling system must be regulated to meet existing operating conditions. One of the principal factors affecting the proper cooling of an engine is the rate of flow of water through the cooling system. The more rapid the flow, the less danger there is of scale deposits and hot spots. This is because the high water velocity has a scouring effect on the metal surfaces of the cooling passages, which causes the heat to be carried away more quickly. As the velocity of the circulating water is reduced, the discharge temperature of the cooling water becomes higher. More heat is therefore carried away by each gallon of cooling water circulated. As the rate of circulation is increased, each gallon of cooling water carries away less heat and the discharge temperature of the cooling water drops. A relatively cool-running engine results. The temperature of engine cooling water may be controlled by one of following two methods:

- In one method, water temperature is controlled by regulating the amount of water discharged by the pump into the engine. This method may be accomplished by means of a manually operated throttle valve as described below.
- The other method involves regulating the amount of water which passes through the freshwater cooler. This method is generally accomplished automatically by means of a thermostatically operated bypass valve as described below.

Manually Operated Throttling Valve

11-22. When the manually operated throttling valve is located in the pump discharge, it may be used to pass the water slowly through the engine for discharge at a high temperature. It may be also used to pass the water rapidly through the engine for discharge at a low temperature. If the pump is driven by an electric motor, these same effects on velocity and discharge temperature of the cooling water can be obtained by increasing or decreasing the speed of the pump. Throttling valves may be used in the seawater circuit to regulate the amount of water passing through the seawater side of the freshwater cooler. An example of temperature control by means of a throttling valve in the seawater circuit is shown in the left portion of Figure 11-1. In some installations of the system shown, engine temperature is controlled by opening or closing the three-way valve. In other installations of the system shown, temperature is controlled automatically by a thermostatically operated bypass valve. Where throttling valves are incorporated in the seawater circuit and a thermostatically operated valve is included in the freshwater circuit, the throttling valves are used only to provide a constant flow of seawater. Temperature is then controlled by the thermostatically operated valve in the freshwater circuit. The throttling valve in the saltwater circuit should be adjusted to maintain the minimum flow of saltwater consistent with maintaining proper temperatures in the freshwater circuit and the lubricating oil system.

Thermostatically Operated Bypass Valves

11-23. In modern marine engine installations, automatic temperature control by thermostatically operated bypass valves is more common than control by throttling valves in the pump discharge or in the seawater circuit. Two types of thermostatic valves (conventional type and three-way proportioning type) are used in the cooling systems of engines. Valves of the latter type are commonly called automatic temperature regulators. Conventional thermostatic valves are generally used in small engines. Temperature regulators are generally used in medium and large engines. The element in both types of thermostatic valves is similar. It is so designed that it will expand or contract, depending on the temperature to which it is exposed. The element of a thermostatic valve may be filled with either a gas or a liquid, or it may be of the bimetal type. In most marine engines, the elements of thermostatic valves are either gas-filled or liquid-filled. The elements of conventional thermostatic valves generally contain a gas. The elements of automatic regulators usually contain a volatile Liquid, such as ether or alcohol. Elements constructed in the form of a sealed bellows may be made of copper, brass, or Monel metal. These metals are used because they are corrosion-resistant and because they will withstand a considerable amount of flexing without fatigue. The element is attached to a valve, which opens and closes as the bellows expands and contracts during variations in the temperature of the coolant. Therefore, the amount of water flowing through the line is automatically regulated. Conventional thermostatic valves may be built into the engine or they may be located outside the engine within the freshwater circuits. The manner in which a conventional thermostatic valve operates is shown in Figure 11-14. As long as the engine and water are cool, the thermostatic valve remains closed. Water from the engine then flows around the bellows, down through the holes in the valve, and out through the bypass outlet to the freshwater pump (see Figure 11-14, view A). Freshwater is therefore bypassed around the cooler until the water gets warm enough to cause the thermostatic valve to open. As the water gets warmer, the increase in temperature causes the element to expand. The valve is then partially opened, a part of the water goes through the cooler, and the rest of the water goes through the bypass outlet (see Figure 11-14, view B). Finally, when the valve is wide open, as a result of the increase in the temperature of the water, the valve seats on the base of the thermostat housing. The flow of freshwater through the bypass outlet is then stopped, and all of the water from the engine passes through the cooler where the temperature of the water is reduced (see Figure 11-14, view C). In many engines, freshwater temperature is regulated by an automatic regulating valve which maintains the freshwater temperature at any desired value by bypassing a portion of the water around the freshwater cooler. An automatic temperature regulator of the type commonly used in the cooling systems of marine engines is shown in Figure 11-15. Although regulators of this type are automatic or self-operated, most installations also have provisions for manual operation in the event of failure of the automatic feature. The temperature regulator consists of a valve and a thermostatic control unit mounted on the valve. The thermostatic control unit consists of two parts (the temperature-control element and the control assembly). The temperature-control element consists of a bellows connected by a flexible armored tube to a bulb mounted in the engine cooling-water discharge line. The temperature-control element is essentially two sealed chambers. One chamber is formed by the bellows and cap sealed together at the bottom and the other chamber is in the bulb. The entire system (except for a small space at the top of the bulb) is filled with a mixture of ether and alcohol, which vaporizes at a low temperature. When the bulb is heated, the liquid increases, forcing the liquid out the bulb and through the tube. The bellows is moved downward and it operates the valve. The control assembly consists of a spring-loaded mechanical linkage which connects the temperature-control element to the valve stem. The force produced by the coil spring in the control assembly balances the force of the vapor pressure in the temperature-control element. Therefore, the downward force of the temperature-control element is balanced at any point by the upward force of the spring. The valve can then be set to hold the temperature of the engine cooling water within the allowable limits. The regulator operates only within the temperature range marked on the nameplate. It may be adjusted for any temperature within this range. The setting is controlled by the temperature-adjusting wheel located under the spring seat. A pointer attached to the spring seat indicates the temperature setting on a scale which is attached to the regulator frame. The scale is graduated from 0 to 9, representing the total operating range of the regulator (Figure 11-16). The location of a temperature regulator in one type of installation is shown in Figure 11-10. The regulator is located in the seawater circuit. In most engines, the regulator is located in the freshwater circuit. When located in the seawater circuit, the regulator controls the amount of seawater flowing through the coolers. When the temperature for which the regulator is set increases, the regulator actuates a valve to increase the flow of seawater through the coolers. When the freshwater temperature drops below the temperature for which the regulator is set, the regulator actuates a valve to decrease the flow of seawater through the cooler. In installations that have the regulator in the freshwater circuit, water is directed to the cooler when the temperature of the water is above the maximum setting of the regulator. After passing through the cooler, where the temperature of the water is decreased, the water returns to the suction side of the freshwater pump to be recirculated. When the temperature of the water is below the maximum setting of the regulator, the water bypasses the cooler and flows directly to the suction side of the pump. Bypassing the cooler permits the water to be recirculated through the engine. In this way, the temperature of the water is raised to the proper operating level. Regardless of whether the regulator is in the freshwater or the seawater circuit, the bulb that causes the regulator to operate is located in the freshwater discharge line of the engine. Temperature regulators are used not only to control the temperature of freshwater but also to control, indirectly, the temperature of oil discharge from the lubricating oil cooler. Control of lubricating oil temperature is possible because the water (freshwater or saltwater) that is passed through the regulator and the freshwater cooler is the cooling agent in the lubricating oil cooler. When the lubricating oil is cooled by seawater, two temperature regulators are installed in the seawater circuit. The temperature regulator bulb of the regulator that controls freshwater temperature is installed in the freshwater circuit. The bulb of the regulator that controls lubricating oil temperature is installed in the lubricating oil system.

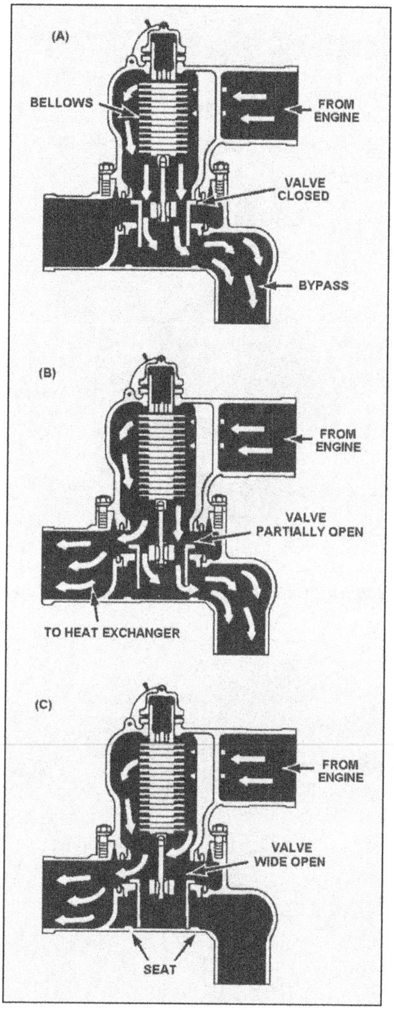

Figure 11-14. Operation of a Conventional Thermostatic Valve

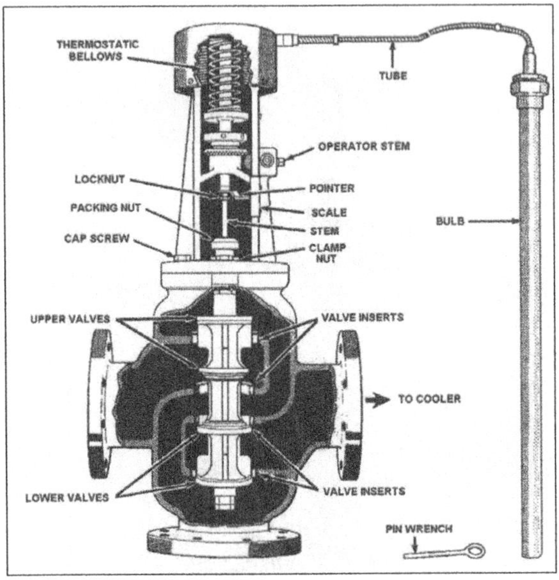

Figure 11-15. Automatic Temperature Regulator

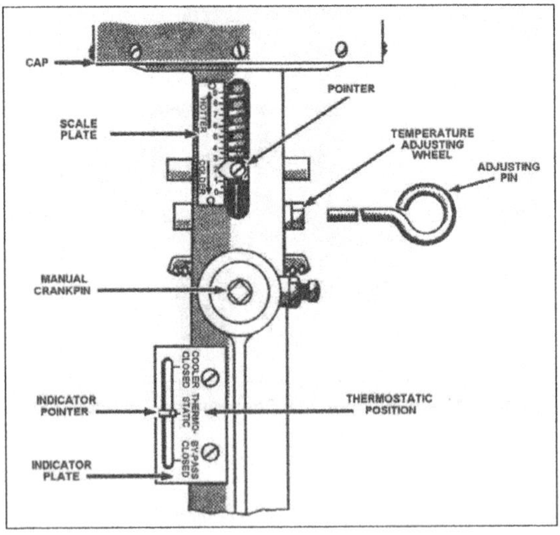

Figure 11-16. Scale and Indicator Plates of Temperature Regulator

MAINTENANCE OF ENGINE COOLING SYSTEM

11-24. The purpose of an engine cooling system is to keep engine parts and working fluids at safe operating temperatures. The system must therefore be in good repair at all times. Corrosion must be prevented and the tendency toward scale formation must be reduced. Heat exchangers and water jackets must be properly cleaned or repaired during a planned maintenance and whenever faulty operation reveals that it is necessary. Circulating pumps must be maintained in the best operating condition. The following information deals with maintenance of heat exchangers and the reduction of scale formation by freshwater treatment.

CARE OF HEAT EXCHANGERS

11-25. If a cooler is to effectively remove excess heat from lubricating oil or freshwater, it must be inspected periodically (usually 30- to 60-day intervals) for excessive scale and foreign material. Excessive scale and accumulations of dirt reduce the efficiency of the cooler. Therefore, unless the cooler is properly cleaned, scale and dirt may accumulate on the tubes to such an extent that the cooler cannot remove enough heat to keep the cooled liquid within prescribed limits. Scale and dirt usually gradually accumulate on the saltwater side of a cooler element. Scale and dirt on the cooler tubes is usually indicated by a gradual increase in oil temperature or in freshwater temperature, depending on the use of the cooler. A gradual increase in the difference between the inlet and outlet pressures of the cooler indicates either excessive accumulations on the tubes or clogged tubes. As the amount of scale increases, the quantity of seawater that must be circulated to obtain the same cooling effect increases. This is due to the insulating effect of the scale coating that forms on the saltwater side of the cooler. When scale formation is suspected, the heat exchanger element should be removed, inspected, and cleaned. Scale will form on the saltwater side of a cooler during normal operation because of the dissolved salts present in the water. One thing which tends to increase the rate of formation of scale is operating the engine with a high seawater temperature. The seawater discharge temperature should be maintained below 130° F. At higher temperatures, the amount of scale formation is considerably greater. Elements of the cooler may become clogged with such things as marine life, grease, or sand. Such clogging greatly reduces cooler capacity. Elements may also become clogged through faulty operation of the seawater strainer and improper lubrication of pumps; or, in oil coolers, a leaky element. Seawater strainers prevent the entrance of seaweed and other debris into the circulating system. These strainers must be replaced or repaired when the screens become punctured or otherwise incapable of preventing entry of dirt into the system. When a seawater strainer is to be cleaned or replaced, the sea suction valve must be secured before the strainer is opened to prevent flooding the space. This precaution must be taken whenever work is to be done on any saltwater piping coming in direct contact with the sea. Many seawater pumps are provided with grease cups for bearing lubrication. Turning such grease cups down too often may result in the grease being squeezed into the cooler element and deposited there. The film of grease deposited will greatly reduce the capacity of the cooler. Seawater pumps should be lubricated as specified in the appropriate manufacturer's technical manual (TM). A hole in the element of the oil cooler will allow some of the lube oil to pass into the water which is used for cooling. Some of the oil introduced into the water may be deposited on the water side of the cooler elements. The film of oil deposited will act in the same manner as the grease film just discussed. Therefore, leaks in oil coolers should be repaired as soon as possible. A vigilant lookout should be maintained for signs of oil or grease in the freshwater system. If oil or grease is present, the source of the contamination should be located and eliminated. When symptoms indicate that excessive amounts of scale or accumulations of dirt are forming on the saltwater side of a cooler, the element should be removed and cleaned. For ordinary cleaning of a cooler element (shell-and tube type), an air lance should be pushed through each tube, the tube sheets washed clean, and all foreign matter removed from the water chests. For more severe fouling, a water lance should be used instead of an air lance. Where there is extreme fouling from oil or foreign material, a rotating bristle brush may be run through each tube. Or, soft rubber plugs (if available) may be driven through the tubes by an air or water gun. A water lance is then used to remove any remaining foreign material from the tubes. No abrasive tools capable of scratching or marring the tube surface are to be used. Wire brushes or metal scrapers should never be used. Air and water lances and other cleaning equipment must be used carefully to avoid damaging the element. A cooler can be cleaned more effectively and easily if it is done before accumulations on the surfaces of the element have had time to dry and harden. The oil sides of shell-and-tube coolers of the removable tube-bundle type can be cleaned by removing the tube bundle and washing it with a jet of hot freshwater. The tube bundle should be dried thoroughly prior to reassembly. However, if the oil is properly purified and filtered, cleaning of the oil side of this type of cooler should seldom be necessary. A chemical rather than a mechanical method is required for cleaning strut-tube coolers. The oil side should be cleaned first. Otherwise, some of the oil will wash out of the tubes and nullify any cleaning already done. The items of equipment necessary to clean the oil side of a cooler vary in size, depending on the size of the unit. With an arrangement such as that shown in Figure 11-17, an approved cleaning agent is pumped through the cooler. The cleaning process can be speeded up considerably by circulating the cleaning agent in the direction opposite to the oil flow. The progress of the cleaning process should be checked frequently. The process is complete when the solution flows freely through the cooler. A hand pump may be used to clean small coolers. First, the element is submerged in a container of cleaning solution. Then the pump is used to force the solution through the element. The oil side of a cooler should be cleaned in the open or in a well-ventilated space if the cleaning agents used give off toxic vapors. Two additional tanks are needed for cleaning the water side of oil coolers. The water side is cleaned chemically by submerging the unit in a weak solution of muriatic acid. The solution should be 1 part muriatic acid to 9 parts cold water. To this is added, for every gallon of muriatic acid, 2 pounds of oxallic acid and 1/4 ounce pyridine. The tank containing the acid must be of earthenware or other acid-resistant material. The second tank is filled with cold freshwater, and the third tank is filled with a 5 percent solution of sodium carbonate. For easier access to the parts of a cooler, the unit should be disassembled and the core removed from the case. The unit to be cleaned is supported by a wire and submerged in the acid solution. Foaming will occur and continue as long as cleaning action is taking place. When foaming ceases, the unit should be placed in the tank of cold water for approximately 1 minute. Then the unit is immersed in the solution of sodium carbonate. Bubbling will occur if all the acid was not removed from the unit by the freshwater. When bubbling occurs, the unit should be left in the solution of sodium carbonate until the bubbling ceases. The unit is then removed from the solution and flushed with freshwater. Warm water should be used. This cleaning procedure can be used to remove any deposits from the cooler case and covers. Corrosion or erosion of the element in a heat exchanger or operation at excessive pressure may cause leaks. These leaks can develop either in the element or in the casing. Leakage from the cooler casing can usually be detected by inspection. However, element leaks are more difficult to detect. Any noticeable decline or rise in the freshwater tank level, with the temperature remaining normal, usually indicates leakage. A leak in the tubes of an oil cooler may be evidenced by water in the lube oil or by slicks in the cooling water. Other indications are the apparent increase in the volume of lube oil in the engine sumps or the loss of oil from the sump without other apparent causes of leakage. A hole made by corrosion in a cooler element indicates that corrosion probably exists throughout the element. A thorough inspection should be made. Corrosion can be prevented to a large extent by doing the following:

- Using the prescribed freshwater treatment.
- Inspecting and replacing zincs as necessary.
- Venting the cooler to remove entrapped air.

11-26. Holes resulting from erosion are generally caused by particles of grit (sand or dirt resulting usually from operation in shallow water) that strike the element at high velocity. For the most part, such grit is so fine that it will pass through the strainer. If the strainer is defective, even larger particles of grit may enter the cooler. Holes in a cooler element may also result from erosion by water at high velocity. Sometimes water flow has to be increased above rated capacity to maintain the desired freshwater temperature. Whenever water flow is greatly increased, the cooler should be cleaned. If the designed maximum operating pressure (indicated on the exchanger nameplate) is exceeded, leaks are apt to result. Excessive pressure is likely to occur in conjunction with clogging because of the added pressure needed to force a given quantity of water through the restricted element. If there is any reason to suspect leaks in the heat exchanger element, the best method of locating them is through a hydrostatic test. The test may be made as follows:

- Remove the element from the casing.
- Block off the discharge side of the element.
- Attach a pressure gauge to the inlet line of the element.
- Supply low-pressure air to the inlet side of the element. Air pressure must not exceed design pressure for the element.
- Immerse the element in a tank of water.

A similar test can be made without immersing the element by filling it with water under pressure.

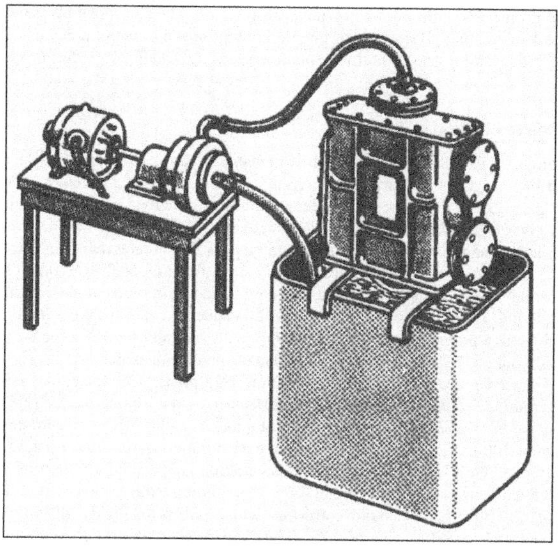

Figure 11-17. Equipment for Cleaning a Heat Exchanger

FRESHWATER TREATMENT AND TESTS

11-27. The purity of water used in an engine's closed circuit cooling system must be maintained to prevent the formation of scale and to control corrosion within the cooling system. These undesirable conditions can be prevented by filling the cooling system with distilled water (zero hardness) and by treating the water to maintain the alkalinity and the sodium chromate and chloride concentrations within specified limits. In the remaining paragraphs of this chapter, we will cover the treatment of water used in the closed circuit of an engine cooling system and with tests used to determine effectiveness of the treatment. All water contains some impurities. Impurities dissolved or suspended in the water of an engine cooling system can cause trouble in the system by forming scale and by causing corrosion. Generally, scale forms only on the hot surfaces in the internal passages, not throughout the system.

Scale

11-28. The formation of scale within the cooling system is caused primarily by certain sulphates of magnesium and calcium. Since these sulphates are present in seawater and since cooling water for shipboard engines is generally distilled from seawater, some slight contamination of the cooling water must be expected. However, the use of distilled water in the cooling system simplifies control of the scale-forming salts. Scale formation must be prevented because scale is a poor conductor of heat. If allowed to accumulate in the cooling system, scale will prevent the proper transfer of heat from the hot engine parts to the cooling water. Improper heat transfer, particularly uneven heat transfer, causes stress in the affected parts which may lead to cracks in cylinder liners, heads, and other parts of the engine. If the water in the cooling system is properly treated, scale formation will be prevented and casualties caused by improper heat transfer will be less likely.

Corrosion

11-29. Unless the water used is properly treated, the internal surfaces of the cooling system may become pitted or eaten away by corrosion. Such corrosion generally results from acidity of the water and the oxygen dissolved in the water. Corrosion in the cooling system may lead to cracks in liners and heads and may cause serious damage to other parts of the cooling system. Corrosion can be controlled by the proper water treatment. The chemical used in treating freshwater in engine cooling systems helps prevent scale and corrosion by maintaining the water within specified limits with respect to alkalinity, chloride content, and chromate concentration.

Terms and Units Related to Water Tests

11-30. The condition of treated water is described in standard terms and units. These are used in recording information regarding the tests and in making reports. The term used to identify the alkalinity of treated cooling water is "pH." The pH unit does not measure alkalinity directly. However, it is related to alkalinity so that a pH number gives an indication of the alkalinity or acidity of the cooling water. The pH scale ranges from 0 to 14 with pH 7 as the neutral point. Solutions having a pH value greater than 7 (pH 8, pH 10) are defined as alkaline solutions; those having pH values less than 7 are defined as acid solutions. The concentrations of chromate and chloride in a test sample of treated water are indicated in terms of parts per million (ppm). This is a weight-per-weight unit denoting the number of parts of a specified substance in relation to 1 million parts of water. For example, 58.5 pounds of salt in 1,000,000 pounds of water represent a concentration of 58.5 ppm. Also note that 58.5 ounces of salt dissolved in 1,000,000 ounces of water or 58.5 tons of salt dissolved in 1,000.000 tons of water represent the same concentration (58.5 ppm). Chromate, pH, and chloride limits must be observed. To minimize serious corrosion and scale deposit and to prolong the life of the cooling system, the treated water must be maintained within the specified limits. The chromate concentration must be maintained between the limits of 700 to 1,700 ppm. A minimum concentration is specified because lower concentrations can result in accelerated corrosion. A maximum concentration is specified to eliminate waste. The pH value must be maintained within the range of 8.25 and 9.75. A minimum value is specified because lower values can result in accelerated corrosion. To avoid corrosion, which occurs in highly alkaline waters, the alkalinity should not be allowed to exceed a value of 9.75. The chloride content must be kept to the lowest practicable value and must never exceed 100 ppm.

WATER TREATMENT

11-31. As previously stated, the chemical treatment of water in the engine cooling system is necessary to maintain the alkalinity and chromate concentration of the water. This greatly reduces scale formation and corrosive action. However, the water treatment discussed in this chapter is a preventive treatment only. It will not remove scale which has already formed in the cooling system. Before the water is treated, the cooling system should be thoroughly cleaned. After cleaning, the system should be thoroughly flushed with freshwater and filled with distilled water. If available instructions do not indicate the proper chemical dosage for a specific engine, the capacity of the complete cooling system (in gallons of water) must be determined. This must be done before the proper dosage of chemicals can be determined. Two different combinations of chemicals are commonly used for the treatment of water for engine cooling systems. The treatment you will use for any particular engine will depend on which set of chemicals is available. The two different combinations are a combination of sodium dichromate and Navy boiler compound and a combination of sodium chromate and disodium phosphate.

Sodium Dichromate and Boiler Compound Treatment

11-32. For each 100 gallons of cooling water to be treated, add 1.5 pounds of sodium dichromate and 3 pounds of boiler compound dissolved in approximately 1 1/2 gallons of warm distilled water. This solution is usually added to the cooling system at the expansion tank. The solution is circulated through the system for at least 10 minutes. Then a 1/2-pint sample of the treated water is drawn off, allowed to cool to at least 80° F, and tested to determine the chromate, alkalinity, and chloride concentration. The procedures for testing, using the test kit provided for this purpose, are discussed later in this chapter. After initial treatment, the cooling water must be tested after each day of operation. When it is apparent that the concentration of chromate and the alkalinity will not drop below the prescribed minimum, the interval between tests may be increased to a maximum of once a month as prescribed by the preventive maintenance schedule. If tests reveal a drop in the chromate concentration below 700 ppm, a solution containing 1 pound of sodium dichromate and 2 pounds of boiler compound should be added for each 100 gallons of cooling water. The addition will bring the chromate concentration within the specified range. If the alkalinity is too high, omit the boiler compound in the solution.

The addition of sodium dichromate alone will reduce the alkalinity. To correct for low pH value, add a solution containing 1.5 pounds of boiler compound for each 100 gallons of cooling water. This addition should bring the pH value within the specified range. If tests reveal that alkalinity is high and chromate concentration is low, add sodium dichromate alone before adjusting the alkalinity. If the alkalinity is too high and the chromate concentration is within the specified range, drain 25 percent of the cooling water from the system and refill with freshwater. Circulate the coolant through the system, retest, and treat as necessary. The concentration of chlorides in the cooling water must not be allowed to exceed 100 ppm. If the test indicates that this limit has been exceeded, the entire system should be drained and the source of chloride contamination should be located and remedied. The system may then be flushed, refilled with freshwater, and chemically treated to proper limits.

Sodium Chromate and Disodium Phosphate Treatment

11-33. When sodium chromate and disodium phosphate are used for cooling water treatment, the procedures for preparing the system, mixing the solution, testing, and controlling the chromate concentration and alkalinity are the same as those used for sodium dichromate and boiler compound treatment. The only differences between the two treatments are the chemicals used and the amounts used. To treat the water using sodium chromate and disodium phosphate, proceed as follows. For each 100 gallons of cooling water to be treated, mix a solution of 1.5 pounds of sodium chromate and 1.5 pounds of disodium phosphate compound dissolved in approximately 1 1/2 gallons of hot distilled water. Add this solution to the cooling system, circulate it in the system, and test. If tests reveal that the chromate concentration is below 700 ppm, a solution containing 1 pound of sodium chromate and 1 pound of disodium phosphate should be added for each 100 gallons of cooling water. The addition will bring the chromate concentration and small pH value within the specified range. This addition should bring the pH value above the specified minimum. If the pH value is too high (above 9.75), drain 25 percent of the cooling water from the system and refill with freshwater. Circulate the coolant through the system, retest, and treat as necessary. If tests reveal that the pH value is low (below 8.25) add a solution containing 1 pound of disodium phosphate for each 100 gallons of cooling water. If tests reveal that chloride concentration exceeds the prescribed limits (above 100 ppm), drain the entire system, locate the source of chloride contamination, and remedy the problem. The system should then be flushed, refilled with freshwater, and chemically treated to proper limits.

CONDUCTING WATER TESTS

11-34. A test kit is provided for testing engine cooling water. It is specifically designed to enable shipboard personnel to conduct chromate, alkalinity (pH), and chloride tests quickly and easily. The kit contains all the equipment and chemicals necessary to do the required tests.

- *Chromate and alkalinity tests.* The tests for chromate and alkalinity are by color comparison. The color of the test sample is compared with the color of two glass disks representing the specified maximum and minimum limits.
- *Chloride test.* The chloride test is different in that the chloride concentration is determined by noting the color and condition of the sample after the addition of a chloride test tablet.

Before taking a test sample of the cooling water, drain water from the drain cock for several seconds. Then draw off a 1 1/2-pint sample. Allow it to cool to at least 80° F before testing. If suspended material is noticed in the sample, filter the water to prevent the suspended material from interfering with the colorimetric tests. To filter the sample, fold a circular filter paper to form a cone. Place the cone in the funnel which is provided. Wet the cone with distilled water and press the upper edge of the cone to the funnel. Place the funnel in the cylindrical sample bottle or other clear, suitable container. Carefully pour the sample water into the cone. As a light source when performing chromate and pH tests, use either natural daylight or a daylight fluorescent lamp suitable for this purpose. Do not use incandescent lighting because the results obtained will be erroneous. When tests are made in natural daylight, the comparator should be held up to the brightest part of the sky, but approximately 30° from the sun. Avoid strong light on the observer's side of the comparator, as reflections will interfere with comparisons. If possible, the observer and comparator should be in a shadow.

Testing for Chromate Concentration

11-35. To test for chromate concentration, proceed as follows:

- Fill the test tube with cooling water sample and place in the center position of the comparator.
- Hold the comparator to a suitable light source.
- If the color of the sample is between the colors of the disks marked 700 and 1700, the chromate concentration is satisfactory. Be careful to note the color rather than the intensity of color. Do not compare the darkness or lightness of the colors. Rather, determine whether the sample is yellower or bluer than the color disk standards.
- If the color is bluer (bluish-green) than the color disk marked 700, the chromate content is too low.
- If the color is yellower (yellowish-green) than the color disk 1700, the chromate content is high. High chromate content is not harmful, but it is wasteful.

Testing for pH Value

11-36. To test for the pH (alkalinity) value, proceed as follows:

- Fill three test tubes to the prescribed mark with sample of cooling water.
- Add 15 drops of pH indicator to one test tube, shake, and place in center position in comparator. Place the other two test tubes in the remaining positions in comparator.
- Hold the comparator to a suitable light source.
- If the color of the sample is between the colors on the color disks marked 8.25 and 9.75, the pH is satisfactory.
- If the color is yellower (yellowish-green) than the color disk marked 8.25, the pH value is low.
- If the color is bluer (bluish-green) than the color disk marked 9.75, the pH is high.

Testing for Chloride Concentration

11-37. To test for chloride concentration, the sample to be tested must contain chromate and be alkaline. The chloride test should therefore be run after the chromate and alkalinity have been determined (and corrected, if necessary). If the chloride test has to be done before the chromate content and alkalinity are adjusted, the sample must have the characteristic yellow color of chromate. To conduct the test to determine the chloride content, proceed as follows:

- Fill the cylindrical sample bottle to the 50-milliliter mark with a sample of the cooling water.
- Add one chloride test tablet, insert a stopper in the bottle, and shake until the tablet is completely dissolved.
- If the sample develops a reddish-brown color, the chloride content is below the maximum safe limit.
- If the sample becomes a cloudy yellow-green, the chloride content is high.

Ships not equipped to conduct the required tests for chromate, alkalinity (pH), and chloride concentration in engine cooling water may submit samples of cooling water to the nearest naval shipyard, tender, or advanced base laboratory for analysis. If test facilities are not available, the cooling system must be drained, flushed, refilled, and chemically treated at intervals of not more than 3 weeks.

SAFETY PRECAUTIONS

11-38. Diesel-driven ships built in recent years have (installed aboard) a new type distilling plant that uses diesel engine jacket water for a heat source. Although the jacket water does not come in direct contact with the distilled water, the chance of a leak is still present. Since chromate chemicals are considered health hazards, they are not to be used in engines that supply jacket water as a heat source to evaporators. In such installations, soluble oil inhibitors are used instead of chromate chemicals. Since chromate chemicals used in water treatment of cooling systems are classified as a health hazard, personnel should avoid any contact of skin or eyes with chromates in either a solid form or a solution. They should also avoid breathing chromate dust or solution spray. Those involved in handling chromate chemicals should use available protective equipment. The protective equipment to be used (such as goggles, face shields, rubber gloves, aprons, and dust respirators) should be consistent with the type and degree of hazard involved. When the skin has come in contact with chromates, the affected areas should be washed with plenty of soap and water immediately after exposure. Suitable precautions must be taken to prevent contamination of the ship's potable freshwater system. Back flow of the engine cooling water through the filling connection must be prevented. Specifications require that an air gap remain between the freshwater supply and fill connection. This arrangement must not be altered. When the engine cooling system must be protected from freezing, glycol base antifreeze is used. The antifreeze solutions specified by the Navy contain their own inhibitors which adequately protect the cooling system from corrosion. Therefore, no additional inhibitors are required. When necessary to change from one type of water treatment to the other, completely drain and flush the system free of any chromates or glycol base antifreeze. This is to prevent any mixing of these materials.

Chapter 12

STARTING SYSTEMS

INTRODUCTION

12-1. The function of a diesel engine starting system is to bring the engine up to sufficient rotational speed so that the events of compression, ignition, and combustion occur in close sequence. In starting a diesel engine, the system has to turn the engine fast enough to obtain ignition through compression. It has to overcome the disadvantage of cold metal in combustion spaces and a certain amount of leakage or blowby, resulting from the loose fit of pistons that have not reached operating temperatures. Diesel engines must start under conditions that often pose special difficulties. Cold weather is probably the hardest starting circumstance. However, other factors such as moisture, salt air, and hard or infrequent use can deteriorate an engine. Also, when a diesel is not maintained and adjusted, it will defy efforts to start it. The three types of starting systems used with internal-combustion engines with which you will be concerned are the electric starting, the hydraulic, and the air starting systems. This chapter discusses these systems and also devices that aid in the ignition in diesel engines.

ELECTRIC STARTING SYSTEMS

12-2. Electric starting systems use direct current (DC). This is because electrical energy in this form can be stored in batteries and drawn upon when needed. The battery's electrical energy can be restored by charging the battery with an engine-driven generator. The following are the main components of an electric starting system:

- Starting motor and drives.
- Generator and storage battery.
- Associated control and protective devices.

STARTING MOTOR AND DRIVES

12-3. The starting motor for diesel engines operates on the same principal as a DC electric motor. The motor is designed to carry extremely heavy loads, but because it draws a high current (300 to 665 amperes), it tends to overheat quickly. To avoid overheating; never allow the starting motor to run for more than 30 seconds at a time. After 30 seconds, allow the motor to cool for 2 or 3 minutes before using again. To start a diesel engine, it must turn over rapidly to obtain sufficient heat to ignite the fuel. The starting motor is located near the flywheel. The drive gear on the starter is arranged so that it can mesh with the teeth on the flywheel when the starting switch is closed. The drive mechanism must function to transmit the turning power to the engine when the starting motor runs; to disconnect the starting motor from the engine immediately after the engine has started; and to provide a gear reduction ratio between the starting motor and the engine. The gear ratio between the driven pinion and the flywheel is usually about 15 to 1. This means that the starting motor shaft rotates 15 times as fast as the engine, or at 1,500 rpm, in order to turn the engine at a speed of 100 rpm. The drive mechanism must disengage the pinion from the flywheel immediately after the engine starts. After the engine starts, its speed may increase rapidly to approximately 1,500 rpm. If the drive pinion remained meshed with the flywheel and also locked with the shaft of the starting motor at a normal engine speed (1,500 rpm), the shaft would be spun at a rapid rate (22,500 to 30,000 rpm). At such speeds, the starting motor would be badly damaged.

Bendix Drive Mechanisms

12-4. The Bendix drive is an example of a starting motor drive mechanism used on such diesel engines as the General Motors Model 268A. The drive provides positive meshing of the drive pinion with the ring gear on the flywheel. Figure 12-1 shows a starting motor equipped with a Bendix drive, friction-clutch mechanism. The pinion of the Bendix drive is mounted on a spiral-threaded sleeve so that when the shaft of the motor turns, the threaded sleeve rotates within the pinion. This moves the pinion outward, causing it to mesh with the flywheel ring gear, thereby cranking the engine. A friction clutch absorbs the sudden shock when the gear meshes with the flywheel. As soon as the engine runs under its own power, the flywheel drives the Bendix gear at a higher speed than the shaft of the starting motor is rotating. This causes the pinion to be rotated in the opposite direction on the shaft spiral. This automatically disengages the pinion from the flywheel as soon as the engine starts. Special switches are required to carry the heavy current drawn by starting motors. Starting motors equipped with a Bendix drive use a heavy-duty solenoid (relay) switch (Figure 12-2) and a hand-operated starting switch. The former is used to open and close the motor-to-battery circuit and the latter to operate the solenoid switch. The starting switch is on the instrument panel and it may be of the push-button or lever type. The solenoid switch is mounted on and grounded to the starting motor housing to allow the heavy-current wires to be as short as possible. When the starting switch is operated, the solenoid is energized and the plunger is drawn into the core to complete the circuit between the battery and the starting motor. Operating precautions on the Bendix drive must be observed. There are times when the engine starts, throws the pinion out of mesh, and then stops. When the engine is coming to rest, it often rocks back part of a revolution. If at that moment an attempt is made to engage the pinion, the drive mechanism may be seriously damaged. Therefore, you must wait several seconds to make certain that the engine is completely stopped before you use the starting switch again. At times the pinion may fail to immediately engage after the starting motor has been energized. When this happens, you will not hear the engine turning over, but you will hear the starting motor develop a high-pitched whine. You should immediately de-energize the motor to prevent its overspeeding. This is because an electric starting motor operating under no-load conditions can quickly overspeed and cause serious damage. If the pinion is to engage and disengage freely, the sleeve and the pinion threads must be free of grease and dirt. The Bendix drive should be lubricated as part of routine maintenance in accordance with instructions in the manufacturer's technical manual (TM).

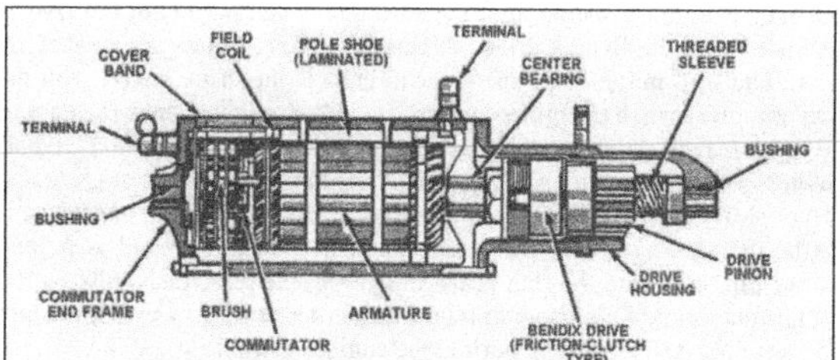

Figure 12-1. A Starting Motor With Bendix Drive

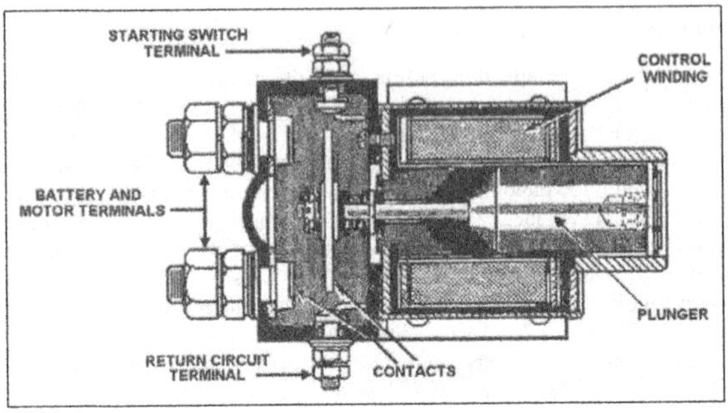

Figure 12-2. Solenoid Switch

Dyer Shift Drive Mechanisms

12-5. A starting motor assembly with the Dyer shift drive mechanism is shown in Figure 12-3. In this system, the drive pinion meshes with the flywheel ring gear before the starting motor switch is closed and before the motor shaft begins to rotate. This eliminates clashing of pinion teeth with the flywheel ring gear and the possibility of broken or burred teeth on either the ring gear or the drive pinion. Two views of the Dyer shift drive (a separate drive assembly and a disassembled mechanism) are shown in Figure 12-4. The upper end of the shift lever is linked to the solenoid switch mounted on top of the starter. The shift lever moves the entire drive mechanism axially along the motor shaft toward the flywheel. At the end of this movement, the drive pinion is meshed with the flywheel ring gear and has come to rest against the pinion stop. When the starting motor shaft begins to rotate, the shift sleeve returns to its original position, out of the way. The instant the engine starts, the flywheel ring gear attempts to spin the pinion faster than the motor shaft is turning. This causes the pinion and pinion guide to spin back out of mesh, and the pinion guide automatically locks the pinion in the disengaged position. Operation of the solenoid switch portion of the Dyer drive is similar to the action of the switch used with the Bendix drive. When the starting switch is operated, the solenoid closes the starting motor switch and also moves the shift lever to engage the drive pinion with the flywheel ring gear. Closing of the starting motor switch energizes the starting motor, which cranks the engine. The starting motor continues to rotate as long as the starting switch is operated. However, the pinion cannot engage the ring gear until the starting motor is stopped and the shift lever is allowed to return to its original position. Four stages of operation of the Dyer shift drive are shown in Figure 12-5. In view A, the mechanism is in the disengaged position. In view B, the starting switch has been operated and the solenoid is pulling the plunger in and beginning to move the pinion toward the ring gear. In view C, the pinion has fully meshed with the ring gear, but the motor shaft has not turned far enough to return the shift sleeve. In view D, the motor shaft has returned the shift sleeve to its original position and is starting the engine.

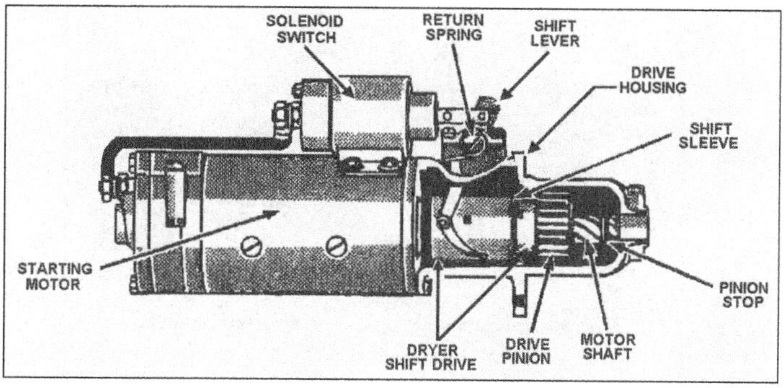

Figure 12-3. A Starting Motor With Dyer Shift Drive

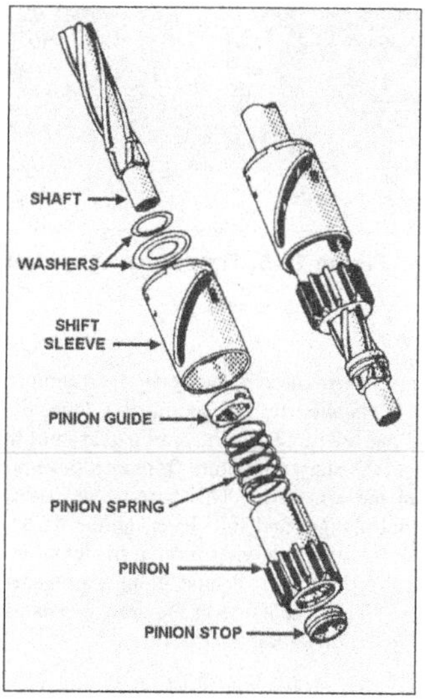

Figure 12-4. Dyer Shift Drive Mechanism

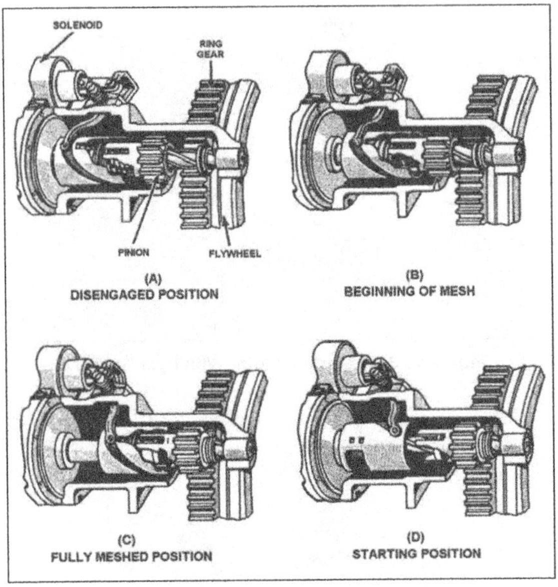

Figure 12-5. Dyer Drive Operation

Sprag Overrunning Clutch Drive

12-6. Another type of drive mechanism used is the Sprag overrunning clutch. This type drive is similar to the Dyer drive in that the pinion is engaged by the action of a lever attached to the solenoid plunger. Once engaged, the pinion will stay in mesh with the flywheel ring gear until the engine is started or until the solenoid switch is disengaged. To protect the starter armature from excessive speeds when the engine starts, the clutch "overruns" or turns faster than the armature. This permits the starter pinion to disengage itself from the flywheel ring gear. The solenoid plunger and shift lever, unlike the Dyer drive, are completely enclosed in a housing to protect them from water, dirt, and other foreign matter. An oil seal installed between the shaft and lever housing and a linkage seal around the solenoid plunger prevents the entry of transmission oil into the starter frame of the solenoid case. The nose house of the drive mechanism can be rotated to obtain a number of solenoid positions with respect to the mounting flange.

GENERATORS AND CONTROLS

12-7. The generator provides a source of electrical current to keep the storage battery charged. The design of a battery-charging generator depends on such factors as:

- Maximum electrical load.
- Design of the charging circuit.
- Type of service.
- Ratio of engine idling time to running time.
- Type of drive.
- Drive ratio (engine rpm to generator rpm).

To keep the battery fully charged, the discharge current must be balanced by a charging current supplied from an external source (such as a battery-charging generator). If the discharge current exceeds the charging current for an appreciable period, the battery will gradually lose its charge and then will not be capable of supplying current to the electrical system. Battery-charging generators sometimes are flange-mounted on the rear of the engine and driven from the timing gear train. However, usually they are cradle-mounted on the side of the engine and driven by a V-belt from the crankshaft pulley. These generators are rated according to the particular application and are designed for clockwise or counterclockwise rotation. They are supplied for use with either 6-, 12-, or 24-volt systems. Battery-charging generators used with the electrical systems in small craft are usually DC generators similar to the one shown in Figure 12-6. The alternating-current (AC) generator (alternator) is installed on some engines because it is designed to produce a constant output over a speed range that varies from idle to top engine speed. A voltage regulator controls the output voltage of the alternator and prevents overcharging the battery. A current-limiting device associated with the regulator protects the alternator and rectifier. An alternator-equipped, battery-charging system is shown in Figure 12-7. The DC battery-charging generator systems also use a control to regulate output of the generator. The type of control used is determined by the design of the generator. Although such controls may control both current and voltage, they are commonly called voltage regulators. Voltage regulators differ considerably in design but all serve essentially the same purposes. When the DC generator is operating and the electrical load is heavy, the generator output is connected through the voltage regulator to the battery. However, when the electrical load is light or there is no load and the battery is fully charged, the voltage regulator operates to prevent the generator's output reaching the battery. The voltage regulator also protects the components of the battery-charging system. It prevents overcharging of the battery. It also prevents the current's flow from the battery to the DC generator when the generator output voltage, which is determined by the speed of the engine, is less than the voltage of the battery. The voltage regulator protects the generator by preventing the output's exceeding the design limits of the generator.

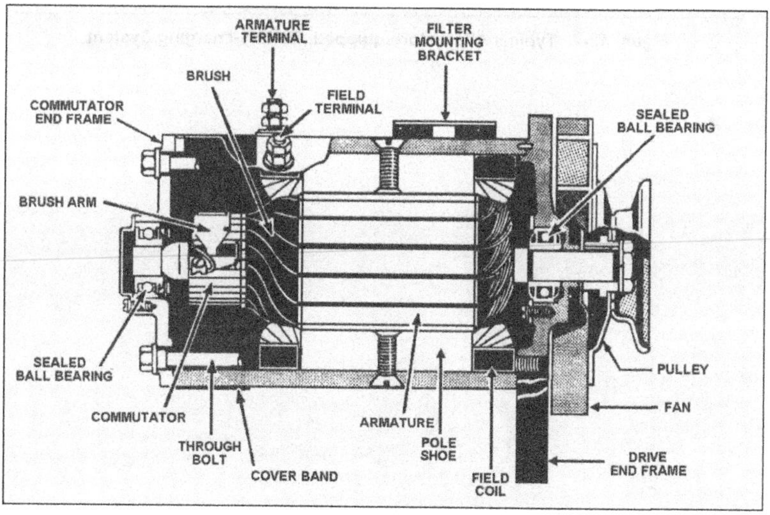

Figure 12-6. Typical DC Generator

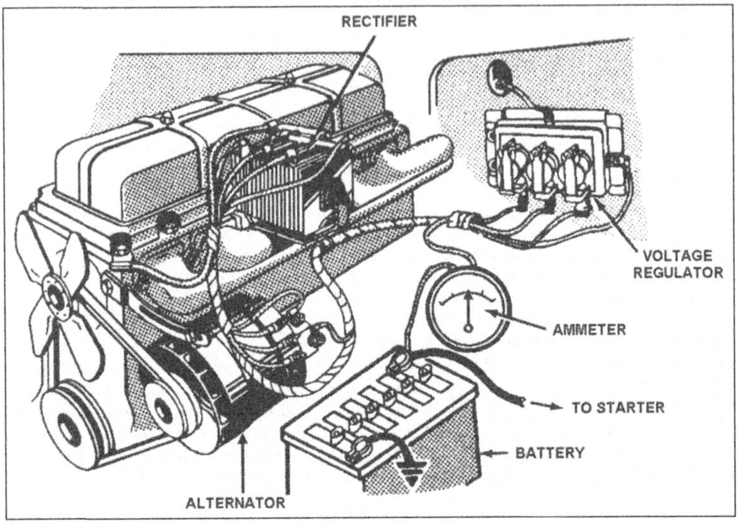

Figure 12-7. Typical Alternator-equipped, Battery-charging System

GENERATOR MAINTENANCE

12-8. For long and satisfactory service of generators, proper maintenance is necessary. The most important factor in maintenance is to keep the equipment clean and free of oil, water, dirt, and other foreign particles. It is also important to keep insulation clean. Dust and dirt tend to block the ventilation passage and increase resistance to the dissipation of heat, thereby causing local or general overheating. If the particles are conducting or forming a conducting paste, the windings may be eventually grounded or short-circuited. Four methods of cleaning generators are by wiping, by using compressed air, by suction, and with solvent. In using compressed air, remember it should always be clean and free of moisture. Cleaning electrical equipment with solvent should be avoided whenever possible. However, its use is necessary at times to remove gummy or greasy substances. Solvents containing gasoline or benzine must not be used in any circumstances. The use of alcohol will injure most types of insulating varnishes. Inhibited methylchloroform (trichloroethylene) is one of the principally approved solvents for cleaning electrical equipment. However, it should be used only in a well-ventilated space. Improper lubrication is a frequent cause of generator failures. Excess grease or oil can be forced through the bearing housing seals and into the commutators, eventually resulting in grounds or short circuits. The excessive quantity and pressure of grease in the bearing housing results in churning, high temperatures, and rapid deterioration of the grease and bearings. The brushes used in generators are one or more plates of carbon bearing against a commutator. They provide a passage of electrical current for an external circuit. The brushes are held in place by brush holders mounted on studs or brackets attached to the brush mounting ring. The brushes should be checked frequently to ensure that they are in good condition and free to move in their holders. The brushes should move smoothly without vibration. If the brushes do not slide smoothly on the commutator, the current output will become erratic. The brushes should be renewed when they are worn to half the original length. Brushes can be accurately seated against the commutator by inserting a strip of number 1 sandpaper, approximately the width of the commutator with the rough side up, between the commutator and the brushes. With the sandpaper held firmly against the commutator and the brushes held in place by normal spring tension, the sandpaper is pulled in the direction of normal rotation of the commutator. When the sandpaper is returned for another pull, the brushes must be lifted. The brushes should be finished with a finer grade of sandpaper, and all dust particles should be removed after the sanding (Figure 12-8). After being used for approximately 2 weeks, the commutator should develop a uniform, glazed, dark brown color on the places where the brushes ride. If a non-uniform or bluish color appears, improper commutation conditions are indicated. Periodic inspections and proper cleaning practices will keep the commutator trouble to a minimum. One of the most effective ways to clean a commutator is to apply a canvas wiper while the commutator is being turned. However, small generators will have to be disassembled for a thorough job. When the generator is disassembled, a toothbrush can be used to clean out the slots. If the commutator is worn or grooved excessively, it can be trued by turning it on a lathe (Figure 12-9).

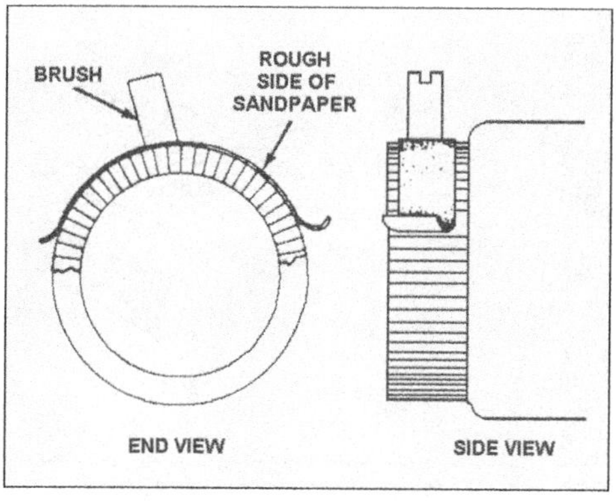

Figure 12-8. Method of Sanding Brushes

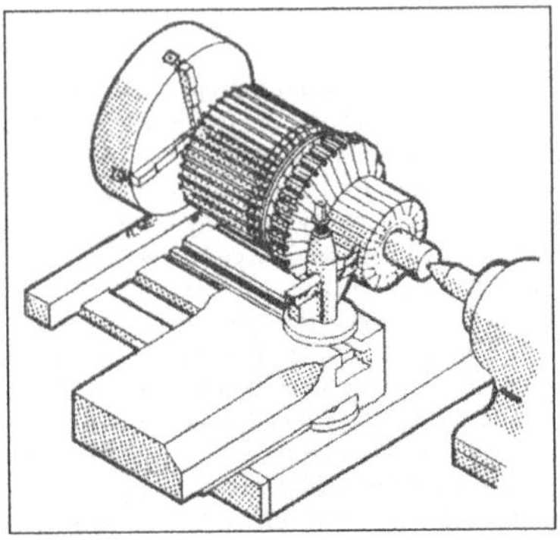

Figure 12-9. Truing Commutator by Turning

HYDRAULIC STARTING SYSTEMS

12-9. Several types of hydraulic starting systems are in use. The system used in most installations consists of a hydraulic starting motor, a piston-type accumulator, a manually-operated hydraulic pump, an engine-driven hydraulic pump, and a reservoir for the hydraulic fluid. One type of hydraulic starting system (Aeroproducts, GM) is shown in Figure 12-10. Hydraulic pressure is obtained in the accumulator by the manually-operated hand pump or from the engine-driven pump when the engine is operating. When the starting lever is operated, the control valve allows hydraulic oil (under pressure) from the accumulator to pass through the hydraulic starting motor, thereby cranking the engine. When the starting lever is released, spring action disengages the starting pinion and closes the control valve. This stops the flow of hydraulic oil from the accumulator. The starter is protected from the high speeds of the engine by the action of an overrunning clutch. Hydraulic starting systems are being used on some smaller diesel engines. This type system can be applied to most engines now in service without modification other than the clutch and pinion assembly. This assembly must be changed when conversion is made from a left-hand to a right-hand rotation.

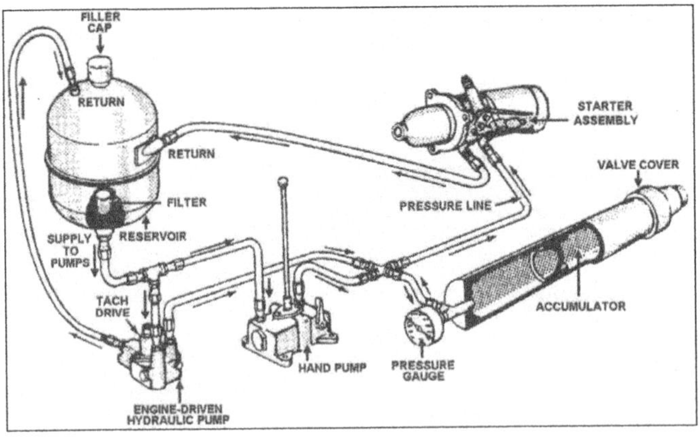

Figure 12-10. Hydraulic Starter System

AIR-STARTING SYSTEMS

12-10. Most modern large diesel engines are started by admitting compressed air into the engine cylinders at a pressure capable of turning over the engine. The process is continued until the pistons have built up sufficient compression heat to cause combustion. The pressure used in air-starting systems ranges from 250 to 600 psi. Some large engines and several smaller engines are provided with starting motors driven by air. These motors are similar to those used to drive such equipment as large pneumatic drills and engine jacking motors. Air-starting motors are usually driven by air pressures varying from 90 to 200 psi.

SOURCE OF STARTING AIR

12-11. Starting air comes directly from the ship's high-pressure, air-service line or from starting-air flasks which are included in some systems to store starting air. From either source, the air on its way to the engine must pass through a pressure-reducing valve which reduces the higher pressure to the operating pressure required to start a particular engine. A relief valve is installed in the line between the reducing valve and the engine. The relief valve is normally set to open at 25 to 50 psi in excess of the required starting-air pressure. If the air pressure leaving the reducing valve is too high, the relief valve will protect the engine by releasing air in excess of the value for which it is set and will permit only air at safe pressure to reach the cylinders. One type of pressure-reducing valve is the regulator shown in Figure 12-11, in which compressed air, sealed in a dome, furnishes the regulating pressure that actuates the valve. The compressed air in the dome performs the same function as a spring used in a more common type of regulating valve. The dome is tightly secured to the valve body, which is separated into an upper flow-pressure outlet) and a lower (high-pressure inlet) chamber by the main valve. At the top of the valve stem is another chamber, containing a rubber diaphragm and a metal diaphragm plate, with an opening leading to the low-pressure outlet chamber. When the outlet pressure drops below the pressure in the dome, air in the dome forces the diaphragm and diaphragm plate down on the valve stem. This partially opens the valve and permits high-pressure air to pass the valve seat into the low-pressure outlet and into the space under the diaphragm. As soon as the pressure under the diaphragm is equal to that in the dome, the diaphragm returns to its normal position and the valve is forced shut by the high-pressure air acting on the valve head. During the starting cycle, the regulator valve is continuously and rapidly adjusting for changes in air pressure by partially opening and partially closing to maintain a safe, constant starting pressure. When the engine starts and there is no longer a demand for air, pressure builds up in a low-pressure chamber to equal the pressure in the dome, and the valve closes completely. High-pressure air entering the valve body is filtered through a screen to prevent the entrance of any particle of dirt which would prevent the valve from seating properly. The screen is held in position around the space under the valve head by the threaded valve seat bushing. This screen should be removed and cleaned periodically to ensure an unrestricted flow of air. If particles of dirt are permitted to accumulate in the screen, the resultant buildup of high air pressure may tear the screen from its position and force it into the working parts, causing damage to the valve seat. To obtain air for the original charging of the dome, two needle valves are opened to the high-pressure chamber of the valve body. As soon as the desired pressure (indicated by the gauge on the discharge side of the regulator) is reached, the needle valves must be closed. The dome will then regulate and maintain the discharge of air at that pressure.

STARTING MECHANISM

12-12. All air-starting systems basically operate similarly and contain the same elements. If you have enough thorough knowledge of the mechanism of one air-starting system, you should be able to understand the principles of operation of other air-starting systems. The following is a description of the system used on General Motors engines. The engine air-starting system used on GM engines (Figure 12-12) is known as the separate distributor type. This is because the starting-air distributor valve (Figure 12-13) is a separate unit for each cylinder. Each distributor valve is individually operated by its cam on the camshaft. Of the 16 cylinders, 8 are air-started (6 in one bank and 2 in the other). However, all of the cylinder heads in both banks are equipped with an air-starting check valve to maintain full interchangeability. On the cylinders that are not air-started, the inlet opening of the check valve is sealed with a removable plug. Air is supplied to the air-starting control valve (Figure 12-12) from the air supply line. When the air-starting control valve is opened by a hand lever, air is admitted to the starting air manifold (a steel pipe extending the full length of the top deck of the engine and located below the exhaust manifold). The starting-air manifold is connected by air lines to each of the starting-air distributor valves. The distributor valves are opened in engine cylinder firing order by their cams on the camshafts. This admits air into the lines that connect each distributor valve to its air-starting check valve (Figure 12-13). When the distributor valve admits air into the lines leading to the air-starting check valve, the pressure opens the check valve, thereby admitting air into the combustion chamber. Air pressure moves the pistons and turns the crankshaft until there is sufficient compression for combustion. Combustion pressure and exhaust gases are kept from backing into the air-starting system by the check valves. As soon as the engine is firing, the hand lever is released and spring pressure closes the air-starting control valve. This shuts off the supply of starting air to the engine. The air-starting control valve is mounted on a bracket bolted to the camshaft drive cover near the hand control lever. It is a poppet valve, opened manually by a lever and closed by a spring. A plug in the valve body holds the spring against the valve head. The valve stem guide is a bronze bushing pressed into the body. A spring and head, placed over the valve stem where it projects from the body, returns the hand lever to the valve's closed position. The hand lever and the operating lever stop are keyed to a shaft in the bracket. A safety device prevents opening of the air-starting control valve while the engine jacking gear is engaged.

Starting-air Distributor Valves

12-13. Each of the eight cylinders that are air-started is equipped with a starting-air distributor valve (Figure 12-13). The starting-air distributor valves (timing valves) are of the poppet type, with forged steel bodies that bolt to the camshaft intermediate covers. The valve is held closed by spring pressure bearing against the top of the valve and is guided in the hollow end of the cam follower which rides on the camshaft air-starting cam. The cam follower is guided in a bronze bushing pressed into the valve body. A lock pin keeps the cam follower in the body. When cam action opens the valve, starting air passes from the air manifold through a chamber in the valve body above the valve head into a line leading to the air-starting check valve in the cylinder head. The cam action opens the valve in the proper valve sequence. The cam follower is lubricated by oil splashed by the cam from the cam pocket.

Air-starting Check Valves

12-14. The air-starting check valve shown in Figure 12-14 is a poppet valve located in the cylinder head. The valve body fits into a recess in the cylinder and is held in place by a capnut that screws into the cylinder head and bears on top of the body. The valve body contains the valve seat and serves as a valve guide. A synthetic rubber seal ring located above the inlet port prevents air leaking to the outside of the valve body. The valve is closed by a spring surrounding an upper portion of the valve stem. This spring fits into a recess in the valve body. It exerts pressure on the spring seat, which is locked to the valve stem by two half-round pieces (valve locks) fitted into a groove on the valve stem. The valve opens into a small chamber (in the cylinder head) with a short, open passage to the cylinder. When the starting-air distributor valve admits air into the line leading to the air-starting check valve, the air passes into a chamber above the valve seat. The pressure of this air opens the check valve and allows the air to pass into the cylinder and move the piston. When the starting-air distributor valve closes, pressure in the valve chamber drops, and spring tension closes the air-starting check valve. When combustion occurs, the air-starting check valve remains closed because the pressure in the combustion chamber is greater than the pressure of the starting air that actuates the check valve. This prevents the backing of exhaust gases and combustion pressures into the air-starting system.

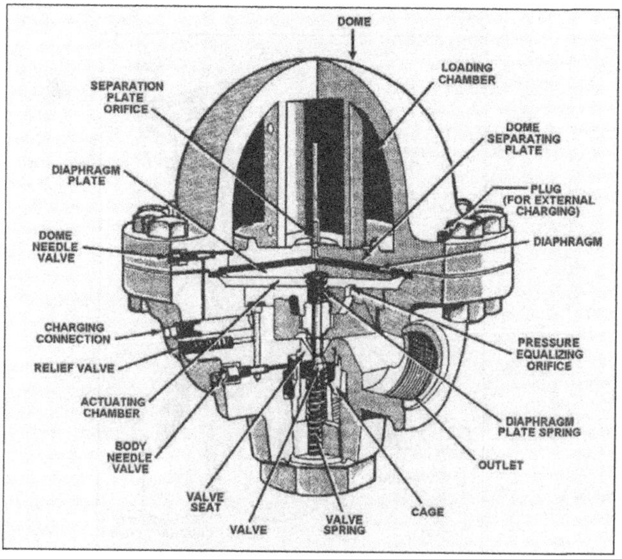

Figure 12-11. Regulator Valve

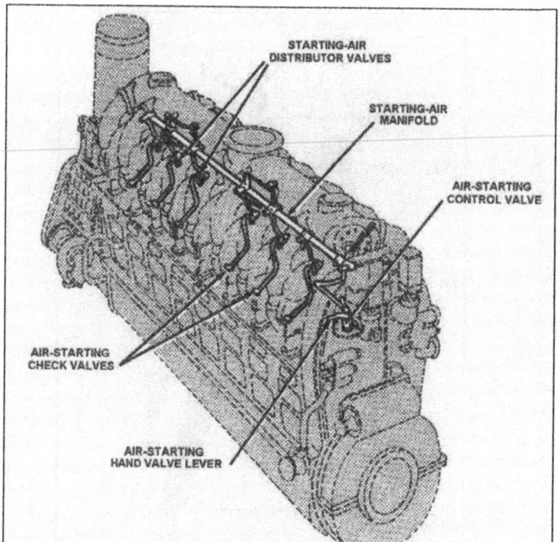

Figure 12-12. GM Engine Air-starting System

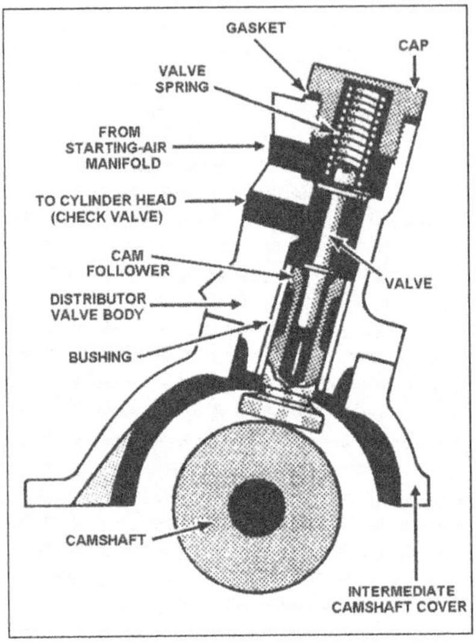

Figure 12-13. Starting-air Distributor Valve - GM

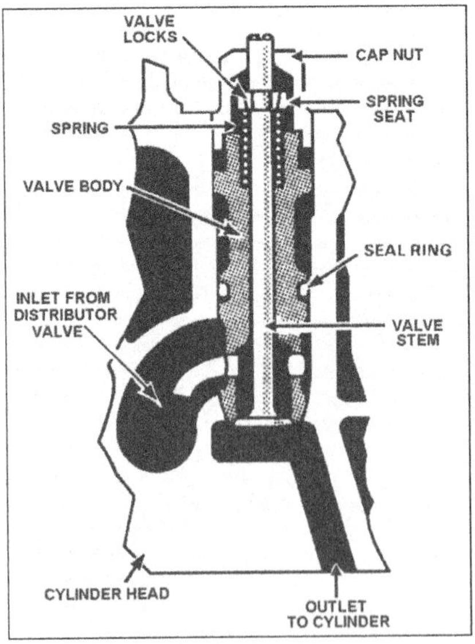

Figure 12-14. Air-starting Check Valve - GM

Chapter 13

TRANSMISSION OF ENGINE POWER

INTRODUCTION

13-1. If the power developed by an engine is to perform useful work, some means must be provided to transmit the power from the engine (driving unit) to such loads (driven units) as the propeller(s) of a ship or boat. This chapter provides general information on how the force available at the crankshaft of an engine is transmitted to a point where it will perform useful work. The combination of devices used to transmit engine power to a driven unit is commonly called a drive mechanism.

TRANSMISSION OF ENGINE POWER RELATED FACTORS

13-2. The fundamental characteristics of an internal-combustion engine make it necessary, in many cases, for the drive mechanism to change both the speed and the direction of shaft rotation in the driven mechanism. Various methods exist by which changes in speed and direction may be made during the transmission of power from the driving unit to the driven unit. However, in most of the installations with which you will be working, the job is done by a drive mechanism consisting principally of gears and shafts. The process of transmitting engine power to a point where it can be used in performing useful work involves a number of factors. The following are the three factors in regards to the transmission of engine power.

TORQUE

13-3. The force which tends to cause a rotational movement of an object is called torque (or twist). The crankshaft of an engine supplies a twisting force to the gears and shafts which transmit power to the driven unit. Gears are used to increase or decrease torque. For example, an engine may not produce enough torque to turn the shaft of a driven machine if the connection between the driving and driven units is direct, or solid. However, if the right combination of gears is installed between the engine and the driven unit, torque is increased and the twisting force is then sufficient to operate the driven unit.

SPEED

13-4. Engine speed is related to torque and to transmission of engine power. For maximum efficiency, an engine must operate at a certain speed. In some installations the engine may have to operate at a higher speed than that required for efficient operation of the driven unit. In other installations, the speed of the engine may have to be lower than the speed of the driven unit. Through a combination of gears, the speed of the driven unit can be increased or decreased so that both driving and driven units operate at their most efficient speeds; that is, so that the proper speed ratio exists between the units.

SPEED RATIO AND GEAR RATIO

13-5. The terms, "speed ratio" and "gear ratio" are frequently used in discussing gear-type mechanisms. Both ratios are determined by dividing the number of teeth on the driven gear by the number of teeth on the driving gear. For example, assume that the crankshaft of a particular engine is fitted with a driving gear which is half as large as the meshing, driven gear. If the driving gear has 10 teeth and the driven gear has 20 teeth, the gear ratio is 2 to 1. Every revolution of the driving gear will cause the driven gear to revolve through only half a turn. Therefore, if the engine is operating at 2,000 rpm, the speed of the driven gear will be only 1,000 rpm. The speed ratio then is 2 to 1. This arrangement doubles the torque on the shaft of the driven unit. Therefore, the speed of the driven unit is only half that of the engine. However, if the driving gear had 20 teeth and the driven gear had 10 teeth, the speed ratio would be 1 to 2 and the speed of the driven gear would be doubled.

The rule applies equally well when an odd number of teeth are involved. If the ratio of the teeth is 37 to 15, the speed ratio is slightly less than 2.47 to 1. In other words, the driving gear will turn through almost two and one-half revolutions while the driven gear makes one revolution. The gear with the greater number of teeth, which will always revolve more slowly than the gear with the smaller number of teeth, will produce the greater torque. Gear trains which change speed, always change torque; when speed increases, the torque decreases proportionally.

TYPES OF DRIVE MECHANISMS

13-6. It may be necessary to change the torque and speed of an engine to satisfy the torque and speed requirements of the driven mechanism. The term, "indirect drive," as used in this chapter, describes a drive mechanism which changes speed and torque. Drives of this type are common to many marine engine installations. Where the speed and torque of an engine need not be changed to drive a machine satisfactorily, the mechanism used is a direct drive. Drives of this type are commonly used where the engine furnishes power for operation of such auxiliaries as generators and pumps.

INDIRECT DRIVE MECHANISMS

13-7. Most engine-powered ships and many boats have indirect drive mechanisms. With this drive, the power developed by the engines is transmitted to the propellers indirectly through an intermediate mechanism which reduces the shaft speed. Speed reduction may be accomplished mechanically, by a combination of gears, or by electrical means. In the propulsion plants of some diesel-driven ships, there is no mechanical connection between the engines and the propellers. In such plants, the diesel engines are connected directly to generators. The electricity produced by such an engine-driven generator is transmitted through cables to a motor. The motor is connected to the propeller shaft either directly or indirectly through a reduction gear. When a reduction gear is included in a diesel-electric drive, the gear is located between the motor and the propeller. The generator and the motor of a diesel-electric drive may be of the AC type or the DC type. However, almost all diesel-electric drives in the Army are of the DC type. Since the speed of a DC motor varies directly with the voltage furnished by the generator, the control system of an electric drive is so arranged that the generator voltage can be changed at any time. An increase or decrease in generator voltage is used as a means of controlling the speed of the propeller. Changes in generator voltage may be brought about by electrical means, by changes in engine speed, and by a combination of these methods. The controls of an electric drive may be in a location remote from the engine, such as the pilothouse. In an electric drive, the direction of rotation of the propeller is not reversed through the use of gears. The electrical system is arranged so that the flow of current through the motor can be reversed. This reversal of current flow causes the motor to revolve in the opposite direction. Therefore, the direction of rotation of the motor and of the propeller can be controlled by manipulating the electrical controls.

DIRECT DRIVE MECHANISMS

13-8. In some marine engine installations, power from the engine is transmitted to the drive unit without a change in shaft speed; that is, by direct drive. In a direct drive, the connection between the engine and the driven unit may consist of a solid coupling, a flexible coupling, or a combination of both. Depending on the type of installation, a clutch may or may not be included in a direct drive. In some installations, a reverse gear is included.

Solid Couplings

13-9. These couplings vary considerably in design. Some solid couplings consist of two flanges bolted solidly together (Figure 13-1).In other direct drives, the driven unit is attached directly to the engine crankshaft. Solid couplings offer a positive means of transmitting torque from the crankshaft of an engine. However, a solid connection does not allow for any misalignment, nor does it absorb any of the torsional vibration transmitted from the engine crankshaft or the drive unit.

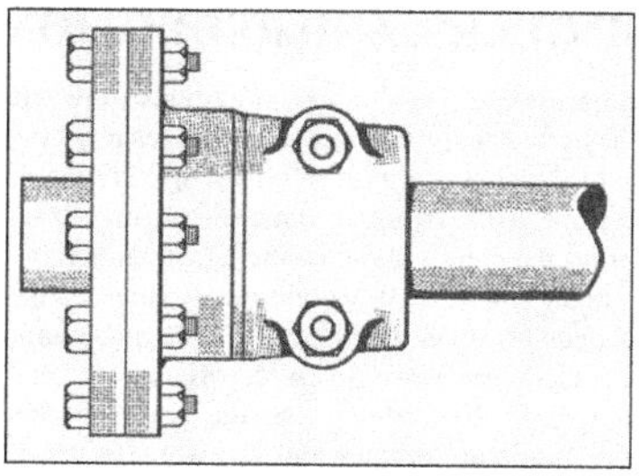

Figure 13-1. Flange-type Solid Coupling

Flexible Couplings

13-10. Since solid couplings will not absorb vibration and will not permit any misalignment, most direct drives consist of a flange-type coupling which is used in connection with a flexible coupling. Flexible connections are common to the drives of many auxiliaries (such as engine-generator sets). Flexible couplings are also used in indirect drives to connect the engine to the drive mechanism. The two solid halves of a flexible coupling are joined by a flexible element. The flexible element is made of rubber, neoprene, or steel springs. Two views of one type of flexible coupling are shown in Figure 13-2. The coupling shown has radial spring packs as the flexible element. Power from the engine is transmitted from the inner ring, or spring holder, of the coupling through a number of spring packs to the outer spring holder, or driven member. A large driving disk connects the outer spring holder to the flange on the driven shaft. The pilot on the end of the crankshaft fits into a bronze bearing on the outer driving disk to center the driven shaft. The ring gear of the jacking mechanism is pressed onto the rim of the outer spring holder. The inner driving disk, through which the camshaft gear train is driven, is fastened to the outer spring holder. A splined ring gear is bolted to the inner driving disk. This helical, internal gear fits on the outer part of the crankshaft gear and forms an elastic drive through the crankshaft gear which rides on the crankshaft. The splined ring gear is split and the two parts are bolted together with a spacer block at each split joint. The parts of the coupling shown in Figure 13-2 are lubricated by oil flowing from the bearing bore on the crankshaft gear through the pilot bearing.

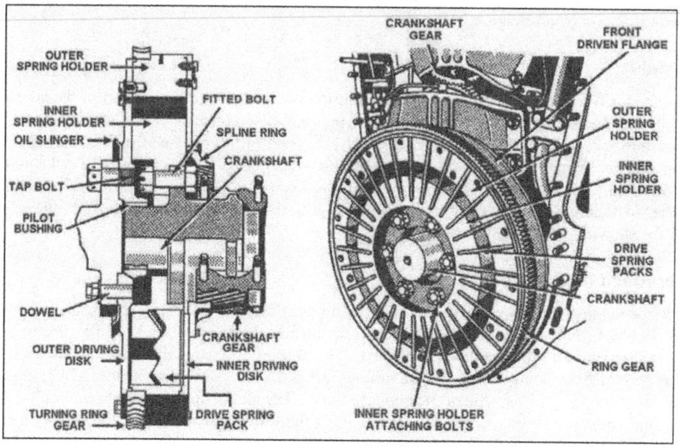

Figure 13-2. Flexible Coupling (GM 278A)

CLUTCHES, REVERSE GEARS, AND REDUCTION GEARS

13-11. Clutches may be used on direct-drive propulsion engines to disconnect the engine from the propeller shaft. In small engines, clutches are usually combined with reverse gears and used for maneuvering the ship. In large engines, special types of clutches are used to obtain special coupling or control characteristics and to prevent torsional vibration. Reverse gears are used on marine engines to reverse the direction of rotation of the propeller shaft when maneuvering the ship without changing the direction of rotation of the engine. They are used principally on relatively small engines. If a high-output engine has a reverse gear, the gear is used for low-speed operation only and does not have full-load and full-speed capacity. For maneuvering ships with large direct-propulsion engines, the engines are reversed. Reduction gears are used to obtain low propeller-shaft speed with a high engine speed. When doing this, the gears correlate two conflicting requirements of a marine engine installation. For minimum weight and size for a given power output, engines must have a relatively high rotative speed. For maximum efficiency, propellers must rotate at a relatively low speed, particularly where high thrust capacity is desired. The Army uses many types of transmissions. In general, this discussion covers the operation and maintenance of transmissions using clutches of the friction, pneumatic, fluid or hydraulic, dog, and electromagnetic type. These may be found on Army marine installations. Information concerning a particular unit can be obtained from the manufacturer's technical manual (TM) for that specific installation.

FRICTION CLUTCHES AND GEAR ASSEMBLIES

13-12. Friction clutches are most commonly used with small, high-speed engines up to 500 hp. However, certain friction clutches, in combination with a jaw-type clutch, are used with engines up to 1,400 hp. Pneumatic clutches with a cylindrical friction surface are used with engines up to 2,000 hp. Friction clutches are of two general styles (disk and band). They can be classified into dry and wet types, depending on whether they operate with or without a lubricant. For engagement of the friction clutches, force-producing friction can be obtained either by mechanically jamming the friction surfaces together by some toggle-action linkage, or through stiff springs (coil, leaf, or flat disk). The following discussion covers operation of friction clutches.

TWIN-DISK CLUTCH AND GEAR MECHANISM

13-13. One of the several types of transmissions used is the Gray Marine transmission (see Figure 13-3). Gray Marine high-speed diesel engines are generally equipped with a combination clutch and reverse and reduction gear unit. This is all contained in a single housing at the after end of the engine. A sectional view of this mechanism is also shown in Figure 13-3.

Clutch Assembly

13-14. The clutch assembly of the Gray Marine transmission is contained in the part of the housing nearest the engine (the left end of Figure 13-3). It is a dry-type, twin-disk clutch with two driving disks. Each disk is connected through shafting to a separate reduction gear train in the after part of the housing. One disk and reduction train is for reverse rotation of the shaft and propeller. The other disk and reduction train are for forward rotation. The forward and reverse gear trains for Gray Marine engines are shown in Figure 13-4. Notice that in Figure 13-3 and Figure 13-4 that the gear trains are different; however, the operation of the mechanisms shown is basically the same.

Clutch Operating Lever

13-15. Since the gears for forward and reverse rotation of the twin-disk clutch and gear mechanism remain in mesh at all times, there is no shifting of gears. In shifting the mechanism, only the floating plate, located between the forward and reverse disks, is shifted. The shifting mechanism is a sliding sleeve which does not rotate but has a loose sliding fit around the hollow forward shaft. A throwout fork (yoke) engages a pair of shifter blocks pinned on either side of the sliding sleeve. The clutch operating lever moves the throwout fork, which in turn shifts the sliding sleeve lengthwise along the forward shaft. When the operating lever is placed forward, the sliding sleeve is forced backward. In this position, the linkage of the spring-loaded mechanism pulls the floating pressure plate against the forward disk and causes forward rotation. When the operating lever is pulled back as far as it will go, the sliding sleeve is pushed forward. In this position, the floating pressure plate engages the reverse disk and back plate for reverse rotation. The clutch has a positive neutral which is set by placing the operating lever in a middle position. Then the sliding sleeve is also in a middle position and the floating plate rotates freely between the two clutch disks. The only control the operator has is to cause the floating plate to bear heavily against either the forward disk or the reverse disk, or to put the floating plate in the positive neutral position so that it rotates freely between the two disks.

Forward Rotation

13-16. The two clutch disks shown in Figure 13-3 are separated from each other by the floating plate. When the floating plate presses against the forward disk, this disk is pressed, in turn, against the front plate, which is bolted to and rotates with the engine flywheel. The friction disk immediately begins to rotate with the front plate at engine speed. The forward disk has internal teeth; the forward sleeve has an integral external gear. The forward-sleeve shaft transmits the rotation to the propeller shaft through the two-gear train (shown in the upper view of Figure 13-4).

Reverse Rotation

13-17. When the floating plate is pressed against the reverse disk, reverse rotation is obtained. In turn, the reverse disk is pressed against the backplate, which is also bolted to the engine flywheel (Figure 13-3). At engine speed, the reverse disk begins to rotate with the back plate. The reverse disk has internal teeth; the reverse sleeve has an integral external gear. The reverse shaft transmits the rotation through the three-gear train (shown in the lower view of Figure 13-4). Notice the presence of an idler gear in the reverse-gear train. This gear reverses the direction of the propeller shaft rotation.

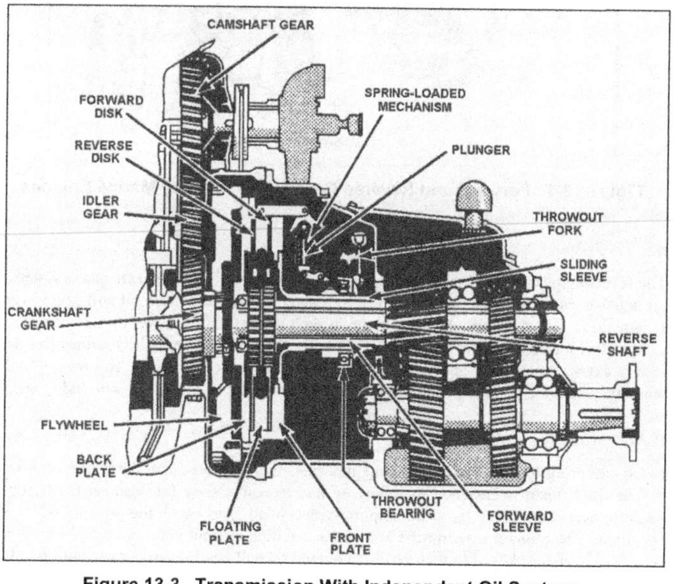

Figure 13-3. Transmission With Independent Oil Systems

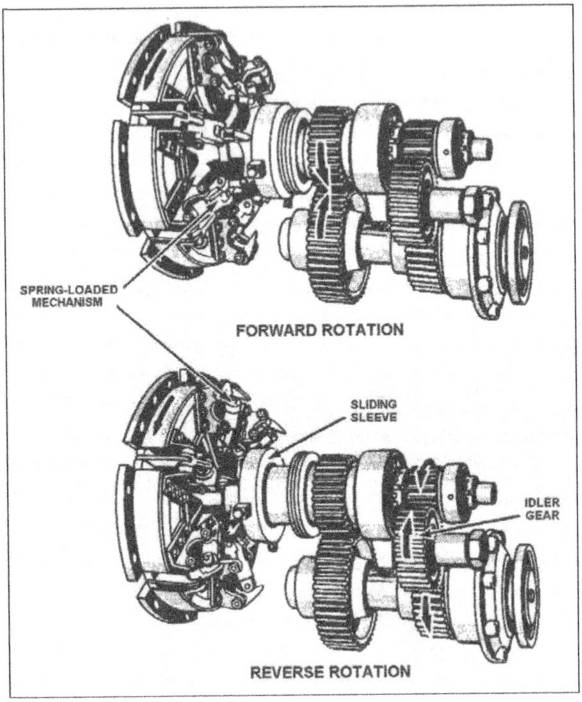

Figure 13-4. Forward and Reverse Gear Trains for Gray Marine Engines

Lubrication

13-18. The reversing gear unit is lubricated separately from the engine by its own splash system. The oil level of the gear housing should never be kept over the high mark because too much oil will cause overheating of the gear unit. The oil is cooled by air which is blown through the baffled top cover by the rotating clutch. Grease fittings are installed for bearings not lubricated by the oil. Do not overgrease the bearings in a dry-type friction clutch because excess grease may get on the friction plates and cause slippage. To prevent binding caused by rust, clutch parts which are not equipped with grease fittings should be lubricated with light machine oil.

Operational Adjustments

13-19. Since the spring-loaded, clutch-operating mechanism is pressure-set at the factory, it is not necessary to adjust it. The mechanism is designed to follow up and to compensate for wear on the friction plates. The simplest way to determine when the disks require replacement is to check the position of the plunger in the engaged position. The plunger is permitted to travel a specified amount in accordance with the manufacturer's instructions. The gears are keyed to their shafts, mounted on ball bearings in permanently fixed centers. They require no adjustment.

FRICTION CLUTCH TROUBLES

13-20. Difficulties which may necessitate repair of friction clutches will vary. These difficulties will depend on the classification of the clutch (wet or dry or disk or band). It also depends on the method of operation (hand, hydraulic, pneumatic, or vacuum). The troubles discussed here (slippage and wear, freezing, and noise) are common to most friction clutches.

Slippage and Wear

13-21. It is difficult to consider slippage and wear separately, since each can be the cause of the other and each intensifies the effect of the other. Slippage generally occurs at high engine speed (when the engine is delivering the greatest torque). It results in lowered efficiency, loss of power, and rapid wear of the clutch friction surfaces. Several possible causes of clutch slippage are:

- Wear.
- Insufficient pressure.
- Overload.
- Fouling.

Over a period of operation, a normal amount of wear results from extended engaging and disengaging of the surfaces. If the surfaces are rough, wear will be excessive. When a friction clutch is being overhauled, the bearing surfaces must not be damaged. Small nicks must be stoned. If the scoring is serious, the damaged surface must be refinished or the part must be replaced. Since water has a deteriorating effect on clutch facings, every effort must be made to keep dry-type clutches free of water. Engaging the clutch while racing the engine may be another cause of excessive wear. It will strain the entire drive system. Some types of friction clutches are not adjustable but depend on the initial compression in the pressure springs. Twin-disk clutches (such as those on Gray Marine engines) are equipped with a spring-pressure system. With this system, it is important to check the springs whenever the clutch is disassembled. Checking should be done with a spring tester. However, if none is available, a check on the free lengths of the springs will give an indication of their condition. The manufacturer's TM should be consulted for the proper values. When an engine is overloaded, torque may be increased to such an extent that slippage will occur. This trouble can be prevented by keeping the load within specified limits. Whenever an engine is fully loaded, watch for symptoms which indicate slippage. A dry-type clutch may slip when the lining surfaces become fouled with oil, grease, or water. Oil and grease usually reach the clutch surfaces because of such careless maintenance procedures as using an excess of grease in lubricating or overfilling the gear case with oil. When oil in a gear case foams, there is leakage from the shaft bearings. Foaming may be caused by air leaks in the oil suction line or by overfilling. When foaming occurs, inspect all lube oil lines for air leaks and check for proper oil level. When filling a reduction gear case, add only enough oil to bring the level up to the FULL mark. Do not add or measure oil when the unit is in operation, since you will not get an accurate oil reading. In a twin-disk clutch installation, leakage of oil from the rear main bearing of the engine may cause oil to appear on the clutch surfaces. The leakage may be caused by the following:

- Excessive bearing clearance.
- Overfilling of the engine crankcase.
- Plugged crankcase breather cap.
- Excessive piston blowby.

The crankcase breather cap must be cleaned periodically to prevent its becoming clogged. Another source of fouling is from grease that may be deposited on a dry-type clutch while it is being overhauled. The parts should not be handled with greasy hands. Any grease that may be deposited should be removed with an approved cleaner. For the pneumatically operated friction clutches where rubber parts are used, only a dry cloth should be used on clutch facings. When clutch slippage becomes apparent, it should be corrected immediately. If slippage occurs, the clutch surfaces are probably worn and the thickness of the clutch linings should be checked. When a lining is worn excessively, it should be replaced. Tightening the adjusting device will not compensate for the excessive surface wear. Instead, such adjustments may lead to scoring of the mating clutch surfaces.

Freezing

13-22. When a clutch fails to disengage, it is said to be frozen. Failure to disengage may be caused by a defective clutch mechanism or by water in the clutch linings. When a clutch becomes frozen, the operating mechanism should be inspected. Check the control rods for obstructions or loose connections. Also check for excessive clearances in the throwout bearing pressure plate, the pivots, and the toggles. In a twin-disk clutch, warped disks will cause the clutch to freeze. Warped disks are caused by extended running in neutral position. If a dry-type clutch is equipped with molded-type clutch linings, moisture will cause the linings to swell and to become soft. When this occurs, many linings tend to stick to the mating surfaces. Every effort should be made to prevent moisture from getting into the clutch linings. If a molded lining becomes wet, it should be permitted to dry in the disengaged position. If allowed to dry in the engaged position, the possibility of sticking increases.

Noise

13-23. Dry-type clutches may produce a chattering noise when being engaged. Excessive clutch chatter may result in damage to the reverse and reduction gears. It also may cause the clutch linings to break loose, resulting in complete clutch failure. The principal cause of clutch chatter is fouling of the linings by oil, grease, or water. Every possible precaution should be taken to guard against oil, grease, or water entering into the unit. When clutch chatter occurs as a result of fouling, replacement of the linings is the only satisfactory means of repair. All metal parts of the clutch must be cleaned in accordance with the manufacturer's instructions.

REDUCTION GEARS

13-24. The main reduction gear is one of the largest and most expensive single units of machinery found in the engineering department. Those that are installed properly and operated properly will give years of satisfactory service. However, any casualty to main reduction gears may put a ship out of operation or force it to operate at reduced speed. Main reduction gear repairs can be very costly, as they usually must be done at a shipyard.

FACTORS AFFECTING GEAR OPERATION

13-25. Proper lubrication is a must for efficient operation of reduction gears. It includes supplying the proper amount of oil to the gears and bearings, plus keeping the oil clean and at the proper temperature. Abnormal noises and vibrations must be investigated and corrected immediately. Gears and bearings must be inspected in accordance with applicable TMs.

Lubrication of Gears and Bearings

13-26. The correct quantity and quality of lubricating oil must be available at all times in the main sump. The oil must be clean and it must be supplied to the gears and bearings at the pressure specified by the manufacturer. To supply the proper quantity of oil, several conditions must be met. The lubricating oil pump must deliver the proper discharge pressure. All relief valves in the lubricating system must be set to function at their designed pressure. Too small a quantity of oil will cause the bearing to run hot. If too much oil is delivered to the bearing, the excessive pressure will cause the oil to leak at the oil seal rings. Too much oil may also cause a bearing to overheat. Lubricating oil must reach the bearing at the proper temperature. If the oil is too cold, insufficient oil will flow. If the oil supply is too hot, some lubricating capacity is lost. For most main reduction gears, the normal temperature of oil leaving the lube oil cooler should be between 120° F and 130° F. For full power operation, the temperature of the oil leaving the bearings should be between 140° F and 160° F. The maximum temperature rise of oil passing through any gear or bearing, under any operating condition, should not exceed 55° F. The final temperature of the oil leaving the gear or bearing should not exceed 185° F. This temperature rise and limitation may be determined by a thermometer or resistance temperature element installed in the oil discharge from the bearings. Lubricating oil must be clean. Oil must be free from impurities such as water, grit, metal, and dirt. Any metal flakes and dirt must be removed when new gears or bearings are wearing in or after they have been opened for inspection. Lint or dirt, if left in the system, may clog the oil spray nozzles. The spray nozzle passages must be open at all times. Spray nozzles should not be altered without proper authorization. Lubricating oil strainers perform satisfactorily under normal operating conditions. However, they cannot trap particles of metal and dirt which are fine enough to pass through the mesh. These fine particles can become embedded in the bearing metal and cause wear on the bearings and journals. These fine abrasive particles passing through the gear teeth act like a lapping compound and remove metal from the teeth.

Effects of Acid and Water in Oil

13-27. Water in the oil is extremely harmful. Even small amounts can cause pitting and corrosion of the teeth. Acid can be an even more serious problem. The oil must be tested frequently for the presence of water and tests should be made periodically for acid content. Corrective measures must be taken immediately when saltwater is found in the reduction gear lubricating oil system. Occasionally gross contamination of the oil by saltwater occurs when a cooler leaks or when leaks develop in a sump which is integral with the skin of the ship. Immediate location and sealing of the leak or removal of the source is not enough. The contaminated oil must be removed from all steel parts. Several instances are known where, because such treatment was postponed (sometimes for a week or less); gears, journals, and couplings became so badly corroded and pitted that it was necessary for naval shipyard forces to remove the gears and recondition the teeth and journals. Burned-out bearings also may result from saltwater contamination of the lubricating oil. Water is always present within the lubrication system in small amounts as a result of condensation. Air which enters the units contains moisture which condenses when it strikes a cooler object and subsequently mixes with the oil. The water displaces oil from the metal surfaces and causes rusting. Water reduces the lubricating value of the oil itself. When the main engines are secured, the oil should be circulated until the temperature of the oil and of the reduction gear casing approximates the ultimate engine room temperature. While the oil is being circulated, the cooler should be operated and the gear should be jacked continuously. The purifier should also be operated to renovate the oil while the oil is being circulated, and after circulation until water is no longer discharged from the purifier. This procedure will eliminate condensation from the interior of the main reduction gear casing and reduce rusting in the upper gear case and gears. Generally, lubricating oil will be kept in good condition if the purifier and settling tanks are used properly. However, if the purifier does not operate satisfactorily and does not have the correct water seal, it will not separate water from the oil. An additional check for the presence of water can be made by taking small samples of oil, in bottles, and allowing the samples to settle for a time. The samples should be taken from a low point in the lube oil system. Samples of lubricating oil should be tested for acid, water, and sediment content. Ships should take advantage of every opportunity to have laboratory tests made of the lubricating oil. The supply of oil in the system should be replenished immediately after the oil has been renovated. The amount of lubricating oil in the sump should never be allowed to drop below the minimum allowable level. An adequate reserve supply of clean oil should always be on hand. With continuous use, lube oil will increase in acidity and the free fatty acids will form a mineral soap which reacts with the oil to form an emulsion. Once oil has emulsified, the removal of water and other impurities becomes increasingly difficult. Also, as the oil emulsifies it loses its lubricating quality. The formation of a proper oil film is rendered impossible, and the oil must be renovated. It sometimes happens that when a ship from the reserve fleet is placed back in commission, the rust-preventive compound is not removed completely. The residue of this compound may cause serious emulsification of the lubricating oil. Operating with emulsified oil may result in wiping of the bearings or in other damage to the reduction gears. Since it is extremely difficult aboard ship to destroy emulsions by heating, settling, and centrifuging, a close check should be made to see that such emulsions do not occur. At the first indication of an emulsion, the plant should be stopped and the oil renovated.

Proper Oil Level

13-28. It is of extreme importance that the oil in the sump be maintained within the prescribed maximum and minimum levels. Too much, as well as, too little oil in the sump can lead to trouble. If the oil level is above the prescribed maximum and the bull gear runs in the oil, the oil foams and heats as a result of the "churning" action. In gear installations where the sump tank extends up around the bull gear and the normal oil level is above the bottom of the gear, an oil-excluding pan (sheet metal shield) is fitted under the lower part of the gear. This is to prevent its running in the sump oil. Under normal conditions, the bull gear comes in contact with only a small quantity of oil. The oil which tends to fill the pan is swept out by the gear and is drained back to the sump. When there is too much oil in the sump, the engines must be slowed or stopped until the excess oil can be removed and normal conditions restored. Routine checks should be made to see that the lubricating oil is maintained at the proper level. Any sudden loss or gain in the amount of oil should immediately be investigated.

UNUSUAL NOISES

13-29. A properly operating gear has a definite sound which the experienced engine operator can easily recognize. The operator should be familiar with the normal operation of the gears aboard a ship at different speeds and under various operating conditions. The readings of lube oil pressures and temperatures may or may not be significant in determining the reasons for abnormal sounds. A burned-out pinion bearing or main thrust bearing may be indicated by a rapid rise in oil temperature for the individual bearing. Noise may indicate the following:

- Misalignment.
- Improper meshing of the gear teeth.
- Gear tooth damage.

When there is either a burned-out bearing or trouble with the gear teeth, the main propeller shaft should immediately be stopped and locked by means of the brake on the line shaft or turning gear motor. An inspection should be made to determine the cause of the abnormal sound or noise. The trouble should be remedied before the reduction gear is placed back in operation. In some cases, conditions of a minor nature may be the cause of unusual noises in a reduction gear which is otherwise operating satisfactorily. If investigation reveals the cause to be minor, the gear should be operated cautiously and under close observation by experienced personnel. A more thorough investigation should be made as soon as practicable to determine the cause of the unusual noise. Upon discovery, the problem should be remedied as soon as possible.

VIBRATION

13-30. If the main reduction gear vibrates, a complete investigation should be made, preferably by a naval shipyard. Vibration is caused by the following:

- Bent shafts.
- Damaged propellers.
- Misalignment between prime mover and gear.
- Bearing wear.
- Improper balance in the gear train.

Remember that when the units are built, the gear wheels are fully balanced (both statically and dynamically). Unbalance in the gear is manifested by unusual vibration and noise or by unusual wear of bearings. When the ship has been damaged, vibration of the main reduction gear may result from misalignment of the engine and main shafting as well as the main gear foundation. When vibration occurs within the main reduction gear, damage to the propeller should be one of the first things to be considered. The vulnerable position of propellers makes them more subject to damage than other parts of the main plant. Bent or broken propeller blades may transmit vibration to the main reduction gear. Propellers may also become fouled with line and steel cable.

MAINTENANCE OF REDUCTION GEARS

13-31. Under normal conditions, all repairs and major maintenance work on main reduction gears should be performed by a naval shipyard. When the services of a shipyard are not available, emergency repairs should be done (where possible) by a repair ship or at an advanced base. Minor inspections, tests, and repairs should be done by the ship's force. It is of utmost importance that the ship retains a complete record of the reduction gears from the time of commissioning. Complete installation data, as furnished by the contractor, should be entered in prescribed records by the ship's engineering personnel when the ship is at the contractor's yard. This should include the following:

- Crown thickness readings and clearances of the original bearings.
- Thrust settings and clearances.
- Backlash and root clearances for gear and pinion teeth.

This information must be available when the alignment is subsequently checked. All repairs, adjustments, readings, and casualties should be reported. All original bearing data, as well as all additional bearing measurements, should be entered in appropriate records. Special tools and equipment are normally provided on board ship for:

- Lifting some reduction gear covers.
- Handling the gear elements when removing or replacing their bearings.
- Making the required measurements.
- Rebabbitting bearings.

The special tools and equipment should be available aboard ship in case repairs have to be made by repair ships or at advanced bases. The manufacturer's TM, which gives detailed information on repairing reduction gears, is also furnished on each ship. Bridge gauges are no longer used to check bearing wear of the main reduction gears. When necessary, the crown-thickness method is used. A bearing shell is classed as having a pressure-bearing half and a non-pressure-bearing half. The non-pressure-bearing half will have a radial scribe line at one end of the geometric center. The pressure-bearing half of every main reduction gear shell has three radial scribe lines on each end of the bearing shell. One of these scribe lines is located at the geometric center of the shell and the remaining lines intersect the center scribe line at a 45° angle. The crown thickness of each shell at these points should be measured with a micrometer at a prescribed distance from the end of the shell. These measurements will be recorded during the initial alignment and should be permanently marked adjacent to each scribe line. The amount of bearing clearance should not be allowed to become sufficiently great to cause incorrect tooth contact. The designed clearances for bearings are given in the manufacturer's TM. These clearances are also shown on the blueprints for the main reduction gears. On a multi-shaft ship, if a main reduction gear bearing is wiped, the preferred procedure is to secure the shaft and reduction gear until the units can be inspected and repaired by a repair activity. The replacement of a bearing in a main reduction gear would be a major undertaking for the ship's force. However, emergency conditions may require action by the ship's force. When this is necessary, many factors must be taken into consideration before repairs are attempted. The first step would be to study the manufacturer's instructions and the blueprints for the reduction gear. The engineer must have a clear understanding of construction details and repair procedures before starting a repair job. He/She must also be able to decide whether or not the repair work should be attempted. Other factors which must be considered are:

- Location of the ship.
- Available repair activities.
- Operational schedule of the ship.

Assume that the after bearing for the inboard pinion has been wiped because of an obstructed oil passageway. When making repairs to this unit, the propeller shaft must be locked rigid and the lubricating oil pumped from the sump before the bearing cap is disturbed. After removing the bearing cap, remove and inspect the upper half of the bearing. Then, with the aid of a special jack, roll out the lower half of the bearing. The jack is used to relieve the weight from the lower half and to properly support the rotating elements when the journal bearings are removed. The journal surface of the shaft and all oil passages or nozzles should be carefully inspected and cleaned. The new bearing should also be inspected and cleaned. The crown thickness, as measured at the factory, is stamped on the new bearing. The measurements of the new bearing should be compared with those of the original bearing and with the specifications in the manufacturer's instructions. The new bearing should also be well oiled. The lower half of the bearing can be rolled into place and the jack removed. The upper half is placed in position. The bearing and its dowel must be in the proper position in accordance with the manufacturer's instructions. The bearing cap, or assembly, can then be lowered into position and securely bolted down. It is possible that the forward bearing for the inboard pinion is also damaged as a result of excessive wear. If one pinion bearing fails, that end of the shaft will tend to move away from the bull gear and an abnormal load will be placed on the other pinion bearing. The other pinion bearing should also be opened and inspected. The bearing should be checked with a micrometer, using the crown thickness method. The readings should be compared with the original readings listed in the TM. If excessive wear is indicated, the bearing should be replaced with a new one. If the measurements indicate no wear of the opposite pinion bearing, the bearing can be reassembled. In determining the condition of bearings, a great deal depends on the type of casualty that has occurred. In a loss of lubricating oil, the pinion bearings are first checked. If these bearings are in good condition, it is possible that the bull gear shaft bearings are in satisfactory condition. However, a close watch should be kept on all bearings after a bearing casualty has been corrected. When the reduction gear is opened, dirt and foreign matter must not be allowed to enter. Repair personnel should remove all loose articles from their clothing. Again, before closing the reduction gear, a careful inspection should be made to see that the inside of the gear is free of all dirt, foreign matter, and tools.

GEAR TEETH

13-32. Gears which have been realigned, or new gears, should be given a wearing-in run at low power before being subjected to the maximum tooth pressure of full power. For proper operation of the gears, the tooth contact or total tooth pressure must be uniformly distributed over the total area of the tooth faces. This is done through accurate alignment and adherence to the designed clearances. The designed center-to-center distance of the axis of the rotating elements should be maintained as accurately as practicable. However, in all cases, the axis of pinions and gear shafts must be parallel. Nonparallel shafts concentrate the load in one end of a helix. This may result in flaking, galling, or pitting, and in a featheredge on teeth, deformation of tooth contour, or breakage of tooth ends. The designed tooth contour must be maintained. If the contour is destroyed, a rubbing contact will occur with consequent danger of abrasion taking place. If proper tooth contact is obtained when the gears are installed, there will be little or no wear of teeth. Excessive wear cannot take place without metallic contact. Metallic contact will not occur if adequate lubrication is provided. An adequate supply of lubricating oil at all times, proper cleanliness, and inspection for scores will prevent the wearing of teeth. After all precautions have been taken, if the lubricating oil supply should fail and the teeth become scored, the gears must be thoroughly overhauled by a shipyard as soon as possible. During the first few months that reduction gears are in service, pitting may occur, particularly along the pitch line. Flakes of metal must not be allowed to remain in the oiling system. Slight pitting does not affect the operation of the gears. Play between the surfaces of the teeth in mesh on the pitch circle is known as backlash. It will increase as the teeth wear. It can increase considerably without causing trouble.

ROOT CLEARANCE

13-33. The designed root clearance with gear and pinion operating on their designed centers can be obtained from the manufacturer's drawing or blueprint. The actual clearance can be found by taking leads or by inserting a long feeler gauge or wedge gauge. The actual clearance should check with the designed clearances. If the root clearance is considerably different at the two ends, the pinion and gear shaft are not parallel. An allowable tolerance is permitted, provided that there is still sufficient backlash and the teeth are not meshed so closely that lubrication is adversely affected, or that the clearance is reduced below allowable limits.

ALIGNMENT OF GEAR TEETH

13-34. When the gear and the pinion are parallel (axis of the two shafts are in the same plane and equidistant from each other), the gear train is aligned. In service, the best indication of proper alignment is good tooth contact and quiet operation. The length of tooth contact across the face of the pinions and gears is the criterion for satisfactory alignment of reduction gears. To static-check the length of tooth contact, about 5 to 10 teeth should be coated with Prussian blue or red lead. The gears should then be rolled together with sufficient torque to cause contact between the meshing teeth and to force the journals into the ahead reaction position in their bearings. After the tooth contact has been determined, the coating should be completely removed to prevent possible contamination of the lubricating oil. If tooth contact is to be checked under operating conditions, the teeth should be coated with red or blue Dyken or copper sulphate.

SPOTTING GEAR TEETH

13-35. Any abnormal conditions, which may be revealed by operational sounds or by inspections, should be corrected as soon as possible. Rough gear teeth surfaces, resulting from the passage of some foreign object through the teeth, should be stoned smooth. If deterioration of a tooth surface cannot be traced directly to a casualty of this nature, give special attention to lubrication and to bearings. Also consider the possibility that a change in the supporting structure has disturbed the parallelism of the rotors. Spotting reduction gear teeth is done by coating the teeth Prussian blue and then jacking the gear in its ahead direction of rotation. As the gear teeth come in contact with the marked pinion teeth, an impression is left on the high part of the gear tooth.

After the gear is rotated one-fourth of a turn or is in a convenient position for stoning, all the high spots, as indicated by the marking, are removed by a small handstone. The bluing on the pinion teeth must be replaced repeatedly. However, if the bluing is too heavily applied, you may obtain false impressions on the gear teeth. A satisfactory tooth contact has been obtained when at least 80 percent of the axial length of the working face of each tooth is in contact, distributed over approximately 100 percent of the face width. Remember that the stoning of gears is useful only to remove a local hump or deformation. No attempt should be made to remove deep pitting or galling.

AIRFLEX CLUTCH

13-36. On larger diesel-propelled ships, the clutch, reverse, and reduction gear unit has to transmit an enormous amount of power. To keep the weight low and the size of the mechanism as small as possible, special clutches have been designed for large diesel installations.

AIRFLEX CLUTCH AND GEAR ASSEMBLY

13-37. The airflex clutch and gear assembly shown in Figure 13-5 consists of two clutches (one for forward rotation and one for reverse rotation). The clutches are bolted to the engine flywheel by means of a steel spacer, so that they both rotate with the engine at all times and at engine speed. Each clutch has a flexible tire (or gland) on the inner side of a steel shell. Before the tires are inflated, they will rotate out of contact with the drums, which are keyed to the forward and reverse drive shafts. When air under pressure (100 psi) is sent into one of the tires, the inside diameter of the clutch decreases. This causes the friction blocks on the inner tire surface to come in contact with the clutch drum, locking the drive shaft with the engine.

Forward Rotation

13-38. The parts of the airflex clutch which give the propeller ahead rotation is shown in the upper view of Figure 13-5. The clutch tire nearest the engine (forward clutch) is inflated to contact and drive the forward drum with the engine. The forward drum is keyed to the forward drive shaft, which carries the double helical forward pinion at the after end of the gear box. The forward pinion is in constant mesh with the double helical main gear, which is keyed on the propeller shaft. By following through the gear train, note that for ahead motion, the propeller rotates in a direction opposite to the engine's rotation.

Reverse Rotation

13-39. The parts of the airflex clutch which give the propeller astern rotation are shown in the lower view of Figure 13-5. The reverse clutch is inflated to engage the reverse drum, which is then driven by the engine. The reverse drum is keyed to the short reverse shaft, which surrounds the forward drive shaft. A large reverse step-up pinion transmits the motion to the large reverse step-up gear on the upper shaft. The upper shaft rotation is opposite to the engine's rotation. The main reverse pinion on the upper shaft is in constant mesh with the main gear. By tracing through the gear train, note that for reverse rotation, the propeller rotates in the same direction as the engine. The diameter of the main gear of the airflex clutch is approximately 2 1/2 times as great as that of the forward and reverse pinion. Therefore, there is a speed reduction of 2 1/2 to 1 from either pinion to the propeller shaft. Since the forward and main reverse pinions are in constant mesh with the main gear, the set that is not clutched in will rotate as idlers driven from the main gear. The idling gears rotate in a direction opposite to their rotation when carrying the load. For example, with the forward clutch engaged, the main reverse pinion rotates in a direction opposite to its rotation for astern motion. Since the drums rotate in opposite direction, a control mechanism is installed to prevent the engagement of both clutches simultaneously.

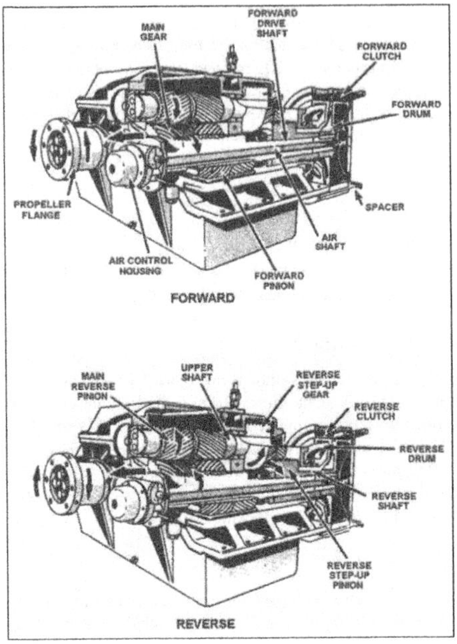

Figure 13-5. Airflex Clutch and Reverse Reduction Gear Assembly

AIRFLEX CLUTCH CONTROL MECHANISM

13-40. The airflex clutch is controlled by an operating level which works the air control housing located at the after end of the forward pinion shaft. The control mechanism (Figure 13-6) with the airflex clutches, directs high-pressure air into the proper paths to inflate the clutch glands (tires). The air shaft, which connects the control mechanism to the clutches, passes through the forward drive shaft. Supply air enters the control housing through the air check valve and must pass through the small air orifice. The restricted orifice delays inflation of the clutch to be engaged for shifting from one direction or rotation to the other. The delay is necessary to allow the other clutch to be fully deflated and out of contact with its drum before the inflating clutch can make contact with its drum. The supply air goes to the rotary air joints in which a hollow carbon cylinder is held to the valve shaft by spring tension. This prevents leakage between the stationary carbon seal and the rotating air valve shaft. The air goes from the rotary joint to the four-way air valve. The sliding-sleeve assembly of the four-way valve can be shifted endwise along the valve shaft by operating the control lever. When the shifter arm on the control lever slides the valve assembly away from the engine, air is directed to the forward clutch. The four-way valve makes the connection between the air supply and the forward clutch, as described here. Eight neutral ports connect the central air supply passage in the valve shaft with the sealed air chamber in the sliding member. In the neutral position of the four-way valve (Figure 13-6), the air chamber is a dead end for the supply air. With the shifter arm in the forward position, the sliding member uncovers eight forward ports which connect with the forward passages conducting the air to the forward clutch. The air now flows through the neutral ports, air chamber, forward ports, and forward passages to inflate the forward clutch gland. As long as the shifter arm is in the forward position, the forward clutch will remain inflated and the entire forward air system will remain at a pressure of 100 psi. At this point, assume that the bridge signals you to reverse the propeller. You should pull the operating lever back to the neutral position and hold it there for 2 or 3 seconds (as a safety factor). Then, pull the lever to the reverse idling position and wait approximately 7 seconds, after which the reverse clutch is fully engaged. Then, you can increase the reverse speed to whatever the bridge has ordered. The reason it was necessary to pause at neutral is that when you shift to neutral, the forward ports are uncovered and the compressed air from the forward clutch and passage vents to the atmosphere. In deflating either clutch, the air is vented through eight ports approximately the same size as the air orifice, so that deflating either clutch actually requires 1 or 2 seconds. Pausing for 2 or 3 seconds at neutral allows sufficient time for the forward clutch to deflate and become disengaged before you start inflating the reverse clutch. The reason it was necessary to pause at the reverse idling position is that when you shift to reverse idling, the air chamber comes over the set of eight reverse ports which open to the central reverse passage in the air shaft. The compressed air begins to inflate the reverse clutch immediately. However, the inflating air must pass through the single air orifice in the supply line, causing a delay of approximately 7 seconds to fully inflate the clutch. When the clutch is in the reverse idling position, wait until the reverse clutch is fully engaged before increasing the speed to prevent damaging the clutch (by slipping). It is impossible to have both clutches engaged at the same time.

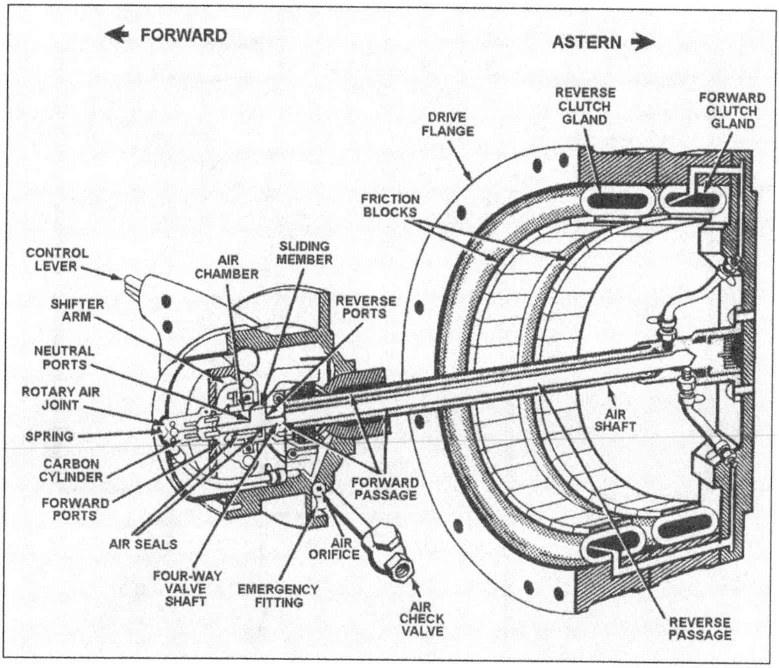

Figure 13-6. Airflex Clutches and Control Valves

LUBRICATION

13-41. On most large gear units, a separate lubrication system is used. Oil is picked up from the gear box by an electric-driven, gear-type lubricating oil pump and is sent through a strainer and cooler. After being cleaned and cooled, the oil is returned to the gear box to cool and lubricate the gears. In twin installations (see Figure 13-7), a separate pump is used for each unit and a standby pump is interconnected for emergency use.

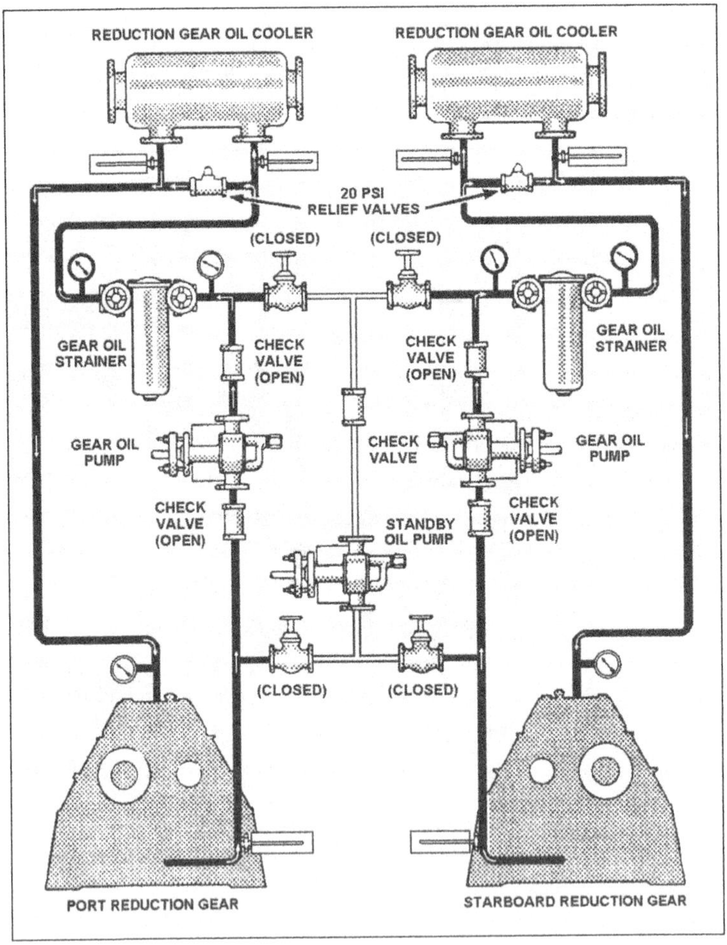

Figure 13-7. Schematic Diagram of Reverse Gear Lubrication System (GM 12-567A)

HYDRAULIC CLUTCHES OR COUPLINGS

13-42. A fluid clutch (coupling) is widely used aboard ships. The use of a hydraulic coupling eliminates the need for a mechanical connection between the engine and the reduction gears. Power is transmitted through couplings of this type efficiently (97 percent) without transmitting torsional vibrations or load shocks from the engine to the reduction gear. The power loss resulting from the small amount of slippage is transformed into heat which is absorbed by the oil in the system. Some slippage is necessary for operation of the hydraulic coupling since torque is transmitted because of the principle of relative motion between the two rotors.

HYDRAULIC COUPLING ASSEMBLY

13-43. The two rotors and the oil-sealing cover of a typical hydraulic coupling are shown in Figure 13-8. The primary rotor (impeller) is attached to the engine crankshaft. The secondary rotor (runner) is attached to the reduction gear pinion shaft. The cover is bolted to the secondary rotor and surrounds the primary rotor. Before proceeding with the discussion on the assembly of rotors and shafts in the coupling housing, study the structure of the rotors themselves. Each rotor is shaped like a doughnut with radial partitions. A shallow trough is welded into the partitions around the inner surface of the rotor. The radial passages tunnel under this trough. When the coupling is assembled, the two rotors are placed facing each other to complete the doughnut. The rotors do not quite touch each other, the clearance between them being 1/4 to 5/8 inches, depending on the size of the coupling. The curved radial passages of the two rotors are opposite each other so that the outer passages combine to make a circular passage except for the small gaps between the rotors. In the hydraulic coupling assembly shown in Figure 13-9, the driving shaft is secured to the engine crankshaft and the driven shaft goes to the reduction gearbox. The oil inlet admits oil directly to the rotor cavities, which become completely filled. The rotor housing is bolted to the secondary rotor and has an oil-sealed joint with the driving shaft. A ring valve, completely encircling the rotor housing, can be operated by the ring valve mechanism to open or close a series of emptying holes in the rotor housing. When the ring valve is opened, the oil will fly out from the rotor housing into the coupling housing, draining the coupling completely in 2 or 3 seconds. Even when the ring valve is closed, some oil leaks into the coupling housing and additional oil enters through the inlet. From the coupling housing, the oil is drawn by a pump to a cooler, then sent back to the coupling.

PISTON COUPLING ASSEMBLY

13-44. Another coupling assembly used on several Army ships is the hydraulic coupling with piston-type, quick-dumping valves. In this coupling, in which the operation is similar to the one described previously, a series of piston valves around the periphery of the rotor housing are normally held in the closed position by springs. By means of air or oil pressure admitted to the valves (see Figure 13-10), the pistons are moved axially to uncover drain ports, allowing the coupling to empty. Where extremely rapid declutching is not required, the piston-valve coupling has the advantages of greater simplicity and lower cost than the ring-valve coupling.

SCOOP CONTROL COUPLING

13-45. Another type of self-contained unit for certain diesel-engine drives is the scoop control coupling (see Figure 13-11). In couplings of this type, oil is picked up by one of two scoop tubes (one tube for each direction of rotation) mounted on the external manifold. Each scoop tube contains two passages:

- A smaller one (outermost) which handles the normal flow of oil for cooling and lubrication.
- A larger one which rapidly transfers oil from the reservoir directly to the working circuits.

The scoop tubes are operated from the control stand through a system of linkages. As one tube moves outward from the shaft centerline and into the oil annulus, the other tube is being retracted. Four spring-loaded centrifugal valves are mounted on the primary rotor. These valves are arranged to open progressively as the speed of the primary rotor decreases. The arrangement provides oil flow for cooling as it is required. Quick-emptying piston valves provide rapid emptying of the circuit when the scoop tube is withdrawn from contact with the rotating oil annulus. Under normal circulating conditions, oil fed into the collector ring passes into the piston valve control tubes. These tubes and connecting passages conduct oil to the outer end of the pistons. The centrifugal force of the oil in the control tube holds the piston against the valve port, thereby sealing off the circuit. When the scoop tube is withdrawn from the oil annulus in the reservoir, circulation of oil is interrupted and oil in the control tubes is discharged through the orifice in the outer end of the piston housing. This releases pressure on the piston and allows it to move outward, thereby opening the port for rapid discharge of oil. Resumption of oil flow from the scoop tube will fill the control tubes and the pressure will move the piston to the closed position.

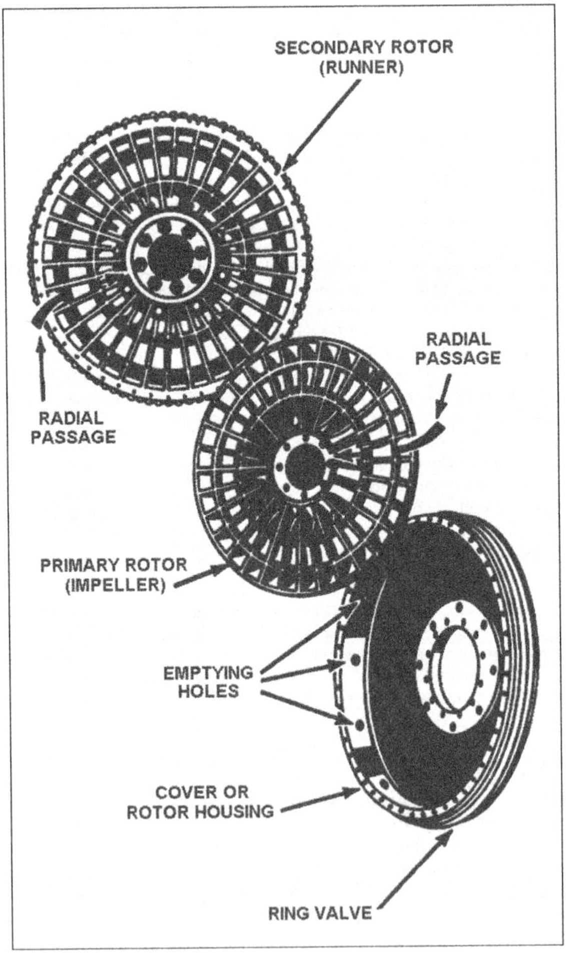

Figure 13-8. Runner, Impeller, and Cover of Hydraulic Coupling

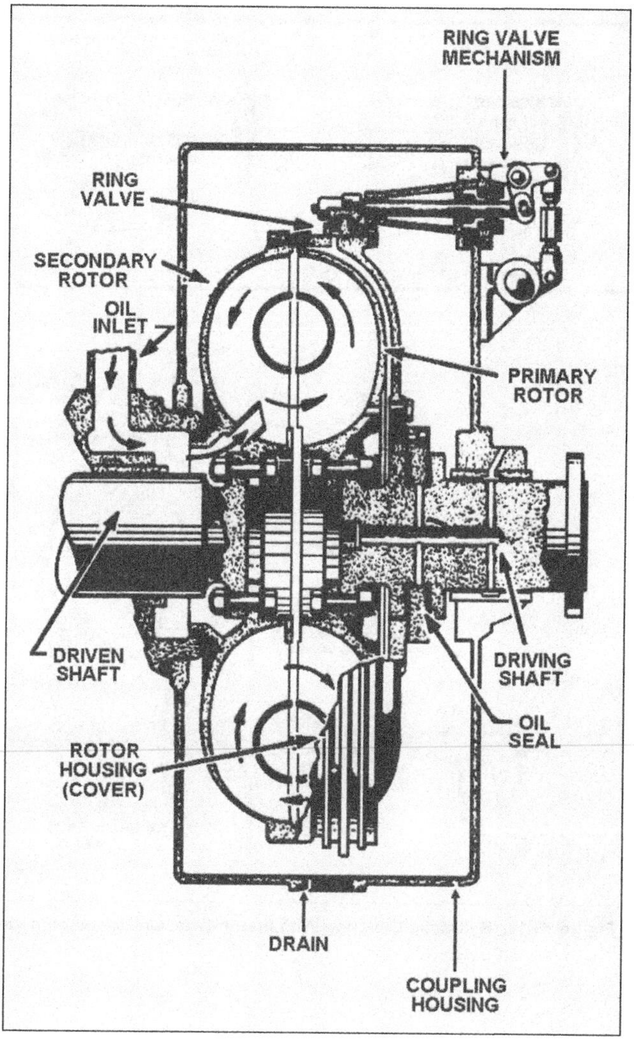

Figure 13-9. Hydraulic Coupling Assembly

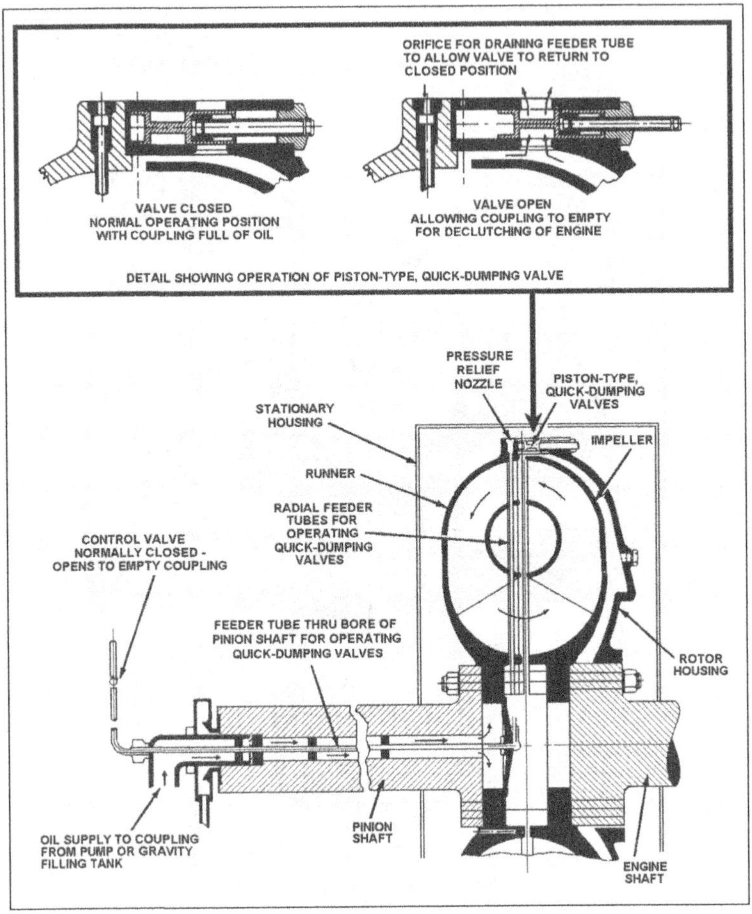

Figure 13-10. Hydraulic Coupling With Piston-type, Quick-dumping Valves

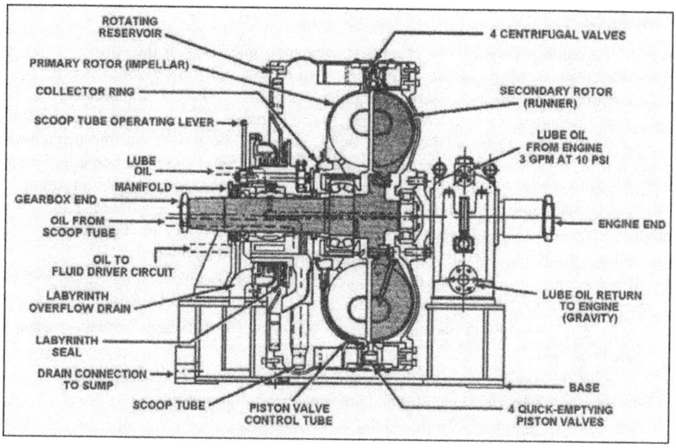

Figure 13-11. Scoop Control Hydraulic Coupling

PRINCIPLES OF OPERATION

13-46. When the engine is started and the coupling is filled with oil, the primary rotor turns with the engine crankshaft. As the primary rotor turns, the oil in its radial passages flows outward under centrifugal force. This forces oil across the gap at the outer edge of the rotor and into the radial passages of the secondary rotor where the oil flows inward. Oil in the primary rotor is not only flowing outward but is also rotating. As the oil flows over and into the secondary rotor, it strikes the radial blades in the rotor. The secondary rotor soon begins to rotate and pick up speed. However, it will always rotate more slowly than the primary rotor because of drag on the secondary shaft. Therefore, the centrifugal force of the oil in the primary rotor will always be greater than that of the oil in the secondary rotor. This causes a constant flow from the primary rotor to the secondary rotor at the outer ends of the radial passages and a constant flow from the secondary rotor to the primary rotor at the inner ends.

HYDRAULIC COUPLING TROUBLES

13-47. Troubles which may be encountered with a quick-dumping hydraulic coupling are dumping while under load and excessive slippage. These troubles are generally caused by plugging of the pressure relief nozzles located in the periphery of the secondary rotors. These pressure relief nozzles consist of drilled Allen setscrews mounted in the secondary rotor at the ends of the radial tubes. They feed air or oil to the dumping valves. The nozzles permit the feeder tubes to drain when the air or oil supply valve is closed. This allows the dumping valve to return to the closed position. The nozzles also permit draining of any oil that has leaked past the control valve when it was shut. Leak-off nozzles are also provided in the periphery of the secondary rotor at the base of the dumping valves. The leak-off nozzles serve as flushing exits for the valves and allow a continual flow of oil past the inlet port of the dumping valves. The oil washes away any particles of foreign matter that may have a tendency to collect as a result of the centrifugal action of the coupling. The best way to prevent the dumping of a hydraulic coupling while under load is to keep the oil system free of all foreign matter. Gasket compound and shredded copper from oil tube packings often cause trouble. Regardless of the source of foreign material, the oil system must be kept clean. To aid in this, the system is equipped with a strainer which effectively catches or traps most of the foreign material. Since a small amount of foreign matter may reach the nozzles, all nozzles must be blown out during each overhaul. To clear nozzles that may have become clogged during operation, operate the dumping control several times. This action may blow the obstruction through the nozzle opening. If this method fails, secure the engine and remove and clean the nozzles.

INDUCTION COUPLINGS

13-48. Induction couplings (Figure 13-12) are used aboard some ships in the Army. In these couplings, the induced current appears as eddy currents. They are used to transmit torque from the prime mover to variable frequency generators. They are also used in propulsion systems. The excitation current is inducted into the coils of the outer members through collector rings and brushes. When used in propulsion systems, the induction coupling provides torsional flexibility between the prime mover and the propeller shafts. Instant disconnecting is done by deenergizing the coupling excitation circuit. The induction coupling limits maximum torque by pulling out of step when excessive torque is applied. On installations having more than one engine per shaft, the induction coupling permits rapid maneuvering by allowing selection of either forward or reverse running engines. The coupling allows for a small amount of misalignment. In operating equipment having induction couplings, the following directions and precautions should be observed:

- Do not attempt to alter plant performance by changing control settings to settings other than those recommended in the manufacturer's TM.
- Ensure that the coupling does not overheat because of insufficient ventilation when the fields rotate at slow speeds.
- Ensure that proper alignment is maintained (although most clutches of this type are capable of operating satisfactorily with a limited amount of misalignment).
- Be thoroughly familiar with the means to mechanically couple the rotating members in the event of total failure of the coupling or excitation system.
- Be thoroughly familiar with any interlocks which serve to prevent—
    - Operation at reduced excitation except when the prime mover is operating at specified reduced speeds.
    - Excitation of the field windings unless the throttle control is in the proper position.
    - Excitation of the field windings unless the shaft turning gear and shaft locking devices are disengaged.
    - Excitation of the field windings at a time when the clutch will turn the driven gear counter to the direction in which it is already being driven by another coupling or clutch.

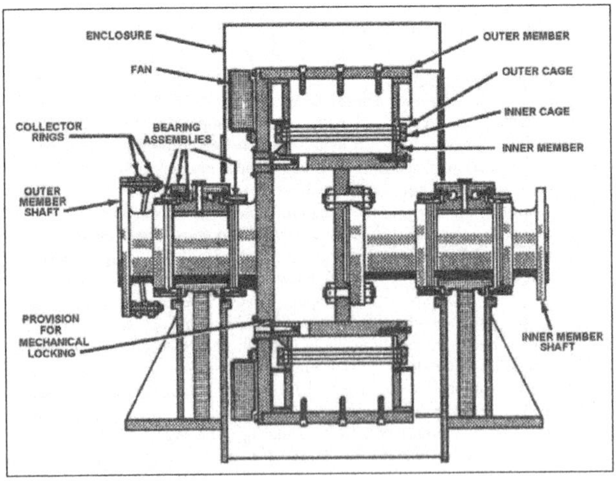

Figure 13-12. Induction Coupling

DOG CLUTCHES

13-49. Dog clutches ensure a positive drive without slippage and with a minimum amount of wear (Figure 13-13). They perform basically the same function as friction clutches in that they allow the engine shaft to be disconnected from the propeller shaft. In some installations, a friction clutch is used in conjunction with a dog clutch. The friction clutch (or synchronizing clutch) is used to synchronize the speed of the shafts, and then the dog clutch is engaged. Engagement of the dog clutch eliminates any slippage and reduces wear on the friction clutch. A combination friction and dog clutch is shown in Figure 13-14. Any difficulty encountered in engaging the dog clutch will probably be due to burrs on the dogs or to misalignment of the mating parts. The dogs may become burred through normal usage. Repair should be made by disassembling the clutch and dressing down the burrs with a small hand grinder. When the damage is extensive and the burrs cannot be satisfactorily removed by grinding, the parts must be replaced. If difficulty in engaging a dog clutch is intermittent, misalignment is probably the cause. When this condition exists, try disengaging the clutch fully and then engaging it again. It may be that the shifting of relative positions of the mating dogs is prevented by the synchronizers used to aid in bringing the two shafts to equal speed. Releasing the pressure on the engaging mechanism will usually permit relative motion between the mating dogs.

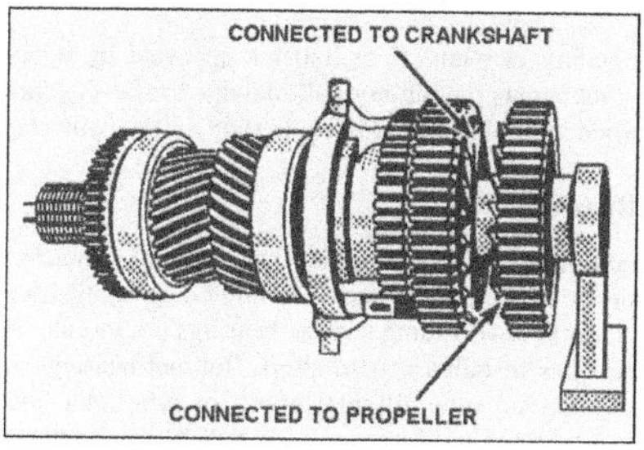

Figure 13-13. Dog Clutches

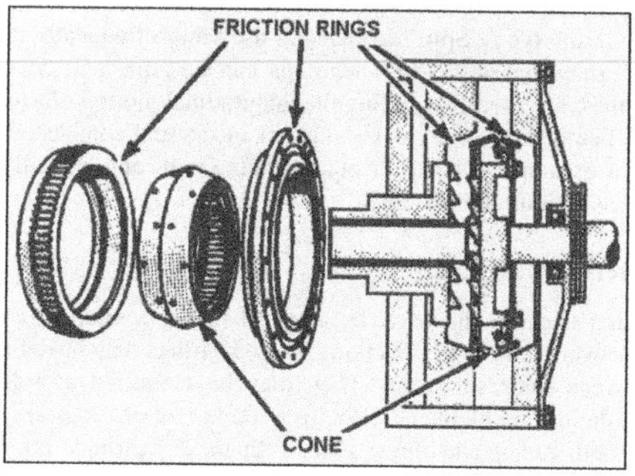

Figure 13-14. Combination Friction and Dog Clutch

BEARINGS

13-50. From the standpoint of mechanics, the term "bearing" may be applied to anything which supports a moving element of a machine. However, this discussion concerns those bearings which support or confine the motion of sliding, rotating, and oscillating parts on revolving shafts or movable surfaces of naval machinery. In view of the fact that naval machinery is constantly exposed to varying operating conditions, bearing material must meet rigid standards. A number of nonferrous alloys are used as bearing metals. In general, these alloys are tin-base, lead-base, base, or aluminum-base alloys. The term babbitt metal is often used for lead-base and tin-base alloys. Bearings must be made of materials which will withstand varying pressures and yet permit the surfaces to move with minimum wear and friction. In addition, bearings must be held in position with very close tolerances permitting freedom of movement and quiet operation. In view of these requirements, good bearing materials must possess a combination of the following characteristics for a given application:

- The compressive strength of the bearing alloy at maximum operating temperature must withstand high loads without cracking or deforming.
- Bearing alloys must have high fatigue resistance to prevent cracking and flaking under varying operating conditions.
- Bearing alloys must have high thermal conductivity to prevent localized hot spots with resultant fatigue and seizure.
- Bearing materials must be capable of retaining an effective oil film.
- Bearing materials must be highly resistant to corrosion.

CLASSIFICATION

13-51. Reciprocating and rotating elements or members supported by bearings may be subject to external loads. These loads can be components having normal, radial, or axial directions, or a combination of the two. Bearings are generally classified as sliding surface (friction) or rolling contact (antifriction).

Sliding Surface or Friction Bearings

13-52. These may be defined broadly as those which have sliding contact between their surfaces. In these bearings, one body slides or moves on the surface of another. Sliding friction is developed if the rubbing surfaces are not lubricated. Examples of sliding surface bearings are journal bearings and thrust bearings (such as the spring or line shaft bearings installed aboard ship). Journal bearings are extensively used aboard ship. Journal bearings may be sub-divided into different styles or types, the most common of which are solid bearings, half bearings, and two-part or split bearings. A typical solid style journal bearing application is the piston bearing, more commonly called a bushing. An example of a solid bearing is a piston rod wristpin bushing (such as that found in compressors). Perhaps the most common application of the half bearing in marine equipment is the propeller shaft bearing. Since the load is exerted only in one direction, they are less costly than a full bearing of any type. Split bearings are used more frequently than any other friction bearing. A good example is the turbine bearing. Split bearings can be adjustable to compensate for wear. Guide bearings, as the name implies, are used for guiding the longitudinal motion of a shaft or other part. Perhaps the best illustrations of guide bearings are the valve guides in an internal combustion engine. Thrust bearings are used to limit the motion of or support a shaft or other rotating part longitudinally. Thrust bearings sometimes are combined functionally with journal bearings.

Rolling Contact or Antifriction Bearings

13-53. These are so called because they are designed so that less energy is required to overcome rolling friction than is required to overcome sliding friction. These bearings may be defined broadly as bearings which have rolling contact between their surfaces. They may be classified as roller bearings or ball bearings according to the shape of the rolling elements. Both roller and ball bearings are made in different types, some being arranged to carry both radial and thrust loads. In these bearings, the balls or rollers generally are assembled between two rings or races, the contacting faces of which are shaped to fit the balls or rollers. The basic difference between ball and roller bearings (see Figure 13-15) is that a ball at any given instant carries the load on two tiny spots diametrically opposite, while a roller carries the load on two lines. Theoretically, the area of the spot or line of contact depends on how much the bearing material will distort under the applied load. Rolling contact bearings must therefore be made of hard materials because if distortion under load is appreciable the resulting friction will defeat the purpose of the bearings. Bearings with small, highly loaded contact areas must be lubricated carefully if they are to have the antifriction properties they are designed to provide. If improperly lubricated, the highly polished surfaces of the balls and rollers will soon crack, check, or pit, and failure of the complete bearing follows. Both sliding surface and rolling contact bearings may be further classified by their function (such as radial, thrust, and angular-contact [actually a combination of radial and thrust]). Radial bearings are designed primarily to carry a load in a direction perpendicular to the axis of rotation. They are used to limit motion in a radial direction. Thrust bearings can carry only axial loads; that is, a force parallel to the axis of rotation. This tends to cause endwise motion of the shaft. Angular-contact bearings can support both radial and thrust loads. The simplest forms of radial bearings are the integral and the insert types. The integral type is formed by surfacing a part of the machine frame with the bearing material. The insert bearing is a plain bushing inserted into and held in place in the machine frame. The insert bearing may be either a solid or a split bushing. It may consist of the bearing material alone or be enclosed in a case or shell. In the integral bearing there is no means of compensating for wear. When the maximum allowable clearance is reached the bearing must be resurfaced. The insert solid-bushing bearing, like the integral type, has no means for adjustment due to wear. It must be replaced when maximum clearance is reached. The pivoted shoe is a more complicated type of radial bearing. It consists of a shell containing a series of pivoted pads or shoes faced with bearing material. The plain-pivot or single-disk thrust bearing consists of the end of a journal extending into a cup-shape housing, the bottom of which holds the single disk of bearing material. The multi-disk thrust bearing is similar to the plain-pivot bearing except that several disks are placed between the end of the journal and the housing. Alternate disks of bronze and steel are generally used. The lower disk is fastened in the bearing housing and the upper one to the journal. The intermediate disks are free. The multi-collar thrust bearing consists of a journal with thrust collars integral with or fastened to the shaft. These collars fit into recesses in the bearing housing which are faced with bearing metal. This type bearing is generally used on a horizontal shaft carrying light thrust loads. The pivoted-shoe thrust bearing is similar to the pivoted-shoe radial bearing except that it has a thrust collar fixed to the shaft which runs against the pivoted shoes. This type bearing is generally suitable for both directions of rotation. Angular loading is generally taken by using a radial bearing to restrain the radial load and some form of thrust bearing to handle the load. This may be done by using two separate bearings or a combination of a radial and thrust (radial thrust). A typical example is the multi-collar bearing which has its recesses entirely surfaced with bearing material. The faces of the collars carry the thrust load and the cylindrical edge surfaces handle the radial load.

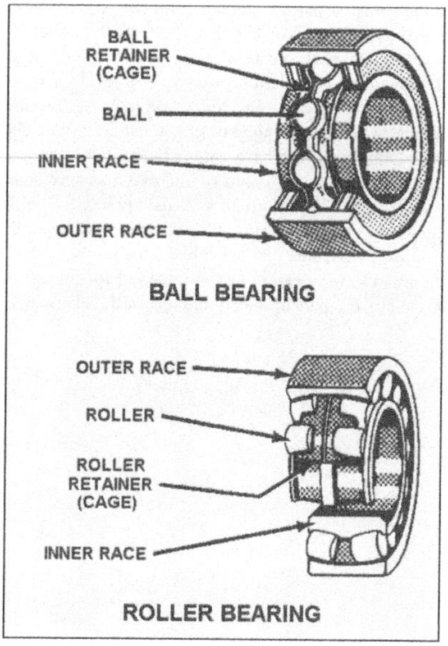

Figure 13-15. Load-carrying Areas of Ball and Roller Bearings

MAIN REDUCTION GEAR BEARINGS

13-54. Reduction gear bearings must support the weight of the gears and their shafts. They must also hold the shafts in place against the tremendous forces exerted by the shafts and gears when they are transmitting power from the turbine shaft to the propeller shaft. Like other radial bearings in main engine installations, these bearings are of the babbitt-lined split type. However, instead of being spherically seated and self-aligning, the reduction gear bearings are rigidly mounted into the bearing housings by dowels or locking screws and washers. The angular direction of forces acting on a main reduction gear bearing, changes with the amount of propulsion power being transmitted by the gear. In some designs, thermocouples or resistance-temperature elements (RTE) for measuring oil temperature are installed in bearings. Or they may be in sight-flow fittings. To avoid wiping, reduction gear bearings must be positioned so that the heavy shaft load is not brought against the area where the bearing halves meet (the split). For this reason, most of the bearings in a reduction gear are placed so that the split between the halves is at an angle to the horizontal plane. Turbine bearings, unlike reduction gear bearings, are self-aligning by means of a spherical seat for the bearing shells. Turbine bearings are pressure-lubricated by the same force-feed system that lubricates the reduction gear bearings.

MAIN THRUST BEARINGS

13-55. The main thrust bearing, which is usually located in the reduction gear casing, absorbs axial thrust transmitted through the shaft from the propeller. Kingsbury or segmental, pivoted-shoe thrust bearings are commonly used for main thrust bearings. This type consists of pivoted segments or shoes (usually six or eight) against which the thrust collar revolves. Ahead or astern axial motion of the shaft, to which the thrust collar is secured, is thereby restrained by the action of the thrust shoes against the thrust collar. These bearings operate on the principle that a wedge-shape film of oil is more readily formed and maintained than a flat film and that it can therefore carry a heavier load for any given size. In a segmental, pivoted-shoe thrust bearing, upper leveling plates on which the shoes rest, and lower leveling plates equalize the thrust load among the shoes. The base ring supports the lower leveling plates. It holds the plates in place and transmits thrust on the plates to the ship's structure via housing members which are bolted to the foundation. Shoe supports (hardened steel buttons or pivots) located in the shoes, separate the shoes and the upper leveling plates. This enables the shoe segments to assume the angle required to pivot the shoes against the upper leveling plates. Pins and dowels hold the upper and lower leveling plates in position. This allows ample play between the base ring and the plates to ensure freedom of movement (oscillation only) of the leveling plates. The base ring is kept from turning by its keyed construction which secures the ring to its housing. Figure 13-16 shows a diagram of the Kingsbury bearing. In a Kingsbury thrust bearing, end play must always be checked with the upper half of the housing solidly bolted down. Otherwise, the base rings could tilt under the freedom given by the leveling plates. This could result in a false reading. A record of end play measurements should be kept and referred to when checking the main thrust bearing. The normal wear of a pivoted-shoe thrust bearing is negligible, even with years of use. However, when a thrust bearing is new, there may be slight settling of the leveling plates. Any noticeable increase in end play indicates that the thrust shoe surfaces should be examined and repaired, if necessary. In most cases, the main thrust bearing cap is removed for inspection. This opening is of such size that it will permit the withdrawal of the pair of ahead and astern thrust shoes located (by means of the turning gear) in line with it.

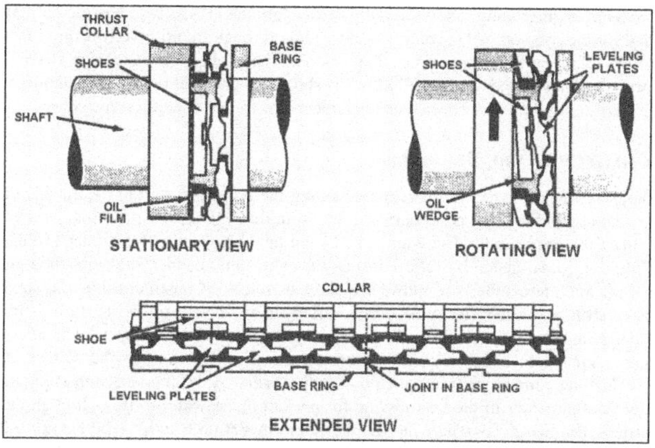

Figure 13-16. Diagrammatic Arrangement of Kingsbury Bearing

CHECKING END PLAY WHILE RUNNING

13-56. The simplest method of checking end play is to use a suitable measuring instrument on any accessible part of the propeller shaft while running the shaft slowly ahead and astern. This would normally be done at the end of a run when the ship is maneuvering to approach the pier, before the machinery and shaft are cold. Speeds should be slow to avoid adding deflections of bearing parts and housing to the actual end play. However, the speeds should be sufficient to overcome the rake of the shaft and ensure that the full end play is actually taken up. End play is measured by a dial indicator mounted on a rigid support close to any convenient coupling flange. Occasionally the shaft may have a shoulder turned on it for the sole purpose of applying a dial indicator. A flange surface must be free of paint, burrs, and rust spots. The flange surface should also be well oiled to prevent damage to the dial indicator.

JACKING ON SHAFT FLANGE

13-57. If it is not feasible to measure end play while running, the next choice is to jack the shaft fore and aft at some convenient main-shaft flange. A dial indicator should be used. Make certain that the shaft movement is free and guard against overdoing the jacking force. The difficulty with the jacking method is finding suitable supports where no structural damage will be incurred when jacking against a main-shaft flange coupling.

MAIN PROPULSION SHAFT BEARINGS

13-58. You will be required to watch and maintain the main propeller shaft bearings. The bearings which support and hold the propulsion shafting in alignment are divided into the following:

- Mainline shaft bearings or spring bearings.
- Stem tube bearings.
- Strut bearings.

MAINLINE SHAFT OR SPRING BEARINGS

13-59. These are of the ring-oiled, babbitt-faced, spherical-seat, shell-type bearings. This bearing is designed primarily to align itself to support the weight of the shafting. In many of the older, low-powered ships, the bearing is not the self-aligning type and it consists only of a bottom half. The upper half of the assembly consists only of a cap or cover (not in contact with the shaft) to protect the shaft journal from dirt. However, the spring bearings of all modern naval ships are provided with both upper and lower self-aligning bearing halves. The brass oiler rings hang loosely over the shaft journal and the lower bearing half, and they are slowly dragged around by the rotation of the shaft. As they glide through the reservoir of oil at the bottom, the rings carry some of the oil along to the top of the shaft journal. The upper bearing half is grooved to accommodate the rings. Spring bearing temperatures and oil levels should be checked hourly while underway. At least once each year, the bearings should be inspected and clearances taken and defects corrected.

STERN TUBE AND STERN TUBE BEARINGS

13-60. The hole in the hull structure for accommodating the propeller shaft to the outside of the hull is called the stern tube. The propeller shaft is supported in the stern tube by two bearings (one at the inner end and one at the outer end of the stern tube). These are called stern tube bearings. At the inner end of the stern tube is a stuffing box packing gland generally referred to as the stern tube gland. The stern tube gland seals the area between the shaft and stern tube, yet allows the shaft to rotate. Construction of the stern tube bearings is similar to that of strut bearings. The stuffing box is flanged and bolted to the stern tube. Its casing is divided into two annular compartments, the forward space being the stuffing box proper. The after space is provided with a flushing connection. This allows a positive flow of water through the stem tube for cooling and flushing. This flushing connection is supplied by the fire main. A drain connection is provided for testing for the presence of cooling water in the bearing and for permitting seawater to flow through the stern tube. This cools the bearing when underway where natural seawater circulation is used. The gland for the stuffing box is divided longitudinally into two parts. The gland bolts are long enough to support the gland when it is withdrawn at least 1 inch clear of the stuffing box. This permits the addition of a ring of new packing, when needed, while the ship is waterborne. Either braided flax packing or special semi-metallic packing is used (ship's engineering drawings show the proper type of packing). This gland is usually tightened to eliminate leakage when the ship is in port. It is loosened (prior to warming up) just enough to permit a slight trickle of water for cooling purposes when the ship is underway.

STRUT BEARINGS

13-61. The strut bearings, as well as the stern tube bearings, are equipped with composition bushings which are split longitudinally into two halves. The outer surface of the bushing is machined with steps to bear on matching landings in the bore of the strut. One end is bolted to the strut. Since it is usually impractical to use oil or grease as a lubricant for underwater bearings, some other material must be used for that purpose. There are certain materials that become slippery when wet. These materials include:

- Natural or synthetic rubber.
- Lignum vitae, a hard tropical wood with excellent wearing qualities.
- Laminated phenolic material consisting of layers of cotton fabric impregnated and bonded with phenolic resin. Strips of this material are fitted inside the bearing.

A rubber composition is the type most used in modern installations.

REDUCTION GEAR INSPECTIONS

13-62. The inspections mentioned here are the minimum required for reduction gears only. Where defects are suspected or operating conditions so indicate, inspections should be made at more frequent intervals. No inspection plates or other fittings of the main reduction gear may be opened without permission of the engineer. Before replacement of an inspection plate or a connection, fitting, or cover which permits access to the gear casing, a careful inspection must be made. This is to ensure that no foreign matter has entered or remains in the casing or oil lines. An entry of the inspections and the name of the engineer who witnesses the closing of the inspection plate should be made in the engineering log. The PMS discussed here are general in scope. Inspection requirements for your ship are listed in the applicable TM. They should be referred to for all maintenance action. Gears should be jacked daily (at anchor) so that the main gear shaft is moved 1 1/4 revolution. The jacking should be done with lubricating oil circulating in the system. The following inspections should be made quarterly:

- Sound with a hammer, hold-down bolts, ties, and chocks to detect signs of loosening of casing fastenings.
- Open inspection plates, inspect gears and oil-spray nozzles. Wipe off oil at different points and note whether the surface is bright, or if already corroded, whether new areas are affected.
- Inspect strainers for the oil-spray nozzles to see that dirt or sediment has not accumulated t herein.
- Take and record main thrust bearing readings.

When conditions warrant or if trouble is suspected, a work request may be submitted to perform a 7-year inspection of the main reduction gears. This inspection includes clearance readings of bearings and journals; alignment checks and readings; and any other inspections, tests, or maintenance work that may be necessary. If the ship's propeller strikes ground or a submerged object, inspect the main reduction gear at the time of occurrence. In this inspection, look for the possible misalignment of the bull gear and its shaft. Where practicable, a shipyard should be requested to check the alignment and concentricity of the bull gear.

REDUCTION GEAR SAFETY PRECAUTIONS

13-63. If churning or emulsification of the oil in the gear case occurs, the gear must be slowed down or stopped until the defect is remedied. If the supply of lubricating oil to the gears fails, the gears should be stopped until the cause can be located and remedied. When bearings have been overheated, gears should not be operated (except in extreme emergencies) until bearings have been examined and defects remedied. If excessive flaking of metal from gear teeth occurs, the gears should not be adjusted (except in case of emergency) until the cause has been determined. Unusual noises should be investigated at once and the gears should be operated cautiously or stopped until the cause of the noise has been discovered and remedied. No inspection plate, connection, fitting, or cover which permits access to the gear casing should be removed without specific authority of the engineer officer. The immediate vicinity of an inspection plate joint should be kept free of paint and dirt. When gear cases are open, precautions should be taken to prevent the entry of foreign matter. The openings should never be left unattended unless satisfactory temporary closures have been installed. Lifting devices should be inspected carefully before being used and should not be overloaded. Naked lights should be kept away from vents while gears are in use (the oil vapor may be explosive). When ships are anchored where there are strong currents or tides, the main shafts should be locked. On a ship with divers, over the side where the rotation of the propeller may result in injury to the diver or damage to equipment, propeller shafts should be locked. When a ship is towed, the propellers should be locked unless it is permissible and advantageous to allow the shafts to trail with the movement of the ships. When a shaft is allowed to turn or trail, the lubrication system must be in operation. The main propeller shaft must be brought to a dead-stop position before an attempt is made to engage or disengage the turning gear. When a main shaft is being locked, the brake should be quickly and securely applied. Where there is a limiting maximum safe speed at which a ship with more than one shaft can steam with a locked propeller shaft, this speed should not be exceeded. When the main gears are being jacked over, the turning gear must be properly lubricated. It should be determined that the turning gear has definitely been disengaged before the main engines are started.

PROPULSION SHAFTING

13-64. The purpose of propulsion shafting (see Figure 13-17) is to transfer torque (turning motion) generated by the main engine to the propeller.

CONSTRUCTION

13-65. The external portion of shafting is usually supported by bearings located in struts or skegs. The inboard shafting is supported on pedestal blocks or spring bearings. The after section of the shafting is exposed to seawater and must be protected from corrosion. The accessible sections of the exposed shaft are protected by coating with the usual hull paint. Shafting which is not accessible for painting and which is not protected by bearing sleeves or metallic casings is properly protected. This can be done by a vulcanized rubber cover, by materials specified by US Coast Guard (USCG) regulations, or by vessel onboard drawings.

SHAFTING MATERIAL

13-66. Propulsion shafting is usually forged in sections from alloy steel ingots. It is hollow-bored from end to end to attain a saving in weight. Aluminum bronze is also used for shafting on Army vessels. Bronze or monel sleeves are shrunk over the shafting for the bearing surfaces. The roundness of the sleeves must be checked frequently as outlined in the applicable TMs. Steel shafts should be kept rust-free, smooth, and even.

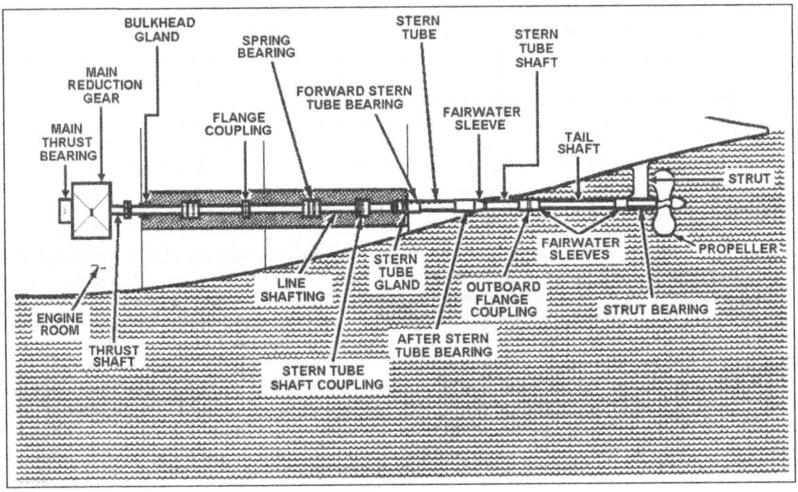

Figure 13-17. Propulsion Shafting

SHAFT COUPLINGS

13-67. Shaft couplings (see Figure 13-18) are used for connecting two separate parts of the shaft or connecting the shaft to the main engine. This is done through connecting nuts and flanges which are locked in place with setscrews, locking keys, or locking plates. Inboard flange couplings are usually set in such bearings as spring, line-shaft, or line-shaft spring bearings.

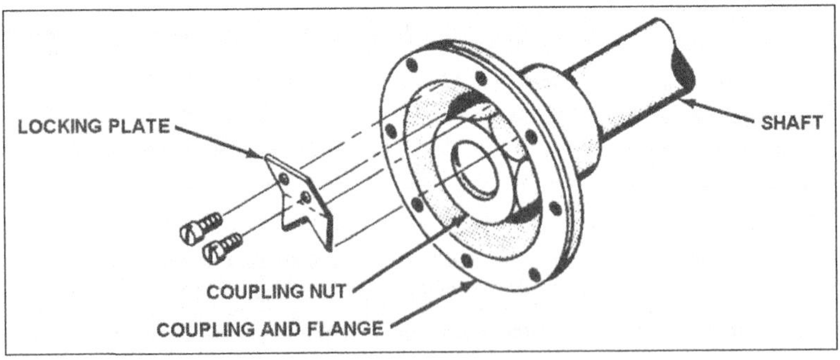

Figure 13-18. Typical Shaft Coupling With Retaining Nut and Flanges

Fitted Flanges

13-68. The ends of the propulsion shafting sections are usually provided with integral flanges for coupling the adjacent shaft sections by means of coupling bolts (see Figure 13-19). The face of each half coupling flange must be finished truly perpendicular to the axis of the shaft. The outer peripheries of each of the matched flange coupling halves must be concentric. For fitting together flanges of two adjacent sections, two different bolting systems may be used. One system is a hex or round-headed (with flats), straight-fitted bolt, backed up by a standard nut and cotter pin. The other or preferred bolting system is a hex or round-headed (with flats) bolt with a body taper of 1/8 inch per foot, backed up by a standard nut or cotter pin. Using these approved designs with upset heads, reduces the danger of pulling headless taper bolts through the flanges when back-driving the vessel. Such action could result in casting the shafting adrift. In all cases when fitting the flange bolts to their particular flanges, a system of permanently marking the bolt location in the flange must be used. This is to ensure the proper fit of the flanges after repair of the shaft.

Couplings Exposed to Seawater

13-69. On shafting exposed to seawater, a coupling known as an outboard cross-key coupling is frequently used. This type of coupling offers little resistance to the flow of water and permits easy uncoupling after long immersion in saltwater. The ends of two shafting sections to be attached by a sleeve cross-key coupling are slightly tapered (see Figure 13-20). These sections are drawn into the sleeve by driving a tapered key through the sleeve and the shafting. The disadvantages in this coupling are high cost, difficulty in fitting the crossed keys that take the astern thrust, and lengthy assembly time.

SHAFT KEYS AND KEYWAYS

13-70. Keys and keyways in shafting, propeller hubs, and couplings (inboard and outboard) must be fitted closely and a press-fit used in installation. Drilling for retainer screws is not required. Drilling for a dowel pin to keep the keys from sliding when installing propellers and couplings is required. Key corners should be chamfered 45°. The keyways should have well-rounded corner fillets; preferably with a radius equal to about 3 percent of the shaft diameter. Key material should be of the same basic material as the shafting and the coupling. Dimensions and sectional areas of the keys should provide sufficient strength to carry the entire load.

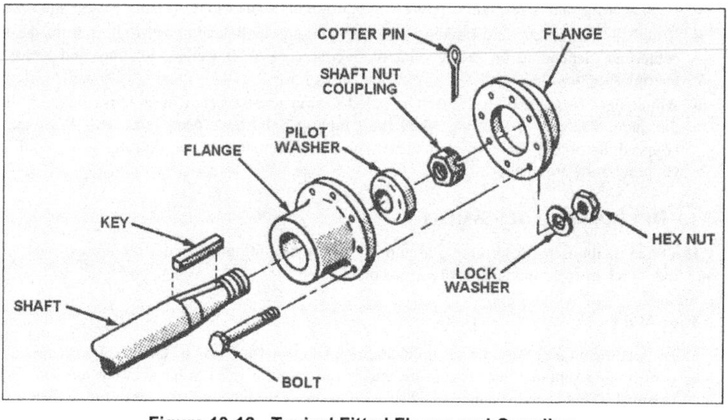

Figure 13-19. Typical Fitted Flange and Coupling

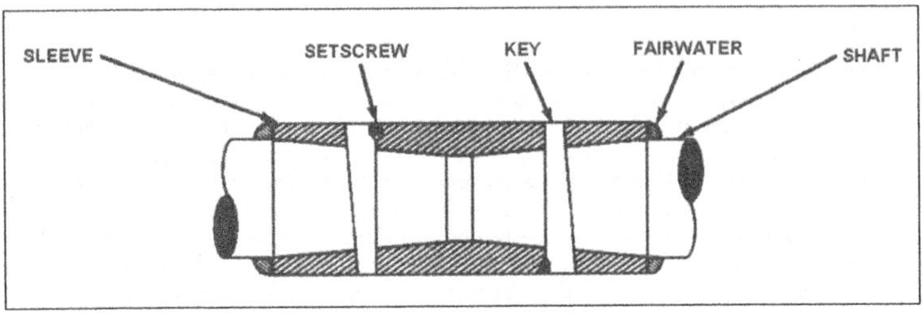

Figure 13-20. Typical Sleeve Coupling

ALIGNMENT OF SHAFTING AND BEARINGS

MAINTAINING ALIGNMENT

13-71. Alignment of shafting and bearings is not permanently fixed. The alignment changes with every docking. This is due to changes in the keel blocking, temperature variations, and direction of the sun's rays relative to the fore and aft line of the vessel. The alignment of shafting and bearings is affected by the temporary removal of machinery attached to the shafting or near the shafting because of the redistribution of weights and stresses. The alignment is not the same when the vessel is waterborne as when it is in dry dock. The final alignment and bolting of the main propulsion shafting should always be done when the vessel is waterborne. The primary purpose of correct alignment is to eliminate shaft-exited vibrations and to prevent an excessive pressure upon any localized portion of the shafting bearing surfaces (journal bearing areas). The longitudinal line connecting the lowest extremities of all shafting journals having the same diameter should form a continuous faired line when the machinery is at operating temperature. When the shafting is correctly aligned at rest, the bottoms of the shaft journals should be in contact with the bearing material. The bearing clearance at the horizontal centerline of the journal should be equally divided. To obtain and maintain acceptable alignment, the fundamentals of long-established practice are as follows:

- Each bearing shall guide and support its proportionate share of the shafting weight and load.
- When shaft couplings are broken, each overhanging shaft length will deflect from the true shaft centerline, depending on the amount of overhanging shaft weight, loading, and location of the bearing supports.
- Alignment-of-sag charts have been prepared for most vessels showing relative flange positions and the angular slopes of shafting when the coupling bolts have been removed. With the bearings adjusted to obtain these measurements, proper alignment of the shafting is ensured when the coupling bolts are secured.

METHODS OF DETERMINING ALIGNMENT

13-72. The three methods for determining alignment are by running a line wire, the optical method, and the flange method. Each of these methods is described below.

Running a Line Wire

13-73. The proper location of the bearings on main propulsion shafting may be checked by running a line wire. This consists of rigging supports just clear of the end of the outer bearings of the set to be aligned. A length of piano wire is stretched between the supports. The supports must be rigid and not subject to deflection when the wire is put under tension. The wire should be attached to the supports so that it can be accurately centered in the end bearings. After the wire has been centered in the end bearings, the wire forms the line of reference (when corrected for sag) for all-the intervening bearings.

Optical Method

13-74. Alignment of shafting by the optical method makes use of the line of sight, which for all practical purposes is a true line. This method consists of boring a large hole in the ends of two boards. One board is fitted at each end bearing of the set to be aligned. A small hole (about 1/16 to 1/8 inch) is drilled in two pieces of thin sheet metal. The sheet metal is placed on the boards and the small holes are adjusted so that they are aligned with the center of the end bearings. Similar boards are prepared for the other bearings. A light is placed behind the board on one end and observed through the hole in the board at the other end. The intermediate boards are adjusted so that the light can be seen through all the holes. The center of these holes serves to establish the reference line.

Flange Method

13-75. When it is suspected that the shaft is out of alignment, it should be checked by slacking the coupling bolts at a coupling near the suspected area on the shaft. Feelers are inserted between the coupling flanges. If there is a greater distance between the faces at one part of the coupling than at another, the shafts are out of alignment at these places.

Chapter 14

DIESEL ENGINE OPERATING PROCEDURES

INTRODUCTION

14-1. This chapter summarizes the theoretical material of the preceding chapters by applying it to the practical problems of operating diesel engines. Since diesel engines differ widely in design, size, and application, no attempt is made to discuss procedures peculiar to specific installations. Descriptions apply generally to various auxiliary and propulsion diesel installations in detailed and specific information. Operating instructions are provided in the manufacturer's manual or applicable technical manual (TM).

OPERATING INSTRUCTIONS FOR DIESEL ENGINES

14-2. Diesel engines are started, operated, and secured under a variety of conditions. The most demanding of these are emergencies and casualties in engine supporting systems. Operation under unusual conditions requires knowledge and understanding of the engine installation, the function of supporting systems, and the reasons for the procedures observed in normal operation. For the normally encountered situations under which engines are operated, checklists have been prepared for each installation. The applicable list of procedures must be followed for each condition of engine operation. Precautions listed in the operating instructions or posted at the operating station must be observed. Diesel engines are brought up to starting speed by hydraulic, electric, or air-starting systems or by air-powered starting motors. The general starting procedure for all types of systems consists of the following:

- Preoperational checks.
- Alignment of supporting systems.
- Cranking of the engine with the starting equipment until ignition occurs and the engine is running.

Steps in the starting procedure differ depending on whether the engine is being started after routine securing, after a brief period of idleness, or after a long period of idleness. First listed are the steps to follow after routine securing.

STARTING AFTER ROUTINE SECURING

14-3. First, the supporting systems are made ready for operation. These are the cooling, lubrication, and fuel systems. The checklist follows:

- Check all valves in the seawater cooling system to ensure that the system is lined up for normal operation.
- Start the separate motor-driven seawater pump if one is provided. If an auxiliary engine is cooled from a ship's saltwater circulating system, ensure that adequate pressure and flow will be available.
- Vent seawater coolers, using the vent cocks or vent valves on the heat exchanger shells. Air or gas can accumulate and reduce the effective surface of a heat exchanger if this is not done.
- Check the level in the freshwater expansion tank. Remember that a cold expansion tank will need a lower fluid level than one that is hot, so leave room for expansion.
- Check the freshwater cooling system. Set all valves in their operating positions, start the motor-driven circulating pump if one is provided, vent the system, and check freshwater level in the expansion tank again. It may have dropped if much air or gas was vented elsewhere in the system.
- Check the lubricating system. Check oil level in the sump; add any necessary oil to bring it to the proper level. Ensure that adequate grease is applied to pump bearings if these require grease lubrication. If oil heaters are installed, raise the lubricating oil to at least 100° F.

- In idle engines, the lube oil film can be lost from the cylinder wall. It is desirable to restore this film before actually starting the engine. Large diesel engines have the necessary features to allow restoration of the film by jacking the engine over without starting it. The pressure in the lube oil system will oil the cylinders and the pistons will distribute the oil film. The lubricating system is pressurized either by starting the motor-driven lubricating oil pump, if there is one, or by operating a hand-operated lubricating oil pump. If the lubricating oil pump is driven by the engine, it will develop pressure when the engine is jacked over. To reduce load on the jacking gear and prevent an accidental start, open any cylinder test valves or indicator cocks. Then turn the engine over using the jacking gear, which may be motor-driven or hand-operated.
- When the preceding operation has been performed, disengage the jacking gear and restore the valves and cocks to their operating positions.
- Line up and prime the fuel system. Check to ensure that there is sufficient clear fuel for the anticipated engine operation.
- Check for malfunction in alarms (such as low-pressure lubricating oil alarm and freshwater high-temperature alarm).
- The engine can now be started with the starting system.
- Follow all proper procedures for the type of starting system in use.
- Once the engine is running, energize the low-pressure, lube-oil alarm and the water temperature alarm. Pay careful attention to all gauges and other indications of engine condition and performance. Diesel engines tend to be noisy, particularly so when cold and idling. Familiarity with the normal sounds of the engine will help avoid unnecessary panic. Lubricating oil pressure is the best indication that a cold engine is operating properly. If it does not rise immediately to the operating pressure, the engine should be shut down and the cause of low pressure determined.
- If possible, avoid placing a load on the engine until it has warmed up. Loading a cold engine will produce carbon in the cylinder heads, cause excessive engine wear, and dilute the lubricating oil. The procedures for placing the engine on the line will depend on the type of installation. In general, it is best to bring it up to speed gradually while being alert to signs of malfunction.

Starting After a Brief Period of Idleness

14-4. Starting a warm engine after it was recently secured and if no unusual conditions are suspected consists of aligning the systems that may have been secured. Ensure valve piping is aligned (such as circulating water, fuel piping, and lube oil piping). Carefully observe lubricating oil pressure. The temperature of coolant may exceed normal operating temperatures for a minute or so until the heat accumulated in the secured engine is removed.

Starting After Overhaul or Long Idle Period

14-5. Some additional checks and inspections should be made when the engine to be started has been idle for a long time or has been overhauled.

- Inspect any parts of the engine system that have been worked on. This is to ensure that the work is complete, that covers have been replaced, and that it is safe to operate any valves or equipment that have been tagged out of service.
- Check all pipe connections to see whether the connections are tight and whether the systems have been properly connected.
- Fill the freshwater cooling system with freshwater if it has been drained. Be sure coolant flows through all parts and components of the system. Vent the system. If possible, apply a hydrostatic test to the cooling system.
- Check the lubrication system thoroughly. Check sump level; fill if necessary. If a separate oil pump is installed, the system can be primed when a slight pressure is registered on the engine oil pressure gauge. Then check visually, with inspection plates removed, to see whether oil is present at all points of the system and in each main bearing. Examine pipes and fittings for leaks. If lubricators are installed, be sure they are filled.

- Inspect air receiver, filter, and blower discharge passages for cleanliness and remove any oil accumulations.
- If the engine has a hydraulic governor, inspect the governor oil level. If an overspeed trip is installed, be sure it is in proper operating condition and position.
- Examine all moving parts of the engine to see that they are clear for running. Check the valve assemblies (including the intake, exhaust, and air-starting valves) for freedom of movement.
- Check fuel injector timing.
- Inspect the fuel oil service tank for the presence of water and sediment. Fill the tank with clean oil if necessary. Start auxiliary fuel pumps, if installed, and see whether fuel pressure gauges are registering properly. Examine fuel oil piping and fittings for leaks, especially fittings and lines inside the engine. Thoroughly vent all air from the fuel system using the vent cocks. Be sure that fuel oil strainers have been cleaned or that new filter elements have been installed. If fuel oil has been standing in the same tank, it should be run through a centrifuge to purify it.
- If the engine has an air-starting system, open the lines on the system and blow them out. Reconnect these lines and pressurize the starting air banks.
- Make a final check to ensure that all parts are in place. Then open all scavenging-air header and exhaust header manifold drains. After making the preparations, the engine is ready to be started, using the procedures for a routinely secured engine.

NORMAL OPERATING PROCEDURES

14-6. Operation of a diesel engine cannot be separated from the operation of the equipment it is driving. Therefore, assume that the operator is fully aware of the complete system he/she is running. Each type of engine and installation has its special requirements of operating routine. A systematic procedure has usually been established, based on these special requirements and on the experience of the engine operators with the particular installations. This procedure should be respected and followed. The following description of procedures is general and should be considered as incomplete in terms of operation of any specific plant. While engines are operating, their performances are monitored and observed for the following two purposes.

- The first is early recognition of unsatisfactory operation or impending malfunctions so that immediate casualty control procedures can be started.
- The second is to develop a comparative record over a given period so that gradually deteriorating conditions can be detected.

For the latter purpose, a complete log of all operating conditions must be kept. The operating pressures and temperatures should be observed and recorded in the log at hourly intervals. Entries over a period of time should be compared and deviations from normal conditions should be noted. Operators should also be alert to changing or unusual noises made by operating machinery. Gradually changing sounds are difficult to detect, especially by inexperienced operators. Often an oncoming watch will detect a new sound of which the present watch was not aware. Unusual operating noise usually indicates conditions for which load, lubrication, cooling, engine speed, or fuel supply are responsible, either directly or indirectly.

LOAD

14-7. The manner of applying a load to an engine and the regulation of the load depend on the type of load and system design. Procedures for loading an engine or placing it on the line will be established by local policies or, in new installations, by system designers. Whenever a cold engine is started, ample time should be allowed to build the load up gradually. Never load the engine heavily until it has warmed up. The manufacturer's instructions should be followed in all but emergency situations. Gradual application of the load will prevent damage to the engine from such conditions as uneven rates of expansion and inadequate lubrication at low temperatures. A diesel engine should not be operated for prolonged periods with less than one-third of its rated load. Combustion at low load is incomplete. Partially burned fuel oil and lubricating oil may cause heavy carbon deposits which will foul the valve stems, the piston rings, and the exhaust systems. A low-load operation could be responsible for the following problems:

- Sticking and burning of exhaust valves.
- Dilution of lubricating oil.
- Scuffing of cylinder liners.
- Increased fuel consumption.
- Excessive smoke when the load is increased.

If an engine must be operated at less than 30-percent power for more than 30 minutes, the load should be increased to above 50-percent power at the first opportunity. Diesel engines are designed to operate up to full-load conditions for prolonged periods. They should never be operated at an overload except in an emergency. This includes both excessive torque and speed loads. Overload may be indicated by excessive temperatures, smoky exhaust, or excessive firing pressures. When conditions indicate an overload, the load should be reduced immediately.

LUBRICATION

14-8. The importance of lubrication is emphasized. The performance of the lubrication system is one of the most important factors of engine operation which the operator can monitor. Indicators continuously show oil temperature and pressure in key parts of the system. While the engine is operating, these indicators and sight glasses should be monitored regularly. An alarm of some sort will usually warn of low pressure, but if one is not installed, oil pressure must be continuously monitored. If the lubrication system uses a wet sump, the level should be checked at regular intervals. Otherwise, the amount of oil in the system should be observed by the available means at regular intervals. Under typical operating conditions, operators should be able to estimate the rate at which the engine burns its lubricating oil and to predict when replenishment will be needed. The condition and the cleanliness of lubricating oil are critical for long engine life. If lubricating oil purifiers are provided, they should be kept running while the engines are running. When the engines are idle, the purifiers should be operated at regular intervals. Metal-edge lubricating oil strainers should be cleared by rotating the cleaning handle two complete revolutions. This is done each watch. The condition of filters is often indicated by the amount of pressure drop across them. Gauges are installed to indicate this differential and should be checked frequently. Equipment is usually available to check viscosity. It should be used daily to test the lubricating oil to determine the percentage of fuel dilution.

PRESSURES AND TEMPERATURES

14-9. All pressures and temperatures must be maintained within the normal operating ranges. If this is not possible, the engine must be secured. All instruments must be checked frequently. The manufacturer's instructions provide detailed information concerning proper operating pressures and temperatures. When this information is not available, the temperature of the lubricating oil as it leaves the engines should be maintained between 140° F and 180° F. The temperature of freshwater should not be less than 140° F or more than 180° F when the water leaves the engine. Temperatures in the saltwater cooling system should not be allowed to go above 130° F. Higher temperatures will cause deposits of salt and other solids in the coolers and piping and will cause corrosion. To ensure efficient operation in engines that are cooled by saltwater, the temperature of the saltwater coolant should never be allowed to drop below 100° F at the engine discharge. Frequent checks of the cooling system should be made to detect any leaks. Coolers and heat exchangers should be vented at least once each watch. The level of freshwater in the expansion tank should be checked frequently and freshwater added as necessary. If the freshwater level gets low enough to cause overheating of the engine, cold water should never be added until the engine has cooled.

CRITICAL SPEEDS

14-10. Vibrations resulting from operation at destructive critical speeds will cause serious damage to an engine. All moving members of machinery have critical speeds. The term applies to certain ranges of speed during which excessive vibration in the engine is created. Every part of the engine has a natural period of vibration or frequency. When impulses set up a vibration which coincides with the natural frequency of the body, each impulse adds to the magnitude of the previous vibration. Finally, the vibration becomes great enough to damage the engine structure. Vibration may be set up by linear impulses from reciprocating parts or by torsional impulses from rotating members. The crankshaft is the member which causes torsional vibrations. Pressure impulse on the piston puts a twist in the crankshaft. When the pressure on the piston is somewhat relieved, the shaft untwists. If pressure impulses, which are timed to the natural period of the shaft, are permitted to continue, the amplitude of vibration will become so great that the shaft may be fractured. However, if the speed of such an engine is changed, pressure impulses will no longer coincide with the natural period of the shaft. The vibration will then cease. Since each engine has a natural period of vibration which cannot be changed by the operator, the only control available to the operator is to avoid operating the engine at critical speeds. If critical speeds exist below the normal speed of the engine, the critical ranges must be passed through as quickly as possible when engine speed is being changed. Detailed information concerning critical speed ranges is provided with each installation. Tachometers must be marked to show any critical speed ranges to make it easier for the operator to keep the engine out of the critical ranges. Tachometers sometimes get out of adjustment. Therefore, they must be frequently checked with mechanical counters.

FUEL

14-11. An adequate supply of the proper type of fuel must be maintained. The fuel system should be frequently checked for leaks. All fuel oil strainers should be cleaned at periodic intervals. Fuel oil filter elements should also be replaced whenever necessary. When diesel fuel oil purifiers are provided, all fuel should be purified before it is transferred to the service tank. The service tank should be frequently checked for water and other settled impurities. This can be done by taking samples from the drain valve at the bottom of the tank. Water and impurities should be drained off.

STOPPING AND SECURING PROCEDURES

14-12. Diesel engines are stopped by shutting off the fuel supply (placing the throttle or the throttle control in the stop position). If the engine installation permits, it is a good idea to let the engine idle without load for a short time before stopping it. This is to allow the engine temperatures to be reduced gradually. It is also good practice to operate the overspeed trip when stopping the engine. This is to check the operating condition of the device. Before tripping the overspeed trip, reduce engine speed to low idling speed. Some overspeed trips reset automatically. However, in some installations, the overspeed trip must be reset manually before the engine can be started again. In addition to the detailed procedures listed in checklists and manufacturer's manuals, the following steps should be taken after an engine has stopped:

- Open the drain cocks on the exhaust lines and those on the scavenging-air inlet headers, if provided.
- Leave open an adequate number of indicator cocks, cylinder test valves, or hand-operated relief valves to indicate the presence of any water in the cylinders.
- See that the air pressure is off. If starting air is left on, the possibility of a serious accident exists.
- Close all sea valves.
- Allow the engine to cool.
- Drain the freshwater when freezing temperatures prevail, unless an antifreeze solution is being used.
- Clean the engine thoroughly by wiping it down before it cools. Clean the floor plates and see that the bilges are dry.
- Arrange to have any malfunctions repaired. No matter how minor they appear to be, repairs must be made and problems must be corrected promptly.

PRECAUTIONS IN OPERATING DIESEL ENGINES

14-13. The specific safety precautions for a given engine must be obtained from its manufacturer's operating instructions. In addition to those of the manufacturer, the following precautions should be observed in operating and maintaining a diesel engine.

RELIEF VALVES

14-14. If a relief valve on an engine cylinder lifts (pops) several times, the engine must be stopped immediately. The cause of the trouble must be determined and remedied. Relief valves must never be locked in a closed position except in an emergency. Pressure-relief mechanisms are fitted on all enclosures in which excessive pressures may develop. Strict compliance with designated adjustments on these mechanisms is essential.

FUEL

14-15. Make sure that fuel is not pumped into a cylinder while valves are being tested or while the engine is being motored. This is because an excessive pressure may be created in the cylinders when combustion of the fuel takes place. When it reaches the injection system, the fuel should be absolutely free of water and foreign matter. The fuel must be centrifuged thoroughly before use. The filters must be kept clean and intact. Fuel oil leakage into the lubricating oil system causes dilution of the lubricating oil with consequent reduction in viscosity and lubricating properties.

WATER

14-16. Do not allow a large amount of cold water, under any circumstances, to enter a hot engine suddenly. Rapid cooling may crack a cylinder liner and head or seize a piston. Stop the engine when the volume of circulating water cannot be increased and the temperature is too high. In freezing weather, all spaces which contain freshwater and which are subject to freezing must be carefully drained unless an antifreeze solution is added to the water.

AIR

14-17. When engines are stopped, all starting-air lines must be vented. Serious accidents may result if pressure is left on. Intake air, air ducts, and passages must be kept as clean as possible.

CLEANLINESS

14-18. Cleanliness is one of the basic essentials in the efficient operation and maintenance of diesel engines. Clean fuel, clean air, clean coolants, clean lubricants, and clean combustion must be maintained. Engines must be kept clean and accumulation of oil in the bilges or in other pockets must be prevented.

EMERGENCY DIESEL GENERATORS

14-19. Large Army ships are equipped with diesel-driven emergency generators. Diesel engines are most suitable for this application because of their self-sufficiency and quick-starting ability. Emergency generators furnish power directly to the electrical auxiliaries, radio, radar, and vital machinery spaces. Emergency generators also serve as a source of power for casualty power installations. All engineering personnel should become familiar with the emergency system aboard their ship. Each emergency generator has its individual switchboard and switching arrangement. This is for control of the generator and for distribution of power to certain vital auxiliaries and a minimum number of lighting fixtures in vital spaces. The capacity of the emergency unit varies with the size of the ship on which it is installed. Regardless of the size of the installation, the principle of operation is the same. Emergency diesel engines are started either by compressed air or by a starting motor. The engines are designed to develop full rated-load power within 10 seconds. In a typical installation, the starting mechanism is actuated when the ship's supply voltage on the bus falls to approximately 80 percent of normal. In a 440-volt system, this would be approximately 350 volts. The generators are not designed for parallel operation. Therefore, when the ship's supply voltage fails, a transfer switch automatically disconnects the emergency switchboard from the main distribution switchboard and connects the emergency generator to the emergency switchboard. With this arrangement, transfer from the emergency switchboard is accomplished manually. Then, the emergency generator must be stopped by hand and manually reset for automatic starting. Since the emergency diesel generators are of limited capacity, only certain circuits can be supplied from the emergency bus. These include such circuits as the steering gear and the interior communication switchboard. If some vital circuit is secured, another circuit may then be cut in, up to the capacity of the generator.

Operating Instructions

14-20. Normally, the emergency diesel generator will start automatically. However, for test purposes, and under other conditions, it may be started and operated manually. The following are the principal points for operation of an air-started diesel generator:

- The engine is started automatically when the ship's supply current fails and causes the solenoid air valve (located between the starting-air tank and the engine) to open, admitting starting air to the engine. The engine then turns over on air until firing begins. As the engine speed increases, the air cutoff governor valve closes and shuts off the starting air. As soon as the normal operating speed is reached and the generator develops normal voltage, the solenoid air valve also closes to shut off the starting air supply. The starting-air tank is charged from the high-pressure air system through a reducing valve. The pressure carried in the starting-air tank varies from 300 psi to 600 psi, depending on the installation.
- To start the engine manually, the solenoid valve must be de-energized. If the ship's supply current is not broken, the switch in the solenoid circuit must be opened. Starting air can then be admitted to the engine by opening the valve manually with the handwheel. After firing begins, the handwheel should be turned to close the valve and cut off the starting air. The handwheel must be turned to the open position of the valve whenever it is desired to leave the generator set available emergency service.
- If the lubricating oil pressure does not build up immediately after the engine starts, shut down the engine and determine the cause of the trouble. Never operate the engine without lubricating oil pressure. At regular intervals, the lube oil pressure, fuel pressure, cooling water temperature, and exhaust temperature should be checked. In addition, the fuel oil and lubricating oil filters should be cleaned regularly.
- To shut down the engine, hold the fuel-control lever to the STOP position. After the lever is released, it automatically returns to the running position to permit the engine to be restarted.

Operating Precautions

14-21. The following operating precautions must be observed:

- Do not operate the engine without lubricating oil pressure. Operating with insufficient lube oil pressure would cause serious damage.
- Do not operate the engine in overloaded or unbalanced condition. Overload condition on one or more cylinders may be indicated by an increase in the exhaust temperature or by smoky exhaust.
- Do not operate the engine with an abnormal water outlet temperature.
- Do not operate the engine after an unusual noise develops; the noise might be an indication of pending trouble. Investigate the noise and correct any problem, particularly if it may prove harmful to the engine.
- If there is danger of freezing during shut-down periods, drain all water jackets.
- If the overspeed device trips and shuts down the engine, investigate the cause of the trouble before restarting the engine.
- Make certain that the fittings of the ventilation system serving the compartment in which the engine is located are open. If the diesel engine was started with the vent system secured, the engine would expend the air in the compartment. Under such conditions, the engine may continue to operate long enough to suffocate the operator. This applies to installations where the engine does not have a direct air supply from the outside to the intake manifold. These precautions are also applicable to emergency diesel fire pumps.

INSPECTION AND MAINTENANCE

14-22. Inspection and maintenance are vital to successful casualty control because they minimize the occurrence of casualties by material failures. Continuous and detailed inspections are necessary to discover partly damaged parts which may fail at a critical time. Also, they serve to eliminate the underlying conditions which lead to early failure (maladjustment, improper lubrication, corrosion, erosion, and other causes of machinery damage). Particular and continuous attention must be paid to the following symptoms of malfunctioning:

- Unusual noises.
- Vibrations.
- Abnormal temperatures.
- Abnormal pressures.
- Abnormal operating speeds.

Operating personnel should thoroughly familiarize themselves with the specific temperatures, pressures, and operating speeds of equipment required for normal operation so that departures from normal will be more readily apparent. If a gauge or other instrument for recording operating conditions of machinery gives an abnormal reading, the cause must be fully investigated. Installation of a spare instrument or a calibration test will indicate whether the abnormal reading is due to instrument error. Any other cause must be traced to its source. Because of the safety factor commonly incorporated in pumps and similar equipment, considerable loss of capacity can occur before any external evidence is readily apparent. Changes in operating speed (from normal for the existing load) of pressure-governor controlled equipment should be viewed with suspicion. Variations from normal pressures, lubricating oil temperatures, and system pressures indicate either inefficient operation or poor condition of machinery. When a material failure occurs in any unit, a prompt inspection should be made of all similar units to determine whether there is any danger that a similar failure might occur. Prompt inspection may eliminate a wave of repeated casualties. Strict attention must be paid to lubrication of all equipment. This includes frequent inspection and sampling to ensure that the correct quantity of the proper lubricant is in the unit. It is good practice to make a daily check of samples of lubrication oil in all auxiliaries. Such samples should be allowed to stand long enough for any water to settle. When auxiliaries have been idle for several hours, particularly overnight, a sufficient sample to remove all settled water should be drained from the lowest part of the oil sump. Replenishment with fresh oil to the normal level should be included in this routine. The presence of saltwater in the oil can be detected by drawing off the settled water by means of a pipette and by running a standard chloride test. A sample of sufficient size for test purposes can be obtained by adding distilled water to the oil sample. It should be shaken vigorously and then the water allowed to settle before draining off the test sample. Because of its corrosive effects, saltwater in the lubricating oil is far more dangerous to a unit than is an equal quantity of freshwater. Saltwater is particularly harmful in units containing oil-lubricated ball bearings.

CORRECTION AND PREVENTION OF CASUALTIES

14-23. The speed with which an engineering casualty is corrected is frequently of paramount importance. This is particularly true when dealing with casualties which affect the propulsion power, steering, and electrical power generation and distribution. If casualties associated with these functions are allowed to become cumulative, they may lead to serious damage to the engineering installation. This damage often cannot be repaired without loss of the ship's operating availability. When possible risk of permanent damage exists, the commanding officer is responsible for deciding whether to continue operation of equipment under casualty conditions. Such action can be justified only when the risk of even greater damage or loss of the ship may be incurred by immediately securing the affected unit. To reemphasize, whenever there is no probability of greater risk, the proper procedure is to secure the malfunctioning unit as quickly as possible even though considerable disturbance to the ship's operations may occur. Although speed in controlling a casualty is essential, action should never be undertaken without accurate information. Otherwise, the casualty may be mishandled and irreparable damage and possible loss of the ship may occur. Speed in handling of casualties can be achieved only by a thorough knowledge of the equipment and associated systems and by thorough and repeated training in the routine required to handle specific predictable casualties.

Casualty Control Training

14-24. Casualty control training must be a continuous step-by-step procedure with constant refresher drills. Realistic simulation of engineering casualties must be preceded by adequate preparation. The amount of advance preparation required is not always readily apparent. You must visualize fully the consequences of any error which may be made in handling simulated casualties originally intended to be of a relatively minor nature. The simulation of major casualties must be preceded by a complete analysis and by careful instruction to all participants. A new crew member must be given an opportunity to become familiar with the ship's piping systems and equipment prior to simulating any casualty which may have other than purely local effects. In the preliminary phases of training, a "dry run" is a useful device for imparting knowledge of casualty control procedures without endangering the ship's equipment by too realistic a simulation of a casualty. Under this procedure, a casualty is announced and all individuals are required to report as though action were taken (an indication must be made that the action is simulated). Definite corrective actions can be made. With careful supervision, the timing of individual actions can appear to be very realistic. Regardless of the state of training, such dry runs should always be carried out before actually attempting to simulate realistically any involved casualty. Similar rehearsals should precede the simulation of relatively simple casualties whenever an appreciable number of personnel involved are new to the ship. This is particularly true after an interruption of regularly conducted casualty control training.

Phases of Casualty Control

14-25. Handling of any casualty can usually be divided into the following three phases:

- Limitation of the effects of the damage.
- Emergency restoration.
- Complete repair.

Limitation of the Effects of the Damage

14-26. This first phase concerns the immediate controlling of the casualty. This first phase prevents further damage to the unit concerned and to prevent the casualty's spreading through secondary effects.

Emergency Restoration

14-27. This second phase consists of restoring, as far as practicable, the services which were interrupted as a result of the casualty. For many casualties, the completion of this phase eliminates all operational handicaps except for temporary loss of standby units which lessens ability to withstand further failure. If no damage to or failure of machinery has occurred, this phase usually completes the operation.

Complete Repair

14-28. This third phase of casualty control consists of making repairs. These repairs will completely restore the installation to its original condition.

Diesel Engine Casualties

14-29. The engineman's duties concerning engineering casualties and their control depend on the type of ship. The engineman will operate engines of various sizes, made by various manufacturers, and intended for different types of services. Detailed information on diesel engine casualty control procedures should be obtained from the manufacturer's instructions, pertinent commander's instructions, and the applicable TMs.

General Safety Precautions

14-30. In addition to following the specific safety precautions listed in the operating instructions for an engine, you must continuously exercise good judgment and common sense in preventing damage to material and injury to personnel. In general, you can aid in preventing damage to machinery by operating engines according to prescribed instructions. Damage may also be prevented by observing all rules of cleanliness when handling the parts of an engine during maintenance or overhaul. Having a thorough knowledge of your duties and being completely familiar with the parts and functions of the machinery being operated and maintained is another way you can aid in preventing damage to equipment. Damage may be prevented by maintaining machinery so that the engines will be ready for service at full power in the event of an emergency and by preventing conditions which are likely to cause fires or explosions. Personnel work safer when they thoroughly know how to perform their duties, how to use their tools and machines, and how to take reasonable precautions around moving parts.

Emergency Starting and Securing

14-31. There is a definite hazard to starting a diesel engine under emergency conditions in that personnel are rushed and they tend to be careless. There is always time to ensure that personnel are clear of external moving parts (such as belt drives and shafts) before the starting gear is actuated. If emergency repairs have been made, be sure that all tools are accounted for before closing up the engine and that all parts have been replaced before starting the engine. An engine can be started and run briefly if it has air and fuel and if the starting system will operate. It will run much longer if it has lubrication and a functional cooling system. With the exception of some boat engines that can be started by towing the boat, there is no backup for the starting system. Usually, sufficient spares and resources are available to restore any casualty to the starting system. However, if the repair is rushed, the danger of careless work increases. In an emergency, an engine may be started by lining up the fuel system and actuating the starter. Before this is done, be sure that there is a supply of air to the engine and engine compartment and that the lubricating system will operate. After starting, establish cooling water flow and review all the normal pre-starting checks as quickly as possible. If an operating engine suffers a casualty, the decision of whether to continue operating or to secure the unit must be made immediately. The condition of the ship's operation is an important factor in this decision. Risk to the ship is present in the following conditions:

- Actual combat situations.
- In severe weather.
- In narrow channels.
- In potential collision situations which include close-formation steaming.

Engines can be operated with casualties to vital auxiliaries if the function of the auxiliary unit can be produced by other means. For instance, cooling water flow can be reestablished from a fire main. An engine can operate for some time with saltwater in its cooling system if it is well rinsed afterward. If the decision is made to secure an engine that has suffered a casualty, the general rule is to stop it as soon as possible. In the case of a propulsion engine, it will usually be necessary to stop the shaft also. This may require slowing the ship until the shaft is stopped and locked with the turning gear, shaft brake, or other means. Engines can almost always be stopped by securing the flow of fuel to them. Occasionally, this does not work since a blower seal leak or similar source permits the engine to run on its own lubricating oil. If the engine cannot be braked to a stop by increasing the load on it, some means of stopping air flow to it must be found. Discharging a CO_2 extinguisher into the air intake is effective, or the air intake can be covered by some other means. If the latter is done, be sure the covering will not be sucked into the blower, causing an additional casualty.

LOCKING MAIN SHAFT

14-32. There are no standard procedures applicable to all types of diesel-driven ships for locking a main shaft. Ships that have main reduction gears, shaft locking by means of the jacking gear is permissible, provided that it has been designed for this purpose as indicated by the manufacturer's instructions. Some ships have brakes that are used for holding the shaft stationary. If no provision has been made for locking the main shaft, it is usually possible to arrange a jury rig, preferably at a flanged coupling, which will hold the shaft. As a precautionary measure, it is best to make up jury rigs in advance of an actual need for locking a shaft.

Chapter 15

TROUBLESHOOTING

INTRODUCTION

15-1. This chapter deals with troubles encountered in starting an engine and those encountered after an engine is started. The troubles discussed in this chapter are those that can be corrected without major overhaul or repair. Troubles that also can be identified by erratic operation of the engine, by warnings indicated by the instruments, and by inspection of engine parts and systems are also discussed. Remember that these troubles are general and may or may not apply to a particular diesel engine. When working with a specific engine, check the applicable technical manual (TM) for that engine.

COMBUSTION ENGINE TROUBLESHOOTER

15-2. Complete failure of a power plant at a crucial moment may imperil both ship and crew. Even comparatively minor engine trouble, if not recognized and corrected as soon as possible, may develop into a major breakdown. Therefore, every operator of an internal-combustion engine must train to be a successful troubleshooter. It may happen that an engine will continue to operate even when a serious casualty is imminent. However, if troubles are impending, symptoms will probably be present. The success of a troubleshooter depends partially upon his/her ability to recognize these symptoms when they occur. A good operator uses most of his/her senses to detect trouble symptoms. He/She may see, hear, smell, or feel the warning of trouble to come. Of course, common sense is also a requisite. Another factor upon which the success of a troubleshooter depends is his/her ability to locate the trouble after once deciding something is wrong with the equipment. Then he/she must be able to determine, as rapidly as possible, what corrective action must be taken. In learning to recognize and locate engine troubles, experience is the best teacher. Instruments play an important part in detecting engine troubles. The engine operator should read the instruments and record their indications regularly. If the recorded indications vary radically from those specified in engine operating instructions, it is a warning that the engine is not operating properly and that some type of corrective action must be taken. Familiarity with the specifications given in engine operating instructions is essential, especially those pertaining to temperatures, pressures, and speeds. When instrument indications vary considerably from the specified values, the operator should know the probable effect on the engine. When these variations occur, before taking any corrective action, the operator should be sure that such variations are not the fault of the instrument. Instruments should be checked immediately when they are suspected of being inaccurate. Periodic inspections are also essential in detecting engine troubles. The following are some of the troubles that can be discovered when performing periodic inspections:

- Failure of visible parts.
- Presence of smoke.
- Leakage of oil, fuel, or water.

Cleanliness is probably one of the greatest aids in detecting leakage. When an engine is secured because of trouble, the procedure for repairing the casualty follows an established pattern if the trouble has been diagnosed. If the location of the trouble is not known, it must be found. To inspect every part of an engine, whenever a trouble occurs, would be an almost endless task. The cause of a trouble can be found much more quickly if a systematic and logical method of inspection is followed. Generally speaking, a well-trained troubleshooter can isolate a trouble by identifying it with one of the engine systems. Once the trouble has been associated with a particular system, the next step is to trace out the system until the cause of the trouble is found. Troubles generally originate in only one system. However, remember that troubles in one system may cause damage to another system or to engine components. When a casualty involves more than one system of the engine, each system must be traced separately and corrections made as necessary. The troubleshooter must know the construction, function, and operation of various systems as well as the parts of each system for a specific engine before he/she can satisfactorily locate and remedy troubles. Many troubles may affect the operation of a diesel engine. However, satisfactory performance depends primarily on the presence of sufficiently high compression pressure and injection of the right amount of fuel at the proper time. Proper compression depends basically on the pistons, piston rings, and valve gear, while the right amount of fuel depends on the fuel injectors actuating mechanism. Such troubles as lack of engine power, unusual or erratic operation, and excessive vibration may be caused by either insufficient compression or faulty injector action.

TROUBLESHOOTING DIESEL ENGINES

15-3. Many troubles encountered by an engine operator can be avoided if prescribed instructions for starting and operating an engine are followed. The lists of troubles which follow are not complete and all of these troubles do not necessarily apply to all diesel engines because of differences in design. A successful troubleshooter generally associates a trouble with a particular system or assembly. The troubles here are discussed according to when they might be encountered, either before or after the engine starts. They are indicative of the system to which they apply. Therefore, further identification is unnecessary.

ENGINE FAILS TO START

15-4. In general, the troubles which prevent an engine from starting may be grouped under the following headings:

- Engine cannot be cranked nor barred over.
- Engine cannot be cranked but can be barred over.
- Engine can be cranked but fails to start.

Table 15-1 gives various conditions which commonly cause difficulties in cranking, jacking over, or starting an engine.

Table 15-1. Troubles Which May Prevent a Diesel Engine From Starting

ENGINE WILL NOT START		
ENGINE CANNOT BE CRANKED NOR BARRED OVER	ENGINE CANNOT BE CRANKED BUT CAN BE BARRED OVER	ENGINE CAN BE CRANKED BUT FAILS TO START
Improperly engaged jacking gear	Depleted air supply	Improper throttle setting
Seized piston	Closed air line valve	Contaminated fuel
Obstructions in cylinder	Engaged jacking gear interlock	Insufficient fuel supply
Improper bearing fit	Faulty air-starting distributor	Improper fuel
	Faulty cylinder air-starting valves	Improper fuel system
		Insufficient compression
		Tripped overspeed device
		Inoperative governor
		Inoperative cold starting device
		Insufficient cranking speed

Engine Cannot Be Cranked Nor Barred Over

15-5. Most pre-starting instructions for large engines specify that the crankshaft should be turned one or more revolutions before starting power is applied. If the crankshaft cannot be turned over, check the turning gear to be sure that it is properly engaged. If the turning gear is properly engaged and the crankshaft still fails to turn over, check to see whether the cylinder test (relief) valves or indicator valves are closed and are holding water or oil in the cylinder. When the turning gear operates properly and the cylinder test valves are open but the engine nevertheless cannot be cranked or barred over, the source of the trouble is probably of a serious nature. A piston or other part may be seized or a bearing may be fitting too tightly. Sometimes the difficulty cannot be remedied except by removing a part or an assembly. Some engines have ports through which pistons can be inspected. If inspection reveals that the piston is defective, the assembly must be removed. Figure 15-1 shows the testing for stuck piston rings through the scavenging-air distributor manifold port. If the condition of an engine without cylinder ports indicates that a piston inspection is required, the whole assembly must be taken out of the cylinder. Engine bearings have to be carefully fitted or installed according to the manufacturer's instructions. When an engine cannot be jacked over because of an improperly fitted bearing, someone probably failed to follow instructions when the unit was being reassembled.

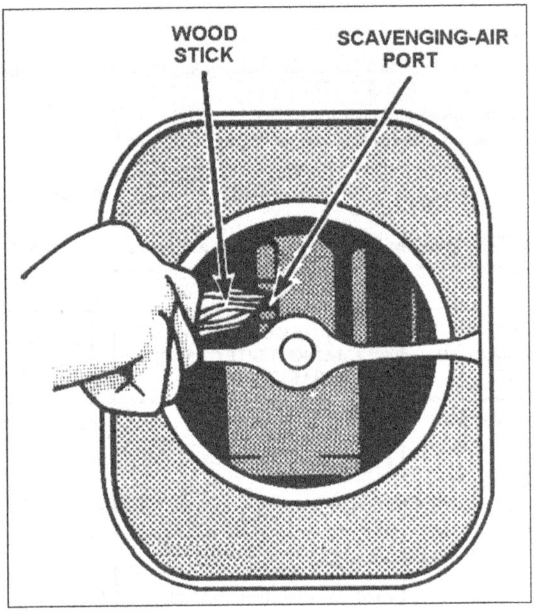

Figure 15-1. Checking the Condition of the Piston Rings

Engine Cannot Be Cranked But Can Be Barred Over

15-6. Most troubles that prevent cranking of an engine but are not serious enough to prevent barring over can be traced to the starting system. However, other factors may prevent an engine from cranking. Only troubles related to starting systems are discussed in this chapter. If an engine fails to crank when starting power is applied, first check the turning or jacking gear to be sure that it is disengaged. If this gear is not the source of trouble, then the trouble is probably with the starting system.

Air-starting System Malfunctions

15-7. Although the design of different air-starting systems varies, the function remains the same. In general, such systems must have a source of air such as the compressor or the ship's air system, a storage tank, and air flasks. An air-timing mechanism may also be used and a valve in the engine cylinder to admit air during starting and to seal the cylinder while the engine is running.

Defect In Timing Mechanism

15-8. All air-starting systems have a unit designed to admit starting air to the proper cylinder at the proper time. The type of unit as well as its name may vary from one system to another. Various names include:

- Timer.
- Distributor.
- Air-starting pilot valve.
- Air-starting distributor.
- Air distributor.

The types of air-timing mechanisms which may be encountered are the following:

- Direct mechanical lift.
- Rotary distributor.
- Plunger distributor valve.

The timing mechanism of an air-starting system is relatively trouble-free except as noted in the following situations. The operation of the direct mechanical lift air-timing mechanism involves the use of cams, push rods, and rocker arms. The mechanism is subject to parts failures similar to those occurring in corresponding major engine parts. Therefore, the causes of trouble in the actuating gear and the maintenance procedures are included with information covering similar parts of major engine systems. Most troubles are a result of improper adjustment. Generally, this involves the lift of the starting-air cam or the timing of the air-starting valve. The starting-air cam must lift the air-starting valve sufficiently to give a proper clearance between the cam and cam valve follower when the engine is running. If proper clearance does not exist between these two parts, hot gases will flow between the valve and the valve seat, causing excessive heating of the parts. Since the starting-air cam regulates the opening of the air-starting valve, those with adjustable cam lobes should be checked frequently to ensure that the adjusting screws are tight. The proper values for lift, tappet clearance, and time of valve opening for a mechanical-lift timing mechanism are given in the manufacturer's TM for the particular engine. Adjustments should be made only as specified. The rotary-distributor timing mechanism requires a minimum of maintenance. However, there may be times when the unit will become inoperative and will have to be disassembled and inspected. Generally, the difficulty is caused by a scored rotor, a broken spring, or improper timing. Foreign particles in the air can cause scoring of the rotor, which results in excessive air leakage. Therefore, the air supply must be kept as clean as possible. Another cause of scoring is lack of lubrication. If the rotor in a hand-oiled system becomes scored because of insufficient lubrication, the equipment could be at fault or lubrication instructions may not have been followed. In either a hand-oiled or a pressure-lubricated system, the piping and the passages must be checked to see that they are open. When scoring is not too serious, the rotor and body should be lapped together. A thin coat of Prussian blue can be used to determine whether the rotor contacts the distributor body. A broken spring may be the cause of an inoperative timing mechanism if a coil spring is used to maintain the rotor seal. If the spring is broken, replacing the spring is the only way to ensure an effective seal. An improperly timed rotary distributor will prevent an engine from cranking. Timing should be checked against information given in the instructions for the specific engine. In a plunger-distributor-valve timing mechanism, the valve requires little attention.

However, it may stick occasionally and prevent proper functioning of the air-starting system. On some engine installations, the pilot air valve of the distributor may not open, while on other installations this valve may not close. The trouble may be caused by the following:

- Dirt and gum deposits.
- Broken return springs.
- Lack of lubrication.

Deposits and lack of lubrication cause the unit valve plungers to bind and stick in the guides. A broken valve return spring prevents the plunger from following the cam profile. A distributor valve that sticks should be disassembled and thoroughly cleaned. Any broken springs must be replaced.

Faulty Air-starting Valves

15-9. Air-starting valves admit starting air into the engine cylinder and then seal the cylinder while the engine is running. These valves may be the pressure-actuated or mechanical lift type.

- *Pressure-actuated valve.* In a pressure-actuated valve, the principal trouble encountered is sticking. The valve may stick open for a number of reasons. A gummy or resinous deposit may cause the upper and lower pistons to stick to the cylinders. This deposit is formed by oil and condensate which may be carried into the actuating cylinders and lower cylinders. Oil is necessary in the cylinders to provide lubrication and to act as a seal. However, moisture should be eliminated. The formation of this resinous deposit can be prevented by draining the system storage tanks and water traps as specified in operating instructions. The deposit on the lower piston may be greater than that in the actuating cylinder. This is because of the heat and combustion gases which add to the formation if the valve remains open. When the upper piston is the source of trouble, sticking can usually be relieved without removing the valve. This is done by using light oil or diesel fuel and working the valve up and down. When this method is used to relieve a sticking valve, be sure that the valve surfaces are not burned or deformed. If this method does not relieve the sticking condition, the valve will have to be removed, disassembled, and cleaned. Pressure-actuated starting valves sometimes fail to operate because of broken or weak valve return springs. Replacement is generally the only solution to this condition. However, some valves are constructed with a means of adjusting spring tension. In such valves, increasing the spring tension may eliminate the trouble. Occasionally the actuating pressure of a valve will not release, and the valve will stick open or be sluggish in closing. The cause is usually clogged or restricted air passages. Combustion gases will enter the air passageways, burning the valve surfaces. These burned surfaces usually have to be reconditioned before they will maintain a tight seal. Keeping the air passages open will eliminate extra maintenance on the valve surfaces.
- *Mechanical-life valve.* The mechanical-lift valve is subject to leakage which, in general, is caused by the valve sticking open. Any air-starting valve that sticks or leaks creates a condition which makes an engine hard to start. If the leakage in the air-starting valve is excessive, the resulting loss in pressure may prevent starting. Leakage in this type valve can be caused by an overtightened packing nut. Overtightening the packing nut is sometimes used to stop minor leaks around the valve stem when starting pressure is applied. However, it may prevent seating of the air valve. As in the pressure-actuated valve, return spring tension may be insufficient to return the valve to the valve seat after admitting the air charge. If this occurs, gases from the cylinder will leak into the valve while the engine is running.

Obstructions such as particles of carbon between the valve and valve seat will hold the valve open, permitting combustion gases to pass. A valve stem bent by careless handling during installation also may prevent a valve from closing properly. If a valve hangs open for any of these reasons, hot combustion gases will leak past the valve and valve seat. The gases burn the valve and seat. This may result in a leak between these two surfaces even though the original causes of the sticking are eliminated. A leaking, valve should be completely disassembled and inspected. It is subject to a resinous deposit similar to that found in a pressure-actuated air valve. A specified cleaning compound should be used for removal of the deposit. Be sure the valve stem is not bent. Check the valve and valve seat surfaces carefully. Scoring or discoloration should be eliminated by lapping with a fine lapping compound. Jewelers rouge or talcum powder with fuel oil may be used for lapping. The preceding discussion tells you that the air-starting system may be the source of many troubles that will prevent an engine from cranking even though it can be barred over. A few of the troubles can be avoided if pre-starting and starting instructions are followed. One such instruction, sometimes overlooked, is that of opening the valve in the air line. With this valve closed, the engine will not crank. Recheck the instructions for such an oversight as a closed valve, an empty air storage receiver, or an engaged jacking gear before starting any disassembly operation.

Electric Start Malfunctions

15-10. Electric starting system malfunctions fall into the following categories:

- Nothing happens when the starter switch is closed.
- Starter motor runs but does not engage the engine.
- Starter motor engages but cannot turn the engine.

The first situation is the result of an electrical system failure. The failure could be an open circuit caused by broken connections or burned out components. Circuit continuity should be tested to ensure that the relay closes and that the battery provides sufficient voltage and current to the starter circuit. If the circuit is complete, there may be resistance through faulty battery connections. Considerable current is needed to operate the solenoid and starter motor. If the starter runs free of engagement, it will produce a distinctive hum or whine. Lack of engagement is usually caused by dirt or corrosion which prevents proper operation of the solenoid or Bendix gears. In some cases, the starter motor may engage the flywheel ring gear but is either not able to turn the engine or cannot turn it quickly enough to obtain starting speed. The cause may be lack of battery power or, more likely, a mechanical problem. If the engine can be barred over, there is excessive friction in the meshing of the starter pinion and the ring gear. Either the teeth are deformed or the starter pinion is out of alignment. Either case would have been preceded by noise the last time the starter was used. A major repair may be necessary.

Engine Can Be Cranked But Fails To Start

15-11. Even when the starting equipment is in operating condition, an engine may fail to start. A majority of the possible troubles which prevent an engine from starting are associated with fuel and the fuel system. However, parts or assemblies which are defective or inoperative may be the source of some trouble. Failure to follow instructions may be the cause of an engine failing to start. Corrective action is obvious for such items as leaving the fuel throttle in the OFF position and leaving the cylinder indicator valves open. If an engine fails to start, follow prescribed starting instructions and recheck the procedure.

Foreign Matter in the Fuel Oil System

15-12. In the operation of an internal-combustion engine, cleanliness is of paramount importance. This is especially true in the handling and care of diesel fuel oil. Impurities are the prime source of fuel pump and injection system troubles. Sediment and water cause wear, gumming, corrosion, and rust in a fuel system. Even though fuel oil is generally delivered clean from the refinery, handling and transferring increase the chances of it becoming contaminated. Corrosion frequently leads to replacement or at least to repair of the part.

- *Water.* Accumulation of water in a fuel system must be prevented, not only to eliminate the cause of corrosion but also to ensure proper combustion in the cylinders. All fuel should be centrifuged and the fuel filter cases should be drained periodically to prevent excessive collection of water. Water in fuel is injurious to the entire fuel system. It will cause irreparable damage in a short time. It not only corrodes the fuel injection pump, where close clearances must be maintained, but it also corrodes and erodes the injection nozzles. The slightest corrosion can cause a fuel injection pump to bind and seize. If not corrected, it will lead to excessive leakage. Water will cause the orifices of injection nozzles to erode until they will not spray the fuel properly, thereby preventing proper atomization. When this occurs, incomplete combustion and engine knocks result.
- *Air.* Air in the fuel system is another possible trouble which may prevent an engine from starting. Even if starting is possible, air in the fuel system will cause the engine to miss and knock, and perhaps to stall. When an engine fails to operate, stalls, misfires, or knocks, there may be air in the high-pressure pumps and lines. In many systems, expansion and compression of such air may take place without the injection valves opening. If this occurs, the pump is air-bound. You can

determine whether air exists in a fuel system by bleeding a small amount of fuel from the top of the fuel filter. If the fuel appears quite cloudy, there are probably small bubbles of air in the fuel.

Insufficient Fuel Supply

15-13. An insufficient fuel supply may result from any one of a number of defective or inoperative parts in the system. Such items as a closed inlet valve in the fuel piping or an empty supply tank are more apt to be the fault of the operator than of the equipment. However, an empty tank may be caused by leakage, either in the lines or in the tank.

- *Lines.* Leakage in low-pressure lines of a fuel system can usually be traced to cracks in the piping. Usually these cracks occur on threaded pipe joints at the root of the threads. Such breakage is caused by the inability of nipples and pipe joints to withstand shock, vibration, and strains. These result from the relative motion between smaller pipes and the equipment to which they are attached. Metal fatigue can also be a cause of breakage. Each system should have a systematic inspection of the installation of fittings and piping to determine whether all parts are satisfactorily supported and sufficiently strong. In some instances, nipples may be connected to relatively heavy parts (such as valves and strainers) which are free to vibrate. Since vibration contributes materially to the fatigue of nipples, rigid bracing should be installed. When practicable, bracing should be secured to the unit itself instead of to the hull or other equipment. Leakage in high-pressure lines of a fuel system also results from breakage. The breakage usually occurs on either of the two end fittings of a line. It is caused by lack of proper supports or by excessive nozzle opening pressure. Supports are usually supplied with an engine and should not be discarded. Excessive opening pressure of a nozzle (generally due to improper spring adjustment or to clogged nozzle orifices) may rupture high-pressure fuel lines. A faulty nozzle generally requires removal, inspection, and repair plus the use of a nozzle tester. In an emergency, high-pressure fuel lines can usually be satisfactorily repaired by silver soldering a new fitting to the line. After making a silver solder repair, test the line for leaks and be sure no restrictions exist. Leakage from fuel lines may also be due to improper replacements or repairs. When a replacement is necessary, always use a line of the same length and diameter as the one removed. Varying the length and diameter of a high-pressure fuel line will change the injection characteristics of the injection nozzle.
- *Tanks.* Although most leakage occurs in the fuel lines, leaks may occasionally develop in the fuel tank. These leaks must be repaired immediately, because of potential fire hazard. The principal causes of fuel tank leakage are improper weld and metal fatigue. Metal fatigue is usually the result of inadequate support at the source of trouble. Excessive stresses develop in the tank and cracks result.
- *Clogged fuel filters.* Another factor that can limit fuel supply to such an extent that an engine will not start is clogged fuel filters. As soon as it is known that clogging exists, the filter elements should be replaced. Definite rules for such replacement cannot be established for all engines. Instructions generally state that elements will not be used longer than a specified time. There are also reasons that an element may not function properly even for the specified interval. Filter elements may become clogged due to the following reasons:
    - Dirty fuel.
    - Too small filter capacity.
    - Failure to drain the filter sump.
    - Failure to use the primary strainer.

  Usually, clogging is indicated by such symptoms as stoppage of fuel flow, increase in pressure drop across the filter, and increases in pressure upstream of the filter. It is also indicated by excessive accumulation of dirt on the element (observed when the filter is removed for inspection). Symptoms of clogged filters vary in different installations. Each installation should be studied for such external symptoms as abnormal instrument indications and engine operation. If external indications are not apparent, visual inspection of the element will be necessary, especially if it is known or suspected that dirty fuel is being used. Fuel filter capacity should at least equal fuel supply pump capacity. A filter with a small capacity clogs more rapidly than a larger one, because the space available for dirt accumulation is more limited. Two standard sizes of fuel filter elements are small and large. The small element is the same diameter as the large but is only half as long. This construction permits substitution of two small elements for one large element. The interval between element changes can be increased through use of the drain cocks on a filter sump. Removal of dirt through the drain cock will make room for more dirt to collect. If new filter elements are not available for replacement and the engine must be operated, you can wash some types of totally clogged elements and get limited additional service. This procedure is for emergencies only. An engine must never be operated unless all the fuel is filtered. Therefore, a washed filter is better than none at all. Fuel must never flow from the supply tanks to the nozzles without passing through all stages of filtration. Strainers are the primary stage in the fuel filtration system. They must be kept in good condition if sufficient fuel is to flow in the system. Most strainers are equipped with a blade mechanism which is designed to be turned by hand. If the scraper element cannot be turned readily by hand, the strainer should be disassembled and cleaned. This minor preventive maintenance will prevent breakage of the scraping mechanism.

- *Inoperative fuel transfer pumps.* If the supply of fuel oil to the system is to be maintained in an even and uninterrupted flow, fuel transfer pumps must be functioning properly. These pumps may become inoperative or defective to the point that they fail to discharge sufficient fuel for engine starting. Generally, when a pump fails to operate, some parts have to be replaced or reconditioned. For some types, it is customary to replace the entire unit. However, for worn packing or seals, satisfactory repairs may be made. If plunger-type pumps fail to operate because the valves have become dirty, submerge and clean the pump in a bath of diesel oil. Repairs of fuel transfer pumps should be made in accordance with maintenance manuals supplied with the individual pump.

Malfunctioning of the Injection System

15-14. The fuel injection system is the most intricate of the systems in a diesel engine. The troubles which may occur depend on the system in use. Since an injection system functions to deliver fuel to the cylinder at a high pressure, at the proper time, in the proper quantities, and properly atomized, special care and precautions must be taken in making adjustments and repairs. If a high-pressure pump in a fuel injection system becomes inoperative, an engine may fail to start. Information on troubles which make a pump inoperative and information for overcoming such troubles is more than can be given in the space available here. Any ship using fuel injection equipment should have available copies of the applicable TM. Regardless of the installation or the type of fuel injection system used, maximum energy obtainable from fuel cannot be gained if the timing of the injection system is incorrect. Early or late injection timing may prevent an engine from starting. If the engine does start, it will not perform satisfactorily. Operation will be uneven and vibration will be greater than usual. If fuel enters a cylinder too early, detonation generally results. This causes pressure to rise too rapidly before the piston reaches top dead center. This in turn causes a loss of power and high combustion pressures. Low exhaust temperatures may be an indication that fuel injection is too early. The following may occur when fuel is injected too late in the engine cycle:

- Overheating.
- Lowered firing pressure.
- Smoky exhaust.
- High exhaust temperatures.
- Loss of power.

An improperly timed injection system should be corrected by following the instructions given in the appropriate TM.

Insufficient Compression

15-15. Proper compression pressures are essential if a diesel engine is to operate satisfactorily. Insufficient compression may be the reason that an engine fails to start. If low pressure is suspected as the reason, compression should be checked with the appropriate instrument. If the test indicates pressures below standard, disassembly is required for complete inspection and correction.

Inoperative Engine Governor

15-16. Many troubles may render a governor inoperative. Those encountered in starting an engine is generally caused by bound control linkage or, if the governor is hydraulic, by low oil level. Whether the governor is mechanical or hydraulic, binding of linkage is generally due to distorted, misaligned, defective, or dirty parts. If binding is suspected, linkage and governor parts should be moved and checked by hand. Any undue stiffness or sluggishness in movement of the linkage should be eliminated. Low oil level in hydraulic governors may be due to leakage of oil from the governor or failure to maintain the proper oil level. Leakage of oil from a governor can generally be traced to a faulty oil seal on the drive shaft or power piston rod or to a poor gasket seal between parts of the governor case. Oil seals should be checked if oil must be added too frequently to governors with independent oil supplies. Depending on the point of leakage, oil seal leakage may or may not be visible on external surfaces. There will be no external sign if leakage occurs through the seal around the drive shaft. Leakage through the seal around the power piston will be visible. Oil seals must be kept clean and pliable. Therefore, the seals must be stored properly so that they do not become dry and brittle or dirty. Leaky oil seals require replacement. Some leakage troubles can be prevented if proper installation and storage instructions for oil seals are followed. Most diesel engine TMs supply information on the governor installed.

Inoperative Overspeed Safety Devices

15-17. Overspeed safety devices are designed to shut off fuel or air when engine speed becomes excessive. Inoperative overspeed devices may cause an engine not to start. They may be inoperative because of improper adjustment, faulty linkage, and a broken spring. The overspeed device may also have been accidentally tripped during the attempt to start the engine. The overspeed device must always be operative before the engine is operated and must be kept operable at all times. If the overspeed device fails to operate when the engine overspeeds, the engine may be secured by manually cutting off the fuel oil or the air supply to the engine. Most engines are equipped with special devices or valves to cut off air or fuel in an emergency.

Insufficient Cranking Speed

15-18. If the engine cranks slowly, required compression temperature cannot be reached. Low starting-air pressure may be the source of such trouble. Slow cranking speed may also be the result of an increase in viscosity of the lubricating oil. This trouble is encountered during periods when air temperature is lower than usual. Oil specified for use during normal operation and temperature is not generally suitable for climate operation.

IRREGULAR ENGINE OPERATION

15-19. The engine operator must be constantly alert to detect any symptoms which might indicate the existence of trouble. Forewarning is often given in the form of sudden or abnormal changes in the supply, temperature, or pressure of the lubricating oil or cooling water. Color and temperature of exhaust afford warning of abnormal conditions and should be checked frequently. Fuel, oil, and water leaks are an indication of possible troubles. Keep the engine clean to make such leaks easier to spot. An operator soon becomes accustomed to the normal sounds and vibrations of a properly operating engine. An abnormal or unexpected change in the pitch or tone of an engine's noise or a change in the magnitude or frequency of a vibration warns the alert operator that all is not well. The occurrence of a new sound such as a knock, a drop in fuel injection pressure, or a misfiring cylinder are other trouble warnings for which an operator should be constantly alert during engine operation. The .following discussion on possible troubles, their causes, and corrective action is general rather than specific. Information is based on instructions for some of the engines used. It is typical of most, though not all, models of diesel engines. A few troubles listed may apply to only one model. Consult the applicable TM for specific information on a particular engine.

Engine Stalls Frequently or Stops Suddenly

15-20. Several troubles which may cause an engine to stall or stop are discussed under starting troubles. These troubles include the following:

- Air in the fuel system.
- Clogged fuel filters.
- Unsatisfactory operation of fuel injection equipment.
- In-correct governor action.

They not only cause starting failures or stalling but also may cause other troubles as well. For example, clogged fuel oil filters and strainers may lead to loss of power, to misfires or erratic or to low fuel oil pressure. Unfortunately, a single engine trouble does not always manifest itself as a single difficulty. It may be the cause of several major difficulties. In addition to those already mentioned, many other factors may cause an engine to stall. Some of these include the following:

- Misfiring.
- Low cooling-water temperature.
- Improper application of load.
- Improper timing.

Also included are such factors as the following:

- Obstruction in the combustion space or in the exhaust system.
- Insufficient intake air.
- Piston seizure.
- Defective auxiliary drive mechanisms.

When an engine misfires or fires erratically or when one cylinder misfires regularly, possible troubles can usually be associated with the fuel or fuel system, worn parts, or the air cleaner or silencer. In determining what causes a cylinder to misfire, follow these prescribed procedures:

- Start the engine and run it at part load until it reaches normal operating temperatures.
- Stop the engine and remove the valve rocker cover.
- Check the valve clearance. Clearance should be .009" (two-valve cylinder head) or .014" (four-valve cylinder head).
- Start the engine and hold the no. 1 injector follower down with a screw driver to prevent operation of the injector. If the cylinder has been misfiring, there will be no noticeable difference in the sound and operation of the engine. If the cylinder has been firing properly, there will be a noticeable difference in the sound and operation when the follower is held down. This is similar to short-circuiting a spark plug in a gasoline engine.
- If cylinder no. 1 is firing properly, repeat the procedure on the other cylinders until the faulty one has been located.
- Providing that the injector operating mechanism of the faulty cylinder is functioning satisfactorily, remove the fuel injector and install a new one as follows:
    - Disconnect and remove the fuel pipes from the injector and fuel connector.
    - Immediately after disconnecting the fuel pipes from the injector, install shipping caps on the filter caps to prevent dirt from entering the injector.
    - Bar the engine or crank the engine with the starting motor, if necessary, to bring the push rod end of the three rocker arms in line horizontally.
    - Loosen the two rocker arm bracket bolts holding the brackets to the cylinder head and swing the rocker arm assembly over away from the valves and injectors.
    - Remove the injector hold-down nut, washer, and injector clamp.

- Free the injector from its seat with the remover and lift it from the cylinder head; at the same time disengage the injector control rack.
- Install the new injector.
- If installation of the new injector does not eliminate the misfiring, compression pressure of the cylinder in question should be checked.

Checking Compression Pressure

15-21. To check compression pressure, use the following procedure:

- Start the engine and run it at approximately one-half rated load until normal operating temperature is reached.
- With the engine stopped, remove the fuel pipes from the injector and fuel connectors.
- Remove the injector from no. 1 cylinder and install an adaptor and pressure gauge from Diagnostic Kit J 9531-01.
- Use one of the two fuel pipes as a jumper connection between the fuel inlet and return manifold to permit fuel to flow directly to the return manifold.
- Start the engine and run it at 600 rpm. Observe and record the compression pressure indicated on the gauge.
- Do not crank the engine with the starting motor when checking the compression pressure.
- To perform the compression pressure check, steps 2 through 5 on each cylinder, repeat all but the first step. The compression pressure in any one cylinder at a given altitude above sea level should not be less than the minimum prescribed for the engines as shown in Table 15-2. In addition, the variation in compression pressure between cylinders of the engine must not exceed 25 psi at 600 rpm.

Table 15-2. Minimum Compression Pressure

	ENGINE					ALTITUDE (Feet above sea level)
	71	71E	71N	71M	71T	
PSI (600 rpm)	390	425	515	425	400	0
	360	395	480	395	370	2,500
	335	365	440	365	340	5,000
	310	340	410	340	315	7,500
	285	315	380	315	295	10,000

If the compression pressure readings of an engine operating at an altitude near sea level were as shown in the following example, it would be evident that no. 3 cylinder should be examined and the cause of the low compression pressure determined and corrected.

Example:

Cylinder	Gauge reading (psi)
1	445
2	440
3	405
4	435
5	450
6	445

Note that all of the compression pressures are above the low limit for satisfactory operation of the engine. Nevertheless, the no. 3 cylinder compression pressure indicates that something unusual has occurred and that a localized pressure leak has developed. Low compression pressures may result from any one of several causes. Piston rings may be stuck or broken and compression may be leaking past the cylinder head gasket, the valve seats, the injector tubes, or through a hole in the piston. To correct any of these conditions, consult the applicable TM that covers the faulty item.

Engine Out of Fuel

15-22. There is a problem in restarting an engine after it has run out of fuel. This stems from the fact that after fuel is exhausted from the fuel tank, fuel is then pumped from the primary fuel strainer and sometimes partially removed from the secondary fuel filter before the supply becomes insufficient to sustain engine firing. Consequently, these components must be refilled with fuel and the fuel pipes rid of air for the system to provide fuel for the injectors. When an engine is out of fuel, perform the following procedure for restarting the engine:

- Fill fuel tank with recommended grade of fuel oil. If only partial filling of the tank is possible, add at least 10 gallons.
- Remove fuel strainer shell and element from strainer cover and fill shell with fuel oil. Install shell and element.
- Remove and fill fuel filter shell and element with fuel oil as with the fuel strainer shell and element above.
- Start engine. Check filter and strainer for leaks.

Note: In some instances, it may be necessary to remove a valve rocker cover and loosen a fuel pipe nut to bleed trapped air from the fuel system. Be sure the fuel pipe is retightened securely before replacing the rocker cover.

Primer J 5956 may be used to prime the entire fuel system. Remove the filler plug in the fuel filter cover and install the primer. Prime the system. Remove the primer and install the filler plug.

Fuel Flow Test

15-23. Refer to the "Engine Operating Conditions" charts of the TM for gallons per minute fuel flow that is applicable to the particular engine being tested. Then proceed as follows:

- Disconnect fuel return tube and hold the open end of tube in a suitable receptacle.
- Start and run engine at approximately 1,200 rpm and measure fuel flow from the return tube for 1 minute.
- Be sure all tube connections between the fuel supply and the pump are tight so that no air will be drawn into the fuel system. Then immerse the end of fuel tube into fuel in the container. Air bubbles rising to the surface of the fuel will indicate a leak on the suction side of the pump.

Clogged Air Cleaners and Silencers

15-24. Sometimes the reason for an engine firing erratically or misfiring is because of clogged air cleaners and silencers. Air cleaners must be cleaned at specified intervals as recommended in the engine manufacturer's TMs. A clogged cleaner reduces intake air, thereby affecting operation of the engine. Clogged air cleaners may cause not only misfiring or erratic firing but also such difficulties as hard starting, loss of power, engine smoke, and overheating. When a volatile solvent is used for cleaning air cleaner element, it is extremely important that the cleaner be dry before it is reinstalled on the engine. Volatile solvents are excellent cleaning agents. However, if they are permitted to remain in the filter, they may cause engine overspeeding or a serious explosion. Oil bath air cleaners and filters are the source of very little trouble if serviced properly. Cleaning directions are generally given on the cleaner housing. Frequency of cleaning is usually based on a specified number of operating hours. However, more frequent cleanings may be necessary where unfavorable conditions exist. When filling an oil bath cleaner, follow the manufacturer's filling instructions. Most air cleaners of this type have a FULL mark on the oil reservoir. Filling beyond this mark does not increase the efficiency of the unit and may lead to serious trouble. When the oil bath is too full, the intake air may draw oil into the cylinders. This excess oil/air mixture, over which there is no control, may cause an engine to "run away", resulting in serious damage.

Low Cooling Water Temperature

15-25. If an engine is to operate properly, the cooling water temperature must be maintained within specified temperature limits. When cooling water temperature becomes lower than recommended for a diesel engine, ignition lag is increased, causing detonation. This results in "rough" operation and may cause an engine to stall. The thermostatic valves that control cooling water temperature operate with a minimum of trouble. Cooling water temperatures above or below the value specified in the TM sometimes indicate that the thermostat is inoperative. However, high or low cooling water temperature does not always indicate thermostat trouble. The engine load may be insufficient to maintain proper cooling water temperature, or the temperature gauge may be inaccurate or inoperative. Check these items before removing a thermostatic control unit. When a thermostat is suspected of faulty operation, it must be removed from the engine and tested. A thermostat may be checked as follows:

- A container which does not block or distort vision is needed. Fill the container, preferably a glass beaker, with water.
- Heat the water to the temperature at which the thermostat is supposed to start opening. This temperature is usually specified in the appropriate TM. Use an accurate thermometer to keep a check on the water temperature. A hot plate or a burner may be used as a source of heat. Stir the water frequently to ensure uniform distribution of heat.
- Suspend the thermostat in such a manner that operation of the bellows will not be restricted. A wire or string will serve as a satisfactory means of suspension.
- Immerse the thermostat and observe its action. Check the thermometer readings carefully to see whether the thermostat begins to open at the recommended temperature. The thermostat and thermometer must not touch the container.
- Increase the temperature of the water until the specified FULL OPEN temperature is reached. The immersed thermostatic valve should be fully open at this temperature.

The thermostat should be replaced for the following reasons:

- No movement (when it is tested).
- There is a divergence of more than a specified number of degrees between the temperature at which the thermostat begins to open or opens fully and the actuating temperatures specified in the manufacturer's TM.

The Fulton-Sylphon automatic temperature regulator is relatively trouble-free. The unit controls temperatures by positioning a valve to bypass some water around the cooler. This system provides for a full flow of water although only a portion may be cooled. In other words, the full volume of cooling water is circulated at the proper velocity, which eliminates the formation of steam pockets in the system. Generally, when the automatic temperature regulator fails to maintain cooling water at the proper temperature, improper adjustment is indicated. However, the element of the valve may be leaking or some part of the valve may be defective.

Failure to follow adjustment procedures is the only cause for improper adjustment of an automatic temperature regulator. Check and follow procedures in the manufacturer's TM issued for the specific equipment. Adjustment consists of changing the tension of the spring (which opposes the action of the thermostatic bellows). This is done with a special tool used to turn the adjusting stem knob or wheel. Increasing the spring tension raises the temperature range of the regulator and decreasing it lowers the temperature range. When a new valve of this type is in service, a number of steps must be taken to ensure that the valve stem length is proper and that all scale pointers make accurate indications. *All* adjustments should be made in accordance with the valve manufacturer's TM.

Obstruction in Combustion Space

15-26. Such items as broken valve heads and valve stem locks or keepers which come loose because of broken valve spring may cause an engine to come to an abrupt stop. If an engine continues to run when such obstructions are in the combustion chamber, the piston, liner, head, and injection nozzle will be severely damaged.

Obstruction in Exhaust System

15-27. This type of trouble is seldom encountered if proper installation and maintenance procedures are followed. When a part of an engine exhaust system is restricted, an increase in the exhaust back pressure results. This may cause high exhaust temperatures, loss of power, or even stalling. An obstruction which causes excessive back pressure in an exhaust system is generally associated with the silencer or muffler. The manifolds of an exhaust system are relatively trouble-free if related equipment is designed and installed properly. Improper design or installation may result in water backing up into the exhaust manifold. In some installations, silencer design may be the cause of water flowing into the engine. The source of water which may enter an engine must be found and eliminated. This may require replacing some parts of the exhaust system with components of an improved design. It may also require relocating such items as the silencer and piping. Inspect exhaust manifolds for water or symptoms of water. Accumulation of salt or scale in the manifold usually indicates that water has been entering from the silencer. Turbochargers on some engines have been known to seize because of saltwater entering the exhaust gas turbine from the silencer. Entry of water into an engine may also be detected by the presence of corrosion or of salt deposits on the engine exhaust valves. If inspection reveals signs of water in an engine or in the exhaust manifold, the problem should be corrected immediately. Check the unit for proper installation. Wet-type silencers must be installed with the proper sizes of piping. If the inlet water piping is too large, an excess of water may be injected into the silencer. If a silencer has no continuous drain and the engine is at a lower level than the exhaust outlet, water may back up into the engine. Dry-type silencers may become clogged with an excessive accumulation of oil or soot. When this occurs, exhaust back pressure increases. This causes troubles such as high exhaust temperatures, loss of power, or possibly stalling. A dry-type silencer clogged with oil or soot is also subject to fire. Clogging can usually be detected by fire, soot, or sparks which may come from the exhaust stack. An excessive accumulation of oil or soot in a dry-type silencer may be due to a number of factors. These are failure to drain the silencer, poor condition of the engine, or improper engine operating conditions. Silencers should be cleaned of oil and soot accumulations when necessary. Even though recommended cleaning periods may be specified, conditions of operation may require more frequent inspections and cleaning. For example, an accumulation of soot and oil is more likely to occur during periods of prolonged idling than when the engine is operating under a normal load. Idling periods should be held to a minimum.

Insufficient Intake Air

15-28. Insufficient intake air may cause an engine to stall or stop. It may be due to blower failure or to a clogged air silencer or air filter. Even though all other engine parts function perfectly, efficient engine operation is impossible if the air intake system fails to supply a sufficient quantity of air for complete combustion of the fuel. Troubles that may prevent a centrifugal blower from performing its function generally involve damage. This could be damage to the following:

- Rotor shaft.
- Thrust bearings.
- Turbine blading.
- Nozzle ring.
- Blower impeller.

Damage to the rotor shaft and thrust bearings usually occurs as a result of insufficient lubrication, an unbalanced rotor, or operation with excessive exhaust temperature. Centrifugal blower lubrication difficulties may be caused by the following:

- Failure of the oil pump to prime.
- Low lube oil level.
- Clogged oil passages or oil filter.
- Defect in the relief valve which is designed to maintain proper lube oil pressure.

Excessive vibration results, if an unbalanced rotor is the cause of shaft or bearing trouble. Unbalance may be caused by a damaged turbine wheel blading or by a damaged blower impeller. Operating a blower when the exhaust temperature is above the specified maximum safe temperature should be avoided. Such operation could cause severe damage to turbocharger bearings and other parts. Causes of excessive exhaust temperature should be found and eliminated before the turbocharger is damaged. Turbine blading damage in a centrifugal blower may be caused by a number of factors. These factors include:

- Operating with an excessive exhaust temperature.
- Operating at excessive speeds.
- Bearing failures.
- Failure to drain the turbine casing.
- Entrance of foreign objects.
- Turbine blades breaking loose.

Damage to an impeller of a centrifugal blower may result from the following:

- Thrust or shaft bearing failure.
- Entrance of foreign bodies.
- Loosening of the impeller on the shaft.

Since blowers are high-speed units and operate with a very small clearance between parts, minor damage to a part could result in extensive blower damage and failure. Although there is considerable difference in principle and construction of the positive-displacement blower (Roots) and the axial-flow, positive-displacement blower (Hamilton-Whitfield), the problems of operation and maintenance are similar. Some of the troubles encountered in a positive-displacement blower are similar to those encountered in centrifugal blowers. However, the source of some troubles may be different because of construction differences. Positive-displacement blowers are equipped with a set of gears to drive and synchronize rotation of the rotors. Many of these blowers are driven by a serrated shaft. Regardless of construction differences, the basic problem in both types of blowers is to maintain the necessary small clearances. If these clearances are not maintained, the rotors and the case will be damaged and the blower will fail to perform its function. Worn gears are one of source of trouble in positive-displacement blowers. A certain amount of gear wear is expected. However, damage resulting from excessively worn gears indicates improper maintenance procedures. During inspections, the values of backlash should be recorded in the material history. This record can be used for the following:

- Establish the rate of increase in wear.
- Estimate the life of the gears.
- Determine when to replace the gears.

Scored rotor lobes and casing may cause blower failure. Scoring of blower parts may be caused by worn gears, improper timing, bearing failure, improper end clearance, or foreign matter. Any of these troubles may be serious enough to cause contact of the rotors and extensive damage to the blower. Timing of blower rotors not only involves gear backlash but also clearances between leading and trailing edges of the rotor lobes and between rotor lobes and casing (see Figure 15-2). Clearance between these parts can be measured with thickness gauges. If clearances are incorrect, check the backlash of the drive gear. If backlash is excessive, the gears must be replaced. Then the rotors must be retimed according to the method outlined in the appropriate manufacturer's TM. Failure of serrated blower shafts may result from failure to inspect the parts or improper replacement of parts. When inspecting serrated shafts, be sure that they fit snugly and that they are not worn. When serrations of either the shaft or hub have failed for any reason, both parts must be replaced.

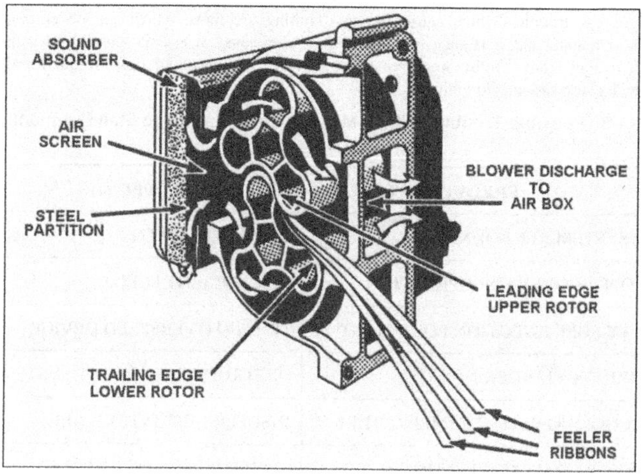

Figure 15-2. Checking Clearances of Positive-displacement Blower Lobes

Piston Seizure

15-29. Piston seizure may be the cause of an engine stopping suddenly. The piston becomes galled and scuffed. When this occurs, the piston may possibly break or extensive damage may be done to other major engine parts. The principal causes of piston seizure are insufficient clearance, excessive temperatures, or inadequate lubrication.

Defective Auxiliary Drive Mechanisms

15-30. Defects in auxiliary drive mechanisms may cause an engine to stop suddenly. Since most troubles in gear trains or chain drives require some disassembly, only the causes of such troubles are discussed. Gear failure is the principal trouble encountered in gear trains. Engine failure and extensive damage can occur because of a broken or chipped gear. If you hear a metallic clicking noise near a gear housing, it is almost a certain indication that a gear tooth has broken. Gears are most likely to fail because of a number of factors. These include the following:

- Improper lubrication.
- Corrosion.
- Misalignment.
- Torsional vibration.
- Excessive backlash.
- Wiped bearings and bushings.
- Metal obstructions.
- Improper procedures.

Gear shafts, bushings and bearings, and gear teeth must be checked during periodic inspections for scoring, wear, and pitting. All oil passages, jets, and sprays should be cleaned to ensure proper oil flow. All gear-locking devices must fit tightly to prevent longitudinal gear movement. Chains are used in some engines for cam shaft and auxiliary drives. In others, they are used to drive certain auxiliary rotating parts. Troubles encountered in chain drives usually result from wear or breakage. Troubles of this nature may be caused by improper tension, lack of lubrication, sheared cotter pins, or misalignment. Possible troubles which may cause an engine to stall frequently or stop suddenly are given in Table 15-3. Some doubt may exist as to the difference between stalling and stopping. In reality, there is none unless we associate certain troubles with each. In general, troubles which cause frequent stalling are those which can be eliminated with minor adjustments or maintenance. If such troubles are not eliminated, it is quite possible that the engine can be started, only to stall again. Failure to eliminate some of the troubles which cause frequent stalling may lead to troubles which cause sudden stopping.

Table 15-3. Possible Troubles Which May Cause an Engine to Stall Frequently or to Stop Suddenly

FREQUENT STALLING OR SUDDEN STOPPING	
FUEL SYSTEM AIR-BOUND	CONTAMINATED FUEL
IMPROPER GOVERNOR OPERATION	INSUFFICIENT FUEL
FAULTY FUEL INJECTION EQUIPMENT	TRIPPED OVERSPEED DEVICE
MISFIRING CYLINDERS	OBSTRUCTED EXHAUST SYSTEM
LOW COOLING WATER TEMPERATURE	INSUFFICIENT INTAKE AIR
IMPROPER LOAD APPLICATION	SEIZED PISTON OR BEARING
IMPROPERLY TIMED VALVES	FAULTY AUXILIARY DRIVES

Engine Will Not Carry Load

15-31. Many of the troubles which can lead to loss of power in an engine may also cause the engine to stop and stall suddenly or may even prevent its starting. Compare the list of some of the troubles that may cause a power loss in the engine in Table 15-4 with those in Table 15-1 and Table 15-3. Such items include insufficient air, insufficient fuel, and faulty operation of the governor. Many of the troubles listed are closely related, these possible troubles are similar to those and the elimination of one may eliminate already discussed in connection with starting others. The operator of an internal-combustion operation may be confronted with additional major difficulties (such as those indicated in Table 15-5). Here again you can see that many of those possible troubles are similar to those already discussed in connection with starting failures and with engine stalling and stopping. The following discussion on irregular engine operation covers only those troubles not previously considered.

Engine Overspeeds

15-32. When an engine overspeeds, the trouble can usually be associated with either the governor mechanism or the fuel control linkage, as previously discussed. When information on a specific fuel system or speed control system is required, check the manufacturer's TM and the special TMs for the particular equipment. These special manuals are available for the most widely used models of hydraulic governors and overspeed trips. They contain specific details on testing, adjusting, and repairing.

Table 15-4. Possible Causes of Insufficient Power in an Engine

IRREGULAR ENGINE OPERATION ENGINE WILL NOT CARRY LOAD OR REACH RATED SPEED (LOSS OR LACK OF POWER)	
LOW COMPRESSION	INSUFFICIENT FUEL
IMPROPERLY TIMED FUEL INJECTION	CLOGGED FUEL NOZZLES OR SPRAY TIPS, FAULTY INJECTION EQUIPMENT
OBSTRUCTIONS IN EXHAUST SYSTEM	IMPROPERLY POSITIONED FUEL CONTROL RACK
IMPROPER GOVERNOR ACTION	MISFIRING CYLINDERS
INSUFFICIENT AIR SUPPLY	ENGINE OVERLOADED

Table 15-5. Additional Causes of Irregular Engine Operation

IRREGULAR ENGINE OPERATION			
ENGINE OVERSPEEDS	**CYLINDER SAFETY VALVES POP FREQUENTLY**	**ENGINE NOT SHUT OFF OR HUNTS**	**ENGINE HUNTS (SPEED VARIES AT CONSTANT THROTTLE SETTING)**
IMPROPER FUNCTION OF GOVERNOR MECHANISM	EXCESSIVE AMOUNT OF FUEL INJECTED INTO CYLINDER	IMPROPER ADJUSTMENT OR MISALIGNMENT OF FUEL CONTROL LINKAGE	IMPROPER FUNCTIONING OF THE ENGINE GOVERNOR
FUEL PUMP CONTROL LINKAGE BINDS	INSUFFICIENT TENSION ON SAFETY VALVE SPRING	STUCK INJECTOR RACKS	LOAD FLUCTUATING TOO UNEVENLY FOR THE GOVERNOR TO MAINTAIN CONSTANT SPEED
TACHOMETER INACCURATE AND READS LOW	BROKEN SUPERCHARGE VALVE (IF APPLICABLE)	FUEL OIL LEAKAGE FROM INJECTORS	
INTAKE AIR OIL BATH CLEANER TOO FULL	CLOGGED OR PARTIALLY OBSTRUCTED EXHAUST PORTS	LUBE OIL LEAKAGE TO THE MANIFOLD AND AIR BOX	

Engine Hunts or Will Not Secure

15-33. Some troubles which may cause an engine to hunt are similar to those which may cause an engine to resist securing efforts. Generally, these two forms of irregular engine operation are caused by troubles originating in the speed control system and fuel control system.

- *Speed control system.* The speed control system of an internal-combustion engine includes those parts designed to maintain the engine speed at some exact value, or between desired limits, regardless of changes in load on the engine. Governors are provided to regulate fuel injection so that speed of the engine can be controlled as load is applied. The governor also acts to prevent overspeeding as in rough seas when the load might be suddenly reduced when the propellers leave the water. If certain parts of the fuel system or governor fail to function properly, the engine may hunt. This means that it may vary at a constant throttle setting or that it may be difficult to stop the engine.
- *Fuel control system.* Fuel control racks that have become sticky or jammed may cause governing difficulties. If the control rack of a fuel system is not functioning properly, engine speed may decrease as the load is removed or the engine may hunt continuously. It may also hunt only when the load is changed. A sticky or jammed control rack may prevent an engine from responding to changes in throttle setting and may even prevent securing. Any such condition could be serious in an emergency situation. Your job is to prevent the occurrence of such conditions if possible. You can check for a sticky rack by securing the engine, disconnecting the linkage to the governor, and then attempting to move the rack by hand. There should be no apparent resistance to the motion of the rack if the return springs and linkage are disconnected. A stuck control rack may be caused by the plunger sticking in the pump barrel or dirt in the rack mechanism. It may also be caused by damage to the rack, sleeve, or gear, or by improper assembly of the injector pump. The cause of sticking or jamming must be determined and damaged parts must be replaced. If sticking is due to dirt, a thorough cleaning of all parts will probably correct the problem. Errors in assembly can be avoided by carefully studying assembly drawings and instructions.
- *Leakage of fuel oil.* Leakage of fuel oil from the injectors may cause an engine to continue to operate when you attempt to shut it down. Regardless of the type of fuel system, results of internal leakage from injection equipment are somewhat the same. Injector leakage will cause unsatisfactory engine operation because of the excessive amount of fuel entering the cylinder. Leakage may also cause the following:
  - Detonation.
  - Crankcase dilution.
  - Smoky exhaust.
  - Loss of power.
  - Excessive carbon formation on spray tips of nozzles and other surfaces of the combustion chamber.
- *Accumulation of lube oil.* Another trouble which may prevent stopping an engine is accumulation of lube oil in the intake air passages (manifold or air box). Such an accumulation creates an extremely dangerous condition. Excess can be detected by removing inspection plates on covers and examining the air box and manifold. If oil is discovered, it should be removed and corrective maintenance should be performed. If oil is drawn suddenly in large quantities from the manifold or air box into the cylinder of the engine, and burns, the engine may run away. The engine governor has no control over the sudden increase of speed that occurs. An air box or air manifold explosion is also a possibility if excess oil is allowed to accumulate. Some engine manufacturers have provided safety devices to reduce the hazards of such explosions. Excess oil in the air box or manifold of an engine also increases carbon formation on liner ports, cylinder valves, and other parts of the combustion chamber. The causes of excessive lube oil accumulation in the air box or manifold varies depending on the specific engine. Generally, the accumulation is due to an obstruction in either the air box or separator drains. In an effort to reduce the possibility of crankcase explosions and runaways, some engine manufacturers have designed a means to ventilate the crankcase. Some engines are ventilated by a passage between the crankcase and the intake side of the blower. In other engines, an oil separator or air maze is provided in the passage between the crankcase and

blower intake. In either type of installation, stoppage of the drains will cause an excessive accumulation of oil. Drain passages must be kept open by being properly cleaned. Oil may enter the air box or manifold from sources other than crankcase vapors. The source of trouble may be from the following:

- Defective blower oil seal.
- Carryover from an oil-type air cleaner.
- Defective oil piping.
- Excessively high oil level in the crankcase.

Under such condition, an oil fog is created in some engines by moving the parts. An oil fog may be caused also by excessive clearance in the connecting rod and main journal bearings. In some types of crankcase ventilating systems, the oil fog will be drawn into the blower. When this occurs, an abnormal amount of oil may accumulate in the air box. Removal of the oil will not remove the trouble. The cause of the accumulation must be determined and repairs made. If a blower oil seal is defective, replacement is the only satisfactory method of correction. When installing new seals, be sure that the shafts are not scored and that the bearings are in satisfactory condition. Special precautions must be taken during installation to avoid damaging oil seals. Such damage is usually not discovered until the blower has been reinstalled and the engine has been put into operation. Be sure an oil seal is lubricated. The oil not only lubricates the seal, reducing friction, but also carries away any heat that is generated. New oil seals are generally soaked in clean, light, lube oil before assembly.

Cylinder Safety Valves Pop Frequently

15-34. On some engines, a cylinder relief (safety) valve is provided for each cylinder. The function of the valve is to open when cylinder pressures exceed a safe operating limit. The valve opens or closes a passage leading from the combustion chamber to the outside of the cylinder. The valve face is held against the valve seat by spring pressure. Tension on the spring is varied with an adjusting nut, which is locked when the desired setting is attained. The desired setting varies with the type of engine and may be found by referring to the manufacturer's TM. Clogging of cylinder ports can be avoided by removing carbon deposits at prescribed intervals. Some engine manufacturers make special tools for port cleaning. Round wire brushes of the proper size are satisfactory for this work. When cleaning cylinder ports, take care to prevent carbon from entering the cylinder. The engine should be barred to such a position that the piston blocks the port.

SYMPTOMS OF ENGINE TROUBLE

15-35. In recognizing symptoms that may help you locate causes of engine trouble, you will find that experience is the best teacher. Even though written instructions are needed for efficient troubleshooting, the information usually given serves only as a guide. It is difficult to describe the sensation that you should feel when checking the temperature of a bearing by hand. It is also difficult to describe the specific color of exhaust smoke when pistons and rings are worn excessively, and, for some engines, the sound that you will hear if the crankshaft counterweights come loose. You must actually work with the equipment to associate a particular symptom with a particular trouble. However, written information can save you a great deal of time and eliminate much unnecessary work. Written instructions can make detection of troubles much easier in practical situations. Symptoms which indicate that a trouble exists may be in the form of an unusual noise or instrument indication. Symptoms may also be smoke or excessive consumption or contamination of the lube oil, fuel, or water. Table 15-6 is a general listing of various trouble symptoms which the operator of an engine may encounter.

NOISES

15-36. Unusual noises which may indicate that a trouble exists or is impending may be classified as pounding, knocking (denotation), clicking, and rattling. Each type of noise must be associated with certain engine parts or systems which may be the source of trouble.

Pounding

15-37. Pounding is a mechanical knock or hammering (not to be confused with a fuel knock). It may be caused by a loose, excessively worn, or broken engine part. Generally, troubles of this nature will require major repairs.

Knocking (Detonation)

15-38. Knocking (detonation) is caused by the presence of fuel or lubricating oil in the air charge of the cylinders during the compression stroke. Excessive pressures accompany detonation. If detonation is occurring in one or more cylinders, an engine should be stopped immediately to prevent possible damage.

Clicking

15-39. Clicking noises are generally associated with an improperly functioning valve mechanism or timing gear. If the cylinder or valve mechanism is the source of metallic clicking, the trouble may be due to a number of following factors:

- Loose valve stem and guide.
- Insufficient or excessive valve tappet clearances.
- Loose cam follower or guide.
- Broken valve springs.
- A valve that is stuck open.

A clicking in the timing gear usually indicates that there are some damaged or broken gear teeth.

Rattling

15-40. Rattling noises are generally due to vibration of loose engine parts. However, other sources of trouble exist when rattling noises occur. These are an improperly functioning vibration damper, a failed antifriction bearing, or a gear pump operating without prime.

Note: When you hear a noise, first make sure that it is a trouble symptom. Each diesel engine has a characteristic noise at any specific speed and load. The noise will change with a change in speed or load. As an operator, you must become familiar with the normal sounds of an engine. Abnormal sounds must be investigated promptly. Knocks which indicate a trouble may be detected and located by special instruments or by the use of a "sounding bar" such as a solid iron screwdriver or bar.

Table 15-6. Symptoms of Engine Trouble

NOISES	INSTRUMENT INDICATIONS			SMOKE	CONTAMINATION OF LUBE OIL, FUEL, OR WATER
	PRESSURE	TEMPERATURE	SPEED		
Pounding (mechanical)	Low lube oil High lube oil pressure	Low lube oil temperature High lube oil temperature	Idling speed not normal Maximum speed not normal	Black exhaust smoke Bluish-white exhaust smoke	Fuel oil in the lube oil Water in the lube oil
Knocking (detonation)	Low fuel oil pressure (in low-pressure fuel supply system)	Low cooling water temperature (fresh)		Smoke arising from crankcase	Oil or grease in the water Water in the fuel oil
Clicking (metallic)	Low cooling water pressure (fresh) Low cooling water pressure (salt)	High cooling water temperature (fresh) Low cylinder exhaust temperature		Smoke arising from cylinder head Smoke from engine auxiliary equipment (blower, pumps, and so on)	Air or gas in the water Metal particles in lube oil
Rattling	High cooling water pressure (salt) Low compression pressure Low-firing pressure High-firing pressure Low scavenging-air receiver pressure (supercharged engine) High exhaust back pressure	High exhaust temperature in one cylinder			

INSTRUMENT INDICATIONS

15-41. An engine operator probably relies on the instruments to warn him/her of impending troubles more than on all the other trouble symptoms combined. Regardless of the type of instrument being used, the indications are of no value if inaccuracies exist. Be sure an instrument is accurate and operating properly. All instruments must be tested at specified intervals, or whenever they are suspected of being inaccurate. The following are three engine problems that are most frequently detected by either high or low instrument readings and the steps taken to correct them.

- Suggested remedy for Excessive Crankcase Pressure—
  - **Step 1.** Check compression pressure and, even if only one cylinder has low compression, remove the cylinder head and replace the head gaskets.
  - **Step 2.** Inspect piston and liner and replace damaged parts.
  - **Step 3.** Install new piston rings.
  - **Step 4.** Clean and repair or replace breather assembly.
  - **Step 5.** Replace blower-to-block gasket.
  - **Step 6.** Replace end plate gasket.
  - **Step 7.** Check exhaust back pressure and repair or replace muffler if an obstruction is found.
  - **Step 8.** Check exhaust back pressure and install larger piping if the piping is too small, too long, or has too many bends.
- Suggested remedy for Low Oil Pressure—
  - **Step 1.** Check oil level and bring it to the proper level on dipstick or correct installation angle.
  - **Step 2.** If wrong viscosity of lubricating oil is being used, consult lubricating oil specifications. Check for fuel leaks at the injector seal ring and fuel pipe connections. Leaks at these points will cause lubricating oil dilution.
  - **Step 3.** If a plugged oil cooler is indicated by an excessively high lubricating oil temperature, remove and clean the oil cooler core.
  - **Step 4.** Remove bypass valve and clean valve and valve seat and inspect valve spring. Replace defective parts.
  - **Step 5.** Remove pressure regulator valve, clean valve and valve seat, and inspect valve spring. Replace defective parts.
  - **Step 6.** Change the bearings. Consult the lubricating oil specifications for the proper grade of oil to use and change oil filters.
  - **Step 7.** Replace any missing plugs.
  - **Step 8.** Check oil pressure with a reliable gauge and replace gauge if found faulty.
  - **Step 9.** Remove and clean gauge line; replace if necessary.
  - **Step 10.** Remove and clean gauge orifice.
  - **Step 11.** Repair or replace defective electrical equipment.
  - **Step 12.** Remove and clean oil pan and oil intake screen. Consult the lubricating oil specifications for proper grade of oil to use and change oil filters.
  - **Step 13.** Remove and inspect valve and valve bore and spring. Replace faulty parts.
  - **Step 14.** Disassemble piping and install new gaskets.
  - **Step 15.** Remove pump. Clean and replace defective parts.
  - **Step 16.** Remove flange and replace gasket.

- Suggested remedy for Abnormal Engine Operating Temperatures—
  - **Step 1.** Clean cooling system with a good cooling system cleaner and thoroughly flush to remove scale deposits.
    - Clean exterior of radiator core to open plugged passages and permit normal air flow.
    - Adjust loose fan belts to the proper tension to prevent slippage.
    - Check for improper size radiator or inadequate shrouding.
    - Repair or adjust inoperative shudders.
    - Repair or replace inoperative temperature-controlled fan.
  - **Step 2.** Check coolant level and fill to filler neck if coolant level is low.
    - Inspect for collapsed or disintegrated hoses. Replace all faulty hoses.
    - If thermostat is inoperative, remove, inspect, and test it. Replace it if found faulty.
    - Check water pump for a loose or damaged impeller.
    - Check flow of water through radiator. A clogged radiator will cause an inadequate supply of water on the suction side of the pump.
    - Clean radiator core.
    - Remove radiator cap and operate engine, checking for combustion gases in the cooling system. The cylinder head must be removed and inspected for cracks and the head gaskets replaced if combustion gases are entering the cooling system.
    - Check for an air leak on the suction side of the freshwater pump.
    - Replace defective parts.
  - **Step 3.** If thermostat is closing, remove, inspect, and test it.
    - Install a new thermostat if necessary.
    - Check for an improperly installed heater.
  - **Step 4.** If continued low coolant operating temperature exists, excessive leakage of coolant past the thermostat seal is a cause. When this occurs, replace the thermostat seal.

SMOKE

15-42. The presence of smoke can be an aid in locating some types of trouble, especially if used in conjunction with other trouble symptoms. The color of exhaust smoke can also be used as a guide in troubleshooting. The color of engine exhaust is a good general indication of engine performance. The exhaust of an efficiently operating engine has little or no color. A dark, smoky exhaust indicates incomplete combustion. The darker the color, the greater the amount of unburned fuel in the exhaust. Incomplete combustion may be due to a number of troubles. Some manufacturers associate a particular type of trouble with the color of the exhaust. More serious troubles are generally identified with either black or bluish-white exhaust colors.

- Suggested remedy for Abnormal Exhaust Colors—
  - **Step 1.** Replace parts causing high exhaust back pressure. High exhaust back pressure or a restricted air inlet causes insufficient air for combustion and will result in incompletely burned fuel. High exhaust back pressure is caused by faulty exhaust piping or muffler obstruction and is measured at the exhaust manifold outlet with a manometer. Clean the items that restrict air inlet to engine cylinders such as clogged cylinder liner ports, air cleaner, or blower air inlet screen. Check emergency stop to see that it is completely open and readjust if necessary.

- **Step 2.** Check for improperly timed injectors and improperly positioned injector rack control levers. Time fuel injectors and perform appropriate governor to correct this condition.
  - o Replace faulty injectors if this condition still persists after timing injectors and performing engine tune-up.
  - o Lugging the engine will cause incomplete combustion and should be avoided.
- **Step 3.** Check for use of an improper grade of fuel. Consult the fuel oil specifications for correct fuel to use.
- **Step 4.** Check for internal lubricating oil leaks and refer to the "high lubricating oil consumption" chart.
- **Step 5.** Check for faulty injectors and replace as necessary. Check for low compression and consult the "hard starting" chart. The use of low cetane fuel will cause this condition and can be corrected by consulting and following the fuel oil specifications.

EXCESSIVE CONSUMPTION OF LUBE OIL, FUEL, OR WATER

15-43. An operator should be aware of engine trouble whenever excessive consumption of any of the essential liquids occurs. The possible troubles signified by excessive consumption depend on the system in question. However, leakage is one trouble which may be common to all. Before starting any disassembly, check for leaks in the system in which excessive consumption occurs.

- Suggested remedy for High Lubricating Oil Consumption—
  - **Step 1.** Tighten or replace parts.
  - **Step 2.** Replace defective gaskets or oil seals.
  - **Step 3.** See "excessive crankcase pressure" chart.
  - **Step 4.** See the "abnormal engine operation" chart.
  - **Step 5.** Remove air inlet housing and inspect blower end plates while engine is operating. If oil is seen on the end plate radiating away from the oil seal, overhaul the blower.
  - **Step 6.** Check engine coolant for lubricating oil contamination. If contaminated, inspect oil cooler core and replace if necessary. Then use a good grade of cooling system cleaner to remove oil from the cooling system.
  - **Step 7.** Replace oil control rings on piston.
  - **Step 8.** Replace piston pin retainer and defective parts.
  - **Step 9.** Remove and replace defective parts.
  - **Step 10.** Check crankshaft thrust washers for wear. Replace all worn and defective parts.
  - **Step11.** Decrease installation angle.
  - **Step 12.** Fill crankcase to proper level only.

Chapter 16

PROPELLERS

INTRODUCTION

16-1. A propeller is essentially a multi-thread screw which rotates in water to produce motion in a vessel. To provide balance, the propeller or screw is equipped with two or more blades. Most propellers rotate with the axis parallel to the direction of vessel motion. As the propeller turns, it acts like a screw advancing through the water. If water had the properties of a solid, the propeller would advance a definite distance with each revolution, as in the case of a bolt which is threaded into a nut. There are also vertical axis propellers, in which the axis of rotation is perpendicular to the line of thrust. This chapter is designed to familiarize you with basic types of propellers and their operation and maintenance.

BASIC TYPES OF PROPELLERS

16-2. Two types of propellers are used within the Army watercraft field. The screw and the vertical axis propellers.

SCREW PROPELLER

16-3. A screw propeller (conventional propeller) has three or more blades projecting from a hub. The use of two or more propellers on a vessel is desirable to reduce the probability of a complete breakdown. An increase in the number of propellers permits the use of smaller propulsion units. Marine propellers usually consist of either three or four blades. Two-blade propellers are practically never used because the diameter is too great. Also, the propeller would be too much out of balance to be used if one blade were lost. The common blade form is elliptical in shape with the greatest width one-half the distance from the center of the hub to the tip of the blade. The elliptical blade is more efficient than a square-tip blade. Square-tip blades absorb more power and give more thrust but with slightly less efficiency than elliptical blades.

Solid Propellers

16-4. The blades and hub of a solid propeller (see Figure 16-1) are formed from a single integral casting and may have either constant or variable pitch. A propeller with a constant pitch will have the same pitch at each radius. A propeller with variable pitch will vary at each radius from the nominal pitch and produce a particular distribution of loading over the propeller radius. The nominal pitch corresponds to the pitch at the 0.7 radius.

Built-up Propellers

16-5. The blades and hub of a built-up propeller are cast separately and may be of different material. The blades are fastened to the hub by studs and nuts. Proper orientation of the blades to the hub is essential to produce the designed propeller pitch.

VERTICAL AXIS PROPELLERS

16-6. The vertical axis propeller differs from the screw propeller in that the axis of rotation is perpendicular to the line of trust. It consists of long airfoil-sectioned blades that project downward from the bottom of the vessel. By varying the pitch of the blades, trust can be exerted in any horizontal direction. The Voith-Schneider and the Dirsten-Boeing propellers consist of an assembly containing long blades of airfoil sections projecting vertically downward from the after part of the vessel's bottom. There may be one or more such assemblies per vessel. Figure 16-2 shows the arrangement of a vertical axis propeller. The blades rotate about a vertical central axis and, at the same time, oscillate around their own axis. By changing the feathering angle, the thrust can be made to act in any direction, even in reverse. As a consequence, no rudder is needed. This results in a highly maneuverable vessel. Vertical axis propellers have not been built in vessels with greater than 2,500 horsepower. Claims for extraordinary efficiency are based on the fact that the rudder, struts, and other vessel parts can be removed from the hull, thereby reducing drag. Currently, vertical axis propellers can only be found installed on the US Army vessel LT. COL. John U.D. Page.

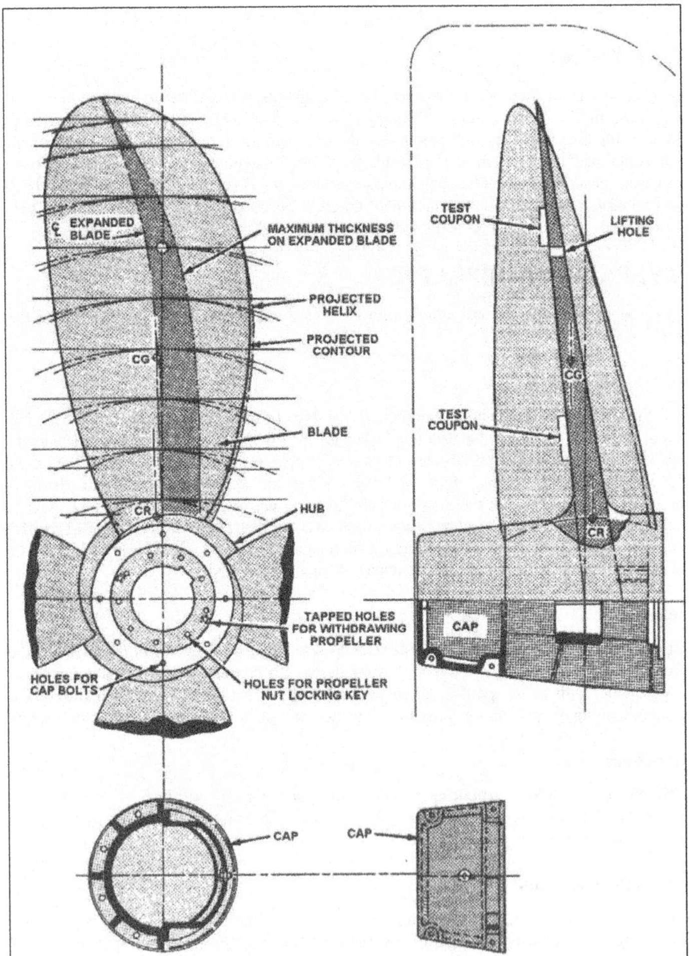

Figure 16-1. Typical Solid Propeller

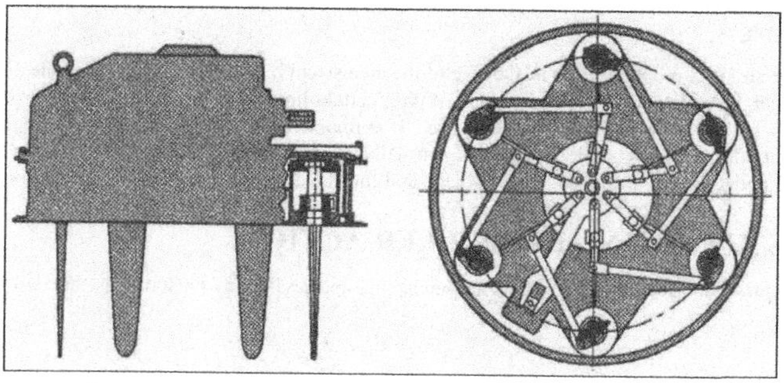

Figure 16-2. Vertical Axis Propeller Assembly With Feathering Device

CLASSIFICATION OF PROPELLERS

16-7. Propellers are classified as being right-hand or left-hand propellers, depending on the direction of rotation. When viewed from astern, with the ship moving ahead, a right-hand propeller rotates in a clockwise direction and a left-hand propeller rotates in a counter-clockwise direction. The great majority of single-screw ships have right-hand propellers. Multiple-screw ships have right-hand propellers to port. Reversing the direction of rotation of a propeller reverses the direction of thrust and, consequently, reverses the direction of the ship's movement.

CONSTRUCTION OF PROPELLERS

MATERIAL

16-8. Propellers may be constructed of manganese bronze, cast steel, or corrosion-resistant steel. Each of these are described below.

Manganese Bronze

16-9. Manganese bronze is an alloy of 16 percent manganese, 68 percent copper, and 16 percent tin. This produces a tough, strong metal which is resistant to saltwater. Manganese bronze propellers are used on most vessels because of this toughness and resistance to saltwater. Manganese bronze is also easier to repair than other metals.

Cast Steel

16-10. Cast steel propellers are composed of approximately 90 percent ingot iron with a small percentage of carbon, manganese, phosphorus, sulfur, and silicon. Cast steel propellers are used on vessels operating where ice is a serious problem. They are also used in freshwater where corrosion is not a problem and on tugs where propellers are likely to hit obstructions in the water.

Corrosion-resistant Steel

16-11. Corrosion-resistant steel is commonly called stainless steel, chromium steel, or chromium-nickel steel. This tough and hard metal may be used in making bearings, shafts, propellers, and drives. Propellers made of corrosion-resistant steel may be used where ice and underwater obstructions are a problem. Corrosion-resistant steel is very hard to bend but will break without causing damage to other blades or to the tail shaft.

Propeller Size

16-12. The size of a propeller (that is, the size of the area swept by the blades) has a definite effect on the total thrust that can be developed on the propeller. Within certain limits, the thrust that can be developed increases as the diameter and the total blade area increase. It is impracticable to increase propeller diameter beyond a certain point. Therefore, propeller blade area is usually made as great as possible by using as many blades as are feasible under the circumstances. Three-blade and four-blade marine propellers are commonly used.

FACTORS AFFECTING PROPELLER ACTION

16-13. Factors that affect propeller action are pitch, slip, and cavitation. Each of these are described below.

Pitch

16-14. The pitch of the propeller is the distance the propeller would advance in one revolution if the water were a solid medium. Propeller pitch is shown in Figure 16-3.

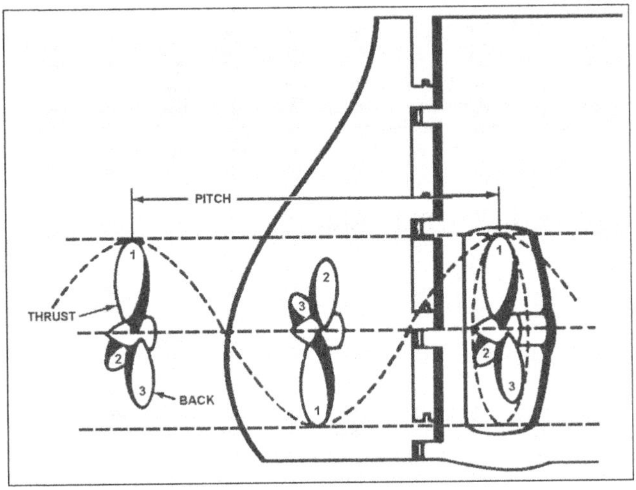

Figure 16-3. Propeller Pitch

Slip

16-15. Since water is not a solid medium, the propeller slips or skids. Therefore, the actual distance advanced in one complete revolution is less than the theoretical advance for one complete revolution. That difference is slip.

Cavitation

16-16. When the blade-tip speed is excessive for the size and shape of the propeller, the vessel is riding high in the water and there is an unequal pressure on the lower and upper blade surfaces. This condition produces cavities or bubbles around the propeller. The result is an increase in revolutions per minute without an equivalent increase in thrust. This results in loss of efficiency. When cavitation is fully developed, it limits vessel speed regardless of available engine power. The speed at which cavitation (see Figure 16-4) begins to occur is different in various types of ships. The turbulence increases in proportion to the propeller rpm.

Specifically, a propeller rotating at a high speed will develop a stream velocity that creates a low pressure. This low pressure is less than the vaporization point of the water, and from each blade tip there appears to develop a spiral of bubbles. The water boils at the low pressure points. As the vapor bubbles of cavitation move into regions where the pressure is higher, the bubbles collapse rapidly and produce a high-pitched noise. The end result of cavitation is a high level of underwater noise, erosion of propeller blades, and vibration with subsequent blade failure from metallic fatigue. There is also an overall loss in propeller efficiency, requiring a proportionate increase in power for a given speed.

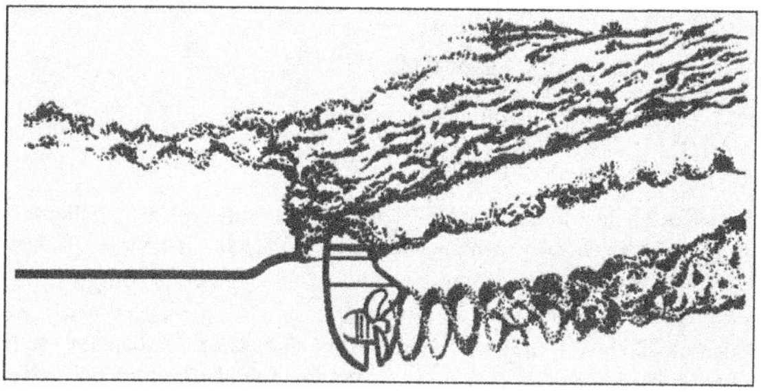

Figure 16-4. Cavitation

COMPUTING SLIPPAGE

16-17. The difference between the theoretical and the actual advance per revolution (called slip) is usually expressed as a ratio. This is a ratio of the theoretical advance per revolution (or the pitch) and the actual advance per revolution.

Therefore:

$$\text{Slip ratio} = \frac{E - A}{E}$$

Where—

E = shaft rpm × pitch = engine distance per minute

A = actual distance advanced per minute.

DETERMINING PROPELLER PITCH

16-18. There are two methods by which the pitch of a propeller can be determined. One is by formula and the other is by actual measurement using a pitchometer.

FORMULA

16-19. The formula method of determining pitch is used while the vessel is underway. Assuming that the vessel is traveling at a speed of 10 knots at 76 rpm with a normal propeller slippage of 10 percent, the pitch may be computed as follows:

$$\text{Pitch} = \frac{\text{speed} \times 6.080 \text{ (feet in a knot)} \times 100 \text{ (percent maximum effective advance)}}{\text{rpm} \times 60 \text{ (minutes)} \times \text{effective advance (100 percent minus slippage factor)}}$$

Therefore:

$$\text{Pitch} = \frac{10 \times 6,080 \times 100}{76 \times 60 \times 90} = 15 \text{ feet}$$

PITCHOMETER

16-20. A pitchometer is an instrument used to determine the pitch of a propeller while in place or when removed. Repair yards are generally equipped with pitchometers. However, if one is not available, the following procedures may be used:

- The propeller cap and nut should be removed.
- A wooden block should be made to fit over the threads of the propeller shaft. If the propeller has been removed, a wooden plug should be used to fit the bore of the propeller in place of a wooden block.
- A stiff batten or straight piece of wood (approximately 6 feet long and 2 1/2 inches wide) is then attached to the block (or plug) at a point corresponding to the center of the shaft, as shown in Figure 16-5. The batten must revolve in a plane at right angles to the axis of the shaft.
- Small wooden strips for guiding a graduated pointer are nailed at spaces of 6 to 12 inches along the batten. These strips should be placed as nearly parallel to the axis of the shaft as possible. The pointer should be graduated in inches and fractions of inches.
- A paper protractor graduated in increments of 5^{0} is then pasted on the block.

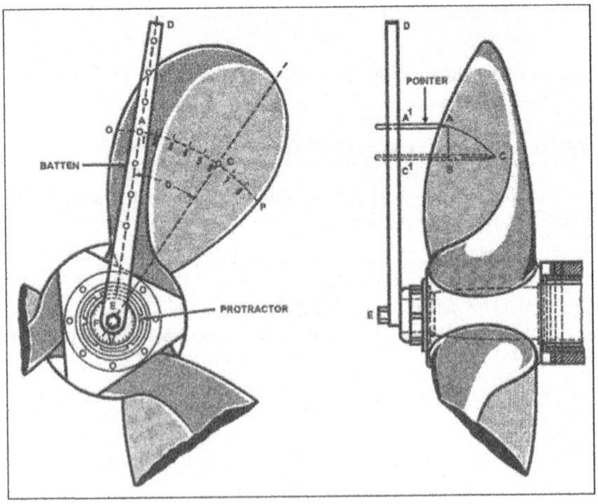

Figure 16-5. Computing Propeller Pitch

Computing Pitch

16-21. In computing the pitch of a propeller, three measurements are necessary:

- The radius at which the pitch is measured.
- The angle corresponding to the distance between two reference points selected on the blade.
- The advance parallel to the shaft.

Note: The advance in inches for an angle of 30° is equal to the pitch in feet.

Measurements

16-22. When taking measurements, use the following procedures:

- Using the distance EA^1 on the batten, an arc is drawn with the pointer, as shown in Figure 16-5.
- Two points are selected on the arc (A and C) so that an angle AEC is a multiple of 5° on the protractor.
- The distance AA^1 and CC^1 is measured using the graduated pointer. The difference (BC) between the measurements is the distance the propeller would move forward while rotating through angle AEC.

$$\text{Pitch (inches)} = \frac{360^{\circ}}{\text{AEC (degrees)} \times \text{BC (inches)}}$$

INSPECTION OF PROPELLERS

PROPELLER MAINTENANCE

16-23. Fouling of a propeller with marine growth has a marked influence on the frictional resistance of the propeller blade surface. This necessitates an increase in engine output to deliver the same number of revolutions as with a clean propeller. Propellers should be cleaned each time the vessel is drydocked. Propellers may be cleaned by divers if the vessel is long out of dock.

PROPELLER VIBRATION

16-24. When vibration difficulties are encountered, the frequency of vibration should be the first thing determined. A vibration with a frequency equal to the revolutions per minute of the shaft is called unbalance vibration. Unbalance vibration is generated by either mechanical unbalance of the propeller or by pitch or surface irregularities between the blades. A vibration having a frequency equal to the product of the number of blades and the shaft rpm is called blade frequency vibration. Blade frequency vibration is generated by hydrodynamic forces acting on the propeller and the hull of the vessel.

Note: A certain amount of vibration is always present aboard ship. However, propeller vibration may also be caused by a fouled blade or by seaweed. If a propeller strikes a submerged object, the blades may be nicked.

Unbalance vibration is usually evidence of variations from propeller design dimensions, excessive runout of the shaft taper, or unbalance of the propeller or cap. The propeller dimensions should be checked with the applicable drawing. The propeller blade pitch and contour should be corrected as necessary and the propeller and cap balanced. Blade frequency vibration, being a function of waterflow around the hull and appendages, may not be caused by any defect in the propellers. It may indicate the necessity for the following:

- Refinement of hull lines.
- Changing the number of propeller blades.
- Other design modifications.

Figure 16-6 shows items to look for during inspections of hub and blades.

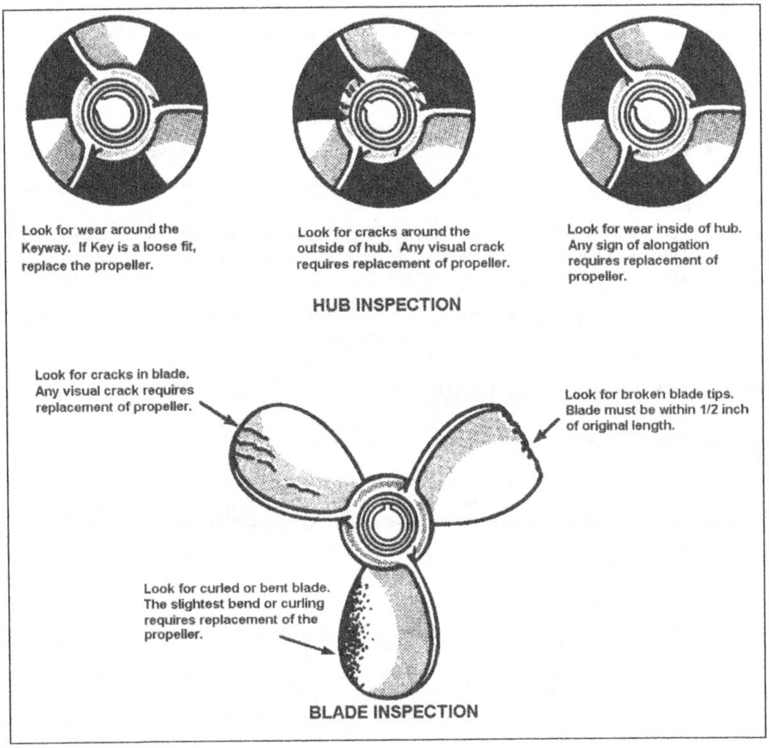

Figure 16-6. Inspection of Hub and Blades

DRYDOCK INSPECTION

16-25. When a vessel is docked, the vessel's engineers should examine the propellers. The results of the examination should be entered in the engineering log. The docking activity should carefully examine the propellers. Any repairs found necessary should be immediately undertaken so that the undocking of the vessel will not be delayed. At each interim and regular overhaul docking, the hub cap should be removed from each propeller. The propeller nut should be examined for proper torque and retightened if necessary to ensure tightness of the propeller hub on the shaft taper. The stenciled hub and blade data should be verified and recorded at each drydocking. Adjustments made, general conditions found, and work performed should be recorded in the engineering records of the vessel. A report should be made of all inspections, repairs, or changes to propellers installed on the vessel, including hub data, pitch measurements, and balancing data. The following hub and blade data is stenciled on the propeller hub:

- Name of vessel.
- Manufacturer's heat number.
- Builder.
- A.B.S. approval.
- Owner's approval.
- Plan number.
- Hull number.
- Diameter and finish weight (pounds).
- Face pitch (at 0.7 radius) for each propeller.

PROPELLER REMOVAL AND INSTALLATION

16-26. To remove and install propellers, certain tools as well as certain procedures are required. Tools and procedures for removing and installing propellers on small and large craft are discussed as follows.

Tools

16-27. After the vessel has been drydocked, no attempt should be made to remove or install a propeller until the proper tools have been assembled. For propellers on large vessels, a clamp, strongback, maul or button set, and soft metal should be obtained for covering fairwater threads. For propellers on small craft, a propeller puller and soft-faced hammer should be obtained.

Removal of Propellers on Small Craft

16-28. The procedure for removing the propeller on a small craft is as follows. Remove the following:

- Lock wire from the lock bolt and fairwater nut and then remove lock bolt.
- Fairwater nut.
- Propeller by using a propeller puller.
- Key from the keyway on the propeller shaft.

Using a Propeller Puller

16-29. Be careful when removing a propeller from the shaft to prevent damage to the shaft threads. Heat can be applied to the hub if necessary. To remove a propeller having tapped drawbolt holes, use a center bolt puller and proceed as follows:

- Remove fairwater nut or cone.
- Loosen propeller nut a few turns.

Note: Do not remove nut unless necessary for use of propeller puller.

- Attach propeller puller with drawbolts (see Figure 16-7).

- Start removing propeller from shaft taper by tightening center bolt or puller.

CAUTION

Each drawbolt must have sufficient thread in the propeller hub or the tapped drawbolt hole will be stripped upon tightening the center bolt.

- Remove propeller puller from propeller hub after loosening propeller, and remove propeller.

Propeller pullers of the type shown can be made for different size propellers. The center hole in the strongback plate does not have to be tapped for the center bolt. A hole large enough to provide ample clearance for the center bolt shank can be drilled in the strongback plate. A nut that fits the center bolt can either be tack-welded to the side of the plate toward the propeller or held there and kept from turning with a wrench.

CAUTION

Propeller usually can be handled by hand, but caution must be exercised to avoid damaging the propeller during handling.

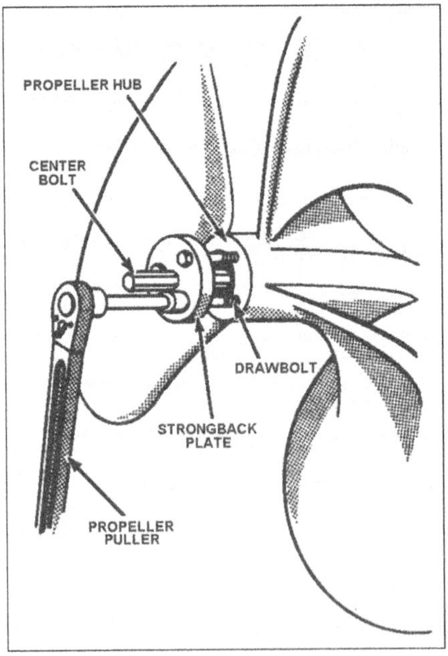

Figure 16-7. Using a Propeller Puller

Installation of Propellers on Small Craft

16-30. Installation of the propeller on a small craft is the reverse order of removal, with one exception. When installing propeller on shaft, the propeller should be pressed or tapped with a soft-faced hammer until flush with the rope guard.

Removal of Propellers on Large Vessels

16-31. On large vessels, to remove a propeller provided with tapped holes, use the following procedures:

- Loosen the propeller nut a few turns, but do not remove.
- Insert drawbolts in tapped holes in the hub face.
- Clamp a drawbolt over the end of the shaft.
- Interpose a pair of steel wedges between the clamp and the shaft end.
- Tighten clamp nuts securely, then drive wedges until the propeller starts.
- If the propeller cannot be made to start, apply heat to expand the hub. Heat is applied by using two soft gas torches on opposite sides, and working the flame around the hub to heat and expand the hub uniformly. The temperature of bronze propellers shall not be permitted to exceed 500° F (260° C). It should be checked at intervals by means of tempil sticks or contact pyrometers. Before applying heat, check to ensure that all propeller filling plugs and cover plates are removed and the adjacent shafting is protected from flame impingement.

To remove a propeller not provided with tapped drawbolt holes, or for some single-screw vessels where the above procedure is not applicable, use the following procedures:

- Install a shore between the after shaft coupling and the forward side of the stern tube to prevent strain on the thrust or main bearings.
- Loosen the propeller nut but do not remove.
- Fill the space between the propeller hub and the sternpost with metal blocking and a pair of metal wedges. Two sets are used (one on top and the other underneath).
- Drive one wedge of each pair from opposite sides.
- If the propeller cannot be made to start, apply heat to expand the hub. Heat is applied by using two soft gas torches on opposite sides, and working the flame around the hub to heat and expand the hub uniformly. The temperature of bronze propellers shall not be permitted to exceed 500° F (260° C). It should be checked at intervals by means of tempil sticks or contact pyrometers. Before applying heat, check to ensure that all propeller filling plugs and cover plates are removed and the adjacent shafting is protected from flame impingement.

Installation of Propellers on Large Vessels

16-32. When installing propellers on large vessels, use the following procedures:

- Properly assemble the propeller on the shaft as this is important to the safety of the vessel. A good fit of the propeller hub on the shaft taper is essential. In addition to the good fit, it is important to exclude all salt air and moisture from the void spaces of the propeller.
- Carefully fit the propeller hub to the shaft taper and check the fit with Prussian blue. A uniformly good fit over the taper should be obtained with a slightly heavier fit on the large end. The standard shaft taper is 1 inch on the diameter per foot of shaft length.
- In final assembly, advance the propeller beyond the mark at which an acceptable fit was obtained. For bronze propellers with standard taper, the movement beyond the mark should be 1/32 inch for each 5 inches of shaft diameter, plus 1 percent for each degree the fitting temperature was below 60° and minus 1 percent for each degree the fitting temperature was above 60°. Propellers with cast iron or cast steel hubs are advanced beyond the mark 1/32 for each 5 inches of shaft diameter. It is not advisable to use grease, oil, or other substance on the taper to ease driving of the propeller. Dry metal-to-metal contact gives a better grip between the hub and shaft.

- Fill cored holes in the hub with rust-preventive compound. This must be done before the propeller is installed on the shaft unless filling and vent holes have been provided in the hub for this purpose. The use of tallow is not approved.
- Fill the space between the end of the shaft liner and the bottom of the propeller hub counterbore with a seamless soft rubber or neoprene ring where propellers are not fitted with an external packing gland. Where an external packing gland is fitted, fill the space between the end of the liner and the bottom of the hub counterbore with rust-preventive compound.
- Fill the annular space between the aft end of the shaft taper and the propeller nut with rust-preventive compound.
- Fill the fairwater cone with rust-preventive compound.
- Measure propellers after repairs are done to verify dimensional accuracy.

Care of Shaft Taper

16-33. Most failures of tail shafts occur under the propeller hub at the large end of the taper, just behind the bronze shaft sleeve. To ensure that saltwater will not enter this region, rubber rings are provided in the propeller gland. The space behind the rubber rings is filled with a mixture of white and red lead and linseed oil. When the propeller is removed, carefully examine the shaft taper, including the keyway, for corrosion or cracks. If the rubber rings have deteriorated, replace them.

PROPELLER CLEANING

16-34. When the propeller fouling rate has become appreciable due to inactivity of the vessel or to water conditions, clean the propellers prior to operation. When cleaning propellers, use the following procedures:

- Remove barnacles by scraping and wire-brushing. This work is more easily done before the growth has dried out. Avoid removing appreciable amounts of metal by grinding or other methods.
- After bronze propellers have been cleaned, polish them with a fine abrasive. If the vessel is to go into immediate service, the propellers may be left bare. If the vessel is to be inactive for an appreciable period, coat the propellers with an approved rust-preventive compound. In case the vessel is to be laid up but will be docked again before being placed in service, paint propellers with the same type of paint and number of coats as the vessel's bottom. Hone-blasting to clean propellers must be controlled so that blade surface roughness or other side effects will not be produced.
- Polish blade surface areas which exhibit roughness after cleaning, because of cavitation or other conditions, by using a 60-grit grinding disk.

PROPELLER REPAIR

STRAIGHTENING BENT BLADES

16-35. Normally a three-man crew is sufficient for propeller repair. Tools required for small craft propeller repair are a blacksmith's forge and anvil, files and an oxyacetylene torch. Straightening and finish work can be conveniently and economically done in the field. Straightening and balancing minimizes out-board bearing maintenance. It also avoids overloading engines, loss of efficiency and speed, and poor performance of misshaped propellers. In straightening moderate bends or curled edges, peen the bend on the concave side with an air hammer and round-edge caulking tool. Follow up the peening by annealing where large arcs are affected. Greater bends must be handled by a heavy sledge, a steam hammer, or a large press. Where practicable, it is recommended that straightening be done cold. Cold-working hardens bronze, and if worked too far, cracks may occur. Large-degree bends should be straightened gradually with frequent passes for annealing (with a torch) and cooling. The repaired area should be annealed after the final bend. On small propellers, a lead mallet and rawhide mallet should be used to straighten bends.

Repairs by Welding

16-36. Faulty and damaged sections of bronze propellers, if not too extensive, may be cut out and rebuilt by bronze welding. Where replacement of large sections is required, bronze plates cast from metal similar to the parent metal should be welded in place rather than the entire section being built up by welding.

PROPELLER MEASURING AFTER REPAIRS

16-37. Propellers should be measured with pitch blocks or a pitchometer. Calipers should be used to verify the accuracy of repairs that have been made. Nominal decrease in blade thickness because of wear and erosion is acceptable. However, the blade section con-tours should remain true.

PROPELLER BALANCING

16-38. Propellers and propeller caps should be balanced to make sure they will not vibrate. A static roll balance is sufficient for propellers that rotate at a relatively low speed and have a small axial length in ratio to diameter. Other propellers can require dynamic balancing. In static roll balancing a propeller, the propeller is mounted on a mandrel. The propeller shaft key should be included in the balancing. With the propeller mounted on the mandrel, the propeller is placed on two rails as shown in Figure 16-8. The rails must be level, straight, and smooth. Material should be removed for balancing from the back or suction side of the blade over as large an area as practicable, but no closer to the blade edge than 10 percent of the blade's width. Weights can be welded to the inside of the propeller cap providing they do not interfere with installation. When the propeller is correctly balanced, it can be stopped in any position of rotation on the balancing mechanism and released, and it will not roll back.

Note: If available, a ball bearing or knife-edge balancer should be used instead of roll rails. The knife-edge balancer has greater sensitivity than roll rails.

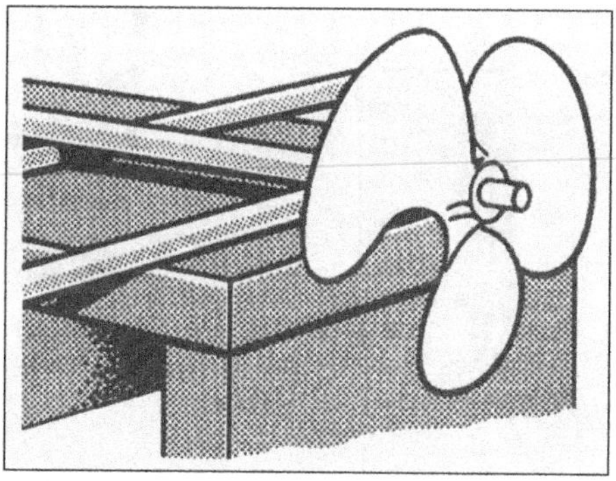

Figure 16-8. Static-balancing a Propeller Using Roll Rails

PROPELLER SHAFT SEALS

16-39. The marine split shaft seal is used on the 1646 class landing craft, utility (LCU). A feature of the marine split shaft seal is the split, inflatable sealing ring. It is located in a recess in the seal housing, just off of the sealing element cavity. The inflatable sealing ring, when not in use, does not contact the shaft surface. When necessary to interchange or replace the sealing element, the inflatable sealing ring is inflated with air from an appropriate source. It forms a restriction around the stationary shaft, restricting seawater ingress and permitting seal repair. After interchange or replacement of sealing element is completed, the inflatable ring is deflated. It recedes back into its recess, losing contact with the shaft. Figure 16-9 shows the shaft inflatable seal and Figure 16-10 shows the split shaft seal.

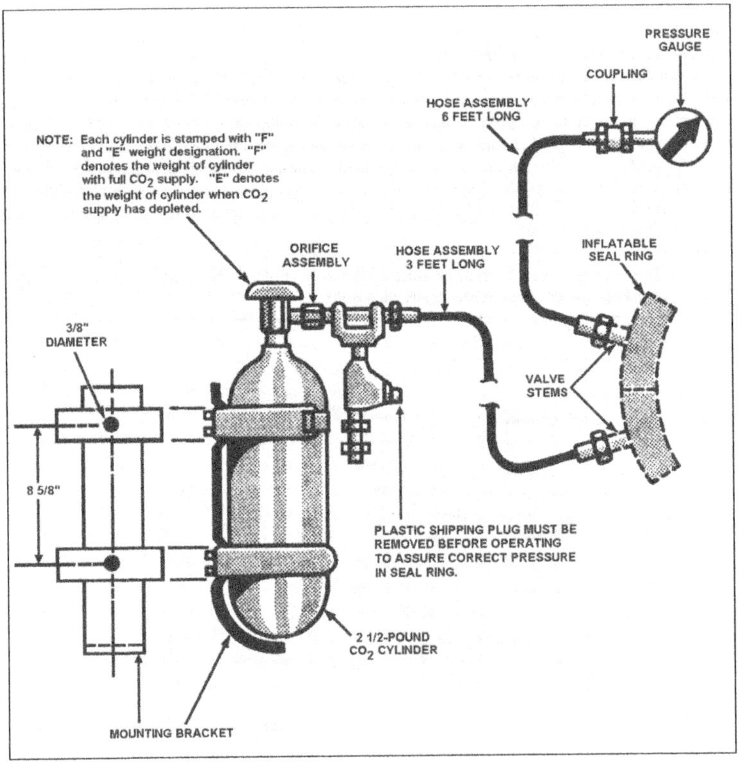

Figure 16-9. Shaft Inflatable Seal

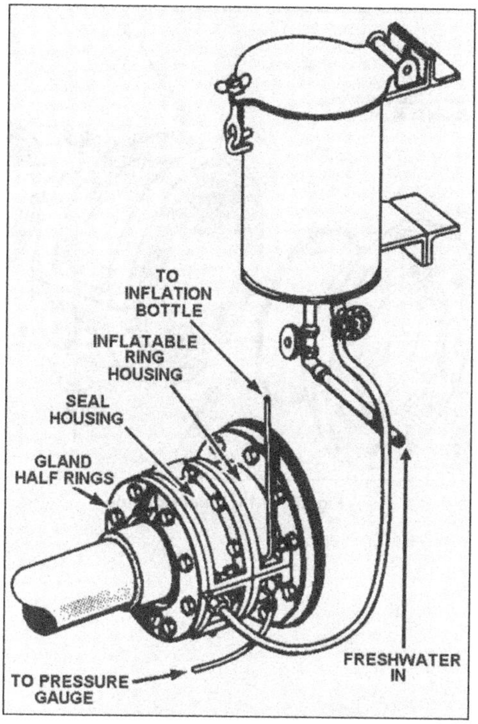

Figure 16-10. Split Shaft Seal

The marine split shaft seal is constructed of a sectional marine bronze housing casting. It houses two flexible sealing elements and a split inflatable sealing ring. The housing is designed for easy installation and simplified maintenance and replacement of sealing elements. Figure 16-11 shows the seal housing. In the forward section of the assembly are installed two split sealing elements. The sealing elements are held in contact with the shaft by encircling garter springs. As a result of the contact, the sealing elements revolve with the shaft. However, actual sealing is done only by the forward sealing vertical element. The vertical surface contacts the lapped face of the seal gland ring. The aft sealing element acts as a spacer ring and is available as a replacement element when interchange is required. The sealing contact is lubricated by a slight flow of clean water, which is injected through the gland ring, into a provided recess in the forward sealing element.

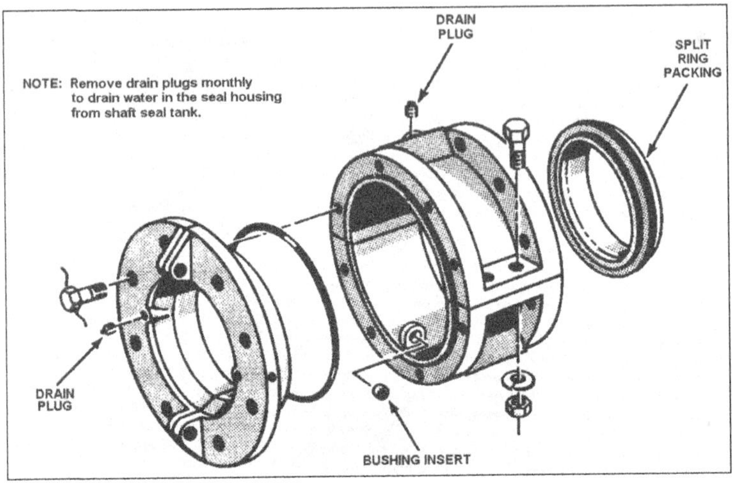

Figure 16-11. Seal Housing

Chapter 17

PUMPS AND VALVES

INTRODUCTION

17-1. A number of pumps and valves are required in engine installations. According to the type of movement that causes the pumping action, pumps may be classed as rotary, centrifugal, propeller, or jet. Valves perform a number of control functions. Those used in engine systems may be generally classed as manually operated and automatic valves. This chapter deals with pumps (Section I) and valves (Section II) that may be found on marine diesel engines.

SECTION I. PURPOSE AND CLASSIFICATION OF PUMPS

An engine installation requires a number of pumps. Some of these pumps deliver fuel to the injection system while others circulate oil through the lubricating system. Still others circulate freshwater or seawater (or both) for engine cooling. Most of these pumps are attached to the engine and are driven by the engine crankshaft through gears, chains, or belts. However, some are separate from the engine and are driven by electric motors. Pump failure may cause failure of the plant or system to which the pump is providing service. Therefore, you must have a knowledge of the operating difficulties which may be encountered and you must know how to perform routine maintenance which will keep pumps in operation. Pumps are classified by their design and operational features and, as stated above, by the type of movement that causes the pumping action. Pumps are also classified by the following:

- Rate of speed.
- Rate of discharge.
- Method of priming.

Some pumps run at variable speed while others run at constant speed. Some have a variable capacity while others have a discharge at a constant rate. Some pumps are self-priming while others require a positive pressure on the suction (intake) side before they can begin to move a liquid. Regardless of classification, a pump must have a power end and a fluid end. Pumps aboard Army ships can be divided into groups according to the principles on which they operate. Most pumps fall into the following four main types:

- Rotary.
- Centrifugal.
- Propeller.
- Jet.

Each type of pump is especially suited for some particular kind of work.

ROTARY PUMPS

17-2. The operating principle of a positive-displacement rotary pump is that rotating screws, lobes, or gears trap liquid in the suction side of the pump casing and force it to the discharge side. Positive-displacement rotary pumps have largely replaced reciprocating pumps for pumping viscous liquids in naval ships. They have a greater capacity for their weight and occupy less space.

Note: Positive displacement means that a definite quantity of liquid is pushed out on each revolution.

Rotary pumps have very small clearances between rotating parts to minimize slippage (leakage) from the discharge side back to the suction side. With close clearances, these pumps must be operated at relatively low speeds to obtain reliable operation and maintain capacity over an extended time. Some rotary pumps operate at 1,750 or higher. Operating the pumps at higher speeds will discharge cause erosion and excessive wear, which will result in increased clearances.

TYPES OF ROTARY PUMPS

17-3. Several types of positive-displacement rotary pumps exist. These types include the following:

- Simple gear.
- Herringbone gear.
- Helical gear.
- Lobe.
- Screw (low pitch and high pitch).

Simple Gear

17-4. The simple gear pump (Figure 17-1) has two spur gears which mesh together. One is the driving gear and the other is the driven gear. Clearances between the gear teeth and the casing and between the gear faces and the casing are only a few thousandths of an inch. When the gears turn, the teeth carry the liquid in the spaces between the teeth at the suction side of the pump. It is carried towards the sides where the liquid is trapped between the tooth pockets and the casing and carried through to the discharge side of the pump. Liquid entering the discharge side cannot return to the suction side because the meshing teeth at the center force the liquid out of the tooth pockets. The small lubricating oil pumps installed on many pumps and other auxiliary machinery are usually simple-gear, positive-displacement, rotary pumps.

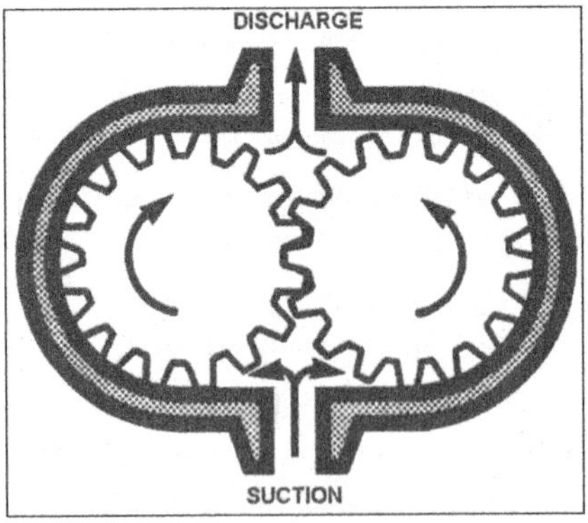

Figure 17-1. Action of Simple Gear Pump

Herringbone Gear

17-5. The herringbone gear pump (Figure 17-2) is a modification of the simple gear pump. In the herringbone gear pump, one discharge phase begins before the previous discharge phase is entirely completed. This overlapping tends to provide a steadier discharge pressure than is obtained with the simple gear pump. In addition, power transmission from the driving gear to the driven gear is smoother. There are no internal driving gears other than the pumping gears themselves. Power-driven pumps of this type are sometimes used for low-pressure fuel oil service, lubricating oil service, and diesel oil service.

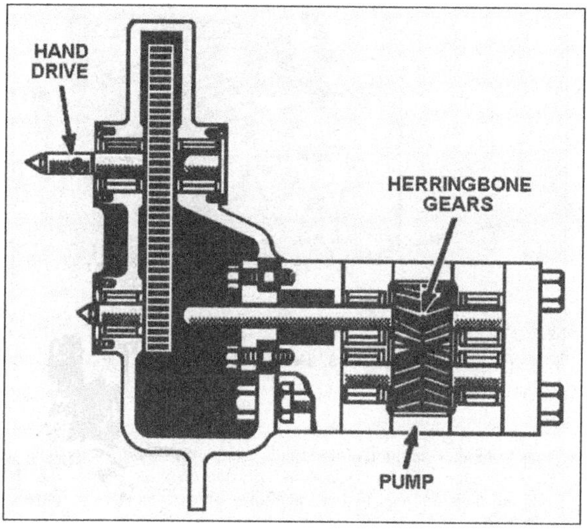

Figure 17-2. Herringbone Gear Pump

Helical Gear

17-6. The helical gear pump (Figure 17-3) is still another modification of the simple gear pump. Because of the helical gear design, the overlapping of successive discharges from spaces between the teeth is even greater than it is in the herringbone gear pump. The discharge flow is even smoother. Since the discharge flow is smoother in the helical gear pump, the gears can be designed with fewer teeth. This allows increased capacity without sacrificing smoothness of flow. The pumping gears of the helical gear pump are driven by a set of timing and driving gears. These gears also function to maintain the required close-tooth clearances while preventing actual metallic contact between the pumping gears.

Note: Metallic contact between the teeth of the pumping gears would provide a tighter seal against slippage. However, it would cause rapid wear of the teeth because foreign matter in the pumped liquid would be present on the contact surfaces.

Roller bearings at both ends of the gear shafts maintain proper alignment and decrease friction loss in power transmission. Stuffing boxes are used to prevent leakage at the shafts. The helical gear pump is used to move nonviscous liquids and light oils at high speeds. In addition, it can be used to pump heavy, viscous material at a lower speed.

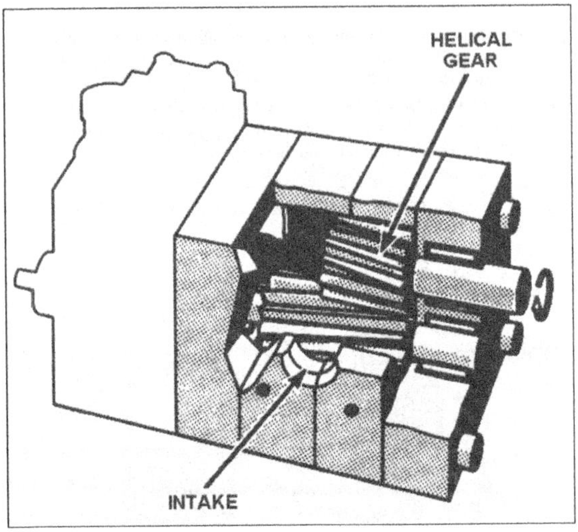

Figure 17-3. Helical Gear Pump

Lobe

17-7. The lobe pump is another variation of the simple gear pump. Modified versions of lobe pumps are used as superchargers on some engines and on vapor compression distilling plants. The lobes are considerably larger than gear teeth, but there are only two or three lobes on each rotor. The rotors are driven by external spur gears on the rotor shafts.

Screw

17-8. Many variations of screw pumps exist. The three main differences among these variations are—

- The number of intermeshing screws.
- The pitch of the screw (low or high).
- The general direction of fluid flow (single or double).

Double-screw, Low-pitch Pump

17-9. The double-screw, low-pitch pump is shown in Figure 17-4. The two pairs of screws, which intermesh with close clearances, are mounted on two parallel shafts. Each pair of screws has opposite threads with respect to the other pair. One shaft drives the other through a set of herringbone timing gears which maintain clearances between the screws as they rotate. All clearances are small. There is no actual contact between the screws or between the screws and the casing. Liquid is trapped between the grooves of the screws and the casing. Meshing of the threads of the two screws forces the liquid along the grooves and toward the center (discharge side) of the pump. In the pump illustrated, liquid enters the thread at both ends of the rotor or screws. Therefore, the axial thrust of each side is balanced.

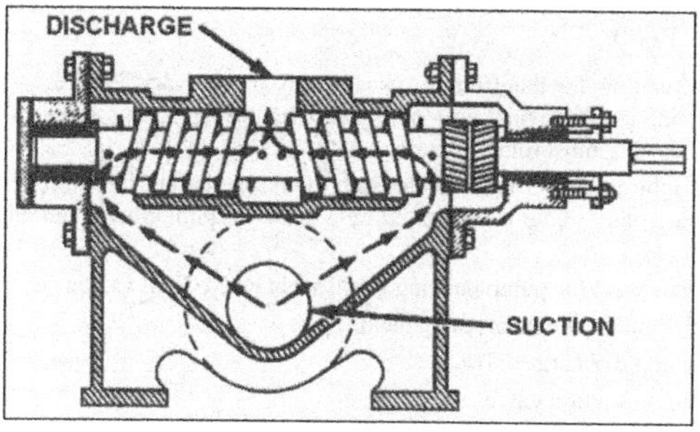

Figure 17-4. Double-screw, Low-pitch Pump

Triple-screw, High-pitch Pump

17-10. The operation of a triple-screw, high-pitch pump (Figure 17-5) is similar to that of the double-screw, low-pitch pump. The pitch of the screws is much longer than in the double-screw, low-pitch pump. This enables the center screw (power rotor) to drive the two outer (idler) rotors directly without external timing gears. Diameters of the idlers are less than that of the power rotor.

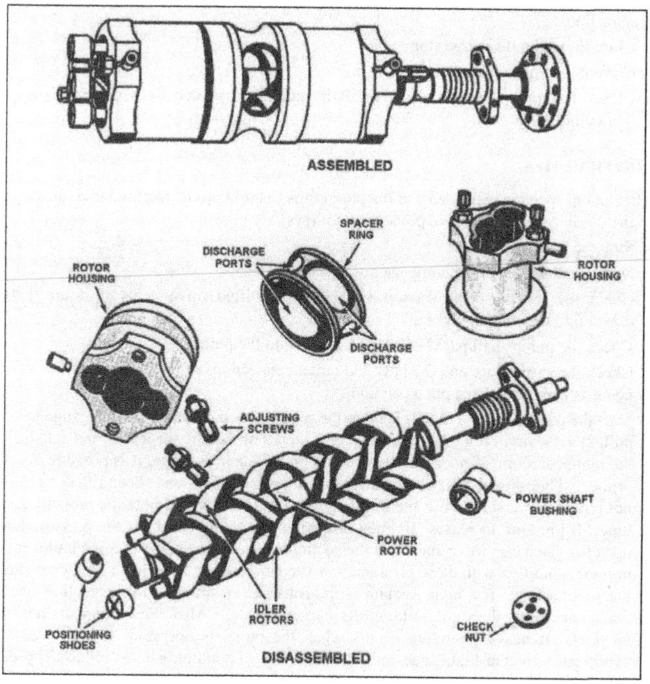

Figure 17-5. Triple-screw, High-pitch Pump

Operating a Rotary Pump

17-11. Operating instructions for the driving unit of rotary pumps vary. Therefore, you must read the posted instructions for the individual driving unit and pump before operating a specific pump. The following instructions apply for starting most rotary pumps:

- Check the lubricating oil level in the sump tank or bearing housing. Fill oil cups or reservoir, if fitted. If the rotary pump is lubricated by a detached pump, open and adjust all delivery and return valves.
- Open the valves on the pump packing gland seals where such valves are fitted.
- Lift all relief and sentinel valves by hand.
- Open the pump discharge valve.
- Open the pump suction valve.
- Start the driving unit (motor or engine).
- Check the lubricating system to see that all bearings are supplied with oil and that the oil is at the correct pressure.
- Check all gauges to see that proper pressures are being developed.
- Adjust the pump-shaft packing glands and gland-sealing needle valve where fitted and where adjustment is needed.

Instructions for stopping and securing a pump are as follows:

- Stop the driving unit. Be sure that the check valve of the pump discharge closes. This is to prevent a backflow, through the pump being secured, from another pump which may be discharging into the same line.
- Close the pump discharge stop valve.
- Close the pump suction valve.
- Close all supply and return valves if the unit is lubricated by a detached pump or by a main lubricating oil system.

Operating Difficulties

17-12. When pump speed is increased and the pump fails to build up the required pressure or fails to discharge fluid when the discharge valve is open, proceed as follows:

- Stop the unit.
- See that all valves in the pump suction lines are open.
- Check the packing of all suction and suction manifold valve stems to ensure that no air is being drawn into the suction piping.
- Check the pump shaft packing for air leakage into the pump.
- Check the spring case and the inlet and outlet connections of the discharge relief valve to ensure that no air is leaking into the pump suction.
- Start the pump again. When it is up to the proper speed, read the suction gauge to see if the pump is pulling a vacuum. If a low vacuum (5 or 6 inches of mercury or less) is indicated, air is leaking into the pump suction. If no vacuum is shown on the suction gauge, it is possible that the pump is not primed. This should rarely occur after the pump casing has once been filled. If the pump still does not build up pressure, close the discharge valve gradually and note the pressure gauge at the same time. If pressure increases, an open discharge line is indicated. If the pressure does not increase, open the discharge valve and close the suction valve, noting if the vacuum builds up. If the pump is in good condition with close clearances, a vacuum ranging from 15 to 25 inches should build up on the suction line. If a high vacuum is indicated, an obstruction in the suction line is probably the reason the pump does not build up discharge pressure. Also, the suction strainer, if fitted, may be clogged. If heavy pounding occurs when the pump is operating at high vacuum, considerable vaporization in the liquid end may be indicated. Pounding can be reduced by decreasing pump speed.

CARE AND REPAIR

17-13. Instructions given in this chapter for maintenance, repair, and operation of pumps are general for all makes and types. Manufacturer's technical manuals (TMs) are furnished for all pump installations except some miscellaneous small, motor-driven pumps. These TMs contain detailed information concerning specific pumps installed. You should study them carefully before attempting to operate or service the pump.

Wearing plates and lines

17-14. Clearances between pump rotors and casing wearing plates and cylinder liners should be maintained according to the manufacturer's plans. On low-pressure, low-suction-lift pumps (such as lubricating-oil, diesel-oil-supply, and tank-drain pumps) pressure drop across the clearance spaces does not generally exceed 50 psi. With these types of pumps, the clearance between parts may wear as much as 0.005 to 0.010 inch without appreciable effect on the capacity of the individual pump. However, when clearances are excessive, parts (wearing plates and liners) must be removed. If a fuel oil tank drain pump, with the suction valve closed, will pull a vacuum of only 16 inches of mercury, parts may require renewal. In each instance, the engineering officer will decide whether the amount of wear or increased clearance requires the renewal of parts.

Bearings

17-15. If the pump bearings are worn excessively, they must be renewed.

Timing gears

17-16. Pumps fitted with timing gears must have correct clearance between the two pumping rotors during operation. To accomplish this, the gears must be securely locked to the rotor shafts in their exact designed position. Be sure that there is no lost motion caused by the looseness of keys or pins holding the rotors in the shafts.

Thrust bearings

17-17. The proper setting of thrust bearings which hold the pumping elements centrally in the pumping casing is important. Examine thrust bearings quarterly and check the position of the rotors. When checking the rotor position, be sure to make enough allowance for expansion of the shaft from the cold condition to the hot running condition.

Couplings

17-18. When the driving unit is connected to the pump by flexible coupling, the coupling is intended to take care of slight misalignment. When misalignment is minor, the coupling should operate satisfactorily without frequent renewal of coupling parts. However, if misalignment is excessive, the coupling parts take severe punishment and the pins, bushings, and bearings must be replaced frequently. Couplings with self-contained lubricant must be kept lubricated. A coupling will wear and eventually fail if its lubricant is lost. Therefore, you must observe the following precautions:

- Inspect the flexible coupling monthly by removing the filler plug to be sure there is a sufficient supply of lubricant.
- Whenever a coupling is dismantled, inspect the teeth to see that they are in good condition. When the coupling is reassembled, check the alignment of the driving unit and pump to prevent excessive coupling wear.

Pump Lubrication

17-19. Lack of proper lubrication is a primary cause of pump failure. Reciprocating engine-driven pumps are usually lubricated by either sight-feed drip cups or wick lubrication. See that oil cups are filled with oil and that an adequate supply is being fed to the bearings. Water pump shafts are usually fitted with water flingers between the pump shaft stuffing-box gland and the bearing housing to prevent the entrance of water from the pump glands to the bearing housing.

Safety Precautions for Rotary Pumps

17-20. You must observe the following precautions in the operation of rotary pumps:

- See that all relief valves are tested at appropriate intervals. Ensure that relief valves function at the designated pressures.
- Never attempt to jack over a pump by hand while the power is on.
- Never operate a positive-displacement rotary pump with the discharge valve closed unless the discharge is protected by a properly set and tested relief valve.
- When working on any vertical, rotary, or centrifugal pump, never rely on the pump coupling to support the weight of the pump rotor assembly. Support the pump rotor assembly in slings or by other acceptable rigging methods.

Variable-Stroke Pumps

17-21. Aboard ship, variable-stroke pumps (positive-displacement) are used largely on electro-hydraulic steering gear and anchor windlasses. In these applications, the variable-stroke pump is referred to as the A-end of the drive system. The hydraulic motor, which is driven by the A-end, is referred to as the B-end. By controlling the pumping action of the A-end, you can run the motor (B-end) in either direction and vary the speed from zero to the maximum rate. On some ships, variable-stroke pumps are also used as in-port fuel oil service pumps. Although variable-stroke pumps are often classified as rotary pumps, they are actually reciprocating pumps. They operate on a principle similar to that of a single-acting reciprocating pump. A rotary motion is imparted to a cylinder barrel or cylinder block in the pump by a constant-speed electric motor. However, the actual pumping is done by a set of pistons reciprocating inside a set of cylinders. The following is a discussion on the way that rotary motion is changed to reciprocating motion (the Waterbury variable-volume pump is an example). The Waterbury variable-volume pump is used as a variable-speed power transmission in the steering gear. It is also used in systems where it is necessary to control pump output within very narrow limits. In the Waterbury pump, the pumping is done by a set of pistons which move back and forth within cylinders. Oil enters and leaves the cylinders through ports or passages in the cylinder heads. The pistons are operated by the cam action of a tilting plate. Therefore, rotary motion is changed to reciprocating (back and forth) motion. The device for making this change (called the demonstrator) consists of a cylinder barrel and a tilting plate attached to a shaft. The tilting plate is fastened to the shaft by a universal joint which permits it to tilt in any direction. The connecting rods (piston rods) rest in sockets in the tilting plate and are attached to pistons which slide up and down in the cylinders. When the handwheel is turned, the complete assembly turns with it. When you place the demonstrator on a sloping surface, the tilting plate will tip or tilt and assume the same angle as the block. The pistons on the low side will be drawn down in the cylinders and those on the high side will be pushed up. When the handwheel is turned to the right, the far side of the tilting plate moves up along the surface of the block and the connecting rods and pistons on that side will be raised. The near side of the plate will be moving down along the block surface and the pistons on this side will move down in the cylinders. The operation is continuous. Each piston moves up during half of each revolution around the shaft and moves down during the other half. The stroke of the pistons (the distance that they move up and down during each revolution of the shaft) depends on the slope or tilt of the tilting plate. If the demonstrator is resting on a level surface and the tilting plate is parallel to the bottom of the cylinder barrel and you turn the handwheel, the tilting plate will not move up and down and the pistons will remain in the same position in the cylinders. The demonstrator so far has been used to show that the pistons will move up and down in the cylinders when the shaft is turned with the angle plate tilted. The device can be converted into an oil pump by adding an adjustable tilting box to control the tilt of the tilting plate and by placing a valve plate on top of the cylinder barrel. The valve plate has two elongated ports which allow oil to flow into the cylinders on one side of the pump and out on the other side. When the tilt of the tilting plate is increased, the stroke of the pistons is increased and more oil passes through the pump. Tilting the tilting plate the opposite direction reverses the direction of the oil flow. The variable-stroke, axial piston pump has the same principal parts as the demonstrator just described. The moving parts are enclosed in a pump case that is kept filled with oil. These pumps are mounted in a horizontal position and are usually driven by an electric motor.

RADIAL-PISTON, VARIABLE-STROKE PUMP

17-22. The pumping action of a radial-piston, variable-stroke pump is similar to the axial-piston pump just described. However, the components are arranged differently. In a radial-piston pump, the cylinders are arranged radially in a cylinder body which rotates around a nonrotating central cylindrical valve or pintle. As the cylinder body revolves, each cylinder contacts either an intake port or an outlet port in the central cylindrical valve. Plungers or pistons which extend outward from each cylinder but against a slipper ring on the inside of a rotating float ring or rotor. The floating ring is so constructed that it can be shifted off-center from the pump shaft. When it is in the neutral position, the pistons do not reciprocate and the pump does not operate, even though the electric motor is still causing the pump to rotate. If the floating ring is forced off-center to one side, the pistons reciprocate and the pump operates. If the floating ring is forced off-center to the other side of the pump shaft, the pump also operates. However, the direction of flow and the amount of flow are both determined by the position of the cylinder body relative to the position of the floating ring.

CENTRIFUGAL PUMPS

17-23. Most modem diesel engines use centrifugal pumps (Figure 17-6) for circulating cooling water. Many types of centrifugal pumps exist, but all operate on the same principal. The operation of centrifugal pumps depends on centrifugal force (which imparts a high velocity to the liquid pumped) produced by the rotation of the impeller at high speed. The liquid is sucked in at the center or eye of the impeller and discharged at the outer rim of the impeller. By the time the liquid reaches the outer rim of the impeller, it has acquired considerable velocity. The liquid is then slowed down by being led through a volute or through a series of diffusing passages. As the velocity of the liquid decreases, its pressure increases. In other words, some of the kinetic energy of the liquid is transformed into potential energy. In the terminology commonly used in the discussion of pumps, the velocity head of the liquid is partially converted to a pressure head.

TYPES OF CENTRIFUGAL PUMPS

17-24. Two types of centrifugal pumps which you will probably encounter aboard ship are the volute and the diffuser.

Volute pump

17-25. In the volute pump, the impeller discharges into a volute (that is, a gradually widening channel in the pump casing). As the liquid passes through the volute and into the discharge nozzle, a considerable portion of its kinetic energy is converted into potential energy.

Diffuser pump

17-26. In the diffuser pump, liquid leaving the impeller is first slowed down by the stationary vanes which surround the impeller. The liquid is forced through the gradually expanding passages of the diffuser ring before entering the volute. The diffuser vanes and the volute reduce the velocity of the liquid in the diffuser pump. In this type of pump there is an almost complete conversion of kinetic energy to potential energy.

17-27. Centrifugal pumps may be classified in several ways. For example, they may be either single stage or multi-stage and horizontal and vertical. A single-stage pump has only one impeller. A multi-stage pump has several impellers housed together in one casing. Each impeller functions separately, discharging to the suction of the next stage impeller. Centrifugal pumps may be classed as horizontal or vertical, depending on the position of the pump shaft.

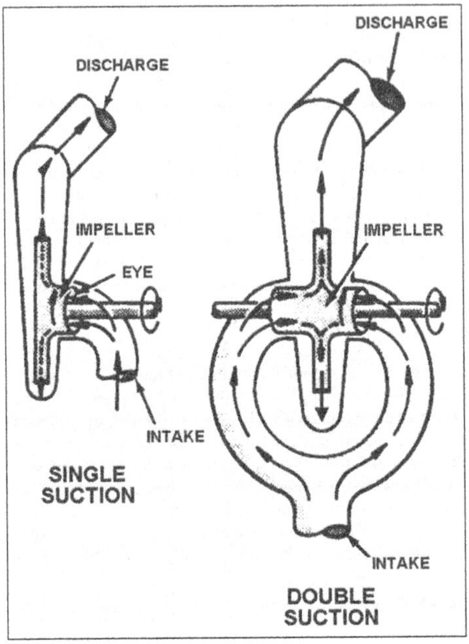

Figure 17-6. Operating Principle of a Centrifugal Pump

CONSTRUCTION OF CENTRIFUGAL PUMPS

17-28. The following information applies in general to most of the centrifugal pumps used with diesel engines. The shaft is protected from excessive wear and corrosion by Monel or corrosion-resistant steel sleeves wherever the shaft comes in contact with the liquid being pumped or with the shaft packing. In many centrifugal pumps, the shaft is fitted with replaceable sleeves. The advantage of having sleeves is that they can be replaced more economically than the entire shaft.

Impellers

17-29. Impellers used on centrifugal pumps may be classified as single suction or double suction. The single-suction impeller allows liquid to enter the eye from one direction only. The double-suction type allows liquid to enter the eye from two directions. Impellers have side walls which extend from the eye to the outer edge of the vane tips. Open impellers do not have these side walls. The impellers are carefully machined and balanced to reduce vibration and wear since they rotate at a very high speed. A close radial clearance must be maintained between the outer hub of the impeller and that part of the pump casing in which the hub rotates. This is to minimize leakage from the discharge side of the pump casing to the suction side. To prevent corrosion of pumps that handle seawater, the impellers and casing of these pumps are made of bronze or gun metal, and the shaft of Monel metal or stainless steel. Because of the close clearance and the high rotational speed of the impeller, the running surfaces of both the impeller hub and the casing at that point are subject to relatively rapid wear. Centrifugal pumps are provided with replaceable wearing rings (Figure 17-7). This is to eliminate the need for renewing an entire impeller and pump casing because of wear. One ring is attached to each outer hub of the impeller. This ring is called the impeller wearing ring. The other ring, which is stationary and attached to the casing, is called the casing wearing ring. Some small pumps with single-suction impellers have only a casing wearing ring and no impeller ring. In this type of pump, the casing wearing ring is fitted into the end plate. Recirculating lines are installed on some centrifugal pumps. This is to prevent the pumps from overheating and becoming vapor-bound in case the discharge is entirely shut off or the flow of fluid is stopped for extended periods. Seal piping is installed for the following reasons:

- To cool the shaft and the packing.
- To lubricate the packing.
- To seal the joint between the shaft and the packing against air leakage.

A lantern ring spacer is inserted between the rings of the packing in the stuffing box. Seal piping leads the liquid from the discharge side of the pump to the annular space formed by the lantern ring. The web of the ring is perforated so that water can flow in either direction along the shaft, between the shaft and the packing.

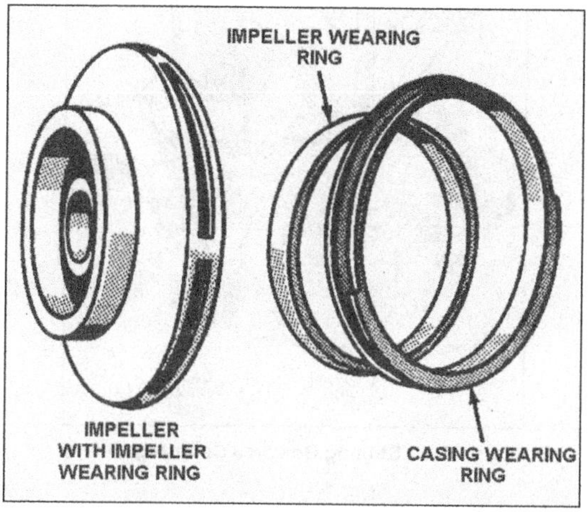

Figure 17-7. Impeller and Wearing Rings for a Centrifugal Pump

Shaft and Thrust Bearings

17-30. Shaft and thrust bearings support the weight of the impeller and maintain the position of the rotor, both radially and axially. Radial bearings may be sleeve or ball type. Thrust bearings may be ball or pivoted segmental type. The power end of a centrifugal pump may be driven by an electric motor or by a diesel engine. Pumps used for continuous service are either turbine or motor driven. Smaller pumps, such as those used for in-port or cruising operation, are generally motor driven. Pumps used for emergency fire main services are generally diesel driven.

Care and Repair of Centrifugal Pumps

17-31. The tests, safety precautions, and maintenance factors for rotary pumps, outlined earlier in the chapter, apply in a general way to centrifugal pumps. However, some of the information which you must have to properly care for and maintain centrifugal pumps is given here. For additional information, as well as for specific information on any one pump, you should consult the appropriate TM.

Stuffing Box Packing

17-32. Packing around the shafts of centrifugal pumps (Figure 17-8) may be the stuffing box type, the labyrinth type, or both. Stuffing box packing should be renewed in accordance with planned maintenance or whenever leakage becomes excessive. The packing should be installed in a uniform thickness all around the shaft sleeves. When installing new packing, pack the stuffing box loosely and set it up lightly on the packing gland. With the pump now in operation, tighten the gland in small steps, with several hours between tightenings, to compress the packing gradually. This procedure will prevent excessive heating and scoring of the shaft or shaft sleeves. A flow of from 40 to 60 drops per minute out of a normal packed stuffing box is required to provide lubrication and to dissipate generated heat. Renew labyrinth packing around worn bushings (bushing wear is due to wearing of the bearings) when the bearings are replaced. Frequently examine the water supply to the packing seals to ensure that it is not obstructed by the packing.

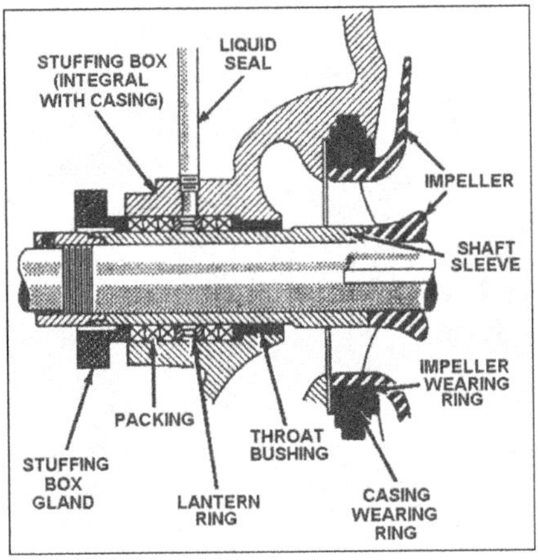

Figure 17-8. Stuffing Box on a Centrifugal Pump

Rotary Seals

17-33. In some pump installations, stuffing box packing is impractical since there must be a certain amount of leak-off for lubrication purposes. It could be rather dangerous to have leaking fluids such as distillate diesels or gasoline. Many gallons of water could also be lost from a freshwater system. In many installations, packing is being replaced by mechanical seal. When the seal requires replacement or when there are signs of abnormal wear or damage to the running surfaces, thoroughly inspect to find the cause. The cause of the trouble must be found and corrected or the failure will recur. Seal failure is often caused by dirt on the running surfaces, by worn bearings, or by bent shafts. Whenever the seal must be replaced, the complete assembly should be replaced in accordance with the manufacturer's instructions. Always be sure the shaft is free of dirt or lint. Then unwrap the seal, being careful not to touch the bearing surface with your fingers. Rinse the seal in an approved cleaning solvent, and allow it to air dry. Never wipe it dry. If a new seal assembly is not available, you may be able to repair the existing seal by lapping the mating surfaces.

Lantern Rings, Sleeves, and Flingers

17-34. When packing a stuffing box fitted with a lantern ring, be sure to replace the packing beyond the lantern ring at the bottom of the stuffing box. In addition, see that the sealing water to the lantern ring is not blanked off by the packing. Sleeves fitted at the packing on the pump shafts must always be tight. These sleeves are usually made secure by shrinking or keying them to the shaft. Be sure that water does not leak between the shaft and shaft sleeves. If the shaft or sleeves are roughened or grooved, turn or grind them to give a smooth surface. If the surface is excessively rough, the sleeves should be renewed. Water flingers are fitted on shafts, outboard of stuffing box glands, to prevent water from following along the shaft and entering the bearing housings. The flingers must be tightly fitted. If the flingers are fitted on the shaft sleeves rather than on the shaft, be sure that no water is allowed to leak under the sleeves. If leakage does occur, fit fiber washers between the ends of the sleeves and the shaft shoulders, and fill all clearances between the shaft and the sleeve with tallow.

Wearing Rings, Impeller, and Casing

17-35. Clearances between the impeller and the casing wearing rings should be maintained as shown on the manufacturer's plans or the Maintenance Requirement Card. When clearances exceed specified figures, the wearing rings must be replaced. This job can be done by the ship's force, but it requires complete disassembly of the pump. If you must undertake this job, follow the manufacturer's instructions carefully. Improper fitting of the rings or incorrect reassembly of the pump can result in serious damage.

Shaft Alignment

17-36. When installing or assembling pumps driven by electrical motors, steam turbines, or diesel engines, be sure that the unit is aligned properly. Misalignment may cause serious operating troubles later. The rotating shafts of the driver and driven units must be in absolute proper alignment as determined by the coupling. Check the shaft alignment of a pump frequently or whenever the pump is opened up or whenever there is noticeable vibration. If the shafts are out of line or inclined at an angle to each other, the unit must be realigned. This is to prevent shaft breakage and renewal of bearings, pump casing wearing rings, and throat bushings. Whenever practicable, check the alignment with all piping in place and with the adjacent tanks and piping filled. Some driving units are connected to the pump by a flexible coupling. Flexible couplings are intended to take care of only slight misalignment. Misalignment should never exceed the amount specified by the pump manufacturer. If misalignment is excessive, the coupling parts are subjected to severe punishment. This requires frequent renewal of pins, bushings, and bearings. When the driving unit is connected, or coupled, to the pump by a flange coupling, the shafting must be frequently realigned. Each pump shaft must be kept in proper alignment with the shaft of the driving unit. Misalignments are indicated by such things as abnormal temperatures, abnormal noises, and worn bearings or bushings. Wedges, or shims, are sometimes placed under the bases of both the driven and driving units (Figure 17-9, view A) for ease in alignment when machinery is installed. Jacking screws may also be used to level the units. When the pump or the driving unit, or both, have to be shifted sidewise to align the couplings, side brackets are welded in convenient spots on the foundation, and large setscrews are used to shift the units sidewise or endwise. When wedges or other packings have been adjusted so that the outside diameters and faces of the coupling flanges run true as they are manually revolved, the chocks are fastened, the units are securely bolted to the foundation, and the coupling flanges are bolted together. These alignments must be checked from time to time and misalignments must be promptly corrected. Some types of installations call for special methods of handling. However, in general practice, three methods are used for checking alignments. These are by use of the following:

- 6-inch scale.
- Thickness gauge.
- Dial indicator.

When using a 6-inch scale to check alignments, check the distance between the faces of the coupling flanges at 90° intervals. Find the distances between the faces at point a, b (on the opposite side), c, and d (opposite point c) (Figure 17-9, view B). This will indicate whether the coupling faces are parallel with each other. If they are not parallel, adjust the driving unit or the pump, or both, with shims until the couplings check true. While measuring the distances, you must keep the outside diameters of the coupling flanges in line. This is done by placing the scale across the two flanges as shown in Figure 17-9, view C. If the flanges do not line up, raise or lower one of the units with shims. Then, if necessary, shift them sidewise, using the jacks welded on the foundation. The scale should be used (as shown in Figure 17-9, view C) at intervals of 90°, as was done in checking between the flange faces (Figure 17-9, view B). The procedure for using a thickness gauge to check alignments is similar to that for a scale. When the outside diameters of the coupling flanges are not the same, use a scale on the surface of the larger flange. Then use "feelers" between the surface of the smaller flange and the edge of the scale. When the space is narrow, check the distance between the coupling flanges with a thickness gauge (see Figure 17-9, view D). Wider spaces are checked with a piece of square key-stock and a thickness gauge. Revolve the couplings one at a time and check at 90° intervals. If the faces are not true, the shaft has been sprung. Many times the shafts must be removed and sent to the ship's machine shop for reworking. When using a dial indicator to check alignments, clamp the indicator to one coupling flange. Then revolve the coupling with the point of the dial indicator on the shaft of the opposite coupling flange. If no variation is shown on the indicator, the coupling is running true. When the coupling with an indicator clamped to it is revolved while the opposite coupling remains still, the degree to which the coupling centers are out of line will be shown. To adjust the centers, loosen bolts at the unit bases and recheck the alignment. When alignment is true, secure the dowel at the unit bases, and insert and fasten the coupling bolts. Use dial indicators whenever possible when aligning a coupling.

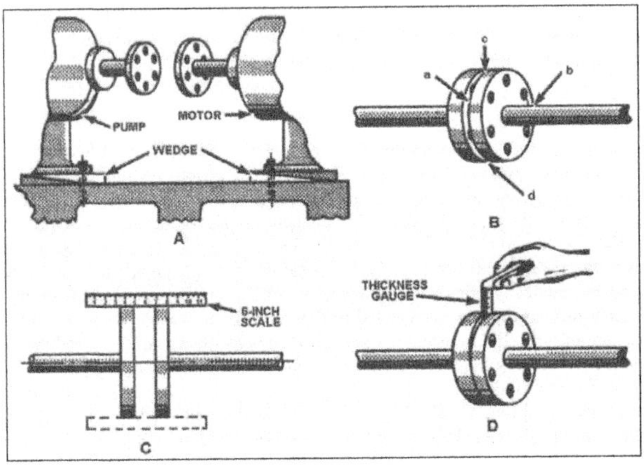

Figure 17-9. Coupling Alignment

MAJOR TROUBLES AND REPAIRS

17-37. A list of the principal troubles that may occur with centrifugal pumps, along with their causes, are given. If the pump fails to build up pressure when the discharge valve is opened and the pump speed is increased, proceed as follows:

- Secure the pump.
- Prime the pump to expel all the air through the air cocks on the pump casing.
- Open all valves on the pump suction line.
- Start the pump again. If the discharge pressure is not normal when the pump is up to its proper speed, the suction line may be clogged, or an impeller may be broken. It is also possible that air is being drawn into the suction line or into the casing. If any of these conditions exist, stop the pump, try to find the source of the trouble, and correct it if possible.

WARNING

Centrifugal pumps are not self-priming, never run a centrifual pump when empty.

The following parts of a centrifugal pump are those that are most frequently in need of repair or replacement:

- Casing wearing and impeller wearing rings.
- Shaft sleeves.
- Bearings.
- Bushings.

Casing Wearing Rings and Impeller Wearing Rings

17-38. The purpose of these rings is to keep internal bypassing of liquid to a minimum. Therefore, clearances should be checked and restored when worn beyond allowable limits, whenever the pump casing is opened up and at least once each year.

Shaft Sleeves

17-39. Operating personnel may have a tendency to take up too hard on the packing in an attempt to prevent stuffing box leakage. This causes scoring of the shaft sleeves. Whenever the pump is opened, the sleeves should be examined. If slightly scored, they should be smoothed. If they are badly scored, they should be replaced.

Bearings

17-40. Worn sleeve bearings cause the rotor to drop, which in turn causes excessive wear of the casing wearing ring and the impeller ring. When a bearing is worn excessively, it should be replaced as soon as possible in accordance with the manufacturer's instruction manual. Whenever a replaceable precision bearing is disassembled, it should be inspected carefully for ridges, scores, and wear. The bearing lining should be firmly anchored to the shell. If the bearing is scored or worn beyond the manufacturer's recommended tolerance, it should be replaced. Journals should be kept free from rust and kept smooth at all times. To remove rust spots, ridges, and sharp edges of scores, the journals should be lapped with an oilstone or with an oilstone powder.

Bushings

17-41. Whenever a pump is disassembled, bushing clearances should be measured. Bearing wear will probably cause bushing wear. Bushings should be renewed if the bearings are restored to their original readings.

Table 17-1 is a troubleshooting chart for centrifugal pumps.

Table 17-1. Troubleshooting Chart for Centrifugal Pumps

PUMP TROUBLES	CAUSES
Failure to deliver water.	Pump not primed; insufficient speed; impeller plugged; wrong direction of rotation (this may occur after motor overhaul).
Short in capacity.	Air leaks in stuffing boxes; insufficient speed; insufficient suction head for hot water; suction strainers fouled; impellers damaged and casing packing defective.
Pressure low.	Insufficient speed; air leak; incorrect discharge valves open in manifold (this may allow the pump to discharge into an open line, causing the pump to operate at other than the design point); mechanical defects (same as for Short in capacity).

Table 17-1. Troubleshooting Chart for Centrifugal Pumps (continued)

PUMP TROUBLES	CAUSES
Pump loses water after starting.	Leaky suction line; water seal plugged; suction lift too high (often caused by fouling of the strainer after the pump is started); air or gases in water.
Pump overloads driver.	Speed too high; liquid of different specific gravity and viscosity too high; rubbing caused by foreign matter in the pump and between the casing rings and the impeller; mechanical defects: rotating element binds, shaft bent, and worn bearings.
Pump vibrates.	Misalignment; poor foundation; impeller partially clogged, causing unbalance; mechanical defects (same as for Pump overloads driver).

PROPELLER PUMPS

17-42. Propeller pumps are used on some ships as circulating water pumps. Propeller pumps closely resemble centrifugal pumps in design and operation, but do not use centrifugal force for their operation. The propeller pump has a propeller closely fitted into a tube-like casing. The propeller pumps the liquid by pushing it in a direction parallel to the shaft. Propeller pumps must be located either below or only slightly above the surface of the liquid to be pumped. This is because they cannot operate with a high suction lift.

JET PUMPS

17-43. All the pumps previously described require motors or turbines to drive them. However, jet pumps have no moving parts. The flow through the pump is maintained by a jet of water or steam which passes through a nozzle at a high velocity. Jet pumps are generally classified as ejectors and eductors.

- Ejectors use a jet of steam to entrain and transport air, water, or other fluid.
- Eductors use a flow of water to entrain and pump fluids.

The basic principles of operation of these two devices are identical. In a simple ejector jet pump, steam under pressure enters the chamber through a pipe which is fitted with a venturi-shaped nozzle. This nozzle has a reduced area which increases the velocity of the steam. The fluid in the chamber in front of the nozzle is driven out of the pump through the discharge line by the force of the steam jet. The size of the discharge line increases gradually beyond the chamber to decrease the velocity of the discharge. As the steam jet forces some of the fluid from the chamber into the discharge line, pressure in the chamber is lowered. Pressure on the surface of the supply fluid forces the fluid up through the inlet into the chamber and out through the discharge line. Therefore, pumping action is established. Jet pumps of the ejector type are occasionally used aboard ship to pump small quantities of drainage overboard. Their primary use on naval ships is to remove air and other non-condensable gases from the main and auxiliary condensers. Figure 17-10 shows a portable eductor of the type found in damage control lockers. The principle of operation is the same as that described for the ejector type of jet pump. However, water is used instead of steam. All of the water which enters the large end of the jet must go out through the small end. Since the exit end is smaller than the entrance end, the water leaving the jet will have a greater velocity than it had upon entering the jet. The venturi shape of the diverging nozzles causes a low-pressure area. This creates suction which draws water through the strainer and entrains it through the diverging nozzle. This ensures a constant flow. Eductors may also be used for salvage work and with fog or foam equipment. Eductors will operate when entirely submerged in a flooded compartment and will discharge against a moderate pressure. Although fire and bilge pumps are still being installed in new ships, fixed eductors are the principal means of pumping water overboard through the drainage system. By the use of eductors, centrifugal fire pumps can serve as drainage pumps without the risk of fouling the pump with debris present in the bilges. This is especially useful when there has been damage to a ship. Because of their simplicity, jet pumps generally require very little maintenance. Since there are no moving parts, only the nozzles will show wear. Erosion causes the nozzles to become enlarged, in which case they are generally renewed. The nozzles are occasionally removed; the strainers (if fitted) are cleaned; and a special reamer is inserted in the nozzles to clean out any rust or scale that may have accumulated.

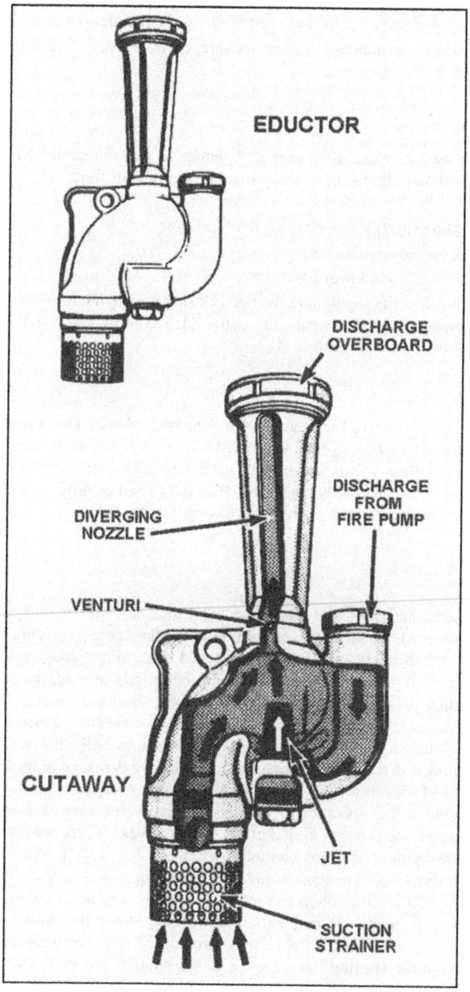

Figure 17-10. Eductor

SECTION II. PURPOSE AND CLASSIFICATION OF VALVES

If a system has a pump, it must also contain devices for controlling the volume of flow, the direction of flow, or the operating pressure of the system. A device that performs one or more of these control functions is a valve. In the various systems of an internal-combustion engine, valves of many types are used to regulate the quantity of liquid flowing to the various components of the systems. As stated earlier, all valves used in engine systems may be grouped into the following two general classifications:

- *Manually operated valves.* Manually operated valves include all valves that are adjusted by hand.
- *Automatic valves.* Automatic valves include check valves, thermostatic valves, and pressure-regulating valves.

STOP VALVES

17-44. Stop valves are used to close off a pipe or opening so that the contained fluid cannot pass through. Some valves can be closed partially to cut down or regulate the flow of fluid. The typical stop valve consists of the following:

- Body, an opening (port) through which the fluid flows.
- Movable disk for closing this port.
- Some means to raise and lower the disk.

In the closed position, the disk fits snugly into the port, closing it completely. When the valve is open, the disk uncovers the port, allowing the fluid to pass through. Each type of stop valve has a different mechanical arrangement for closing the port in the valve.

GLOBE VALVES

17-45. Globe valves generally derive their name from their body shape. Other types of valves may also have globular bodies; therefore, the name may tend to be misleading. A cross-sectional view of a globe stop valve is shown in Figure 17-11. Globe valves are widely used throughout the engineering plant for a variety of services. These valves may be used partly open as well as fully open or fully closed. They are suitable for use as throttling valves. The moving parts of a globe valve consist of the following:

- A disk.
- A valve stem.
- A handwheel.

The stem, which connects the handwheel and the disk, is threaded. It fits into threads in the valve bonnet. When you turn the handwheel, the stem moves up or down in the bonnet, carrying the disk with it. The valve is closed by turning the handwheel clockwise and is opened by turning it counterclockwise. The valve should never be jammed in the open position. After a valve has been fully opened, the handwheel should be turned toward the closed position one-half turn. Unless this is done, the handwheel is likely to freeze in the open position and it will be difficult to close the valve. Many valves have been damaged in this manner. Another reason for not leaving globe valves fully open is that it is sometimes difficult to tell whether a valve is open or closed. If a valve is jammed in the open position, the stem may be damaged or broken by someone who thinks that the valve is closed and tries to force it open. Valves that are exceptions to the above rule are called back-seating valves. Sometimes the operation of a back-seating valve may require that it be fully opened. Whenever this is so, special instructions to that effect will be given. These valves are so designed that, when fully open, the pressure being controlled cannot reach the valve stem packing. This thereby eliminates possible leakage past the packing. The edge of the port, where the disk touches it, is called the valve seat. The edge of the disk and the seat are machined and ground together to form a tight seal. The rate at which fluid flows through the valve is regulated by the position of the disk. When the valve is closed, the disk fits firmly against the valve seat. When it is open, fluid flows through the space between the edge of the disk and the seat. Packing is placed in the stuffing box or space that surrounds the valve stem and is held in place by a packing gland. With continued use of the valve, the stem will gradually wear the packing away and a leak may develop. You can generally stop a slow leak by tightening the gland a turn or two. If this fails, pressure should be removed from the valve, and the packing should be renewed.

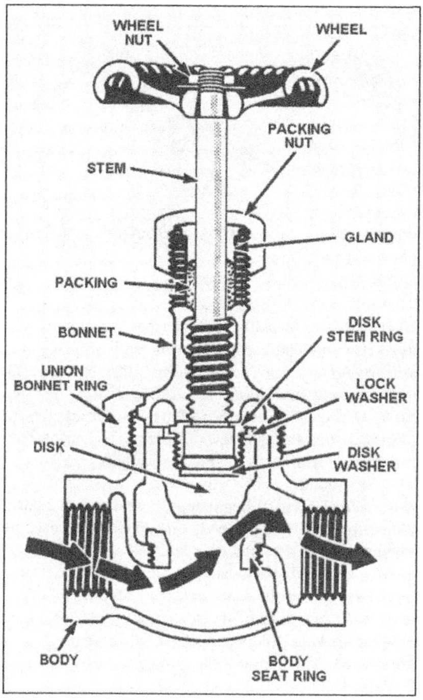

Figure 17-11. Cross-sectional View of Globe Stop Valve

Repair of Globe Valves

17-46. The repair of globe valves (other than routine renewal of packing) is generally limited to refinishing the seat and disk surfaces. When this work is being done, certain precautions should be observed. When refinishing the valve face and seat, do not remove any more material than necessary. Valves that do not have replaceable valve seats can be refinished only a limited number of times. Before doing any repair to the seat and disk of a globe valve, check to be sure that the valve disk is secured rigidly to, and is square on, the valve stem. Also check to be sure that the stem is straight. If the stem is not straight, carefully inspect the valve disk for evidence of wear, for cuts on the seating area, and for improper fit of the disk to the seat. If the disk and the seat appear to be in good condition, spot-in to find out whether they actually are in good condition.

Spotting-in Globe Valves

17-47. Spotting-in is the method used to visually determine whether the seat and the disk make good contact with each other. To spot-in a valve seat, first apply a thin, even coating of Prussian blue over the entire machined face surface of the disk. Then, insert the disk into the valve and rotate it a quarter turn, using a light, downward pressure. The Prussian blue will adhere to the valve seat at those points where the disk makes contact. Figure 17-12 shows a correct seat when it is spotted-in. It also shows various kinds of imperfect seats. Wipe all the Prussian blue off the disk face surface after you have noted the condition of the seat surface. Apply a thin, even coat of Prussian blue to the contact face of the seat, and again place the disk on the valve seat and rotate the disk a quarter turn. Examine the resulting blue ring on the valve disk. The ring should be unbroken and of uniform width. If the blue ring is broken in any way, the disk is not making a proper fit.

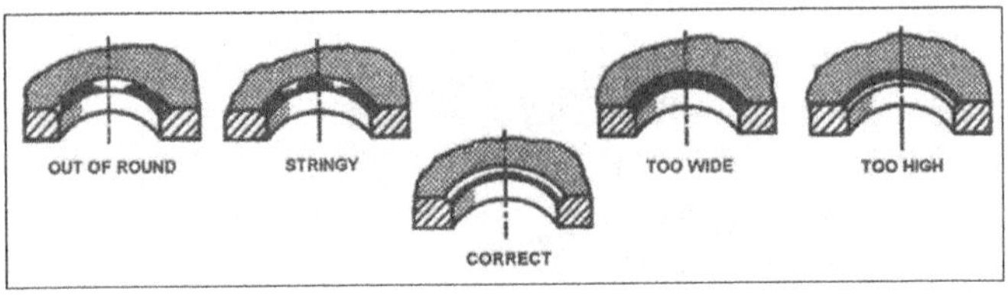

Figure 17-12. Examples of Spotted-in Valves

Grinding-in Globe Valves

17-48. Grinding-in is the manual process used to remove small irregularities from the contact surfaces of the seat and the disk of a valve. Do not confuse grinding-in with refacing processes in which lathes, valve reseating machines, or power grinders are used to recondition the seating surfaces. To grind-in a valve, first apply a small amount of grinding compound to the face of the disk. Then insert the disk into the valve and rotate the disk back and forth about a quarter turn. Shift the disk-seat relation from time to time so that the disk will be moved gradually, in increments, through several rotations. During the grinding process, the grinding compound will gradually be displaced from between the seat and disk surfaces. Therefore, you must stop every minute or so to replenish the compound. When you do this, wipe both the seat and the disk clean before applying the new compound to the disk face. When it appears that the irregularities have been removed, spot-in the disk to the seat in the manner previously described. Grinding is also used to follow up all machining work on valve seats or disks. When the valve seat and disk are first spotted-in after they have been machined, the seat contact will be very narrow and will be located close to the bore. Grinding-in, using finer and finer compounds as the work progresses, causes the seat contact to become broader. The contact area should be a perfect ring, covering approximately one-third of the seating surface. Be careful that you do not over grind a valve seat or disk. Over grinding tends to produce a groove in the seating surface of the disk. It also tends to round off its straight, angular surface. Machining is the only process by which over grinding can be corrected.

Lapping Globe Valves

17-49. When a valve seat contains irregularities that are slightly larger than can be satisfactorily removed by grinding-in, the irregularities can be removed by lapping. A cast-iron lapping tool (lap) of exactly the same and shape as the valve disk is used to true valve-seat surface. Two lapping tools are shown in Figure 17-13. The most important points to remember while using the lapping tool are as follows:

- Do not bear heavily on the handle of the lap.
- Do not bear sideways on the handle of the lap.
- Change the relationship between the lap and the valve seat so that the lap will gradually and slowly rotate around the entire seat circle.
- Keep a check on the working surface of the lap. If a groove develops, have the lap refaced.
- Always use clean compound for lapping.
- Replace the compound often.
- Spread the compound evenly and lightly.
- Do not lap more than is necessary to produce a smooth, even seat.
- Always use a fine grinding compound to finish the lapping job.
- Upon completion of the lapping job, spot-in and grind-in the disk to the seat and remove any traces of grinding or lapping compound.

Only approved abrasive compounds will be used for reconditioning valve seats and disks. Four grades of compounds are used for lapping and grinding valve disks and seats. A coarse-grade compound is used when extensive corrosion or deep cuts and scratches are found on the disks and seats. A medium-grade compound is used to follow up the coarse grade. It may also be used to start the reconditioning process on valves which are not severely damaged. A fine-grade compound should be used when the reconditioning process nears completion. A microscopic fine grade is used for finish lapping and for grinding-in.

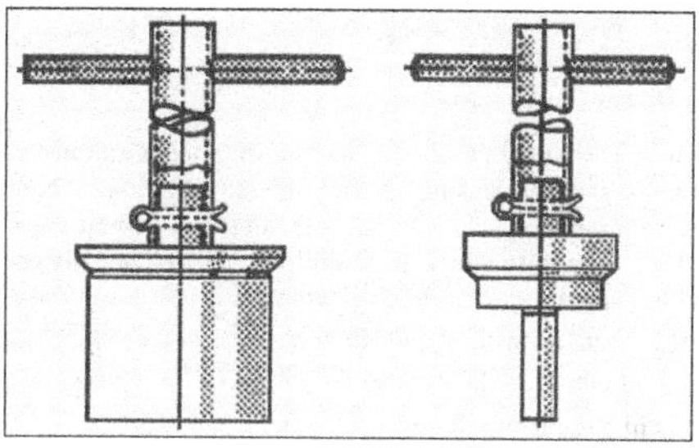

Figure 17-13. Lapping Tools

Refacing Globe Valves

17-50. Badly scored valve seats may be refaced on a lathe, with a power grinder, or with a valve reseating machine. The lathe, rather than the reseating machine, should be used for refacing all valve disks and all hard surfaced valve seats. Work that must be done on a lathe or with a power grinder should be turned over to shop personnel. The discussion here applies only to refacing valve seats with a reseating machine. To reface a valve seat with a reseating machine, attach the correct facing cutter to the machine. With a fine file, remove high spots on the surface of the flange on which the chuck jaws must fit.

Note: A valve reseating machine can be used only with a valve that has the inside of the bonnet flange bored true with the valve seat. If this condition does not exist, the valve must be reseated in a lathe and the inside flange must be bored true.

Before placing the chuck in the valve opening, open the jaws of the chuck wide enough to rest on the flange of the opening. Now tighten the jaws lightly so that the chuck securely grips the sides of the valve opening. Tap the chuck down with a wooden mallet until the jaws rest firmly and squarely on the flange; then tighten the jaws. Adjust and lock the machine spindle in the cutting position and start the cutting by turning slowly on the crank. Feed the cutter slowly so that very light cuts are taken. After some experience, you will learn by the "feel" whether the tool is cutting evenly all around. Remove the chuck and determine if enough metal has been removed from the face of the seat. Be sure the seat is perfect. Then remove the cutter and face off the top part of the seat with a flat cutter. Dress the seat down to the proper dimensions as shown in Table 17-2.

Table 17-2. Proper Dimensions of Dressing the Seat Down

WIDTH OF SEAT	SIZE OF VALVE
1/16 inch	1/4 to 1 inch
3/32 inch	1 1/4 to 2 inches
1/8 inch	2 1/2 to 4 inches
3/16 inch	4 1/2 to 6 inches

After the refacing, grind in the seat and disk. Spot-in as necessary to check the work. A rough method of spotting-in that you may use is to place pencil marks on the bearing surface of the seat or disk. Then, place the disk on the seat and rotate the disk about a quarter turn. If the pencil marks on the seating rub off, the seating is considered satisfactory.

Repacking Valve Stuffing Boxes

17-51. If the stem of a globe is in good condition, stuffing-box leaks can usually be stopped by tightening up on the gland. If this does not stop the leakage, repack the stuffing box. The gland must not be set up on or packed so tightly that the stem binds. If the leak persists, a bent or scored valve stem may be the cause of the trouble. Installing of packing does not require great skill, but a mechanically correct job demands care. When installing packing in moving parts, be sure of the following:

- Proper material is used. Consult the packing chart.
- Rod or stem is not bent, scored, or rusty.
- Packing gland is in alignment with the rod or stem.
- Packing gland is not cocked.
- Old packing has been removed.
- Threads on the gland studs are not burred so as to prevent setting up of the gland nuts.

Cut the end of the packing rings square and leave about a 1 1/16-inch space between the ends to allow for elongation when the gland is tightened. Place the packing in the stuffing box so that the joints will be staggered. Never fill the stuffing box to the extent that the gland cannot enter. Set the gland nuts up evenly and permit some leakage while the packing is adjusting itself to the stem. Never jam the packing glands tight with a wrench.

GATE VALVES

17-52. The manner in which a gate valve (see Figure 17-14) is used has a great deal to do with the service life of the valve. Gate valves should always be used either wide open or fully closed. They should not be used in a partially opened position. When a gate valve is partly open, the gate is not held securely. Therefore, it swings back and forth with the pulsation of flow. As the gate swings, it strikes the valve body and the finished surfaces, nicking and scoring them. When these surfaces are imperfect, the valve gate cannot seat accurately and seal off the flow. A gate valve should never be installed in any position where a throttling or flow-regulating valve is required. For such service, a globe valve should be used. Lapping is the best method for correcting gate valve defects such as light pitting or scoring and imperfect seat contact. The lapping process is the same for gate valves as it is for globe valves, except that the lap is turned by a handle which extends through the end of the valve body. The lapping tool, without its handle, is inserted into the valve so that it covers one of the seat rings. Then, the handle is attached to the lap and the lapping is begun. The wedge gate can be lapped to a true surface, using the same lap that is used on the seat rings. No more material should be removed than is necessary. It is possible to resurface a gate valve only a limited number of times. By removing too much material each time, the total number of times the surfaces can be renewed is decreased. The overall life of the valve will thereby be shortened. On large gate valves, when the seat rings become so deteriorated that they cannot be repaired by lapping, the seat rings can be removed from the valve body and replaced with new ones. It is not advisable to attempt to repair a gate valve without removing it from the piping system. Removing the valve simplifies the repair job and gives more assurance that a good job can be done. Leakage around the stem of a gate valve is caused by troubles similar to those encountered in leaking globe valves. The procedure for stopping leakage around the valve stem is the same for both types of valves.

CAUTION

Do not use the gate as a lap.

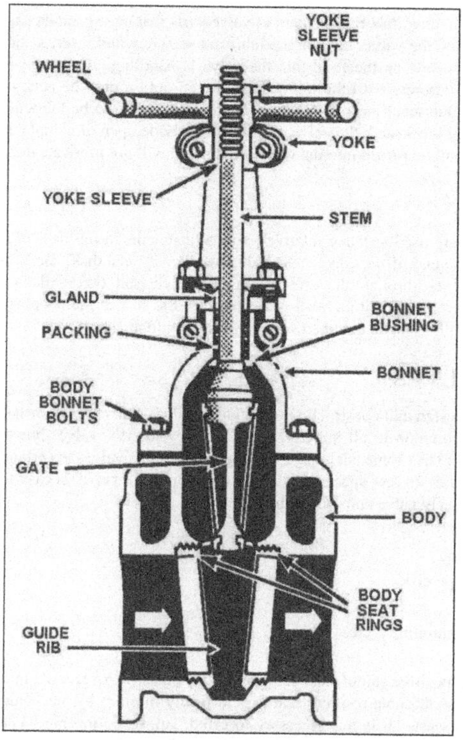

Figure 17-14. Sectional View of Gate Stop Valve

PLUG VALVES

17-53. Plug valves (sometimes referred to as plug-cocks) are frequently used in gasoline and oil feed pipes as well as in water drain lines. The ordinary petcock is a good example of the valve. The body of a plug valve is shaped like a cylinder with holes or ports in the cylinder wall in line with the pipes in which the valve is mounted. Either a cone-shape or a cylindrical plug attached to the handle fits snugly into the valve body. A hole bored through the plug is in line with the ports in the valve body. Turning the plug valve handle (which is in line with the hole in the plug) lines up the hole in the plug with the ports in the valve body so that fluid can pass through the valve. The flow can be stopped by turning the plug 90° (one-quarter turn) from the open position. Some plug valves are designed as three- or four-way selector valves. Three or more pipes are connected to a single valve in line with the same number of ports in the cylinder wall. Two or more holes drilled in the plug provide a variety of passages through the valve. When a valve of this kind is located in a fuel line, the liquid may be drawn from any one of two or three tanks by setting the handle in different positions. A hard lubricant in stick form is used to effectively seal and lubricate the rotating plug. Proper lubricating ensures tightness, maximum life, and ease of operation. Improper lubrication may cause the valve to stick or to leak and may cause excessive wear of the rotating plug. Excessive lubrication should be avoided as it may cause grease to be deposited in the cooling system components. Instructions regarding proper lubrication of plug-cock valves must be strictly followed if valve trouble is to be avoided. Many times only lubricating the valve is needed to eliminate leakage or sticking of the plug. However, if the valve has not been properly lubricated for a long time, replacement of valve parts that have been damaged by lack of lubricant may be required. To lubricate the valve, remove the lubricant setscrew and insert a stick of lubricant. The type of lubricant to be used depends on the fluid that the valve is handling. Lubricant is forced into the plug-cock valve by the lubricant setscrew until the lubricant is forced out around the neck or the stem of the valve. A check valve within the lubricant passage allows the plug-cock valve to be lubricated under pressure. When a plug-cock valve is being lubricated, the valve must be either wide open or completely closed. If this precaution is not taken, lubricant will be forced into the water stream and will not lubricate the valve.

NEEDLE VALVES

17-54. Needle valves are used to make relatively fine adjustments in the flow of fluid. A needle valve has a long tapered point at the end of the valve stem. This needle acts as a disk. Because of the long tapered point, part of the needle passes through the opening in the valve seat before the needle actually seats. This arrangement permits a very gradual increase or decrease in the size of the opening. Therefore, it allows more precise control of flow than can be obtained with an ordinary globe valve.

BUTTERFLY VALVES

17-55. Although the design and construction of butterfly valves can vary somewhat, a butter-fly disk and some means of sealing are common to all butterfly valves. The butterfly valve shown in Figure 17-15 (although relatively new in service) has some advantages over gate and globe valves in certain applications. The butterfly valve is lightweight, takes up less space than a gate valve or globe valve, is easy to overhaul, and is relatively quick-acting. The butterfly valve consists of the following:

- Body.
- Resilient seat.
- Butterfly-type disk.
- Stem.
- Notched positioning plate.
- Handle.

This valve provides a positive shutoff and can be used as a throttling valve set in any position from fully open to fully closed. The replaceable resilient seat is held firmly in place by mechanical means. Neither bonding nor cementing is necessary. It is not necessary to grind, lap, or do machine work to replace the valve seat. Therefore, overhaul of the valve is relatively simple. The resilient seat is under compression when it is mounted in the valve body. This forms a seal around the periphery of the disk and around both upper and lower points where the stem passes through the seat. Packing is provided to form a positive seal around the stem in the event that the seal formed by the seat becomes damaged. When closing the valve, the handle needs only to be turned a quarter turn to rotate the disk 90°. The resilient seat exerts positive pressure against the disk, assuring a tight shutoff. Butterfly valves can be designed to meet a wide variety of applications. Classes of butter-fly valves used for various services are indicated in Table 17-3.

Table 17-3. Classes of Butter-Fly Valves

CLASS A	Freshwater and Seawater
Series 50	50 psi maximum working pressure (22 inches per second (IPS) and above)
Series 150	150 psi maximum working pressure (22 inches IPS and below)
Series 200	200 psi maximum working pressure (20 inches IPS and below)
CLASS B	**JP-5 fuel**
Series 150	150 psi maximum working pressure (20 inches IPS and below)

Table 17-3. Classes of Butter-Fly Valves (continued)

CLASS C	Oil, fuel
Series 150	150 psi maximum working pressure (20 inches IPS and below)
CLASS D	**Oil, diesel, or lubricating**
Series 150	150 psi maximum working pressure (20 inches IPS and below)

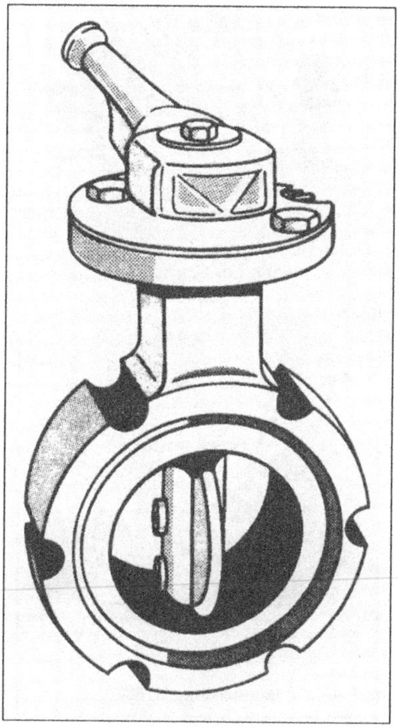

Figure 17-15. Butterfly Valve

CHECK VALVES

17-56. Check valves are designed to permit flow through a line in one direction only. They are used in drain lines (for example, where it is important that there be no reversal of flow). Valves of this type must be properly installed. Most of them have an arrow or the word "inlet" cast on the valve body to indicate direction of flow. If a valve lacks any such indication, a close check must be made to ensure that the flow of fluid in the system will operate the valve in the proper direction. The port in a check valve can be closed by a disk, a ball, or a plunger. The valve opens when pressure on the inlet side is greater than that on the outlet side, and it closes when the reverse is true. All such valves open and close automatically. These valves are made with threaded, flanged, or union faces, with screwed or bolted caps, and for specific pressure ranges. The disk of a swing check valve is raised as soon as the line pressure of fluid entering below the disk is of sufficient force. While the disk is raised, continuous flow takes place. If for any reason the flow is reversed or if back pressure builds up, this opposing pressure forces the disk to seat, thereby stopping the flow. The operation of a lift check valve is similar to that of a swing check valve. The exception is that the valve disk moves in an up-and-down direction instead of through an arc. This difference is shown in Figure 17-16.

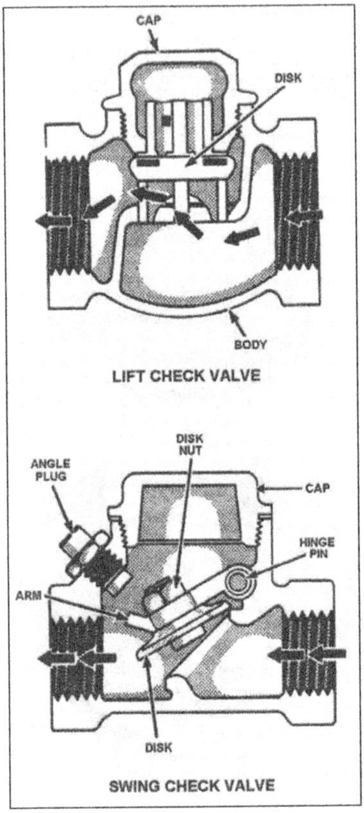

Figure 17-16. Check Valves

STOP CHECK VALVES

17-57. Most valves can be classified as being either stop valves or check valves. However, some valves can function as either a stop valve or a check valve, depending on the position of the valve stem. These valves are known as stop check valves. A stop check valve is shown in cross section in Figure 17-17. The flow and operating principles of this type valve closely resemble the check valve. However, the valve stem is long enough so that when it is screwed all the way down it holds the disk firmly against the seat, thereby preventing any flow of fluid. In this position, the valve acts as a stop valve. When the stem is raised, the disk can be opened by pressure on the inlet side. In this position, the valve acts as a check valve, allowing the flow of fluid in only one direction. The maximum lift of the disk is controlled by the position of the valve stem. Therefore, the position of the valve stem can limit the amount of fluid passing through the valve even when the valve is operating as a check valve. Stop check valves are used in many drain lines, on the discharge side of many pumps, and as exhaust steam valves on auxiliary machinery.

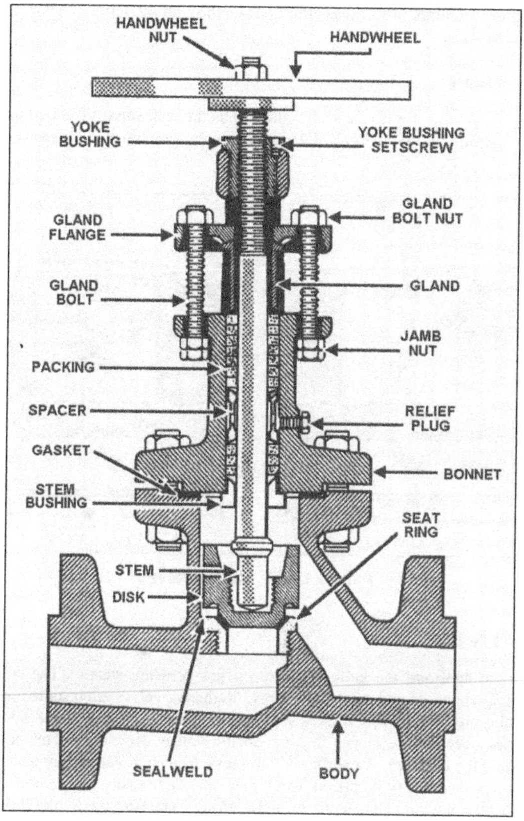

Figure 17-17. Stop Check Valve

RELIEF VALVES

17-58. Relief valves (Figure 17-18) are installed in water, air, and oil lines and on various units of machinery aboard ship. They open automatically when the pressure within the line becomes too high. Relief valves protect piping much the same as fuses protect electrical equipment and wiring in the home. Many types of relief valves exist. Most have either a disk or steel ball acting against a coil spring.

DISK-TYPE RELIEF VALVE

17-59. This type of valve consists of a valve body, a valve disk, and a stem. The steel spring pushes down on the disk and keeps the valve closed. The force of the spring is generally adjusted by setting an adjusting nut on top of the spring. The inlet side of the valve is connected to the system to be protected. When the force on the bottom of the disk, exerted by the pressure of fluid in the line, becomes greater than the compression of the spring, the disk is pushed off the seat. This opens the valve. The valve outlet may be opened to the atmosphere in compressed air or steam lines. When the valve is used to protect a pump, its outlet is connected to the suction line leading to the pump. The excess fluid passes through the relief valve and back to the inlet side of the pump.

BALL-TYPE RELIEF VALVE

17-60. This type of valve operates on the same principal as the disk valve. The ball valve is generally used on lube oil lines. The operating pressure is regulated by adjusting the threaded plug (not shown) which holds the spring in place.

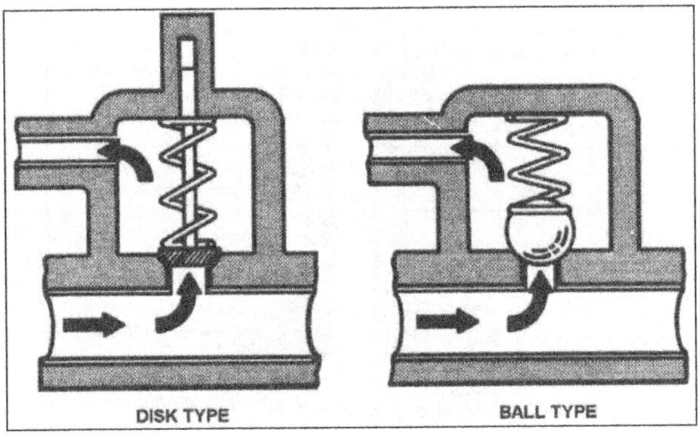

Figure 17-18. Relief Valves

REDUCING VALVES

17-61. The heating system and the galley operate on low-pressure steam. The source of steam for these systems is the boilers, which are under high pressures. Reducing valves are installed in the steam lines to these systems to reduce the pressure. These valves will hold a constant pressure in the delivery lines, even if the boiler pressure varies over a wide range. Most reducing valves depend on a balance between the outlet or operating pressure and the pressure of a spring or compressed air in a sealed chamber. Although some of these valves are quite complicated, the principle on which they operate is easily understood. The simplified reducing valve in Figure 17-19 has a main valve, a piston, and a spring. The compression of the spring pushes the piston to the left and opens the valve. When the steam is turned on, it passes through the open valve and builds up pressure in the outlet chamber. Whenever the force exerted on the piston becomes greater than the force exerted by the spring, the piston moves to the right and closes the valves. During the operation, the outlet steam pressure and spring force remain in balance with the valve partly open. Any slight variation in outlet pressure will upset this balance. The piston will move and increase or decrease the size of the valve opening and restore the original outlet pressure. In some valves, the spring is replaced by a sealed chamber which contains compressed air. The air pressure acts on a diaphragm instead of on a piston. The valve can be set to maintain any desired pressure by adjusting the air pressure in the sealed chamber. Sometimes the diaphragm is located between two chambers (one of them opened to the inlet and the other to the outlet steam). The action of the diaphragm operates a valve which in turn regulates the steam pressure on a piston connected to the main valve. In all reducing valves, the outlet pressure controls the rate at which the inlet steam is permitted to pass the valve.

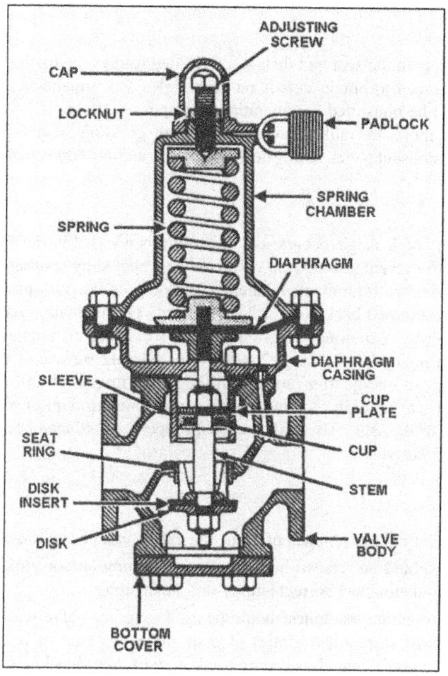

Figure 17-19. Reducing Valve

INSTALLING VALVES

17-62. The best position in which to install a valve is with the stem pointing straight up. When the stem points downward, the bonnet acts as a pocket for scale and other foreign matter in the line. Such matter can interfere with valve operation by cutting and eventually destroying inside stem threads. The recommended position for double-disk gate valves is with the stem upright. In liquid lines subject to freezing temperatures, an upside-down position for valves is undesirable because liquid trapped in the bonnet can freeze and rupture it. Even when installed upright, valves in such lines should have drain plugs in the body as a precaution against freezing. The proper method for installing a globe valve is determined by the purpose for which it is used. The general rule should be that, unless otherwise required, a globe valve will give more satisfactory service when installed with pressure below the seat. This type of installation permits the replacement of the valve stem without isolating the entire system. Install all check valves so that the disk will open with the flow. To ensure closing of the disk when occurs, the position of check valves in line must permit closure of the disk by gravity.

CAUTION

When a stop valve has been modified to a stop check valve, a plate or disk should be inserted under the handwheel nut and marked -- STOP CHECK CAUTION: INSTALL IN CORRECT DIRECTION. Failure to make such marking can result in installation of valve in reverse direction, causing possible damage or malfunction.

Reseating Valves

17-63. The depth of defects in the seat and disk of valves determines the methods to be used for repair. Valve drawings should be referenced to obtain details on seat angles and dimensions. Technical manuals for valve reseating machines should be consulted for operating instructions. In some cases, replacement of a valve can be more economical and more desirable than repair. When grinding seating surfaces, avoid overheating to prevent changes in dimensions, changes in physical properties, and possible cracking after cooling.

Very Minor Defects

17-64. If a small amount of leakage occurs and if there are no visible defects in the seat or disk and no working or other defects to prevent seating, the valve can be reseated by spotting-in the seat and disk. The seat and disk are ground together with an abrasive such as grinding compound, powdered emery, or ground glass mixed with oil. The disk is turned back and forth on the seat. Occasionally, the disk is lifted from its seat and its position is shifted slightly. Continue the grinding until a bearing all around is obtained. As a test of the work, put pencil marks at intervals of about 1/2 inch on the bearing surface of the disk or seat. Drop the disk on the seat and rotate it about one-quarter turn. If all the pencil marks rub off, the seating is satisfactory. For globe valves with tapered seats and disks, remove only a minimum amount of metal, as this grinding operation reduces the effectiveness of the seal. The ability to seat properly is reduced because the wedging action of the tapered faces is partially destroyed.

Minor Defects

17-65. When defects do not require renewal of parts, the valves can be as follows:

- **Disks.** Disks should be refaced in a lathe by machining and/or grinding. Make sure that disks are properly centered and have correct angles and dimensions.
- **Seats.** Valve reseating machines should be used to reface valve seats in place. These machines are available for most sizes and varieties of both globe and gate valves. Ensure that machines of the proper size, seat angle, and abrasive are used. Use of these machines usually results in labor savings and usually avoids rewelding of welded-in valves. A cast-iron dummy disk and grinding compound can also be used to remove minor defects. The disk must have the proper seat angle and be remachined occasionally to maintain this angle and remove shoulders. A jig to guide and support this dummy disk should be used. Seats of valves removed for repair can be machined and/or ground by valve reseating machines, lathes, or other suitable equipment. Spotting-in and grinding-in are accomplished as discussed previously in this section.

Major Defects

17-66. To avoid replacing complete valves in the event of major damage to the seating surfaces, the following process applies.

- **Replace Parts.** Disks, seats, and inserts can be replaced in some valves.
- **Replace Lost Metal.** Welding, plating, and metal spraying can be used to build up seating surfaces which have been eroded, scored, or otherwise damaged. This operation is followed by machining, grinding, and spotting-in.

VALVE REPAIR

17-67. Refer to the troubleshooting chart (Table 17-4) for the probable causes of and remedies for valve troubles and repair. The following paragraphs describe procedures for shipboard repair of a valve.

PACKING STUFFING BOX

17-68. The stuffing box is packed by placing successive turns of packing in the space around the rod of the valve stem. Where string packing is used, it is simply coiled around the rod. The ends are beveled to make a smooth seating for the bottom of the gland. The gland is then put on and set up by the bonnet nut or the gland bolts and nuts. To prevent string packing from folding back when the gland nut is tightened down, the packing should be wound to the right or in the same direction that the gland nut is turned. This will ensure that there are no joints in the packing and will reduce the possibility of leakage. Where successive rings are used, the ends of the rings should be cut square and even with ends butted to make a level joint. If rings are put in place by sticks, splitting of the packing should be avoided.

REPACKING VALVES UNDER PRESSURE

17-69. Certain types of gate, globe, angle, and stop check valves, for some pressures, are designed to back-seat the stem against the valve bonnet when fully open. This allows the stem stuffing box to be repacked under pressure when necessary. In addition, high pressure valves usually have a pipe plug as a leak-off to the cooling chamber. This pipe plug should be removed when repacking the valve under pressure. For valves where discharge cannot be permitted during replacing under pressure, the discharge will be blanked off securely before the valve is opened for back-seating and repacking.

Table 17-4. Troubleshooting Chart for Valve Troubles and Repairs

TROUBLE	PROBABLE CAUSE	REMEDY
Main valve fails to open when operated electrically.	Lack of electrical power.	With operating switch closed, check for proper voltage to no. 1 solenoid.
	Solenoid fails to position pilot valve properly.	Check linkage for binding.
	Burned out solenoid coil.	Check for visible evidence of burn. Check coil for direct shorts or breaks, using an ohmmeter.
	Stuck pilot valve.	Attempt to turn pilot valve manually. If hard to turn, remove for repair. (If stuck, pilot valve is not removed, solenoid cannot turn valve. This will cause solenoid to overheat and perhaps burn out.)
	Plugged drain line.	If main valve fails to open when pilot valve is properly positioned, inspect drain line to ensure that it is plugged. (Remember that fluid should discharge from drain line each time main valve opens.)

Table 17-4. Troubleshooting Chart for Valve Troubles and Repairs (continued)

TROUBLE	PROBABLE CAUSE	REMEDY
	Main valve stuck in closed position.	Position control to open valve and remove tubing from the control to main valve cover. If main valve still fails to open, it must be disassembled and freed.
	Ruptured diaphragm.	This condition will be apparent when valve is disassembled.
Main valve opens slowly.	Pilot valve partially stuck in open position.	Check linkage for binding; slowly try to turn pilot valve manually; if it is hard to turn, it must be removed and repaired.
	Drain line partially plugged.	Inspect drain line and remove obstruction.
Main valve fails to close electrically.	Lack of electrical power.	Ensure that proper voltage is being supplied to no. 2 solenoid.
	Solenoid fails to position pilot valve properly.	Check linkage for binding.
	Burned out solenoid coil.	Check for visible evidence; check coil for direct short or breaks using an ohmmeter; if necessary, replace coil.
	Stuck pilot valve.	Turn pilot valve manually; if it is hard to turn, remove for repair. (Remember that a stuck pilot valve can cause solenoid to burn out.)
	Plugged supply line.	Loosen pipe plug in top of main valve indicator and, with pilot positioned to close main valve, watch for a steady flow. If flow stops, the main valve inlet to pilot valve, tubing from pilot valve cover, or pilot valve itself is plugged. It will then be necessary to work back from main valve cover to locate point of stoppage.
Main valve fails when operated manually.	Any of the probable causes listed above, except when electrical power is required.	Any of the remedies listed above, except where electrical power is required.

PRECAUTIONS FOR INSTALLING AND OPERATING REDUCING AND REGULATING VALVES

17-70. Reducing valve installation should conform to the valve manufacturer's recommendations. In general, they should be installed in a horizontal run of pipe with the superstructure in a vertical position. They should be located so that they are not in a piping pocket where water and dirt can accumulate. Their location should also permit removal of the valve superstructure and working parts without removing the valve from the line. Reducing valves have an arrow cast on the valve body to indicate the direction of flow through the valve. To ensure proper operation, this marking should be observed when installing a valve. Each steam-reducing station should consist of the following:

- An inlet cutout valve.
- A steam strainer.
- A reducing valve.
- A tapered increaser.
- An outlet cutout valve.

A line bypassing these components should contain a cutout valve and a throttling valve. On the discharge of the reducing station, a pressure-relief valve and a pressure gauge should be provided. On some installations of reducing valves in parallel, only one bypass line, pressure gauge and relief valve are provided. Reducing valves must be warmed up and drained before they are adjusted. Cutout valves in the inlet and outlet lines of a reducing station should be fully open when the reducing station is in use. Reducing station relief valves must be kept in good operating condition. They should be lifted by hand weekly. Swing and stop check valves in main and secondary drainage systems will be dismantled and inspected annually for cleanliness, freedom from foreign matter, and integrity of the valve seat.

Chapter 18

PIPING, SEALS (GASKETS AND PACKING), AND INSULATION

INTRODUCTION

18-1. As an engineer, you will be working with piping, fittings, seals (packing and gaskets), and insulation. You will be responsible for routine maintenance of this equipment in the engine room and throughout the ship. The machinery of a system cannot work properly unless the piping and valves that make up the system are in good working order. The information in this chapter, as it is throughout the book, is of a broad and general nature. You should refer to the appropriate technical manuals (TMs) and/or ship's plans, information books, and plant or valve manuals for specific problems with individual equipment.

SECTION I. PIPING AND TUBING

Piping is defined as an assembly of pipe or tubing, valves, fittings, and related components forming a whole or a part of a system for transferring fluids (liquids and gases).

TUBULAR PRODUCT IDENTIFICATION

18-2. In commercial usage, there is no clear distinction between pipe and tubing since the correct designation for each tubular product is established by the manufacturer. If manufacturers call a product pipe, it is pipe; if they call it tubing, it is tubing. However, in the Army a distinction is made between pipe and tubing based on the way the tubular product is identified as to size. The following are the three important dimensions of any tubular product:

- Outside diameter (OD).
- Inside diameter (ID).
- Wall thickness.

A tubular product is called tubing if its size is identified by actual measured OD and by actual measured wall thickness. A tubular product is called pipe if its size is identified by a nominal dimension called iron pipe size and by reference to a wall thickness schedule designation. The size identification of tubing is simple enough, since it consists of actual measured dimensions. However, the terms used for identifying pipe sizes may require some explanation. A nominal dimension such as iron pipe size is close to, but not necessarily identical with, an actual measured dimension. For example, a pipe with a nominal pipe size of 3 inches has an actual measured OD of 3.50 inches. A pipe with a nominal pipe size of 2 inches has an actual measured OD of 2.375 inches. In the larger sizes (about 12 inches) the nominal pipe size and the actual measured OD are the same. For example, a pipe with a nominal pipe size of 14 inches has an actual measured OD of 14 inches. Nominal dimensions are used to simplify the standardization of pipe fittings, pipe taps, and threading dies. The wall thickness of pipe is identified by reference to wall thickness schedules established by the American Standards Association. For example, a reference to schedule 40 for a steel pipe with a nominal pipe size of 3 inches indicates that the wall thickness of the pipe is 0.216 inches. A reference to schedule 80 for a steel pipe of the same nominal pipe size (3 inches) indicates that the wall thickness of the pipe is 0.300 inches. You have probably seen pipe identified as standard (Std), extra strong (XS), and double extra strong (XXS). These designations, which are still used to some extent, also refer to wall thicknesses. However, pipe is manufactured in a number of different wall thicknesses and some pipe does not fit into the Std, XS, and XXS classifications. The wall thickness schedules are being used increasingly to identify the wall thickness of pipe because they provide identification of more wall thicknesses than can be identified under the Std, XS, and XXS classifications.

PIPING MATERIALS

18-3. Many different kinds of pipe and tubing are used in shipboard piping systems. Nonferrous pipe and nonferrous tubing are used for many shipboard piping systems and for most shipboard heat exchangers. Nonferrous materials are used chiefly where their special properties of corrosion resistance and high heat conductivity are required. Various types of seamless copper tubing are used for refrigeration installations, plumbing and heating systems, lubrication systems, and other shipboard systems. Copper-nickel alloy tubing is available in composition 70-30 (70 percent copper and 30 percent nickel) and in composition 90-10 (90 percent copper and 10 percent nickel). The 70-30 composition is generally used in heat exchangers. Many other kinds of pipe and tubing besides those mentioned here are used in shipboard piping systems. Remember that design controls the selection of any particular pipe or tubing for any particular system. Although many kinds of pipe and tubing look almost exactly alike from the outside, they may respond very differently to pressures, temperatures, and other service conditions. Therefore, each kind of pipe and tubing can be used only for specified applications.

PIPING FITTINGS

18-4. Piping sections of the proper size and material are connected by various standard fittings. These include several types of threaded, bolted, welded, and silver-brazed fittings and expansion joints.

THREADED JOINTS

18-5. Threaded joints are the simplest type of pipe fittings. Threaded fittings are not widely used aboard modern ships except in low-pressure water piping systems. The union fittings are provided in piping systems to allow the piping to be taken down for repairs alterations. Unions are available in many different materials and designs (see Figure 18-1) to withstand a wide range of pressures and temperatures. The union is used a great deal for joining piping up to 2 inches in size. The pipe ends connected to the union are threaded, silver-brazed, or welded into the tail pieces (union halves). Then the two ends are joined by setting up (engaging and tightening up on) the union ring. The male and female connecting ends of the tail pieces are carefully ground to make a tight metal-to-metal fit with each other. Welding or silver-brazing the ends to the tail pieces prevent contact of the carried fluid or gas with the union threading.

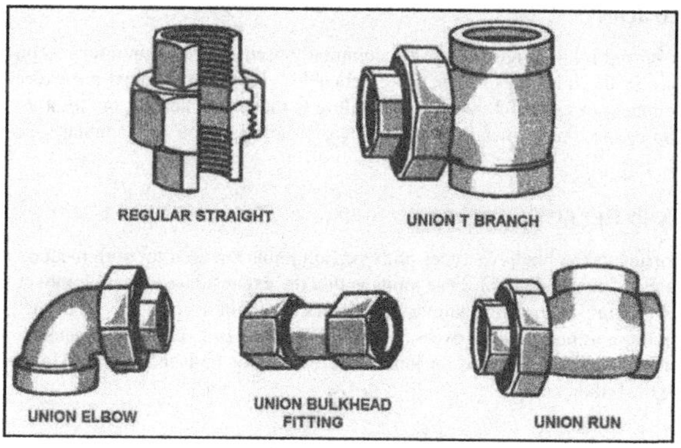

Figure 18-1. Unions

Bolted Flange Joints

18-6. Bolted flange joints are suitable for all pressures now in use. The flanges are attached to the piping by welding, brazing, screw-threading (for some low-pressure piping), or rolling and bending into recesses. Figure 18-2 shows the most common types of flange joints used. Flange joints are manufactured for all standard fitting shapes such as the tee, cross, elbow, and return bend.

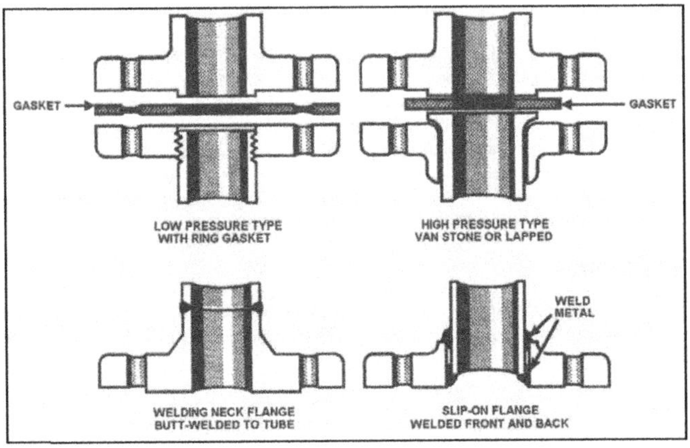

Figure 18-2. Four Types of Bolted Flange Piping Joints

Welded Joints

18-7. The majority of joints found in subassemblies of piping systems are welded joints (Figure 18-3), especially in high-pressure piping. The welding is done according to standard specifications which define materials and techniques. The three general classes of welded joints are butt-weld, fillet-weld, and weld.

Silver-Brazed Joints

18-8. Silver-brazed joints (Figure 18-4) are commonly used for joining nonferrous piping when pressure and temperature in the lines make their use practicable. Temperature must not exceed 425° F; for cold lines, pressure must not exceed 3,000 psi. The alloy is melted by heating the joint with an oxyacetylene torch. This causes the molten metal to fill the few thousandths of an inch annular space between the pipe and the fitting.

Corrugated and Bellows Joints

18-9. The corrugated and bellows types of expansion joints are used for both medium and high pressures and temperatures. The principle of these joints is that the expansion-contraction movement is absorbed by the changing curvature of the corrugations or bellows (as in an accordion). The internal sleeves, free to slide axially in these joints, serve to excessive turbulence and erosion of the expansion parts. Figure 18-5 shows a corrugated bulkhead expansion joint. It provides for both radial and axial movement of piping with respect to the bulkhead.

Flange Safety Shields

18-10. A fuel fire in the fire room or engine room can be caused by a leak at a fuel oil or lube oil pipe flange connection. Even the smallest leak can spray fine droplets of oil on nearby hot surfaces. To reduce this possibility, spray shields (Figure 18-6) are provided around piping flanges of flammable liquid systems, especially in areas where the fire hazard is apparent. Spray shields are usually made of aluminized glass cloth and are simply wrapped and wired around the flange.

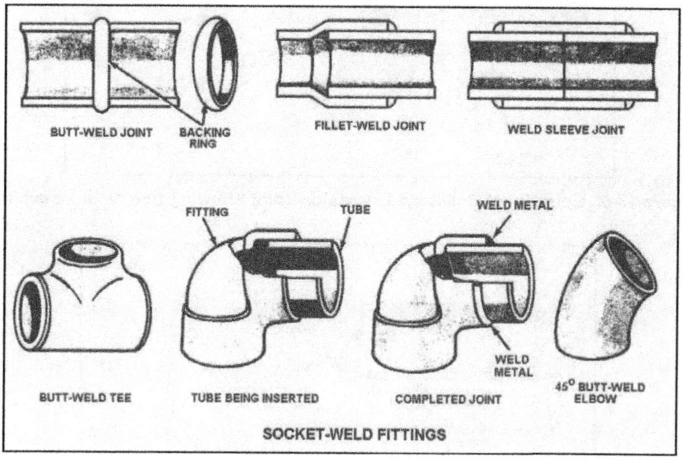

Figure 18-3. Various Types of Welded Joints

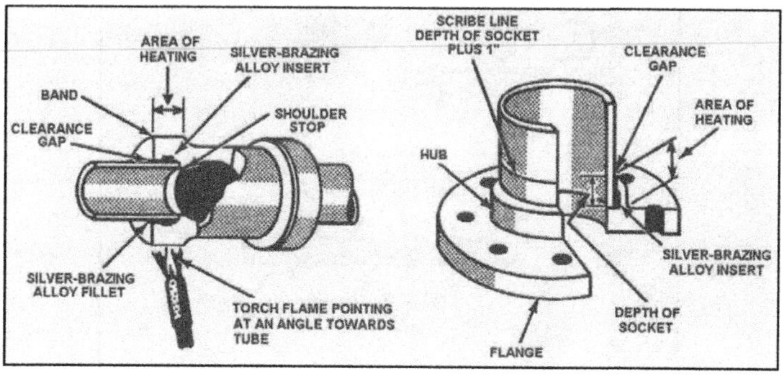

Figure 18-4. Silver-brazed Joints

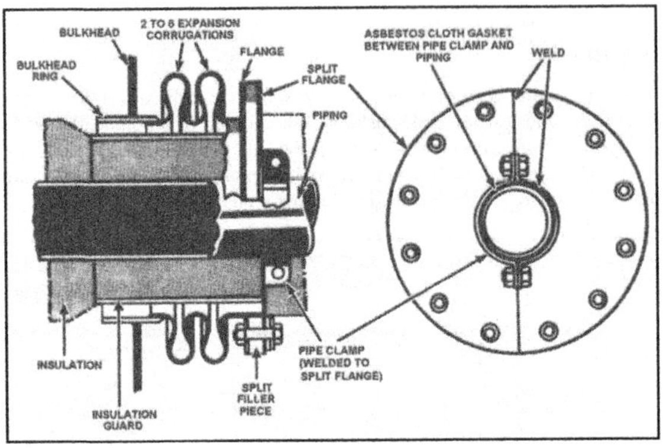

Figure 18-5. Corrugated Bulkhead Expansion Joint Showing Details of Installation

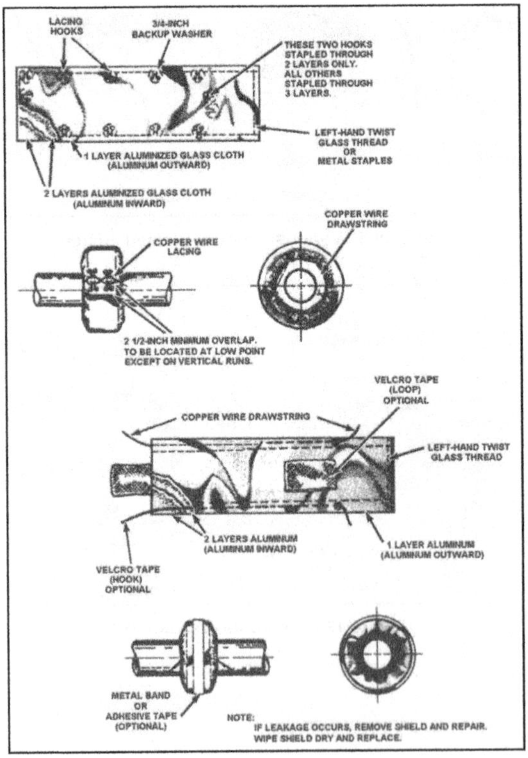

Figure 18-6. Flange Safety Shield

PIPING CARE AND MAINTENANCE

18-11. Reasonable care must be given the various piping assemblies as well as the machinery connected by the piping. All joints, valves, and cocks in the lines must be examined frequently and kept tight. Where piping passes through decks or bulkheads and a possibility exists for movement of one with respect to the other, stuffing boxes or flexible bulkhead connections are provided to take up the movement. This is true if expansion bends or other offsets are not provided in the piping. The external surfaces of uncovered and ungalvanized steel or iron piping should be kept properly painted and free of moisture. Copper and brass piping is seldom painted.

LEAKAGE AT JOINT

18-12. Continual leakage at a joint where a branch line joins another line is usually due to improper for expansion in one of the lines or to excessive vibration. Oftentimes, these leaks can be corrected by slightly altering the anchorages, connections, hangers, or leads of the piping to allow the required expansion to prevent strain. The leaks can also be repaired by fitting supports to the piping to prevent vibration. Leaky joints may also be due to poor alignment of the piping or to movement of decks or bulkheads. Realignments should be made so that flanges or screw threads meet properly without being forced. Sometimes flange joints may need to be refaced or distance (shims) pieces may need to be fitted. Small leaks in gaskets should be taken up immediately, before a dangerous blowout results from progressive growth of the leak.

PIPE THREAD LEAKS

18-13. Pipe thread leaks should be repaired promptly. Leaky screwed joints that cannot be tightened with a reasonable amount of tightening should be repaired. They should be taken apart, cleaned, examined for bad thread conditions, recoated with the appropriate (if any) compound and carefully reassembled to avoid any other thread damage.

PERMANENT REPAIRS

18-14. Copper or brass piping may be permanently repaired by brazing. Small holes may be plugged with a rivet or a screw plug. Leaky piping sections may be semi-permanently repaired by wrapping the piping with tightly drawn wire that is soldered or brazed as it is applied. Several layers of wire securely bonded give a strong, tight repair.

SALTWATER PIPING

18-15. The life of saltwater piping may be lengthened by operating the systems with the minimum practical water velocities and by eliminating grounds from electrical systems, especially direct current (DC) circuits. Saltwater piping may also last longer by eliminating air from the saltwater systems and by promptly correcting any leaks. Insulating with sheet rubber, the hangers which support piping other than that made of wrought iron or steel, can also lengthen the life of saltwater piping. Eliminating wire-drawing by fully opening valves where throttling is unnecessary can prolong the life of saltwater piping.

PIPING IDENTIFICATION SYSTEM

18-16. Shipboard piping will be identified by a color coding (Figure 18-7) and marking as specified in TB 44-0143. Color codes will be applied to valve handwheels only. Valve stems, threads, and tags will not be painted. Fire hose racks will be painted red. Color-coded arrows indicating the direction of flow will also be stenciled on the piping. The liquid within the piping system will be identified by lettering which will be applied when two or more fluids fall within a color. For example, hydraulic oil and lube oil tank vent and fill valves will be color-coded.

Note: Potable water lines will be painted light blue or stripped with 6-inch light blue bands at fittings or each side of a partition, deck or bulkhead, and at intervals of not more than 15 feet in all spaces.

WATERCRAFT COLOR CODES

COLOR	APPLICATION
GREEN	BILGE SYSTEM, SEAWATER SYSTEM
PURPLE	REFRIGERANT SYSTEMS
GRAY	COMPRESSOR AIR SYSTEMS
YELLOW	DIESEL FUEL SYSTEMS GASOLINE CONTAINERS NAPHTHA CONTAINERS } PAINT ENTIRE CONTAINER
ORANGE	LUBRICATING OIL HYDRAULIC OIL
RED	FIRE EXTINGUISHING SYSTEMS
BLACK	STEAM AND HOT WATER HEATING SYSTEMS
DARK BLUE	NONPOTABLE FRESHWATER SYSTEMS
LIGHT BLUE	POTABLE FRESHWATER SYSTEMS

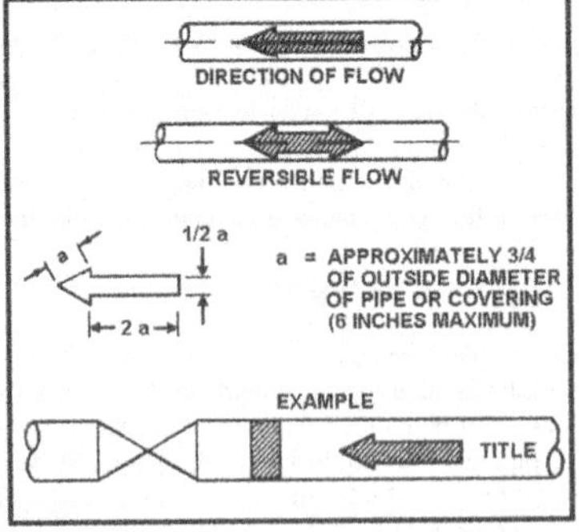

Figure 18-7. Color Coding for Piping System

The lettering along with arrows indicating the direction of flow will be located at the following points:

- Where entering or leaving mechanical equipment.
- Where appearing or disappearing through a deck or bulkhead.
- At all tee, cross, or branches of systems.
- At all valves.
- At points which will clarify a complicated system.

Note: Where a straight line runs through a relatively small compartment, it may be marked only once, and not at each bulkhead.

Nonpotable freshwater line outlets will be labeled as being unfit for drinking.

PIPING SAFETY PRECAUTIONS

18-17. Some general safety precautions to be observed concerning piping systems are as follows:

- To prevent water hammer, drain water from steam piping before admitting steam.
- Before opening large steam valves, open bypasses to warm lines and equalize pressures. If there are no bypasses, "crack" (just slightly back the valve off its seat) the valves.
- Open trap bypasses when admitting steam to piping.

When breaking a flange joint, particularly in steam and hot water lines or in those saltwater lines that have a possible direct connection with the sea, special precautions should be taken as follows:

- Make sure that there is no pressure on the line.
- Be sure that valves that cut pressure off the part of the line undergoing repair are secured (such as tagging and locking) so that they cannot be accidentally opened.
- Ensure that the line is completely drained.
- Make sure that two of the flange-securing nuts (diametrically opposite if possible) remain in place while the others are removed. The two remaining nuts should then be slacked sufficiently to allow breaking of the joint. If the line is clear, all the nuts may be removed. This precaution is necessary to prevent accidents such as scalding personnel or flooding compartments.
- After remaking a steam joint, tighten the nuts according to prescribed methods.
- Never use piping for handholds or footholds.
- Never secure chain falls to piping.
- Never use piping for supporting weights.
- Secure copper and brass piping so that contact with bilges is avoided. The brackets should be lined with insulating material to prevent direct contact between piping and any of the ship's structure.

SECTION II. SEALS

Seals are used to prevent the loss of fluid and pressure. Seals fall into two major classes (static and dynamic).

- Static seals are used to prevent the flow of fluid across a static joint between two parts of a pressure vessel. A typical example is the gasket between the cylinder head and cylinder block of an engine. Static seals are referred to as gaskets.
- Dynamic seals are used to prevent the flow of fluid across a sliding or rotating joint between two members of a hydraulic or pneumatic device. Typical examples are the piston packing in a cylinder and the shaft packing of a rotary pump. Dynamic seals are referred to as packing.

SEAL CLASSIFICATION

18-18. Seals may be classified as metallic and nonmetallic. Examples of metallic seals are sheet metal gaskets and soft copper or aluminum washers. Examples of nonmetallic seals are sheet cork, leather, and rubber. Sub-classifications of nonmetallic seals are organic and inorganic. Typical examples of organic seals are leather, flax, jute, and cotton. Typical examples of inorganic seals are asbestos and synthetic rubber. Seals may also be classified as molded, sheet, or braided packing. Typical molded packing includes cup packings, V-rings, U-cups, and O-rings. Sheet packing is usually rolled from a mass of granulated or shredded packing material in a suitable binder. Seals may also be a mixture of metallic and nonmetallic materials and are available in many shapes and sizes. Figure 18-8 shows some of the seals that are produced commercially.

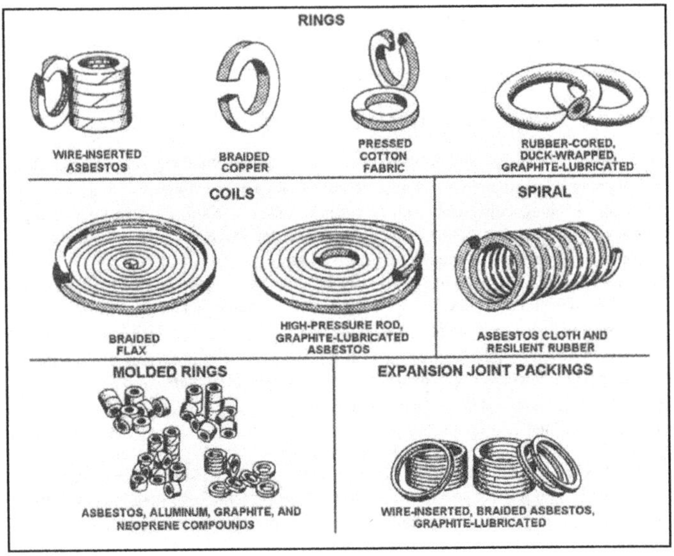

Figure 18-8. Commercially Prepared Seals and Packing

SIMPLE GASKETS (STATIC SEALS)

18-19. Gaskets are made of sheet material that is soft, pliable, and easily compressed to form a good seal. Sheet material may be broken or torn f not handled with care. Some of the most common types of sheet material are:

- Neoprene.
- Oil-treated paper.
- Cork.
- Lead foil.
- Asbestos.

Gaskets made from sheet material must be hand-fabricated. Soft metal gaskets of copper or annealed aluminum are used on surfaces that are serrated with sharp 60^{0} vees directly opposite each other. These serrations sink into the soft metal to form a seal. O-ring seals are used on equipment that requires frequent disassembly. O-rings are used in grooves so that assembly of the parts will not pinch the ring. Both surfaces against the O-ring must be as smooth as possible. Cylinder sleeves should have a bevel and rounded edge to prevent the O-ring from being cut during assembly. O-ring seals are recommended for pressure up to 1,500 psi. At higher pressures, the O-ring should be supplemented with backup washers. O-rings are never to be used more than once.

HAND FABRICATION OF SHEET GASKETS

18-20. Gasket cutters are used to cut gaskets from sheets of gasket material such as cork, rubber, leather, fiber, and composition. To use these cutters, the material is placed on a wood surface and the two cutters ere adjusted; one for the inside cut and the other for the outside cut. The shank is placed in a brace and the center pin is inserted in the gasket material. The brace is then rotated to form gasket. If the gasket is to have bolt holes, the holes can be punched out using a gasket punch. Figure 18-9 shows a gasket cutter and punch. Gasket punches are available in different sizes. Holes should be punched larger than the bolts. Material should be cut so that the inside diameter is the same as the inside diameter of the pipe at the flange surface. Gaskets may also be cut with shears, snips, or a sharp knife. When using any of these tools, the material must be marked before cutting out the gasket. Marking the gasket may be done by laying the material over the flange and marking the cutting limits by light blows with a ball peen hammer. Another way to mark a gasket is to chalk the face of the flange, then lay the gasket material over the flange and apply pressure. This procedure will transfer the chalked impressions of the flange to the gasket material.

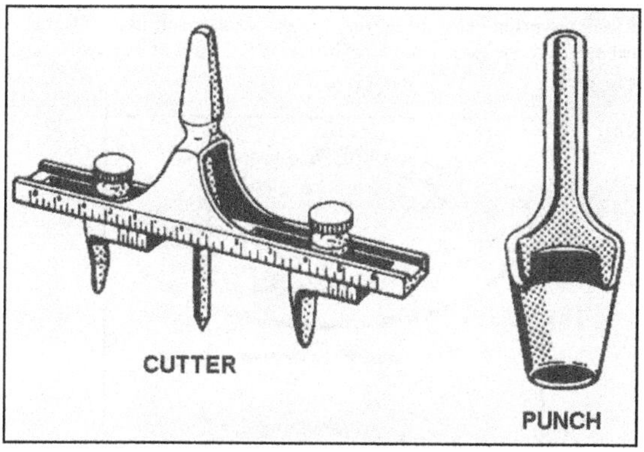

Figure 18-9. Gasket Cutter and Punch

PREFABRICATED GASKETS

18-21. Prefabricated gaskets may be obtained through normal supply channels. They may be ordered in sets or individually by size and type of material. Three types of prefabricated gaskets used on fixed joints of steam lines are flat ring, serrated face, and asbestos-metallic, spiral wound.

Flat Ring

18-22. Flat ring or plain-face gaskets are made of Monel metal or soft iron to specified shapes and sizes. A flat ring gasket and a variation of this type are shown in Figure 18-10.

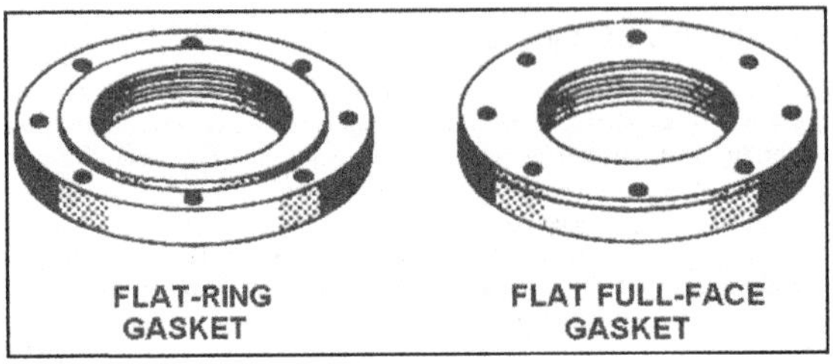

Figure 18-10. Flat-ring Gaskets

Serrated Face

18-23. Serrated face gaskets are also made from Monel metal or soft iron. The raised serrations help to make a better seal and give the gasket some resiliency. Two types of serrated-face gaskets are shown in Figure 18-11.

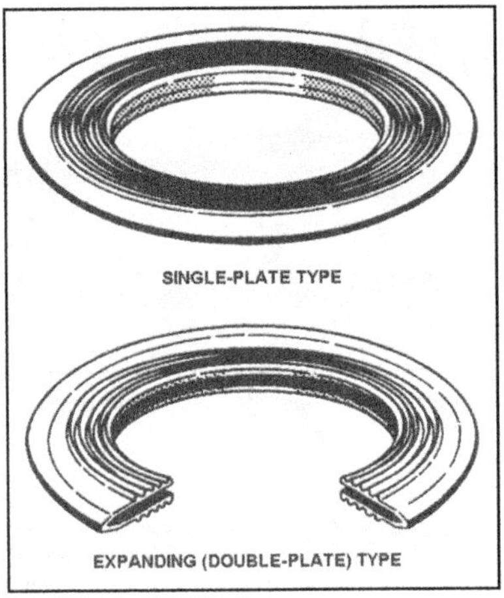

Figure 18-11. Serrated-face Gaskets

Asbestos-metallic, Spiral-wound

18-24. This gasket is composed of alternate layers or plies of dovetailed metal ribbon and strips of asbestos felting spirally wound, ply upon ply. A steel ring is attached to the outer periphery of the gasket. This ring acts as a bolt guide and a centering ring, as well as a reinforcement. An asbestos-metallic, spiral-wound gasket is shown in Figure 18-12.

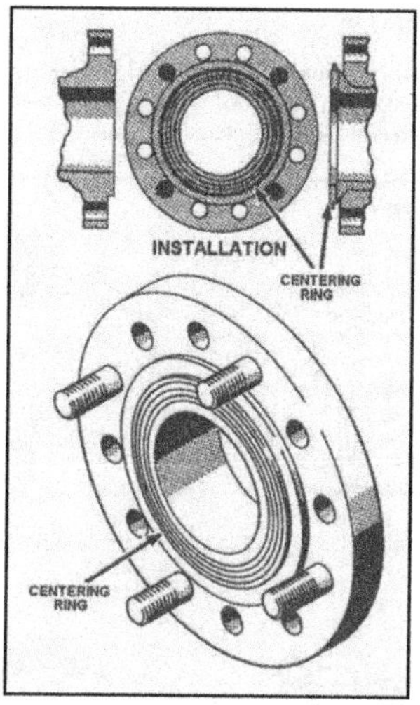

Figure 18-12. Asbestos-metallic, Spiral-wound Gasket

REMOVAL AND INSTALLATION OF GASKETS

18-25. Before a flange joint, be sure there is no pressure on a line. Cutoff valves should be secured to prevent accidental operation during repair. Two of the flange securing nuts (diametrically opposite if possible) should remain in place while the others are being removed. The two remaining nuts should then be slacked sufficiently to allow breaking the joint. After the line is proved to be clear, all nuts should be removed, and then the gasket. If the flange or joint is subject to removal periodically, the face of the gasket should be covered with graphite compound to prevent the gasket from sticking. Most of the trouble experienced with leaky joints in piping is due to poor alignment or to improper allowance for expansion. Either of these faults tends to put excessive strain on the joints. If the joints do not line up properly, the piping should be realigned so that flange joints or screw threads meet properly without using force. On flange joints, it may be necessary to reface the flange. Screw-in joints should be coated with a waterproof compound to prevent leakage and to act as a lubricant. On most flange-type joints, gaskets can be used dry provided space is allowed for expansion of the joint. After the gasket and flange securing nuts have been installed, the cutoff valves should be opened and the joint checked for leakage.

PACKING (DYNAMIC SEALS)

18-26. The primary purpose of packing is to seal joints in machinery and equipment against leakage. Packing is used with four principal types of joints:

- Sliding joints (such as piston rods, pistons, and expansion joints).
- Rotating joints (such as shafts).
- Helical and intermittent joints (such as valve stems).
- Fixed joints (such as flanges and bonnets).

Moving joints offer the greatest difficulty in packing. They must be sealed to prevent leakage without causing excessive friction, undue wear to the moving part, and early packing failure.

Stuffing Boxes and Materials

18-27. Packing is inserted in stuffing boxes (Figure 18-13) which consist of cavities located around valve stems, rotating shafts, or reciprocating pump rods. The packing material is compressed as necessary and held in place by flanged and bolted or threaded gland bushings.

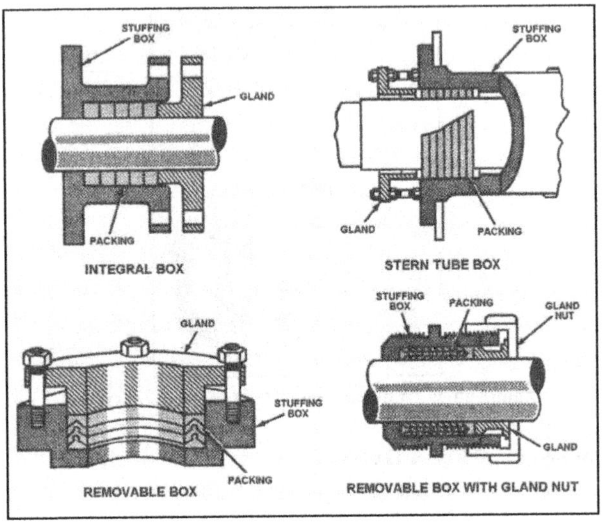

Figure 18-13. Typical Stuffing Boxes

Reciprocating Packing

18-28. Square-braided asbestos packing and plastic nonmetallic asbestos are used for sealing such moving steam joints as rods and valve stems. The square-braided asbestos packing is composed of 90 percent asbestos with brass or copper wire, inserted yarns, and a high-temperature lubricant. The plastic nonmetallic asbestos packing is composed of a plastic core of asbestos fibers, graphite, and a binder encased in a braided Monel wire jacket. These metal inserts or jackets tend to act as a bearing surface for the packing.

Hydraulic Packing

18-29. For high-pressure hydraulic service (such as steering gear and elevator shafts), V-type packing is used. This type packing is composed of laminated plies of fabrics impregnated with a heat- and oil-resisting compound. It is so designed that the shape permits the plies to under pressure.

Rotary Packing

18-30. When sealing rotating joints, it is possible to create enough friction to prevent the machine or apparatus from operating. The heat of friction created by the packing will build up on the faces of the packing and shaft unless dissipation is provided. This is done by use of packing which has high heat conductivity and by leakage. The type packing generally used in rotating joints of pumps and low-speed steam turbines is flexible-metallic and plastic-metallic. In the rotating joints of high speed turbines, labyrinth glands and carbon-ring packing are used.

PACKING TOOLS

18-31. When stuffing boxes need repairing, all necessary tools should be assembled before starting. The tools that may be used for removing and installing packing are an awl or ice pick and a corkscrew, hook, or knife.

- **Awl or Ice Pick.** An awl or ice pick is used to push the packing down to the right position for an even fill of the stuffing box. The point on the awl or ice pick should be blunted with a file before being used.
- **Corkscrew.** A corkscrew is used for removing the old packing. The corkscrew is twisted into the packing and then pulled up to loosen it.
- **Hook.** A hook may also be used for removing packing.
- **Knife.** A knife may be used to cut away old packing and to shape new packing for the stuffing box.

PACKING A TYPICAL STUFFING BOX

18-32. The following procedures should be followed when packing a typical stuffing box.

Securing the Equipment

18-33. Secure the equipment (such as the engine pump) in accordance with the applicable TM.

Packing Removal

18-34. Remove the packing gland (which is sometimes in halves) to expose the packing. No other parts are required to be dismantled. Unscrew followers so the packing can be removed. Ball joints can be dismantled should the faces require taking up.

Packing Requirements

18-35. Use the following as a guide in selecting the correct type of packing for the more common pumping service:

- **Cold water to 220° F (104° C).** Use class 1, nonreinforced asbestos, or class 3, semi-metallic plastic, or a combination of these types.
- **Water over 220° F (104° C).** Use class 1, reinforced asbestos.
- **Petroleum to 450° F (232° C).** Use class 2, crinkled or wrapped Babbitt combination set of Babbitt end rings and semi-metallic or plastic center rings.
- **Petroleum over 450° F (232° C).** Use class 2, crinkled aluminum foil.

Packing Installation

18-36. First make sure the box is clean and all old packing is removed. If the box has a seal cage (lantern ring), be sure that it is located opposite the sealing-liquid inlet tap. Measure the depth of the box and the location of the sealing-liquid tap to determine the proper location of the seal cage. Check to see that the shaft or shaft sleeve has not been damaged. Use all new packing. Place enough rings or packing in the bottom of the box so that when the packing is compressed, the seal cage will be in the proper position. Stagger the ring joints. Woven and braided packing does not require that each individual ring be compressed. Dip each ring in oil and push into the bottom of the box. When the box is half full, draw up snugly, and then back off gland until finger-tight. Since packing expands with heat, a box that is more than finger-tight when cold will generally smoke when started. Plastic and metallic packing must have each ring compressed individually. This can be done with split packing rings. Dip each ring into cylinder oil before inserting in the box. Each ring should be drawn up tight using the split packing rings and glands. Give the shaft a few turns by hand to gloss packing. Combination sets should always be installed according to the packing arrangement sheet supplied with the packing. Combination sets usually consist of one or more rings of metallic or semi-metallic with intermediate rings of plastic or other types of soft packing.

Operation and Maintenance

18-37. Drip leakage is necessary to assure proper lubrication throughout the box. In services where nonlubricating, corrosive, or highly volatile liquids are pumped, an external means of lubrication must be provided. This may be done by using a grease seal or injection of a suitable sealing liquid into a seal cage located at an intermediate point in the stuffing box. For volatile hydrocarbons, oil injection or circulation is used. Watch stuffing boxes carefully when starting. At first sign of heating, shut unit down and allow boxes to cool off before restarting. Several starts may be necessary before leakage breaks through and the box runs cool. Do not back off glands on a hot stuffing box. This could result in leakage around the outside of the packing. Packing is essentially a close-fitting bearing and therefore requires lubricating. Packing cannot run without leakage or a substitute means of lubrication. Keep packing adjusted so that there is a small amount of leakage. Always turn off smothering water (if smothering glands are used) when observing packing leakage. Draw up land bolts one flat at a time. Allow sufficient time between tightening flat for pressure to be transmitted through packing. Observe the effect on leakage.

PACKING OF MOVING JOINTS

18-38. The packing of moving joints offers the most difficulty. The seals must prevent leakage without causing excessive friction, undue wear of the moving part, or rapid deterioration of the packing. Packing is inserted in stuffing boxes that consist of annular chambers located around valve stems, rotating shafts, and reciprocating pump rods. The packing material is compressed to the extent necessary and held in place by gland nuts or other devices. Types of packing used for sealing moving joints depend primarily on whether the seal is for a sliding or a rotating joint. The packing of a sliding joint may be one of a variety of types. High-pressure asbestos rod packing was formerly used exclusively for sealing steam joints such as rods and valve stems. However, this type packing has been superseded to a large extent by wire-inserted, square-braided asbestos (for pressures up to 400 psi and temperatures up to 700° F). Plastic nonmetallic asbestos encased in a braided wire covering (for pressures up to 650 psi and temperatures to 850° F) is also largely used now. Sealing rotating joints is more difficult than sealing sliding joints. With the rotating joint, it is possible for the packing to create enough friction to prevent the machine from operating. In the sliding joint, the heat of friction created by the packing is dissipated through the moving part of the joint. This does not happen in the rotating joint where friction heat builds up on wearing faces of the packing and the shaft unless some way is provided for its dissipation. Packings composed of materials with high-heat conductivity, along with an allowance for leakage, take care of this heat dissipation in rotating joints. However, pressure applied to the packing must be maintained at the minimum which keeps leakage within allowable limits. A relatively new material used for sealing the packing glands of all freshwater pumps, including main feed pumps, is asbestos impregnated with Teflon (TFE), commonly referred to as Teflon packing.

CAUTION

Teflon packing has an operating temperature limit of 500° F. Above this temperature, the material may break down and give off a poisonous gas.

When installing rod packing, proper packing material must be used and any deficiencies in the joint itself must be detected and corrected. The best grade of packing cannot seal a rod effectively if the—

- Rod is bent, scored, or rusty.
- Gland is cocked.
- Stuffing box and gland are scored or nicked.

- Gland is out of alignment with the shaft.
- Old, hard, dry packing is not removed.
- Threads on the gland studs are burred to the extent that setting up of the gland nuts is prevented.

Whenever a stuffing box is broken down, the box, gland, rod, and studs should be carefully inspected to determine whether any of the aforementioned conditions exist. The packing of any moving joint should never be jammed tight with a wrench. This increases the friction and causes wear of both the packing and the rod. When installing packing rings in stuffing boxes of moving rods, be sure that the ends of the rings are cut square, not beveled. Leave enough clearance between the ends of the rings to allow for elongation when the packing is set up. After all possible causes of faulty sealing have been corrected; install the cut rings in the box one at a time with the joints staggered. Insert the gland, draw it up with a wrench, and then back it off until it is finger-tight. A slight leakage will occur during the time the packing is adjusting itself to the rod and box. As the packing expands, further backing off on the nuts may be necessary. Hydraulic rod packing (such as tuck and rock hard) must be soaked in water for approximately 12 hours to allow for swelling, before cutting and fitting in a pump plunger. Step-type joints are best. When necessary to use this packing dry, be sure to allow enough clearance to provide for swelling which will occur after the packing has absorbed moisture.

Note: Regardless of how good packing material may be, if the surface of the shaft passing through the stuffing box is scored or damaged in any way, the packing will not last long. When replacing packing, carefully inspect the shaft in the area where is passes through the stuffing box as well as the interior of the stuffing box itself. Operating requirements and time permitting ensure that the packing will make contact with the straightest, smoothest possible surface. You may have to have the shaft repaired and refinished or replaced.)

PACKING PRECAUTIONS

18-39. On hydraulic lifts, rams, and accumulators, use a V-type packing or O-ring. For water, this packing should be frictioned with crude, reclaimed, or synthetic rubber. For oils, the packing should be frictioned with oil-resistant synthetic rubber. Regarding the use of packing, the following general precautions should be observed. Do not use—

- Metallic or semi-metallic packing on bronze or brass shafts, rods, plungers, or sleeves. If these materials are used, scoring may occur. Use a braided packing that is lubricated throughout or a nonmetallic plastic packing in the center of the box with an end ring of the braided packing at each end of the box.
- A packing fractioned with rubber or synthetic rubber of any kind on rotary or centrifugal shafts. Such packing will overheat.
- Braid-over-braid packing on rotary or centrifugal shafts. The outer layer will wear through quickly and eventually the packing will become rags.
- Packing with a rubber binder on rotary compressors. It will swell and bind, thereby developing excessive frictional heat. The use of flexible metallic packing is recommended or a lead-base plastic packing alternated with the flexible metallic packing can be used.
- A plastic packing alone on worn equipment or out-of-line rods; it will hold. A combination of end rings of plain braided asbestos or of flexible metallic packing may be used for temporary service until defective parts can be repaired or replaced.
- A soft packing against thick or sticky liquids or against liquids having solid particles. This packing is too soft to hold back such liquids as cold boiler fuel oil. It usually gets torn. Some of the solid particles which may be suspended in these liquids embed themselves in the soft packing. Thereafter they act as an abrasive on the rod or shaft. Flexible metallic packing is best for these conditions.

SECTION III. INSULATION

The purpose of insulation is to retard the transfer of heat from piping that is hotter than the surrounding atmosphere or to piping that is cooler than the surrounding atmosphere. Insulation helps to maintain the desired temperatures in all systems. In addition, it prevents sweating of piping that carries cool or cold fluids. Insulation also serves to protect personnel from being burned by coming in contact with hot surfaces. Piping insulation represents the composite piping covering which consists of insulating material, lagging, and fastening. The insulating material offers resistance to the flow of heat.. The lagging, usually of painted canvas, is the protective and confining covering placed over the insulating materials. The fastening attaches the insulating material to the piping and to the lagging. Insulation covers a wide range of temperatures, from extremely low temperatures of refrigerating plants to very high temperatures of the ship's boilers. No one material can be used to meet all the conditions with the same efficiency.

INSULATION MATERIALS

18-40. To standardize various insulating materials, the material should possess the following qualities:

- Have low heat conductivity.
- Be noncombustible.
- Be lightweight.
- Have easy molding and installation capability.
- Be moisture repellent.
- Be noncorrosive, insoluble, and chemically inactive.
- Composition structure and insulating properties to be unchanged by temperatures at which it is to be used.
- Once installed, material should not cluster, become lumpy, disintegrate, or build up in masses from vibration.
- Be vermin proof.
- Be hygienically safe to handle.

INSULATING CEMENTS

18-41. Insulating cements are composed of a variety of materials, differing widely among themselves as to heat conductivity, weight, and other physical characteristics. Typical of these variations are the diatomaceous cements and the mineral and slag wool cements. These cements are less efficient than other high-temperature insulating materials. However, they are valuable for patchwork emergency repairs and for covering small irregular surfaces (valves, flanges, and joints). The cements are also used for a surface finish over block or sheet forms of insulation, to seal joints between the blocks, and to provide a smooth finish over which glass-cloth lagging may be applied.

REMOVABLE INSULATION

18-42. Removable insulation is usually installed in the following locations:

- Manhole covers, inspection openings, turbine casing flanges, drain plugs, strainer cleanouts, and spectacle flanges.
- Flanged pipe joints adjacent to machinery or equipment that must be broken when units are opened for inspection or overhaul.
- Valve bonnets of valves larger than 2 inches, iron pipe size, that operate at 300 psi and above, or at 240° F and above.
- All pressure-reducing and pressure-regulating valves, pump pressure governors, and strainer bonnets.

The entire insulation may be made removable and replaceable for a small unit of machinery or equipment such as an auxiliary turbine. It would be difficult to install both permanent insulation over the casing and removable and replaceable covers over the casing flanges. Covers should fit correctly and should project over adjacent permanent insulation (see Figure 18-14).

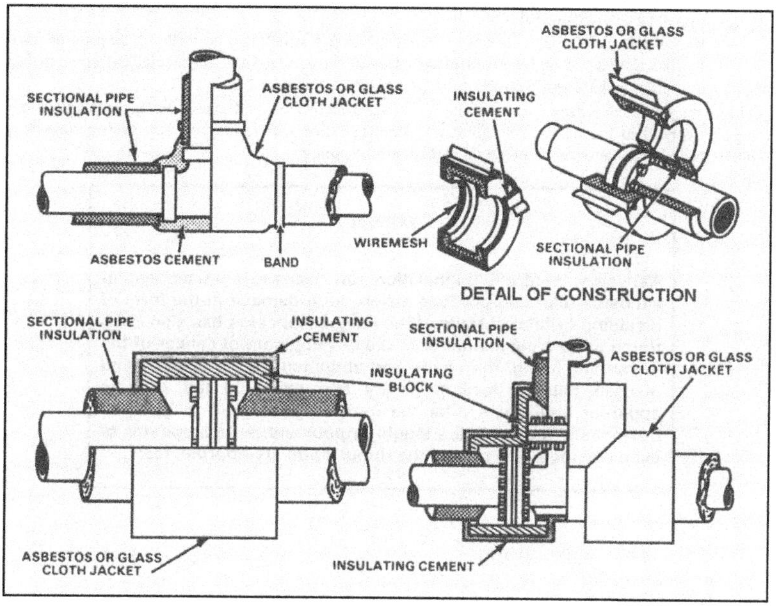

Figure 18-14. Permanent Insulation of Pipe, Flanges, and Valves

GENERAL INSULATION PRECAUTIONS

18-43. The following general precautions should be observed when applying and maintaining insulation:

- Fill and seal all air pockets and cracks. Failure to do this will cause ineffectiveness of the insulation.
- Seal the ends of the insulation and taper off to a smooth, airtight joint. At joint ends or other points where insulation may be damaged, use sheet metal lagging. Flanges and joints should be cuffed with 6-inch lagging.
- Keep moisture out of all insulation work. Moisture is an enemy of heat insulation as much as it is of electrical insulation. Any dampness increases the conductivity of all heat-insulating materials.
- Insulate all hangers and other supports at their point of contact from the pipe or other unit they are supporting. Otherwise, a considerable quantity of heat will be lost by conduction through the support.
- Sheet metal covering should be kept bright and not painted unless the protecting surface has been damaged or has worn off. The radiation from bright-bodied and light-colored objects is considerably less than from rough and dark-colored objects.

- Once installed, heat insulation requires careful inspection, upkeep, and repair. Lagging and insulation removed to make repairs should be replaced just as carefully as when originally installed. When replacing insulation, make certain that the replacement material is of the same type as had been used originally.
- Insulate all flanges with easily removable forms which can be made up as pads of insulating material, wired or bound in place. Cover the whole with sheet metal casings which are in halves and are easily removable.

The main steam, auxiliary steam, auxiliary exhaust, feedwater, and steam heating piping systems are lagged to hold in the heat. The circulating drainage, fire, and sanitary piping systems are lagged to prevent condensation of moisture on the outside of the piping.

WARNING

Asbestos control: Inhalation of excessive quantities of asbestos filler can produce severe lung damage in the form of disabling or fatal fibrosis of the lungs. Asbestos has also been found to be a causal factor in the development of cancer of the membrane lining, the chest, and abdomen. Lung damage and disease usually develop slowly and often do not become apparent until years after the initial exposure. You have no business doing asbestos lagging ripout without a respirator or even staying in the area of the ripout if you are unprotected.

Chapter 19
INSTRUMENTS

INTRODUCTION

19-1. A variety of instruments, indicators, and alarms are used in the engineering plant aboard ship. Those discussed in this chapter are included to give you an idea of the wide range and type used and their importance in the engineering plant.

PRESSURE GAUGES

19-2. Pressure gauges include a variety of Bourdon tube gauges, bellows, and diaphragm gauges, and manometers. Bourdon tube gauges are generally used for measuring pressure of 15 psi and above. Bellows and diaphragm gauges and manometers are generally used for measuring pressures below 15 psi.

BOURDON TUBE GAUGES

19-3. The most commonly used Bourdon tube gauges are the—

- Simplex.
- Vacuum.
- Compound
- Duplex.

They operate on the principle that pressure in a curved tube has a tendency to straighten out the tube. The tube is made of bronze for pressures under 200 psi and of steel for pressures over 200 psi. Figure 19-1 shows a Bourdon pressure gauge case. The Bourdon tube is in the shape of a C and is welded or silver-brazed to the stationary base. The free end of the tube is connected to the indicating mechanism by a linkage assembly. The threaded socket, welded to the stationary base, is the pressure connection. When pressure enters the Bourdon tube, the tube straightens out slightly and moves the link connected with the toothed-gear sector. The teeth on the gear sector mesh with a small gear on the pinion to which the pointer is attached. Therefore, when pressure in the tube increases, the gear mechanism pulls the pointer around the dial and registers the amount of pressure being exerted in the tube. The simplex gauge may be used for measuring the pressure of the following:

- Steam.
- Air.
- Water.
- Oil.
- Similar fluids or gases.

The Bourdon tube vacuum gauge shown in Figure 19-2 is commonly used on auxiliary condensers to indicate vacuum in inches of mercury. Vacuum gauges indicate pressure below atmospheric pressure, where pressure gauges indicate pressure above atmospheric pressure. The duplex Bourdon tube gauge shown in Figure 19-3 has two separate gear mechanisms within the same case. A pointer is connected to the gear mechanism of each tube. Each pointer operates independently of the other. Duplex gauges are commonly used for such purposes as showing the pressure drop between the inlet side and the outlet side of lube oil strainers. If the pressure reading for the inlet side of the strainer is much greater than that for the outlet side, the strainer is most likely very dirty and is therefore restricting the flow of lube oil through the strainer. The compound Bourdon tube gauge shown in Figure 19-4 uses a single Bourdon tube. This tube has such great elasticity that it can measure vacuum (in inches) to the left of the zero point and pressure (in psi) to the right of the zero point.

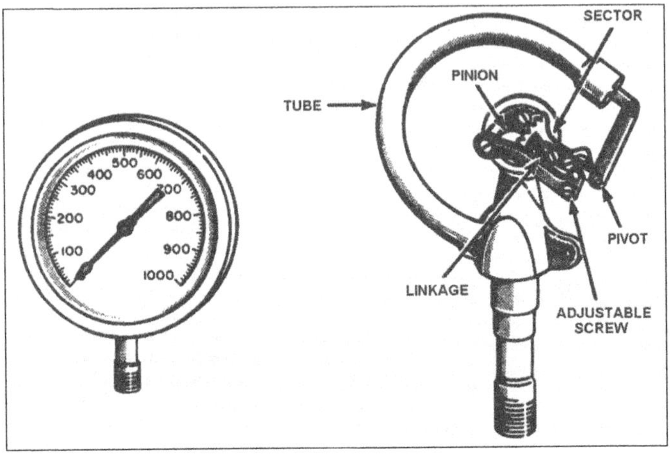

Figure 19-1. Bourdon Pressure Gauge

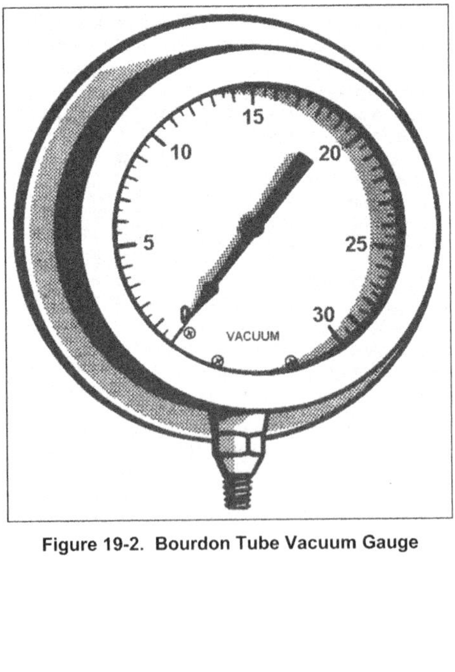

Figure 19-2. Bourdon Tube Vacuum Gauge

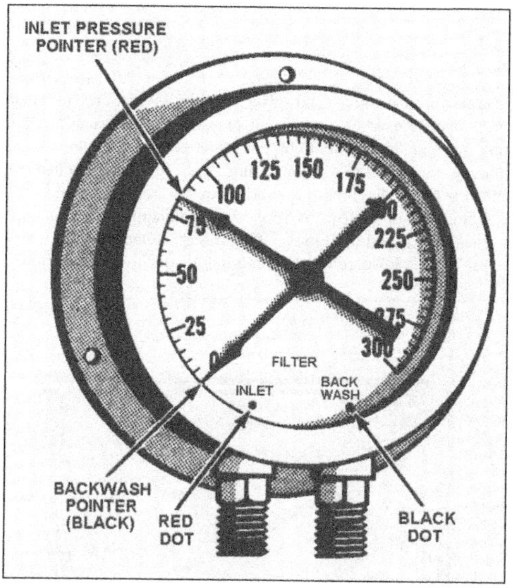

Figure 19-3. Duplex Bourdon Tube Gauge

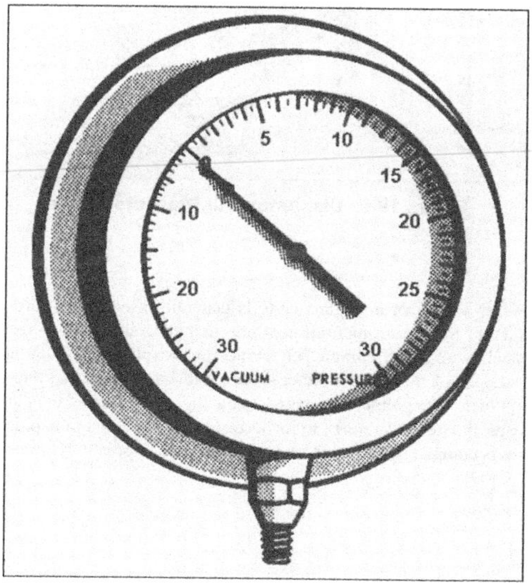

Figure 19-4. Compound Bourdon Tube Gauge

DIAPHRAGM GAUGES

19-4. Diaphragm gauges are sensitive and give reliable indications of small differences in pressure. Diaphragm gauges are generally used to measure air pressure in the space between the inner and outer boiler casings. The indicating mechanism of diaphragm gauge, as shown in Figure 19-5, consists of a tough, pliable, neoprene rubber membrane connected to a metal spring. This spring is attached by a simple linkage system to the gauge pointer. One side of the diaphragm is exposed to the pressure being measured, while the other side is exposed to the atmosphere. When pressure is applied to the diaphragm, it moves upward, pushing the metal spring ahead. As the spring is pushed up it moves the pointer, connected to it with a chain, to a higher reading on the dial. When the pressure is lowered the diaphragm pulls the pointer back toward the zero point.

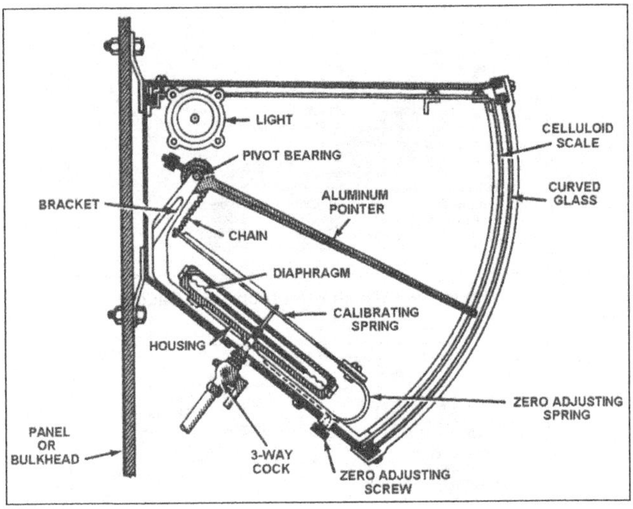

Figure 19-5. Diaphragm Air Pressure Gauge

BELLOWS GAUGES

19-5. A bellows gauge, as shown in Figure 19-6, is generally used to measure pressures less than 15 psi. Some types are satisfactory for measuring draft pressures and general low pressures. The bellows of the gauge is made of stainless steel, brass, beryllium-copper, Monel, or phosphor bronze. When pressure increases in the sensing line to a bellows gauge, the bellows increases in length and operates a system of gears and levers which are connected to the pointer. The pointer then registers the higher reading on the dial. When the pressure to the bellows gauge decreases, the bellows returns to its normal length, returning the pointer toward zero. When the pressure to the bellows is completely removed, the hairspring of the pointer returns the pointer all the way back to zero.

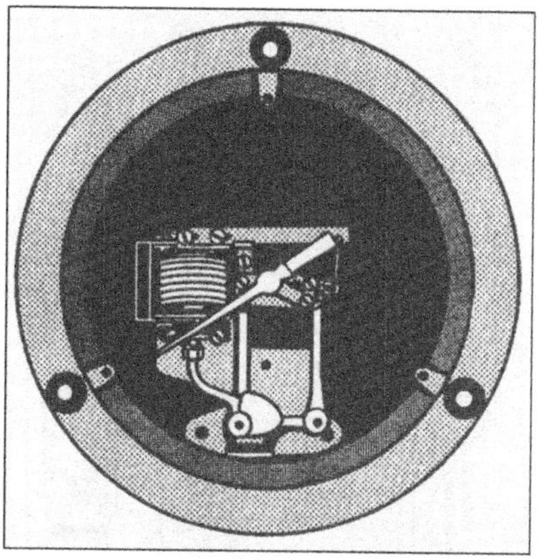

Figure 19-6. Indicating Mechanism of a Bellows Pressure Gauge

MANOMETERS

19-6. A manometer is perhaps the most accurate, least expensive, and simplest instrument for measuring low pressure or low pressure differentials. In its simplest form, a manometer consists of a glass U-tube of uniform diameter filled with a liquid. The most commonly used liquids are water, oil, and mercury. One end of the U-tube is open to the atmosphere and the other end is connected with the pressure to be measured. Manometers are available in many different sizes and designs. However, they all operate on the same principle. Two common types of manometers are shown in Figure 19-7. The U-tube manometer is a primary measuring device indicating pressures or vacuum by the difference in the height of two columns of fluid. Connect the manometer to the source of pressure, vacuum, or differential pressure. When the pressure is imposed, add the number of inches one column of fluid travels up to the amount the other column travels down to get the pressure (or vacuum) reading. The height of a column of mercury is read differently than that of a column of water. Mercury does not wet the inside surface. Therefore, the top of the column has a convex (meniscus) shape. Water wets the surface and therefore has a concave meniscus. A mercury column is read by sighting horizontally between the top of the convex mercury surface and the scale. A water manometer is read by sighting horizontally between the bottom of the concave water surface and the scale. Should one column of fluid travel further than the other column, due to minor variations in the inside diameter of the tube or to the pressure imposed, the accuracy of the reading obtained is not impaired. Figure 19-8 shows a comparison of column height for mercury and water manometers. The manometer reading may be converted into other units of measurement by use of the following pressure conversion chart.

1" water	=	.0735" mercury
1" water	=	.0361 psi
1" mercury	=	.491 psi
1" mercury	=	13.6" water
1 psi	=	27.7" water
1 psi	=	2.036" mercury

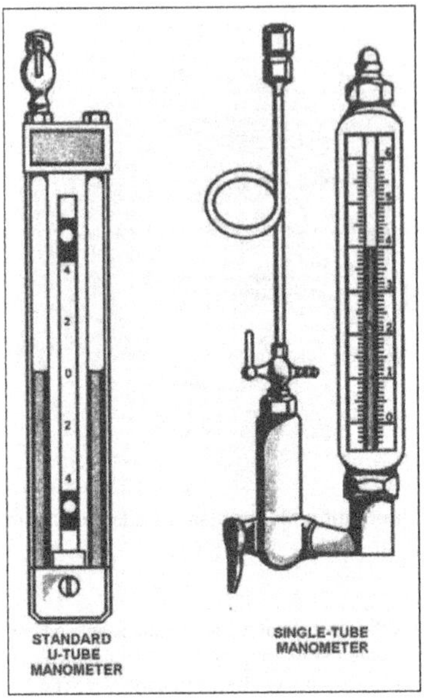

Figure 19-7. Manometers

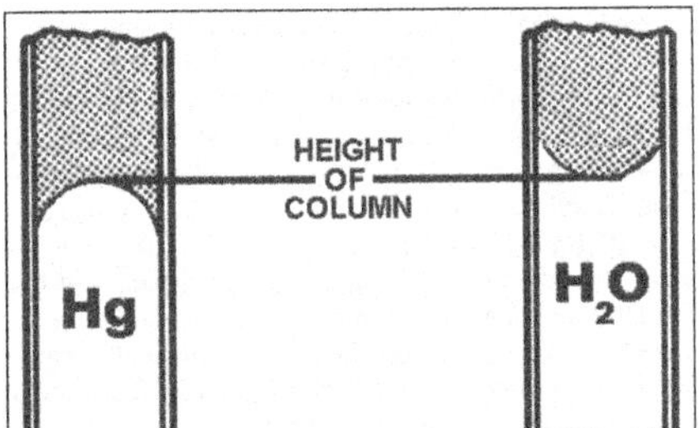

Figure 19-8. Comparison of Column Height for Mercury and Water Manometers

THERMOMETERS

19-7. A thermometer is an instrument which measures temperature. The temperature measured may be of the following:

- Steam to the main engines.
- Brine in an ice machine.
- Oil and bearings in the main engines.
- Substances in many other locations.

In general, a thermometer measures changes in temperature by using the effect of heat on the expansion of a liquid or a gas. The following are the designs of thermometers:

- Direct-reading, liquid-in-glass.
- Bimetallic.
- Distant-reading, indicating-dial thermometers.

Liquid-in-Glass Thermometers

19-8. These thermometers are filled with mercury, ethyl alcohol, benzine, water, or some other liquid suitable for the temperature range involved. Most of the liquid-in-glass thermometers used in engineering spaces are filled with mercury. When a thermometer is exposed to a temperature to be measured, heat causes the liquid in the bulb to expand and rise in the glass stem. Cold, or the absence of heat, causes the liquid to contract and fall. Liquid-in-glass thermometers are designed so that the face will be in the best position for reading when the thermometer is installed. Some thermometers with different angles are shown in Figure 19-9. Mercury-filled thermometers must not be misused or improperly handled. Mercury produces .a highly .toxic vapor which is hazardous to personnel if breathed in excessive concentration.

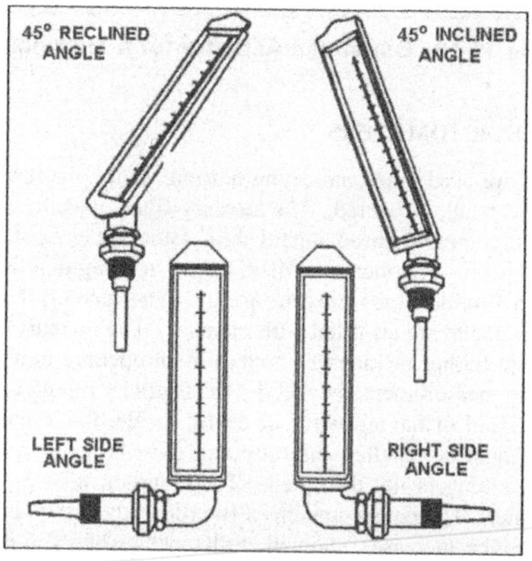

Figure 19-9. Thermometers With Angle Sockets

Bimetallic Dial Thermometers

19-9. Bimetallic dial thermometers use a bimetal element for indicating temperature changes on a circular dial. The bimetallic actuator (Figure 19-10) is a single-helix coil fitted closely to the inside of the stem tube. A rise in temperature causes the actuating element to expand. The element is composed of two thin strips of different metals with each metal having a different rate of expansion. Therefore, the element expands by unwinding instead of by expanding against the sides of the tube. The pointer shaft is secured to the free end of the element. It also registers the amount of element movement on the dial face.

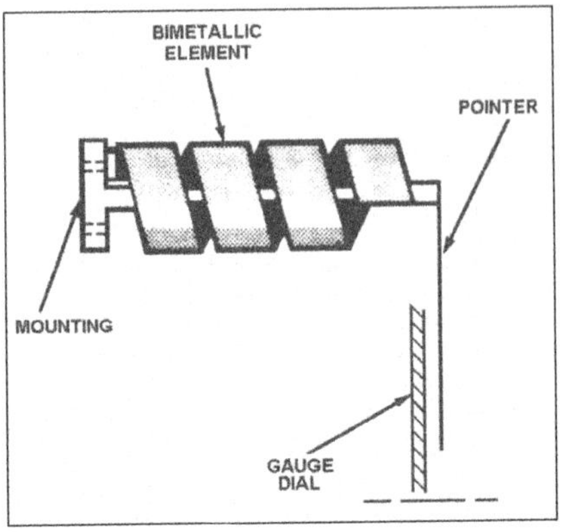

Figure 19-10. Bimetallic Actuator for a Thermometer

DISTANT-READING DIAL THERMOMETERS

19-10. These thermometers are used when indicating portions of the instrument must be placed at a distance from where the temperature is being measured. The mercury-filled, distant-reading dial thermometer shown in Figure 19-11 is the type most commonly used aboard ship. Other types exist, but they are not commonly used aboard ship and are not discussed. The mercury-filled, distant-reading dial thermometer consists of a mercury bulb, capillary tubing, and a Bourdon tube pressure gauge. The mercury bulb, the capillary tubing, and the Bourdon tube in the pressure gauge are all filled with mercury. The mercury bulb is the sensing element. It is inserted and fastened securely (either threaded or bolted) in an opening in a pipe or turbine pump casing, for instance, where a temperature measurement is desired. The capillary tubing is constructed similarly to armored electric cable. However, instead of having wires or cables inside, the armor contains a small flexible metal tube. The capillary tubing must be handled carefully and must be kept free of kinks and twists so that the mercury column within the tubing is not disrupted. The Bourdon tube pressure gauge is the same as that discussed earlier in this chapter. The gauge contains a Bourdon tube which has a tendency to straighten out as the pressure in the Bourdon tube increases. Something happens when you put together the three major units (mercury bulb, capillary tubing, and pressure gauge) of the distant-reading thermometer. Remember that all three of these units are filled with mercury and that when the temperature rises; the increase causes the mercury in the bulb to expand. The expansion of the mercury in the bulb causes a pressure to be built up in the capillary tubing. This same pressure is transmitted through the capillary tubing into the Bourdon tube in the pressure gauge. The Bourdon tube straightens out and, through an assembly of gears and levers, causes a pointer to move around a dial which has previously been calibrated to show corresponding temperatures. The expansion of the mercury in the bulb results in a movement of the pointer that is directly proportional to the temperature applied to the bulb.

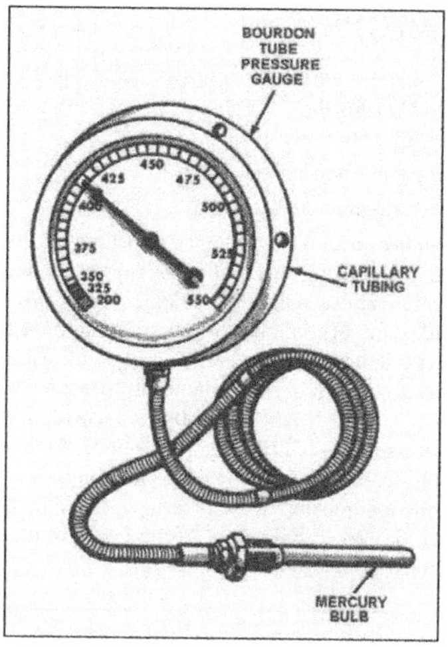

Figure 19-11. Distant-reading Dial Thermometer

PYROMETERS

19-11. Pyrometers are used to measure a wide range of temperatures, generally between 300° F and 3,000° F. They are used aboard ship to measure temperatures in heat-treatment furnaces, in exhaust temperatures of diesel engines, and for other similar temperatures. The pyrometer includes a thermocouple and a meter. The thermocouple is made of two dissimilar metals joined at one end. It produces an electric current when heat is applied at its joined end. The meter, calibrated in degrees, indicates the temperature at the thermocouple. The operating principle of the thermocouple of the pyrometer is shown in Figure 19-12.

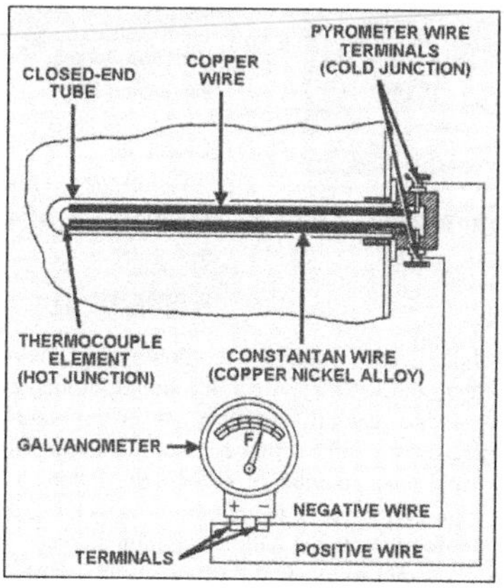

Figure 19-12. Arrangement of a Thermocouple

LIQUID-LEVEL INDICATORS

19-12. In the engineering plant aboard ship, operating personnel must know the level of various liquids in several locations. These include the following:

- Level of water in the ship's boilers.
- Fuel oil level in storage tanks and service tanks.
- Water level in the deaerating feed tank.
- Lubricating oil level in the oil sumps and reservoirs of pumps and other auxiliary machinery.

Liquid-level indicators, other than sounding rods and tapes, are constructed in a variety of designs and sizes. Some are simple and some are relatively complex. Some indicators measure liquid level directly by measuring the height of a column of liquid. One of the most common liquid-level indicators is the static head gauging system (Figure 19-13). This type indicator is generally used to measure the liquid level in fuel oil storage tanks aboard ship. This system uses a mercury manometer to balance a head of liquid in the tank against the column of liquid in the manometer. The balance chamber is located so that its orifice (opening) is near the bottom of the tank. A line connects the top of the balance chamber to the mercury-filled bulb of the indicator gauge. Another line connects the space above the mercury column to the top of the tank. Since the height of the liquid in the tank has a definite relationship to the pressure exerted by the liquid, the scale can be calibrated to show the height (or liquid level). When the height of the liquid in the tank is known, the measurement of height can be readily converted to volume (gallons).

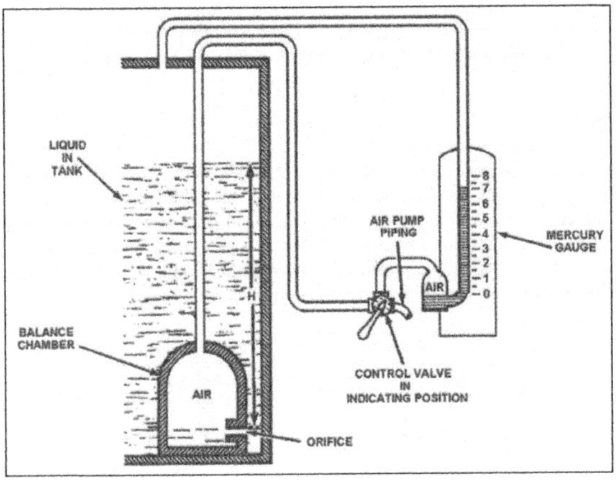

Figure 19-13. Static Head Gauging System

FLUID METERS

19-13. Fluid meters are used aboard ship to measure the flow rate of fuel oil, water, and other liquids. You cannot get an accurate measurement of the rate of flow of a liquid through a meter unless the recorded amount is read correctly. The dial readings are generally in US gallons unless otherwise specified on the dial face. To read the straight-reading meter, read from left to right and add the number indicated by the small pointer. As the pointer turns, so does the right-hand, numbered roller. Even though the next higher number is partly exposed, always read the lesser number, which is the number disappearing from sight. When the small pointer is at 0, all the numbers in the straight-reading dial will be centrally aligned. When the small pointer is at 8 or 9, the next larger number on the numbered roller is almost completely exposed, but the lesser number must be read. To read the round-reading dial face, take the lesser number of the two between which the hand points in each circle. Each circle indicates tens, hundreds, thousands, tens of thousands, and so on. Place the number taken from the circle marked "tens" in the units position, place the number taken from the circle marked "hundreds" in the tens place; and continue similarly for the remaining circles. Each division of any circle stands for one-tenth of the whole number indicated by that circle. If a hand is on or near a number, read that number instead of the next lower number when the hand in the next lower circle is on or past 0.

REVOLUTION COUNTERS AND INDICATORS

19-14. Measurements of rotational speed are necessary for operation of pumps, forced draft blowers, main engines, and other machinery or equipment of the engineering plant. As a result, various instruments are used to indicate the shaft speed or to count the number of turns a shaft makes. One type of instrument is the revolution counter, which is mounted on the throttle board in the engine room. It counts the total number of turns that are made by a main propulsion shaft. These counters are similar to the speedometer of an automobile. The most commonly used instrument in the engineering plant aboard ship to measure rotational speed is the tachometer. For most shipboard machinery, rotational speed is expressed in rpm. Tachometers generally used aboard ship are centrifugal and chronometric.

CENTRIFUGAL TACHOMETERS

19-15. Centrifugal tachometers may be either portable (single and multiple range) or be permanently mounted. The portable multi-range tachometer has three ranges:

- Low (50 to 500 rpm).
- Medium (500 to 5,000 rpm).
- High (5,000 to 50,000 rpm).

Do not shift from one range to another while the portable centrifugal tachometer is in use. Normally, permanently mounted centrifugal tachometers operate off the governor or speed-limiting assembly. The tachometer continuously records the actual rotational speed of the machinery shaft. The portable centrifugal tachometer is operated manually. A small shaft which protrudes from the tachometer case is applied manually to a depression or projection on the end of a rotating shaft of a pump, motor, or other machinery. The centrifugal or rotating movement of the machinery shaft is converted to instantaneous values of speed on the dial face of the tachometer. Figure 19-14 shows portable tachometers.

CHRONOMETRIC TACHOMETERS

19-16. Chronometric portable tachometers, like the one shown in Figure 19-14, are a combination watch and revolution counter. This type measures the average number of revolutions per minute of a motor shaft or pump shaft. This tachometer has an outer drive shaft which runs free when applied to a rotating shaft until a starting button is depressed to start the timing element. Note the starting button. The chronometric tachometer retains readings on its dial after its drive shaft has been disengaged from a rotating shaft, until the pointers are returned to zero by the reset button (usually the starting button). The range of a chronometric tachometer is usually from 0 to 10,000 rpm and from 0 to 3,000 feet per minute (fpm). Each portable centrifugal or chronometric tachometer has a small, rubber-covered wheel and a number of hard rubber tips. The appropriate tip or wheel is fitted on the end of the tachometer drive shaft and held against the shaft to be measured. Portable tachometers of the centrifugal or chronometric type are used for intermittent readings only and are not used for continuous operation.

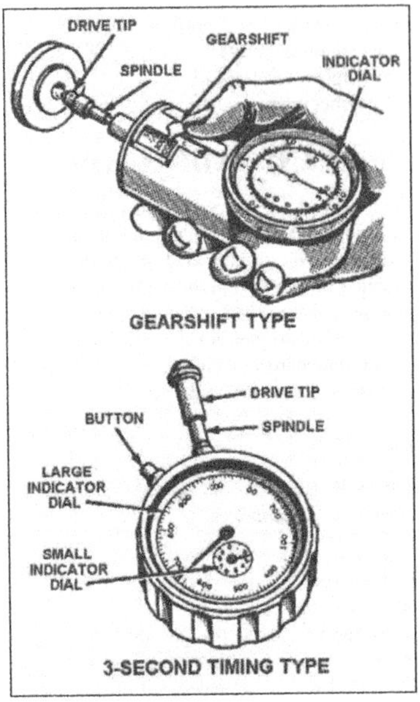

Figure 19-14. Portable Tachometers

OTHER ENGINEERING INSTRUMENTS

19-17. Additional instruments and indicators are used in the engineering plant. This discussion will acquaint you with those instruments and indications which you most likely will be seeing or working with in the engine room.

SALINITY INDICATOR

19-18. Electrical salinity indicating cells (Figure 19-15) are installed throughout distilling plants to maintain a constant check on distilled water. An electrical salinity indicator consists of a number of salinity cells in various locations in the plant (such as in the evaporators, condensate pump discharge, and air-ejector condenser drain). These salinity cells are all connected to a salinity indicator panel. Since the electrical resistance of a solution varies according to the amount of ionized salts in solution, it is possible to measure salinity by measuring the electrical resistance. The salinity indicator panel is equipped with a meter calibrated to read directly either in equivalents per million (epm) or in grains per gallon (gpg). The newer type salinity indicators are calibrated in epm.

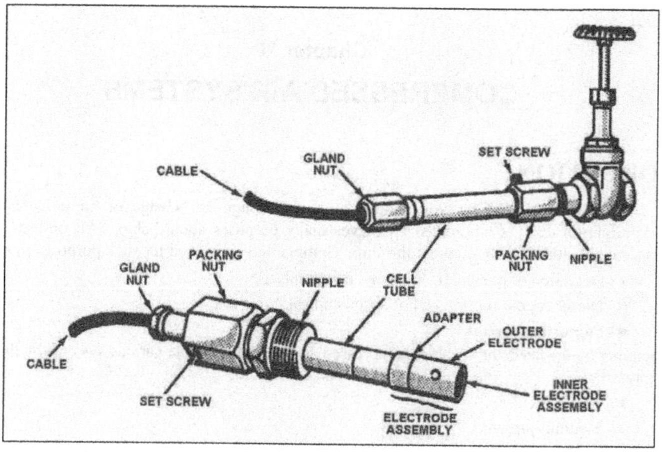

Figure 19-15. Salinity Cell and Valve Assembly

Lube Oil Pressure Alarm

19-19. Lube oil pressure alarms are installed on all generators and main propulsion engines to signal when the lube oil pressure to the bearings is dangerously low. Low lube oil pressure can cause a number of casualties that will impair the operating condition of the machinery involved. The lube oil pressure, which is either a rapidly ringing bell or loud siren, receives its signal from the bearing that is the most remote from the lube pump. When the alarm is sounded, the affected machinery must be stopped immediately and the cause determined and corrected.

Engine Order Telegraph

19-20. The engine order telegraph (speed indicator) relays the speed requested by the officer of the deck while underway to the throttleman in the engine room. The engine order telegraph is generally mounted on the throttle board next to the throttle valves so that it is readily visible to the throttleman.

Chapter 20

COMPRESSED AIR SYSTEMS

INTRODUCTION

20-1. As a marine engineer you should have a thorough knowledge of air compressors and their construction and care. Compressed air serves many purposes aboard ship. Air outlets are installed in various suitable locations throughout the ship. Compressed air is used for such purposes as the following:

- Operation of pneumatic tools and equipment.
- Diesel engine starting and/or speed control.
- Propulsion control.

Depending on the needs of the ship, compressed air is supplied to the various systems by the following air compressors:

- High-pressure.
- Medium-pressure.
- Low-pressure.

Reducing valves may be used to reduce high-pressure air to a lower pressure for a specific use. This chapter introduces you to some of the more common types of air compressors used aboard ship, to compressed air supply systems, and to safety precautions to be observed when dealing with pressurized air.

AIR COMPRESSORS

20-2. The air compressor is the heart of any compressed air system. The compressor takes in atmospheric air, compresses it to the pressure desired, and "pumps" it into supply lines or into storage for later use. Air compressors vary in design, construction, and method of compressions.

AIR COMPRESSOR CLASSIFICATION

20-3. Air compressors are classified in various ways. A compressor may be duplex, single-stage or multi-stage, horizontal, angle, or vertical (as shown in Figure 20-1). In a duplex compressor (Figure 20-1, view D), the two pistons travel together (parallel). Both pistons (compression elements) are either discharging or taking a suction at the same time. A compressor may be designed so that either only one stage of compression takes place within one compressing element or so that more than one stage takes place within one compressing element. In a single-stage compression of air from suction pressure to final discharge pressure is completed in a single compression element; that is, a single cylinder. In a multi-stage compressor, air is compressed in two or more compression elements to reach final discharge pressure. A compressor may also be either single-acting (compression takes place in only one stroke per revolution) or double-acting (compression takes place on both strokes per revolution).

Note: This means that compression is taking place on both sides of the piston and that each end of the cylinder is fitted with suction and discharge valves.

In general, compressors are classified according to capacity (high or low) of the compressing element, source of driving power, method by which the driving unit is connected to the compressor (belt-driven, direct-connected), pressure developed, and whether air delivered is oil-free or non-oil-free.

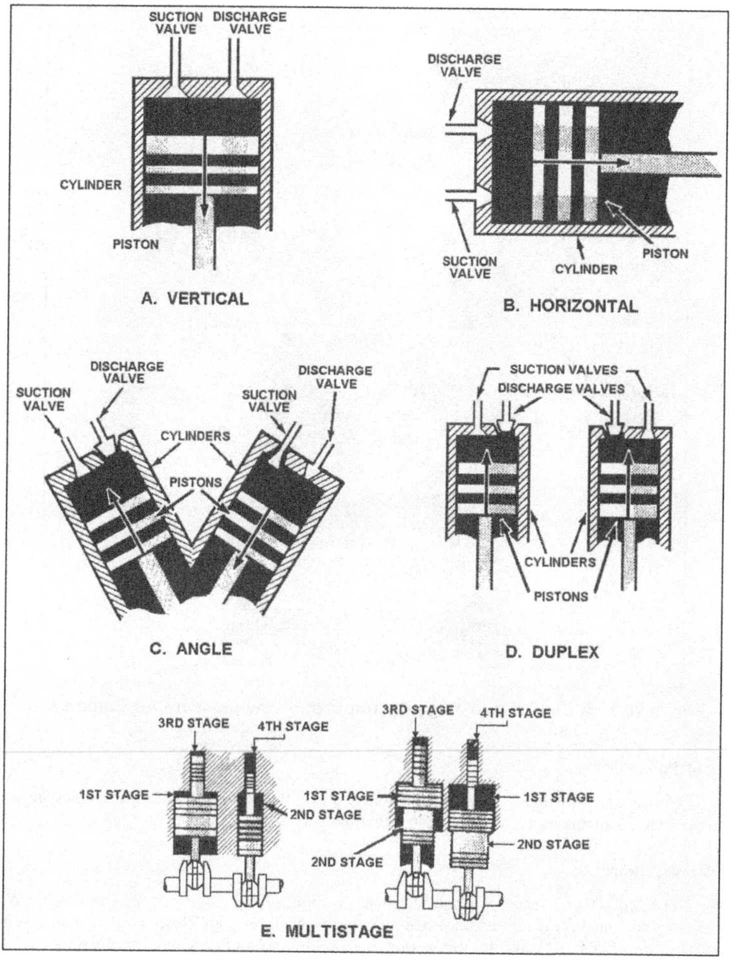

Figure 20-1. Types of Air Compressors

Types of Compressing Elements

20-4. Shipboard air compressors may be centrifugal, rotary, or reciprocating. The reciprocating type is generally selected for capacities from 200 to 800 cubic feet per minute (cfm) and for pressures of 100 to 5,000 psi. The rotary lobe type is for capacities up to 8,800 cfm and for pressures no -more than 20 psi. The centrifugal type is for 800 cfm or greater capacities (up to 2,100 cfm in a single unit) and for up to 125 psi. Most general-service-use air compressors aboard ship are "reciprocators" (Figure 20-2). In this type compressor, air is compressed in one or more cylinders, very much like the compression that takes place in an internal-combustion engine.

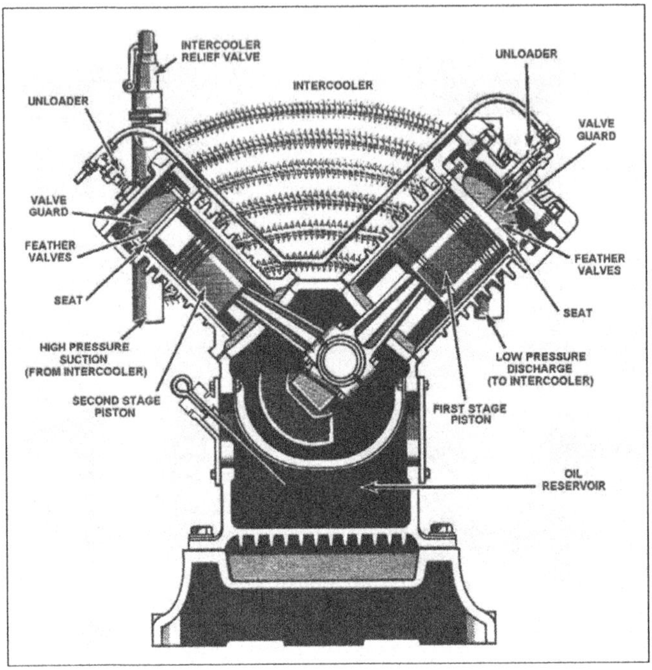

Figure 20-2. A Simple Two-stage Reciprocating Low-pressure Air Compressor

Sources of Power

20-5. Compressors are driven by electric motors or steam-turbines. Aboard ship, most low-pressure and high-pressure air compressors are driven by electric motors.

Device Connections

20-6. The driving unit may be connected to the compressor by one of several methods. When the compressor and the driving unit are mounted on the same shaft, they are close-coupled. Close-coupling is usually restricted to small capacity compressors that are driven by electric motors. Flexible couplings are used to join the driving unit to the compressor when the speed of the compressor and the speed of the driving unit are the same. This is called a direct-couple drive. V-belt drives are commonly used with small, low-pressure, motor-driven compressors and with some medium-pressure compressors. In a few installations, a rigid coupling is used between the compressor and the electric motor of a motor-driven compressor. In a steam-turbine drive, compressors are usually but not always driven through reduction gears, or, in the case of centrifugal (high-speed) compressors, through speed increasing gears.

Pressure Classification

20-7. Compressors are classified as follows:

- **Low-pressure.** These compressors have a discharge pressure of 150 psi or less.
- **Medium-pressure.** These compressors have a discharge pressure of 151 psi to 1,000 psi.
- **High-pressure.** These compressors have a discharge pressure above 1,000 psi.

Low pressure compressors

20-8. Most low-pressure, reciprocating air compressors are of the two-stage type with either a vertical "V" (Figure 20-2) or a vertical "W" (Figure 20-3) arrangement of cylinders. Two-stage, two-cylinder, V-type, low-pressure compressors have one cylinder for the first (lower pressure) stage of compression and one cylinder for the second (higher pressure) stage of compression. Three-cylinder, W-type compressors have two cylinders for the first-stage of compression and one cylinder for the second stage. This arrangement is also shown in Figure 20-4, view A. Notice that the pistons in the lower pressure stage (1) have a larger diameter than the piston in the higher pressure stage (2).

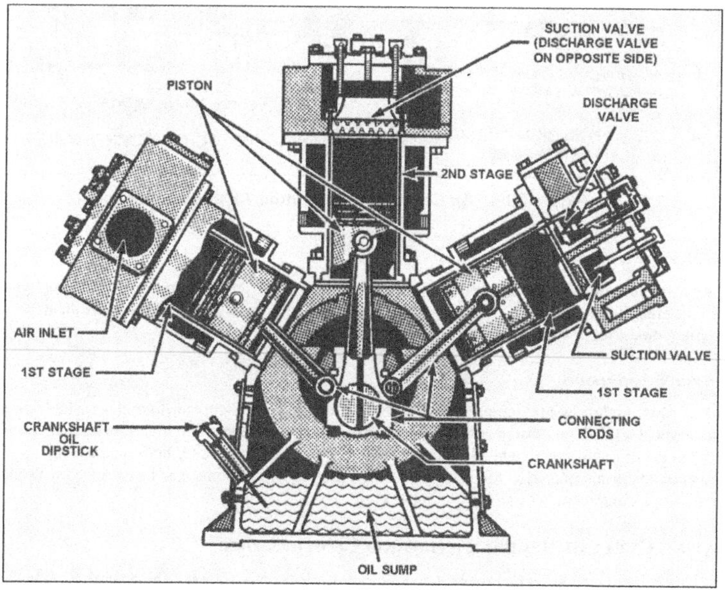

Figure 20-3. Low-pressure, Reciprocating Air Compressor, Vertical W Configuration

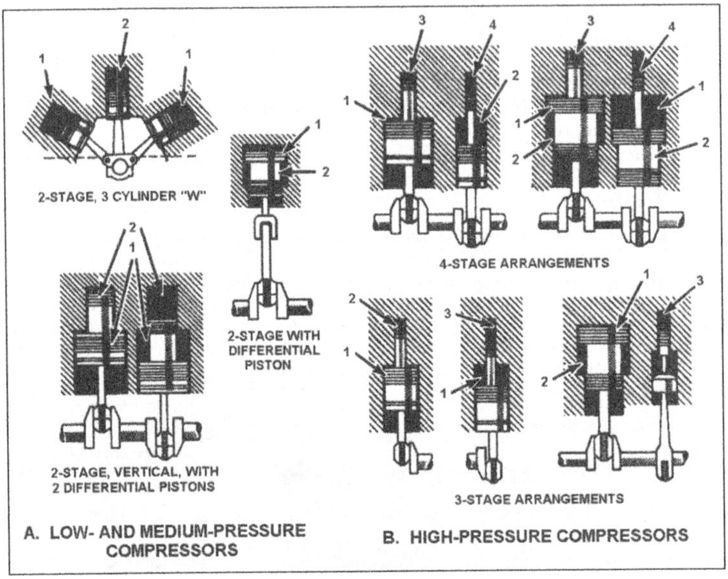

Figure 20-4. Air Compressor Cylinder Arrangements

Medium-pressure compressors

20-9. Medium-pressure air compressors are of the two-stage, vertical, duplex, single-acting type. Many medium-pressure compressors have differential pistons. This type of piston has more than one stage of compression during each stroke of the piston (Figure 20-4, view A).

High-pressure compressors

20-10. Most high-pressure compressors have motor-driven, liquid-cooled, four-stage, single-acting units with vertical cylinders. An example of the cylinder arrangements for high-pressure air compressors installed aboard ship are shown in Figure 20-4, view B. Small-capacity, high-pressure air systems may have three-stage compressors. Large capacity, high-pressure air systems may be equipped with four-, five-, or six-stage compressors.

OPERATING CYCLE OF RECIPROCATING AIR COMPRESSORS

20-11. Reciprocating air compressors are similar in design and operation. The operating cycle during one stage of compression in a single-stage, single-acting compressor is discussed here. The cycle of operation, or compression cycle, within an air compressor cylinder (Figure 20-5) includes two strokes of the piston (a suction stroke and a compression stroke).

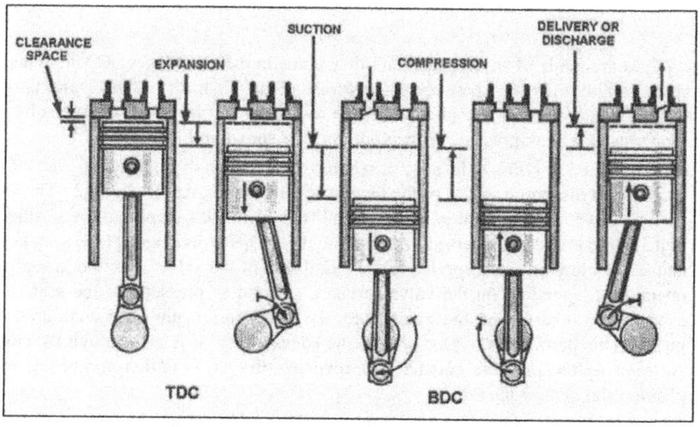

Figure 20-5. The Compressor Cycle

Suction Stroke

20-12. The suction stroke begins when the piston moves away from top dead center (TDC). Air under pressure in the clearance space (above the piston) expands rapidly until the pressure falls below the pressure on the opposite side of the inlet valve. At this point, the difference in pressure causes the inlet valve to open, and air is admitted to the cylinder. Air continues to flow into the cylinder until the piston reaches bottom dead center (BDC).

Compression Stroke

20-13. The compression stroke starts as the piston moves away from BDC and continues until the piston reaches TDC again. When the pressure in the cylinder equals the pressure on the opposite side of the air inlet valve, the inlet valve closes. Air is compressed as the piston moves toward TDC. This continues until the pressure in the cylinder becomes great enough to force the discharge valve against the discharge line pressure and the pressure of the valve springs.

Note: The discharge valve opens shortly before the piston reaches TDC.

During the remainder of the compression stroke, air which has been compressed in the cylinder is discharged at almost constant pressure through the open discharge valve. The basic operating cycle just described is completed twice per revolution of the crankshaft in double-acting compressors (once on the downstroke and once on the upstroke).

Components of Reciprocating Air Compressors

20-14. Reciprocating air compressors consist of a system of connecting rods, crankshaft, and flywheel. These are used to transmit power developed by the driving unit to the pistons and the lubrication, cooling, control, and unloading systems.

Compressing Element

20-15. A compressing element of a reciprocating compressor consists of the cylinder, the piston, and the air valves.

Valves

20-16. The valves are made of special steel and they come in different types. Opening and closing of the valves is caused by the difference between the pressure of the air in the cylinder and the pressure of the external air on the intake valve or the pressure of the discharged air on the discharge valve. Two types of valves commonly used in high-pressure air compressors are shown in Figure 20-6.

- *Strip or Feather Valve.* The strip or feather valve shown in Figure 20-6, view A is used for the suction and discharge valves of the lower pressure stages, that is, 1 and 2. The valve shown is a suction valve. The discharge valve assembly is identical except that the positions of the valve seat and the guard are reversed. At rest, the thin strips lie flat against the seat, covering the slots and sealing any pressure applied to the guard side of the valve. In either a suction or discharge operation (depending on the valve service), as soon as pressure on the seat side of the valve exceeds the pressure on the guard side, the strips flex against the contoured recesses in the guard. This permits air to pass around the edges of the strip and through the slots in the guard. As soon as the pressure equalizes or reverses, the strips unflex and return to their original position flat against the seat.
- *Disk-type Valve.* The disk-type valve in Figure 20-6, view B is used for the suction and discharge valves of the higher pressure stages, that is, 3 and 4. The fourth stage assembly is shown. These valves are of the spring-loaded, dished-disk type. At rest, the disk is held against the seat by the spring and is sealed by pressure applied to the keeper side of the valve. In either a suction or discharge operation (depending on the valve service), as soon as the pressure on the seat side of the valve exceeds the pressure on the keeper side, the disk lifts against the stop in the keeper. This compresses the spring and permits air to pass through the seat around the disk and through the openings in the sides of the keeper. As soon as the pressure equalizes or reverses, the spring returns the disk to the seat.

Cylinders

20-17. Various designs of cylinders are used, depending primarily on the number of stages of compression required to produce the maximum discharge pressure. Several common cylinder arrangements for low- and medium-pressure air compressors are shown in Figure 20-4, view A. Several arrangements for cylinders and pistons of high-pressure compressors are shown in Figure 20-4, view B. The stages are numbered 1 through 4, and a 3-and a 4-stage arrangement are shown. In 5- and 6-stage compressors, the same basic stage arrangement is followed.

Pistons

20-18. The pistons may be of two types (trunk or differential). Trunk pistons (Figure 20-7, view A) are driven directly by the connecting rods. Since the upper end of a connecting rod is fitted directly to the piston (also referred to as wrist or trunk pins), there is a tendency for the piston to develop a side pressure against the cylinder walls. To distribute the side pressure over a wide area of the cylinder walls or liners, pistons with long skirts are used. This type of piston minimizes cylinder wall wear. Differential pistons (Figure 20-7, view B) are modified trunk pistons having two or more different diameters. These pistons are fitted into special cylinders which are arranged so that more than one stage of compression is achieved by one piston. The compression for one stage takes place over the piston crown. Compression for the other stages takes place in the annular space between the large and small diameters of the piston.

Lubrication system

20-19. Except for the oil-free nonlubricated compressors; high-pressure, air compressor cylinders are generally lubricated by means of an adjustable mechanical force-feed lubricator. This is driven from a reciprocating or a rotary part of the compressor. Oil is fed from the cylinder lubricator by separate lines to each cylinder. A check valve is installed at the end of each feed line to keep the compressed air from forcing the oil back into the lubricator. Each feed line is equipped with a sight-glass oil flow indicator.

Lubrication begins automatically as the compressor starts up. The amount of oil that must be fed to the cylinder depends on the following:

- Cylinder diameter.
- Cylinder wall temperature.
- Viscosity of the oil.

Figure 20-8 shows the lubrication connections for the cylinders. The type and grade of oil used in compressors is specified in the equipment technical manual (TM). It is vital to the operation and reliability of the compressor. The running gear is lubricated by an oil pump which is attached to the compressor and is driven from the compressor shaft. This pump (usually of the gear type) draws oil from the reservoir (oil sump) (see Figure 20-8) in the compressor base and delivers it through a filter to an oil cooler (if installed). From the cooler, the oil is distributed to the top of each main bearing, to spray nozzles for reduction gears, and to outboard bearings. The crankshaft is drilled so that oil fed to the main bearings is picked up at the main bearing journals and carried to the crank journals. The connecting rods contain passages which conduct lubricating oil from the crank bearings up to the wrist pin bushings. As oil leaks out from the various bearings, it drips back into the oil sump (in the base of the compressor) and is recirculated. Oil from the outboard bearings is carried back to the sump by the drain lines. A low-pressure air compressor lubrication system is shown in Figure 20-9. This system is similar to the running gear lubrication system for the high-pressure air compressor. Nonlubricated reciprocating compressors have lubricated running gear (shaft and bearings) but no lubrication for the pistons and valves. This design produces oil-free air.

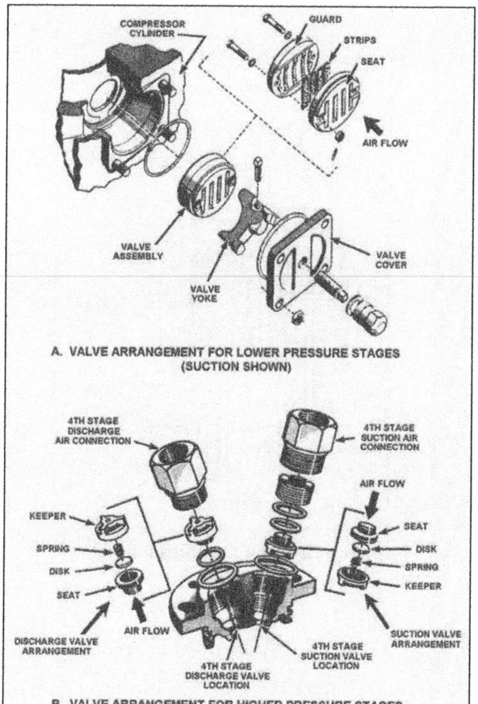

A. VALVE ARRANGEMENT FOR LOWER PRESSURE STAGES (SUCTION SHOWN)

B. VALVE ARRANGEMENT FOR HIGHER PRESSURE STAGES (BOTH SUCTION AND DISCHARGE SHOWN)

Figure 20-6. High-pressure Air Compressor Valves

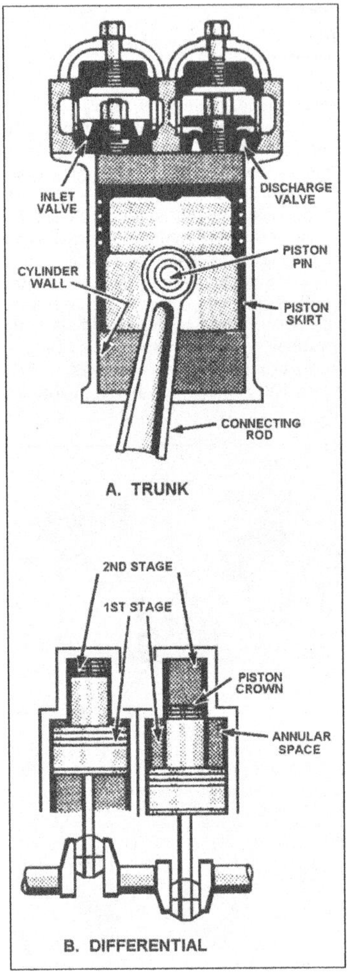

Figure 20-7. Air Compressor Pistons

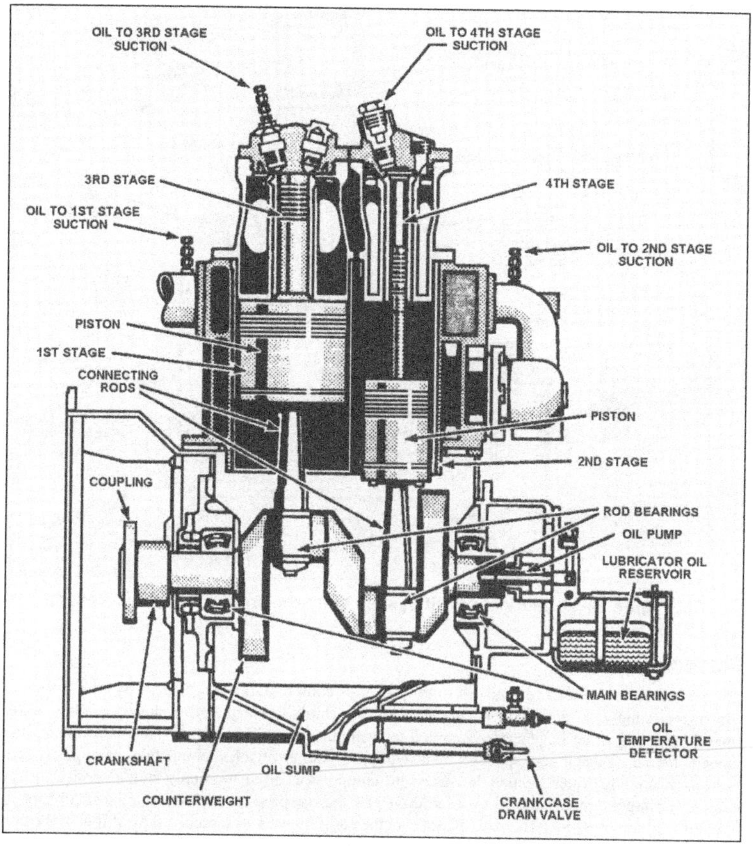

Figure 20-8. High-pressure Air Compressor

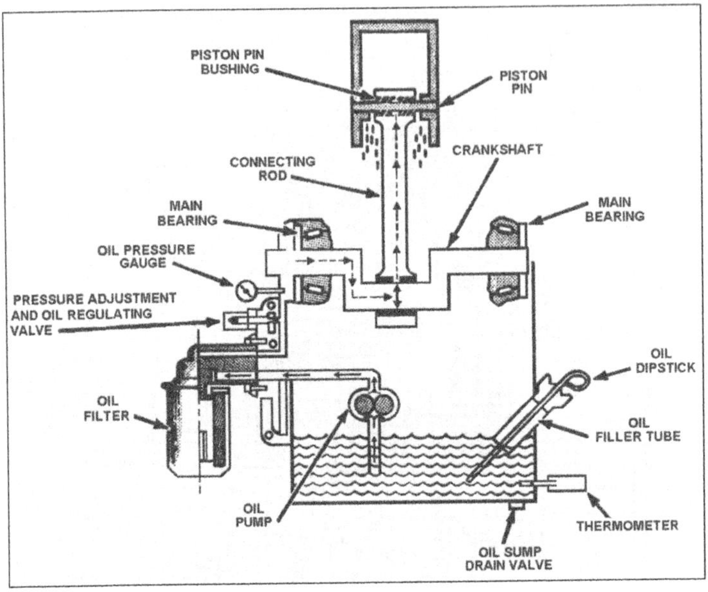

Figure 20-9. Lubricating Oil System of a Low-pressure Air Compressor

Cooling systems

20-20. Most high-pressure and medium-pressure compressors are cooled by the ship's auxiliary freshwater supply or by seawater supplied from the ship's fire main or machinery cooling water service mains. Cooling water is generally supplied to each unit from either of two sources. Compressors located outside the larger machinery spaces may be equipped with an attached circulating water pump as a standby source of cooling water. Small low-pressure compressors used to supply ship's service air and diesel engine starting air are air-cooled by a fan driven by the compressors and so are some small-capacity, high-pressure air compressors. The path of water in the cooling water system of a typical four-stage compressor is shown in Figure 20-10. Not all cooling water systems have an identical path of water flow. However, in systems equipped with oil coolers it is important to use the coldest water available for circulation through the cooler. Valves are usually provided to control the water to the cooler independently of the rest of the system. Therefore, oil temperature can be controlled without harmful effects to other parts of the compressor. Cooling the air in the intercoolers and aftercoolers is very important as well as cooling the cylinder jackets and heads. The amount of cooling water required depends on the capacity in cfm and pressure. High-pressure air compressors require more cooling water for the same cfm than low-pressure units. When seawater is used as the cooling agent, all parts of the circulating system must be of corrosion-resistant materials. The cylinders and heads are therefore composed of a bronze alloy with water jackets cast integral with the cylinders. Each cylinder is generally fitted with a liner of special cast iron or steel to withstand the wear of the piston. Wherever practicable, cylinder jackets are fitted with handholes and covers so that the water spaces can be inspected and cleaned. Jumpers are usually used to make water connections between the cylinders and heads because they prevent possible leakage into compression spaces. However, in some compressors, water passes directly through the joint between the cylinder and the head. With this latter type, the joint must be properly gasketed to prevent leakage which, if allowed to continue, will damage the compressor. The intercoolers and aftercoolers remove heat generated during compression and promote condensation of any vapor that may be present. Figure 20-11 is a diagram of a basic cooler and separator unit showing the collected condensate in the separator section. The collected condensate must be drained at regular intervals to prevent carryover into the next stage. Accumulation at low points may cause water hammer, freezing, bursting of pipes in exposed locations, faulty operation of pneumatic tools, and possible damage to electrical apparatus when air is used for cleaning. The removal of heat is also required for economical compression. During compression, the temperature of the air increases. This causes the air to expand to a larger volume. This in turn requires a corresponding increase of work to compress it. Therefore, multi-staging with interstage cooling of the air reduces the power requirement for a given capacity. Interstage cooling reduces the maximum temperature in each cylinder and thereby reduces the amount of heat which must be removed by the water jacket at the cylinder. The resulting lower temperature in the cylinder also ensures better lubrication of the piston and the valves. Figure 20-12 shows the pressures and temperatures through a stage compressor. The intercoolers and the aftercoolers (on the output of the final stage) are of the same general construction except that the aftercooler is designed to withstand a higher working pressure than the intercoolers. Water-cooled intercoolers may be of the straight tube and shell type (Figure 20-11) or, if size dictates, of the coil type. In coolers with an air discharge pressure below 250 psi, air may flow either through the tubes or over and around them. In coolers with an air discharge pressure above 250 psi, air generally flows through the tubes. In tubular coolers, baffles deflect air or water in its course through the cooler. In coil-type coolers, air passes through the coil and water flows around the coils. Air-cooled intercoolers and aftercoolers may be of the radiator type. However, they may consist of a bank of finned copper tubes located in the path of cooling air supplied by the compressor cooling fan. Each intercooler and aftercooler is generally fitted with relief valves on both the air and water sides. Water relief valves are usually set 5 psi above the maximum working pressure that may be applied to the system. The air relief valves must be set in accordance with directions given in the applicable TM. Intercoolers and aftercoolers are normally fitted with moisture separators on the discharge side to remove the condensed moisture and oil from the airstream. The separators are of a variety of designs. Liquid is removed by centrifugal force, impact, or sudden changes in velocity and/or direction of flow of the airstream. Drains on each separator remove the water and oil. Oil coolers are of the coil type, of the tube and shell type, or of a variety of commercial types. External oil coolers are generally used. However, some compressors are fitted with a base-type oil cooler in which cooling water is circulated through a coil placed in the oil sump. As with the intercoolers and aftercoolers, the materials of the tubes, coils, or cores of these coolers are made of copper-nickel alloy with shell and tube sheets of bronze composition. On all late model compressors, the circulating water system is arranged so that the quantity of cooling water passing through the oil cooler can be regulated without disturbing the quantity of water passing through the cylinder jackets, intercoolers, or aftercoolers. Thermometers or other temperature-measuring devices are fitted to the circulating water inlet and outlet connections, to the intake and discharge of each stage of compression, to the final air discharge, and to the oil sump.

Control systems

20-21. The control system of a reciprocating air compressor may include one or more devices for start-stop control, constant-speed control, speed-pressure governing, and automatic, high-temperature, shut-down devices. Control or regulating systems for air compressors in use by the Navy are largely of the start-stop type. In this type, the compressor starts and stops automatically as the receiver pressure falls or rises within predetermined setpoints. On electrically driven compressors the system is simple. The receiver pressure operates against a pressure switch that opens when the pressure upon it reaches a given limit and closes when the pressure drops a predetermined amount. Centrifugal compressors do not have automatic start-stop controls mainly because of their high horsepower. An automatic load/unload control system is used. On electrically driven units required to start at either of two pressures (as in some of the medium-pressure systems) one of two pressure switches is selected. This switch has a three-way valve or cock that admits pressure from the air accumulator to the selected pressure switch. Another method is to direct the air from the receiver through a three-way valve to either of two control valves set for the respective range of pressures. A line is run from each control valve to a single-pressure switch which may be set at any convenient pressure since the setting of the control valve selected will determine the operation of the switch. The constant-speed control regulates the pressure in the air receiver by controlling the output of the compressor without stopping or changing the speed of the unit. This control prevents frequent starting and stopping of compressors when there is a fairly constant, but low demand, for air. Control is provided by directing air to unloading devices through a control valve set to operate at a predetermined pressure. The automatic high-temperature shutdown devices are fitted on all recent designs of high-pressure air compressors. Therefore, if the cooling water temperature rises above a safe limit, the compressor will stop and will not restart automatically. Some compressors are fitted with a device that will shut down the compressor if the temperature of the air leaving any stage exceeds that of a preset valve.

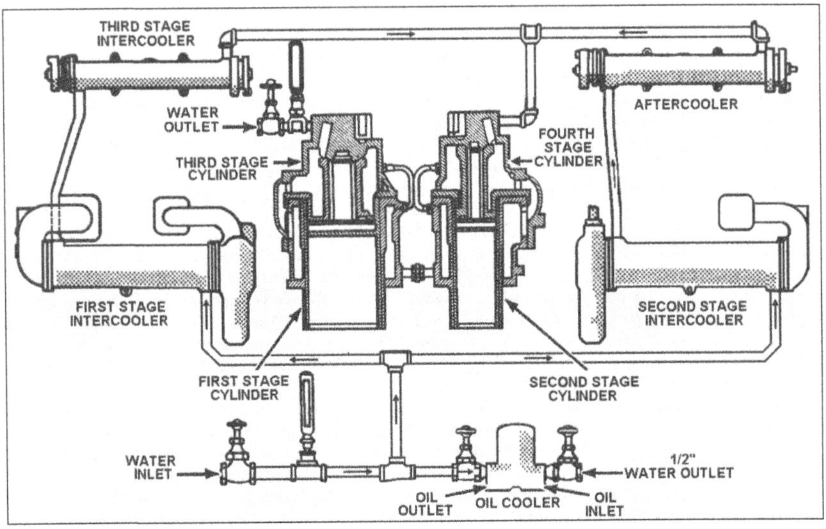

Figure 20-10. Cooling Water System in a Typical Multi-stage Air Compressor

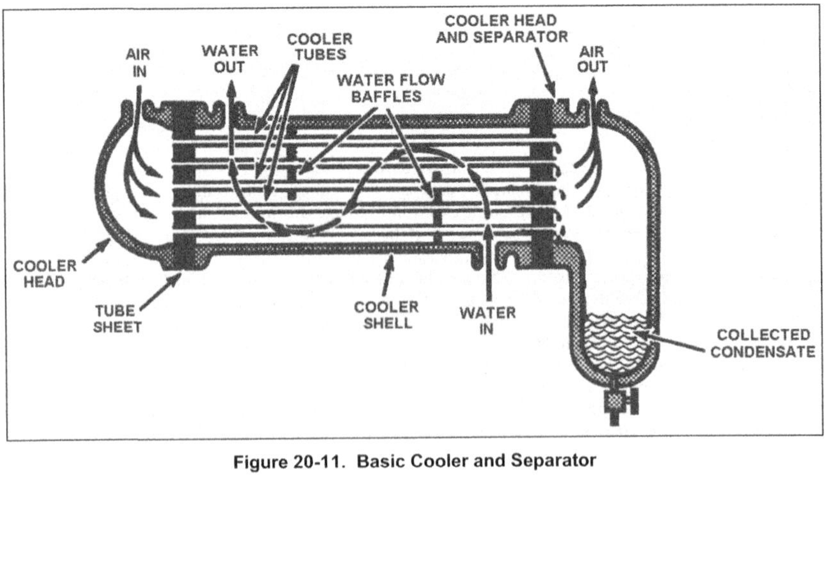

Figure 20-11. Basic Cooler and Separator

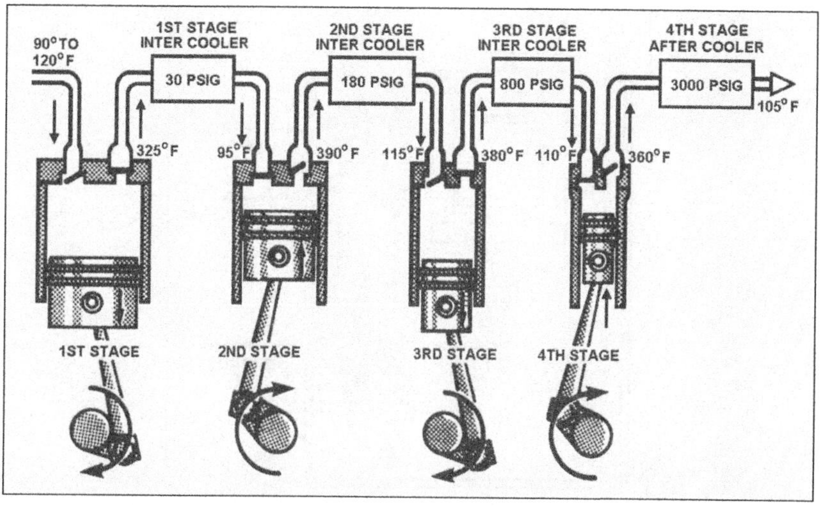

Figure 20-12. Cooling Water System in a Typical Multi-stage Air Compressor

Unloading systems

20-22. Air compressor unloading systems are installed for the removal of all but the friction loads on the compressors. That is, they automatically remove the compression load from the compressor while the unit is starting and they automatically apply the load after the unit is up to operating speed. For units having start-stop control, the unloading stem is separate from the control system. For compressors equipped with constant-speed control, the unloading and control systems are integral parts of each other. A detailed explanation for every type of unloading device used to unload air compressor cylinders cannot be given here. However, you should know something about several of the loading methods which you will probably encounter. These include:

- Closing or throttling the compressor intake.
- Holding intake valves off their seats.
- Relieving intercoolers to the atmosphere.
- Relieving the final discharge to the atmosphere (or opening a bypass from the discharge to the intake).
- Opening up cylinder clearance pockets.
- Using miscellaneous constant-speed unloading devices.
- Using various combinations of these methods.

As an example of a typical compressor unloading device, consider the magnetic unloader. Figure 20-13 shows the unloader valve arrangement. This unloader consists of a solenoid-operated valve connected with the motor starter. When the compressor is at rest, the solenoid valve is de-energized, admitting air from the receiver to the unloading mechanism. When the compressor reaches near-normal speed, the solenoid valve is energized, releasing pressure from the unloading mechanism and loading the compressor again. For detailed information on the various unloading devices, refer to the pertinent manufacturer's TMs for compressors installed on your ship.

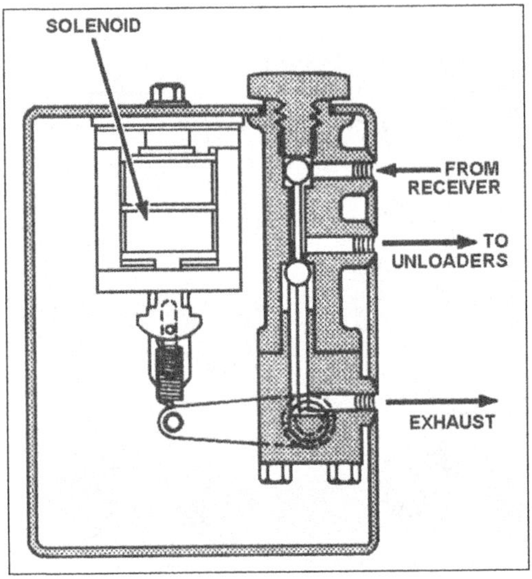

Figure 20-13. Magnetic Unloader

ROTARY-CENTRIFUGAL AIR COMPRESSORS

20-23. A nonreciprocating air compressor that may be found aboard ship is variously referred to as a rotary compressor, a centrifugal compressor, or a "liquid piston" compressor. The unit is actually something of a mixture, operating partly on rotary principles and partly on centrifugal principles. Perhaps, most accurately, it might be called a rotary-centrifugal compressor. The rotary-centrifugal compressor is used to supply low-pressure compressed air. Since it is capable of supplying air that is completely free of oil, it is often used as the compressor for pneumatic control systems and for other applications where oil-free air is required. The rotary-centrifugal air compressor (see Figure 20-14) consists of a round, multi-blade rotor which revolves freely in an elliptical casing. The elliptical casing is partially filled with high-purity water. The curved rotor blades project radially from the hub. The blades, along with the side shrouds, form a series of pockets or buckets around the periphery. The rotor, which is keyed to the shaft of an electric motor, revolves at a speed high enough to throw the liquid out from the center by centrifugal force. This will result in a solid ring of liquid revolving in the casing at the same speed as the rotor but following the elliptical shape of the casing. This action alternately forces the liquid to enter and recede from the buckets in the rotor at high velocity. To follow through a complete cycle of operation, start at point A in Figure 20-14. The chamber (1) is full of liquid. The liquid, because of centrifugal force, follows the casing, withdraws from the rotor and pulls air in through the inlet port. At (2), the liquid has been thrown outward from the chamber in the rotor and has been replaced with atmospheric air. As the rotation continues, the converging wall (3) of the casing forces the liquid back into the rotor chamber, compressing the trapped air and forcing it out through the discharge port. The rotor chamber (4) is now full of liquid and ready to repeat the cycle, which takes place twice in each revolution. A small amount of water must be constantly supplied to the compressor to make up for that which is carried over with the compressed air. The water which is carried over with the compressed air is removed in a refrigeration-type dehydrator.

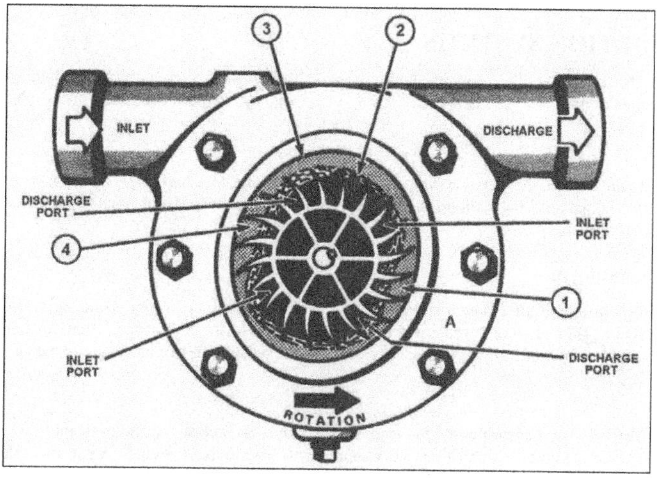

Figure 20-14. Rotary-centrifugal Air Compressor

COMPRESSED AIR RECEIVERS

20-24. An air receiver is installed in each space that houses air compressors (except centrifugal and rotary lobe types). The receiver is an air storage tank. If demand is greater than compressor capacity, some of the stored air is supplied to the system. If demand is less than the compressor capacity, the excess is stored in the receiver or accumulator until the pressure is raised to its maximum setting, at which time the compressor unloads or stops. Therefore, in a compressed air system, the receiver functions to minimize pressure variations in the system and to supply air during peak demand. This minimizes start-stop cycling of air compressors. Air receivers may be horizontal or vertical. Vertically mounted receivers have convex bottoms to permit proper draining of accumulated moisture, oil, and foreign matter. All receivers have such fittings as the following:

- Inlet and outlet connections.
- Drain connections and valves.
- Connections for operating a line to compressor regulators.
- Pressure gauges.
- Relief valves (set at approximately 12 percent above normal working pressure of the receiver).
- Handhole or manhole plates (depending on the size of the receiver).

The discharge line between the compressors and the receiver is as short and straight as possible. This is to eliminate vibration due to pulsations of air and to reduce pressure losses due to friction. In high-pressure air systems, air receivers are called air flasks. These are usually cylindrical in shape, having belled ends, and female-threaded necks. The flasks are also constructed in shapes to conform to the hull curvature for installation between hull frames. One or more air flasks connected together constitute an air bank.

DISTRIBUTION SYSTEMS

20-25. The remainder of a compressed air system is the piping and valves which distribute compressed air to the points of use.

HIGH-PRESSURE AIR

20-26. Figure 20-15, view A is an illustration of the first part of a high-pressure air system aboard ship. The 3,000/150 psi reducing station is used for emergencies or abnormal situations to provide air to the low-pressure air system.

LOW-PRESSURE AIR

20-27. Low-pressure air (sometimes referred to as LP ship's service air) is the most widely used air system aboard ship. Figure 20-15, view B is an illustration of the first part of a low-pressure air system. Many of the low-pressure air systems are divided into the subsystems of vital and nonvital air.

Vital Air

20-28. Vital air is used primarily for engineering purposes such as water level controls and air pilot-operated control valves. Vital air is also supplied to electronics systems. Vital air systems are split between all main machinery groups with cross-connect capability.

Nonvital Air

20-29. Nonvital air is provided for many different purposes (such as tank-level indicating systems and air hose connections). Air for a nonvital air system is supplied through a priority valve. This valve will shut automatically to secure air to nonvital components when pressure in the air system drops to a specified set point. It will reopen to restore nonvital air when pressure in the system returns to normal. This system gives vital air first priority on all the air in the low-pressure system.

MOISTURE REMOVAL

20-30. The removal of moisture from compressed air is an important part of compressed air systems. If air at atmospheric pressure, with even a very low relative humidity, is compressed to 3,000 psi or 4,500 psi, it becomes saturated with water vapor. Some moisture is removed by the intercoolers and aftercoolers. Air flasks, receivers, and banks are also provided with low-point drains to periodically drain any collected moisture. However, many shipboard uses of air require air with an even smaller moisture content than is obtained through these methods. In addition, moisture in air lines can create other problems which are potentially hazardous (such as the freezing up of valves and controls). For example, this can occur if very high-pressure air is throttled to a very low pressure at a high flow rate. The Venturi effect of the throttled air produces very low temperatures which will cause any moisture in the air to freeze into ice. This makes the valve (especially an automatic valve) either very difficult or impossible to operate. Droplets of water in an air system with a high pressure and high flow rate can also cause serious water hammer within the system. For these reasons, air dryers or dehydrators are used to dry the compressed air. Two basic types of air dehydrators are in use (the desiccant type and the refrigerated type).

Desiccant-type Dehydrators

20-31. A desiccant is a drying agent. More practically, a desiccant is a substance with a high capacity to remove (absorb) water or moisture. It also has a high capacity to give off that moisture so that the desiccant can be reused. In compressed air system dehydrators, a pair of desiccant towers (flasks full of desiccant) are used. One is on service dehydrating the compressed air while the other is being "reactivated." A desiccant tower is normally reactivated by passing dry, heated air through the tower being reactivated in the direction opposite to normal dehydration airflow. The hot air evaporates the collected moisture and carries it out of the tower to the atmosphere. The purge air is heated by electrical heaters.

Once the tower that is reactivating has completed the reactivation cycle, it is placed on service to dehydrate air and the other tower is reactivated. Another type of desiccant dehydrator in use is the Heat-Les dehydrator. These units require no electrical heaters or external source of purge air to operate. Figure 20-16, view A shows the compressed air entering at the bottom of the left tower, passing upward through the desiccant where it is dried to a very low moisture content. The dry air passes through the check valve to the dry air outlet. Simultaneously, a small percentage of the dry air is passed through the orifice between the towers. It flows down through the right tower, reactivates the desiccant, and passes out through the purge exhaust. At the end of the cycle, the towers are automatically reversed, as shown in Figure 20-16, view B.

Refrigerated-type Dehydrators

20-32. Another method of removing moisture from compressed air is refrigeration. If the compressed air is passed over a set of refrigerated cooling coils, oil and moisture vapors will condense from the air. They can be collected and removed via a low point drain. Some installations may use a combination of a refrigerated dehydrator and desiccant dehydrator to purify the compressed air.

Compressed Air Plant Operation and Maintenance

20-33. Any air compressor or air system must be operated in strict compliance with approved operating procedures. Compressed air is potentially very dangerous. Cleanliness is of utmost importance in all maintenance that requires opening compressed air systems.

SAFETY PRECAUTIONS

20-34. Many hazards are associated with pressurized air, particularly high-pressure air. Serious explosions have occurred in high-pressure air systems because of diesel effect. If a portion of an unpressurized system or component is suddenly and rapidly pressurized with high-pressure air, a large amount of heat is produced. If the heat is excessive, the air may reach the ignition temperature of the impurities present in the air and piping (oil, dust). When ignition temperature is reached, a violent explosion will occur as these impurities ignite. In addition, ignition temperatures may result from rapid pressurization of a low-pressure, dead-end portion of the piping system, malfunctioning of compressor aftercoolers, leaky or dirty valves, and many other causes. Every precaution must be taken to have only clean, dry air at the compressor inlet. Air compressor accidents have also been caused by improper maintenance procedures such as disconnecting parts while they are under pressure, replacing parts with units designed for lower pressures, and installing stop valves or check valves in improper locations. Improper operating procedures have also caused air compressor accidents with resulting serious injury to personnel and damage to equipment. To minimize hazards inherent in the process of compression and in the use of compressed air, safety precautions must be strictly observed. Some of these hazards and precautions are as follows:

- Explosions may be caused by dust-laden air or by oil vapor in the compressor or receiver. The explosions are triggered by abnormally high temperatures. These temperatures may be caused by leaky or dirty valves, excessive pressurization rates, and faulty cooling systems.
- Never use distillate fuel oil or gasoline as a degreaser to clean compressor intake filters, cylinders, or air passages. These oils vaporize easily and will form a highly explosive mixture with air under compression.
- Secure a compressor immediately if you see that the temperature of the air discharged from any stage rises unduly or exceeds the maximum temperature recommended.
- Never leave the compressor station after starting the compressor unless you are sure that the control, unloading, and governing devices are operating properly.
- To prevent damage due to overheating, compressors must not run at excessive speeds. Cooling water must be properly circulated.
- If the compressor is to remain idle for any length of time and is in an exposed position in freezing weather, the compressor circulating water system should be thoroughly drained.

- Before working on a compressor, be sure that it is secured and cannot start automatically or accidentally. The compressor should be blown down completely. Then, all valves (including control or unloading between the compressor and the receiver should be secured. Appropriate tag-out procedures for the compressor control and the isolation valves must be followed. Leave the pressure gauges open at all times.
- When cutting air into the whistle, siren, or a piece of machinery, be sure the supply line to the equipment has been properly drained of moisture. When securing the supply of air to the affected equipment, be sure all drains are left open.
- Prior to disconnecting any part of an air system, be sure that the part is not under pressure. Pressure gauge cutout valves should always be left open to the sections to which they are attached.
- Avoid rapid operation of manual valves. The heat of compression caused by sudden flow of high pressure into an empty line or vessel can cause an explosion if oil or other impurities are present. Valves should be slowly cracked open until flow is noted. They should be kept in this position until pressures on both sides have equalized. The rate of pressure rise should be kept under 200 psi per second.

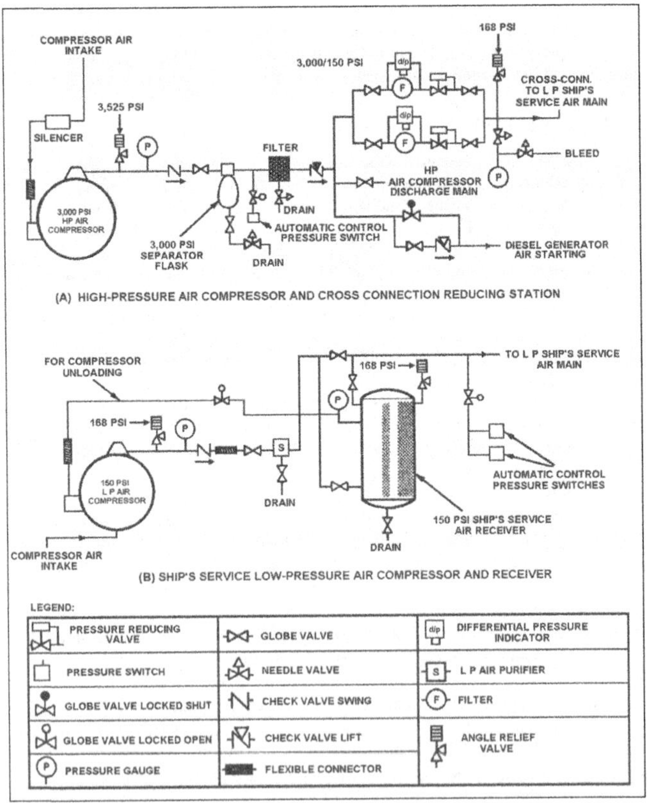

Figure 20-15. High-pressure and Low-pressure Air Compressor Piping Arrangements

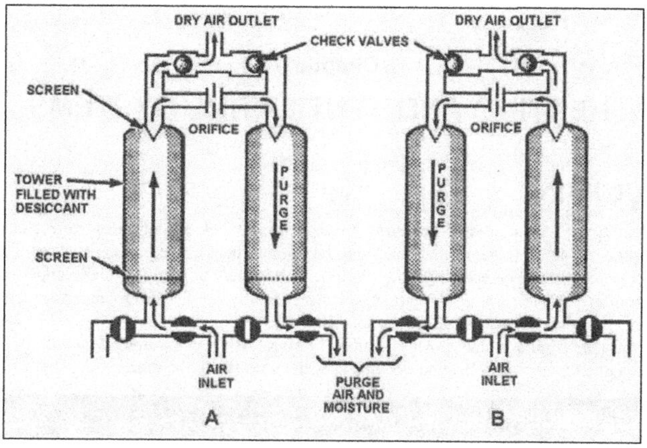

Figure 20-16. Heat-Les Dehydrator

Chapter 21

HEATING AND VENTILATING SYSTEMS

INTRODUCTION

21-1. Heating of Army vessels pertains to that process of adding heat to maintain the desired temperature. Any one of several methods may be used to heat a compartment or space aboard vessels. Some compartments may be heated for the machinery to operate properly. Other compartments are heated for crew comfort. Heads and washroom spaces are generally not heated because they are surrounded by and draw air from heated spaces. It is generally desirable to heat the engine room to keep it warm while in port and while the engine is shut down. Sections I through III of this chapter deal with boilers and section IV discusses other elements of heating systems as well as ventilation systems.

SECTION I. BOILER SPECIFICATIONS

The Code of Federal Regulations (CFR) specifies the design and construction of steam and hot water boilers used for heating and hot water supply. Steam and hot water boilers are constructed of steel plates or cast iron. This section describes the construction, limitations, and boiler fittings and appliances found on a typical marine boiler aboard an Army vessel.

PRESSURE AND TEMPERATURE LIMITATIONS

21-2. The maximum allowable pressure for steel plate steam and hot water, low-pressure boilers shall not exceed 30 psi. Cast iron, low-pressure boilers are limited to 15 psi for steam and 30 psi for water. The maximum allowable temperature for each type of low-pressure boiler is 250° F or 121° C.

BOILER INSPECTION AND ACCESS OPENINGS

21-3. A manhole is an opening in the shell or drum of a boiler to provide access for cleaning, inspection, and repair. Manholes may be elliptical or circular. An elliptical manhole opening is not less than 11 X 15 inches or 10 X 16 inches in size. A circular manhole is not less than 15 inches in diameter. Manhole openings are closed with a cover plate and are sealed against leakage by use of a gasket. The minimum width of bearing surface for a gasket on a manhole opening is 11/16 inches. No gasket for use on a manhole or handhole of any boiler can have a thickness greater than 1/4 inches when compressed. Handholes are small openings in the heads or shells of boilers, tanks, and other vessels. They are used for inspecting, cleaning, and repairing purposes. Vertical-fire-tube or similar-type boilers having gross internal volume of more than 5 cubic feet have at least three handholes or washout plugs in the lower tube sheet. Boilers having gross internal volume not over 5 cubic feet have at least two washout openings in the lower part of the waterleg and at least one washout opening near the line of the lower tube sheet. A handhole opening is not less than 2 3/4 inches by 3 1/2 inches. However, it is recommended that where possible larger sizes be used. Handholes are closed with metal plates secured in place by clamping devices called dogs. Gaskets are placed between handhole plates and openings to prevent leakage. A washout opening is provided in the shell's mud drum of the boiler to permit washing out sludge and other foreign matter. It is used for return pipe connections. The washout plug is placed in a tee so that the plug is directly opposite and close as possible to the opening in the boiler. Washout plugs are no smaller than 1 1/2-inch pipe size for boilers having a gross internal volume of more than 5 cubic feet and 1 inch pipe size for boilers having a gross internal volume of not more than 5 cubic feet. Internally fired boilers, in which the least furnace dimension is 28 inches or over, are equipped with an access opening of not less than 11 by

15 inches, 10 by 16 inches, or 15 inches in diameter. The minimum size of access door used in a boiler setting is 12 by 16 inches or equivalent area, the least dimension being 11 inches.

BOILER FITTINGS AND APPLIANCES

21-4. All boilers must have certain safety controls. These are valves, fittings, and regulators.

SPRING-LOADED SAFETY VALVE

21-5. This valve is fitted with a spring which normally holds the valve disk in a closed position. The spring allows the valve to open and close at predetermined pressures. Spring-loaded safety valves are characterized by pop action. Each steam-heating boiler has one or more approved safety valves set to discharge at a pressure not to exceed 30 psi. Boilers of more than 500 square feet of water heating surface, but without superheaters, are fitted with at least two safety valves. Boilers with integral superheaters have at least two safety valves attached to the drum and one safety valve fitted to the superheater outlet. The steam safety valve capacity for each boiler is such that the pressure cannot rise more than 6 percent above the maximum allowable pressure of the boiler. Accumulation tests are conducted to determine the adequacy of protection provided by the safety valves. During these tests, all steam outlet connections, except the safety valves and other valves necessary to operate the boiler, are closed. The fires are increased to their maximum capacity for a period of 15 minutes for fire tube boilers and 7 minutes for water tube boilers.

RELIEF VALVE

21-6. This valve is used to relieve excess pressure in the boiler. Each hot-water boiler has one or more relief valves. The relief valve is set to relieve at a pressure not to exceed 30 psi. Its capacity for each boiler is such that the pressure cannot rise more than 3 psi above the maximum allowable pressure of the boiler. It is installed with the spindle vertical and it may be connected directly to the boiler or to a fitting connected to the boiler. No shutoff valves are installed in the relief valve line.

WATER PRESSURE GAUGE

21-7. This gauge is connected to the boiler flow connection. Pressure gauges have a lever or T-handle cock located near the gauge to allow shutoff of the boiler's water to the gauge. Pressure gauge connections are of nonferrous material when smaller than 1-inch pipe size and longer than 5 feet between the gauge and the point of connection of the pipe to the boiler. When smaller than 1/2-inch pipe size and shorter than 5 feet between the gauge and the point of connection of the pipe to the boiler, pressure gauge connections are also of nonferrous material.

THERMOMETER

21-8. Each hot-water boiler has a thermometer located near the water pressure gauge to allow the temperature and water pressure to be read at the same time. The thermometer indicates the temperature in degrees Fahrenheit in the boiler or near the boiler outlet.

STACK SWITCH

21-9. As referred to in this text, a stack switch is actually a temperature combustion regulator which performs various functions that protect the boiler system from damage when a malfunction or a dangerous condition exists. Examples are flame failure, ignition failure, excessive water temperature, and low water.

BOTTOM BLOWOFF

21-10. Each boiler has a blowoff pipe fitted with a valve or cock and connected to the lowest water space available. The water pressure of the boiler is used to blow out the sediment that collects at the bottom of the water space through the bottom blowoff.

AUTOMATIC LOW-WATER FUEL CUTOFF

21-11. These cutoffs are found on automatically fired boilers. They shut off the fuel supply when the water falls below the safe water level. If a water-feeding device is installed in conjunction with the low-water fuel cutoff, it is constructed so that the water inlet cannot feed water into the boiler through the float chamber. The fuel or feed-water control devices are attached directly to the boiler.

BOILER TESTS, INSPECTIONS, AND STAMPINGS

21-12. Each boiler is subjected to a hydrostatic pressure test of not less than 60 psi by the manufacturer. After installation, the boiler is hydrostatically tested annually at twice the pressure at which the relief valve is set to open. The hydrostatic test is conducted by filling the boiler and heating system with water and sealing off the relief valve. Pressure is applied to the system by some external source such as water pressure or compressed air. Then the system is checked for leaks. New boilers that have been properly inspected and found acceptable are stamped with the information in the data plate shown in Figure 21-1. Each manufacturer uses a different data plate format, but all data plates must have at a minimum, the following information:

- The manufacturer's name and boiler serial number.
- A Coast Guard number, assigned to the boiler by the Officer in Charge, Marine Inspection (OCMI), who makes the final tests and inspections. The Coast Guard number is composed of an OCMI (CGN) serial number by which a particular boiler can be identified. The first two digits identify the calendar year the number was assigned; the remaining digit or digits identify the number of boilers tested by the inspector of that location on that particular date.
- The Coast Guard symbol applied to the data plate after the boiler has passed inspection. This symbol is also shown in Figure 21-1.
- The maximum allowable working pressure (MAX AWP) of hot water in pounds and the maximum temperature.
- The heating surface area in square feet.
- The minimum relief valve relieving capacity in pounds.

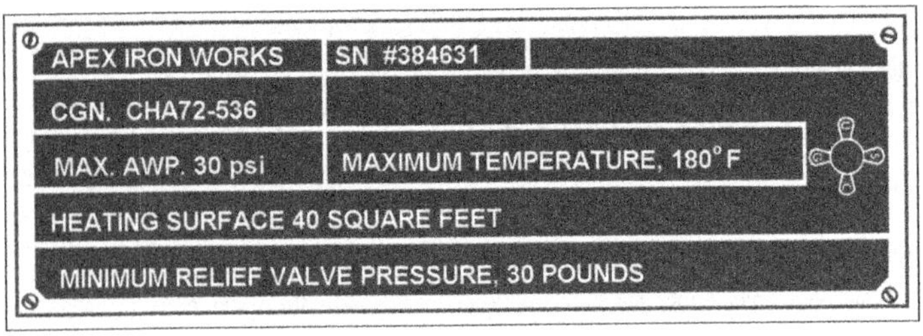

Figure 21-1. Inspection Data Plate

DESCRIPTION OF A TYPICAL BOILER

21-13. A typical boiler used on Army vessels is an oil-fired, fully automatic, central-heating boiler. It is used to supply hot water to radiators, convectors, or other heating elements. An automatic oil burner assembly is used to fire into a center combustion chamber. The boiler shown in Figure 21-2 is of the vertical fire-tube, two-pass, top-fired design. Figure 21-3 shows how the center combustion chamber makes up the first pass; the return fire tubes make up the second pass. The base of the boiler is insulated from the deck by an airspace and wet bottom section. The top cover on which the oil burner is mounted is insulated from the boiler. The top cover may be removed for periodic cleaning and inspection of the boiler tubes and combustion chamber. Domestic hot water is heated by the boiler through a side arm heater

(Figure 21-2). Hot water from the boiler is piped to a coil, in the side arm heater, which is surrounded by water from the domestic hot water tank. A thermostatically controlled circulating pump in the domestic hot water piping system keeps the domestic hot water flowing through the side arm heater as heat is needed. The expansion tank is used as a cushion for the expanding water. Water is, for all practical purposes, incompressible. Therefore, some means of taking up the expansion must be provided to protect the system and to prevent the relief valve from relieving too frequently. The expansion tank is usually located above the boiler and is the highest fixture in the heating system. Its position enables it to collect air from the system. The oil burner is of the fully automatic, high-pressure, atomizing type. It consists of a motor, a blower, and a two-stage fuel unit. The motor and blower provide air for combustion. The two-stage fuel unit supplies oil under controlled pressure to the oil atomizing nozzle. Automatic ignition of the oil is provided by a high-tension spark obtained from an ignition transformer. The ignition spark is turned off by the safety stack control after ignition has taken place. The stack switch also shuts the burner off if ignition has not taken place within 90 seconds. The normal on-and-off operation of the burner is controlled by an adjustable temperature control device. This device is commonly called an aquastat and is mounted on the boiler. A room thermostat may also be used to control operation of the burner. However, it only controls the temperature of the space in which it is located.

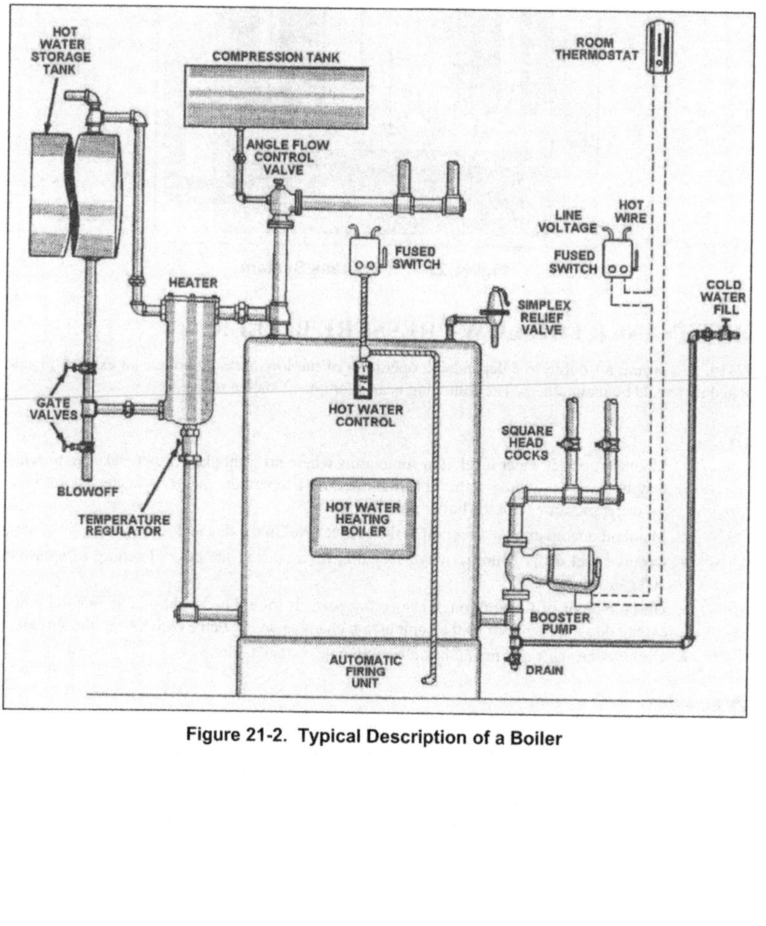

Figure 21-2. Typical Description of a Boiler

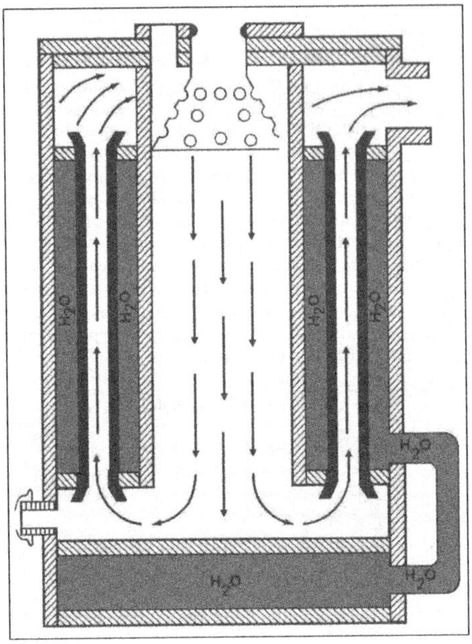

Figure 21-3. Two-pass System

MAINTENANCE FOR LOW-PRESSURE BOILER

21-14. To ensure a trouble-free dependable operation of the low-pressure boiler, an explicit maintenance schedule should be maintained. The following is an example of such a schedule:

DAILY—

- Maintain correct water level. On some units where no sight glass or petcocks are provided, this may be done by adding water until the pressure/temperature gauge indicates approximately 25 psi water pressure with the boiler shut down.
- Maintain uncontaminated fuel oil at the proper level in the day tank.
- Observe fuel oil pressure to make sure that it remains at the desired setting of approximately 100 psi.
- Observe color of flame through inspection port. It should be bright orange in color and should extend down to the floor of the combustion chamber so the flame ends sweep the surface.
- Observe the stack and maintain a clear smokeless stack outlet.

WEEKLY—

- Secure the boiler long enough to drain off some of the boiler water from the bottom blow connection to remove accumulations of sludge from the wet bottom and/or water column.
- The blowdown valve on the low-water cutoff column should be opened weekly. Flush the bowl and fillings to ensure that the control is in working order.
- Clean external boiler parts and inspect and service allied pumps as found necessary.

MONTHLY—

- Test safety devices.
- Relief valve. Increase water pressure inside boiler by adding more water from the domestic system until the valve operates at its preset value (30 psi).
- Low-Water Cutoff. Secure boiler and drain off enough water to uncover low-water cutoff, then reset boiler. It should not operate until you reestablish water level.

QUARTERLY—

- Inspect DC brushes and commutator; blow out with compressed air.
- Inspect AC brushes and slip rings; blow out with compressed air.
- Blow out air passages of all heaters with compressed air.

SEMIANNUALLY—

- Clean the boiler thoroughly every 6 months or at the beginning or end of each heating season. If the oil burner has been allowed to operate for extended periods with a smoky fire, clean the fire tubes and main combustion chamber more frequently than every 6 months.
- Remove the burner, top mounting cover, and insulation to provide access for cleaning. Punch the tubes with a flue-cleaning brush. Scrape down any soot or carbon that has formed in the center firing tube. Inspect and clean the combustion cone. Any accumulation of soot or carbon that has been punched through the tubes or main firing chamber can be cleaned out through the bottom port. The cleaning operation can be done with a vacuum cleaner if one is available.
- Remove the pipe plugs on the crosses of the low-water cutoff column and punch the fillings clean.
- Normally the oil burner is cleaned every 6 months. However, the nozzle lining one of the most important parts of the oil burner will require cleaning any time it becomes clogged. Boiler smoking or erratic operation usually indicates a partially clogged nozzle.
- Inspect and thoroughly clean the helix of the stack switch.
- The motor on the oil burner is equipped with either sleeve bearings or presealed ball bearings. If the motor has sleeve bearings, oil the bearings occasionally with light lubricating oil. Do not over lubricate.

BOILER TROUBLESHOOTING TIPS

21-15. In the case of an LCU working a beach with heavy surf, the boiler should be secured and the stack covered prior to the landing. On all vessels, if boiler develops a tendency to go into flame safety lockout sporadically, check the following:

- Air supply.
- Fuel system.
- Water level.
- Electrodes.

If all is well here, change transformer. If boiler starts with a rumble, or pulsates, or operates in a smoky condition, check the following:

- Air damper setting.
- Fan blade.
- Fuel pressure.
- Color and pattern of flame.

On some installations it is possible to create a partial vacuum in the boiler room by securing all ports and other openings when engines are running. This causes the engines to use up the available air at a rate which exceeds the inflow of fresh air. If the boiler cycles on at this time, the flame will probably be drawn up through the fan and damper inlet. Serious damage may be the end result. Therefore, do not operate boiler with vessel "buttoned up."

SECTION II. WAY-WOLFF BOILER

The Way-Wolff boiler is classified as a packaged boiler. The components of the Way-Wolff boiler are shown in Figure 21-4. This is because it is completely assembled at the factory and delivered ready for operation, pending connection of suitable water pipes, wiring, and fuel lines. It is further classified as a two-pass fire tube boiler because of the following:

- Products of combustion pass down through the central cylindrical combustion chamber.
- Expand throughout the bottom combustion chamber.
- Pass up through the vertical fire tubes to the top where they pass through duct work and emerge up the stack.

The Way-Wolff boiler is found on such Army vessels as the Q-boat, LCU, and some 65-foot tugs. This section covers the construction, installation, and maintenance of the Way-Wolff boiler.

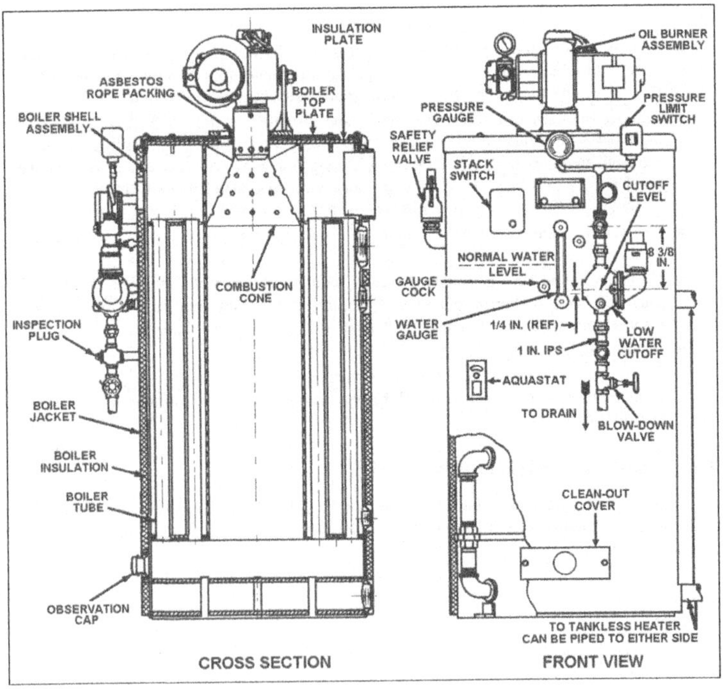

Figure 21-4. Way-Wolff Boiler Components

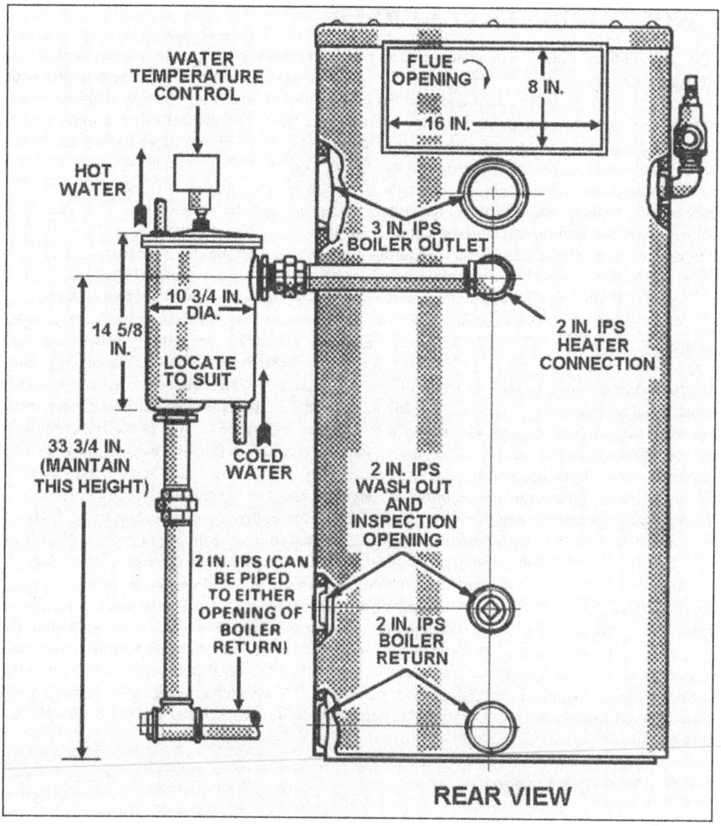

Figure 21-4. Way-Wolff Boiler Components (continued)

CONSTRUCTION OF THE WAY-WOLFF BOILER

21-16. The oil burner unit is mounted on top of the boiler and secured to the top plate by three 3/8-inch cap screws threaded into nuts welded to the underside of the top plate. The top plate is secured to the boiler main shell by four 3/8-inch studs welded to the shell. The top plate is insulated from the shell by 1/2-inch asbestos mill board as shown in Figure 21-4. The boiler has a 1-inch thick, 4-ply asbestos aircel insulating jacket protected by the outer steel jacket. The burner fires into the heat-resisting, perforated combustion cone. After passing through the center firing tube, the gases return through the thirty-six 1 1/2-inch outside diameter (OD) fire tubes to the 10- X 4-inch stack outlet. The bottom of the boiler is fitted with a removable wet bottom section to which the return connection must be made. A combustion cleanout and observation port is provided in the base of the boiler. Another observation port is located on top of the boiler so the oil spray pattern can be observed during operation. A 3/4-inch iron pipe size inlet, brass-body, hot-water safety-relief valve is provided to relieve pressure in excess of 30 psi. The boiler water temperature and pressure are indicated on the combination gauge mounted in the 1/2-inch iron pipe size connection on the front of the boiler.

WAY-WOLFF BOILER OIL BURNER

21-17. The oil burner unit is operated automatically by the stack switch, aquastat, and manual starter. It is an integral unit consisting of a special motor, flange-mounted on the burner case. It drives the burner blower mounted on the motor shaft and also the fuel unit by means of the flexible coupling. The fuel unit incorporates a two-stage pump, strainer, and pressure regulating and shutoff valve in unit. An electrically operated magnetic valve is also provided in the fuel line from the fuel unit to the burner fuel pipe assembly for additional safety. The special burner motor inverter operates on 115 volts DC and has, in addition to the DC commutator, a set of slip rings on the armature which supplies 110 volts, 60 cycles. This AC supply is fed into the primary of the ignition transformer through the stack switch. The secondary or high-tension side of the ignition transformer supplies 10,000 volts at 23 milliamperes to the ignition electrodes connected to the ignition transformer. During the start-up or ignition, an electric spark between the ignition electrodes ignites the oil which is atomized by the nozzle tip assembly. The atomizing nozzle tip is protected by a 100-mesh Monel strainer in the nozzle adapter. Oil is delivered to the atomizing nozzle at approximately 100 psi through the fuel tube. Air for combustion is drawn in through the air inlet housing assembly. It is controlled by the butterfly valve damper mounted on the butterfly valve shaft. This can be secured in any position by the knurled head screw in the butterfly valve lever. The air inlet housing may be connected to a source of outside fresh air when the space in which the boiler is located is subjected to varying negative pressures. These can be due to exhaust fan, diesel engine intake, or other sources. The combustion air is forced through the blower tube and directed to the outside of the blower tube by the air deflector, mounted on the fuel tube so that maximum is attained as the air leaves the angular vanes of the air diffuser. This turbulence permits the air to thoroughly mix with the atomized oil spray from the nozzle tip. Part of the air for combustion is blown through the holes in the side of the blower tube above the air diffuser into the space between the center firing tube of the boiler and combustion cone. This air, termed secondary combustion air, is preheated by the stack gases surrounding the upper portion of the center firing tube. This secondary combustion air is preheated by the incandescent outer surface of the combustion cone. It is then injected into the combustion cone through the series of concentric holes in the combustion cone to provide further progressive mixing of air and atomized oil. The secondary air also serves to normalize the temperature of the combustion cone which is fabricated of heat-resisting metal. Because of its low mass and specific heat, the combustion cone comes up to full firing temperature within seconds. In conjunction with the secondary air, introduced as described above, it produces a stable and efficient fire without any trace of smoke immediately from a cold start. A 3/8-inch diameter asbestos rope gasket on the outside of the blower tube and between the burner mounting bracket and the boiler top plates seals the burner opening.

CIRCULATING PUMPS

21-18. Three 1 1/4-inch iron pipe size brass body, water-circulating pumps are provided as follows:

- One is connected in the heating system return line. It may be connected to either of the two lower 1 1/4-inch coupling connections on the boiler. The circulating pump is controlled automatically by the room thermostat through a manual starter. The hot water outlet to the heating system may be made from either of the two upper 1 1/2-inch iron pipe size connections on the boiler.
- A second water circulating pump is for circulating boiler water through the heating coils of the potable water, side arm heater. It is controlled automatically by the potable hot water tank temperature control through another manual starter.
- The third pump is connected to the potable water circulating return line for circulating potable hot water throughout the vessel. It is controlled manually by another starter. A typical circulating pump is shown in Figure 21-5.

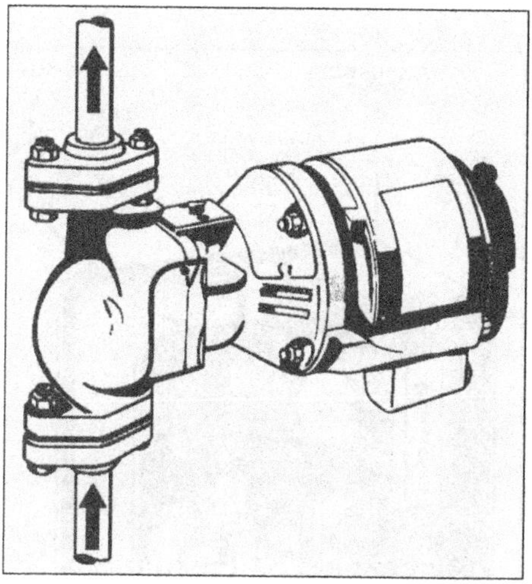

Figure 21-5. Typical Circulating Pump

STACK SWITCH

21-19. The stack switch (Figure 21-6) is mounted in an opening on the front of the boiler. It is provided with a set screw for securing it in place. The helix of this control projects into the upper return flue and is operated by the temperature of the stack gases. This control ensures that the ignition is "ON" when the burner starts. After the temperature of the stack gases has increased about 200° F, the helix of this control operates a contact to open the ignition circuit. During the preliminary ignition period, the current through the main relay in this control is shunted through a bimetal lockout warp switch resistor. If stack gases have not increased approximately 100° F before the preset lockout time of approximately 90 seconds, the burner motor stops, thereby shutting off the oil. The control is then manually reset by pressing the button in the cover of this control after allowing approximately 2 to 5 minutes for the safety lockout warp switch to cool. In normal operation, the stack gas temperature increases the required approximate 100° F, within 15 to 30 seconds after a start. At this time, the helix operates a contact to open the circuit to the lockout warp switch resistor. Then, after increasing another approximate 100° F, the ignition circuit is opened. In case of power failure or momentary interruption during operation, the control shuts the burner down and automatically restarts after the stack gases have cooled to allow the cold or starting contact of the stack switch to make contact. In case of flame failure during operation, the stack temperature and helix cool, thereby making the cold or starting contact. The burner will lock out after the preset time of approximately 90 seconds.

AQUASTAT

21-20. The aquastat (Figure 21-7) governs the normal "ON" and "OFF" operation of the burner according to the setting of an adjustable control. This control is adjustable through a range of 100° F to 240° F. An adjustable differential of 5° to 45° is provided. The contacts of this control operate the relay of the stack switch. This occurs so that the full load of the burner motor and ignition circuit is made through the heavy contacts of the stack switch. The heat-sensitive thermo element of the aquastat (which is immersed in the boiler water) will contract due to this cooling effect of the entering water. This will cause another set of electrical contacts to close, causing a flow of electricity to the stack switch.

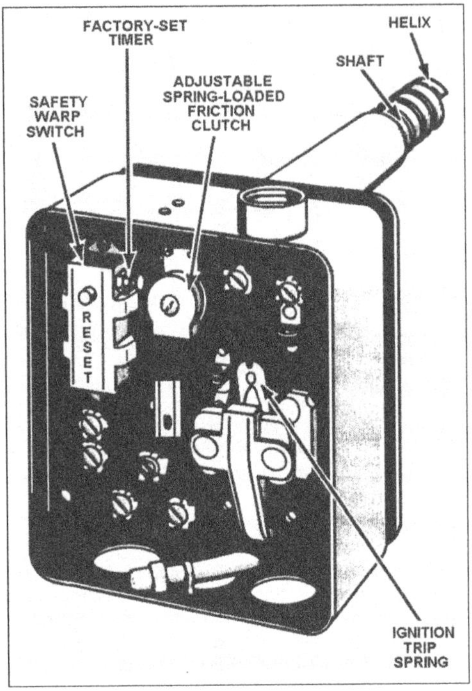

Figure 21-6. Stack Switch

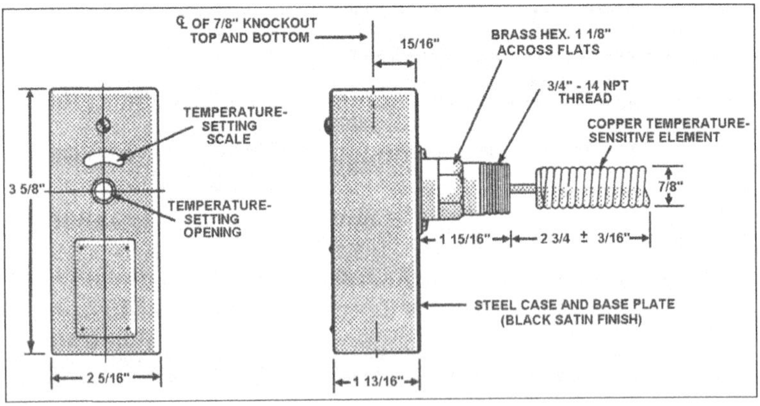

Figure 21-7. Aquastat

OTHER PARTS OF A WAY-WOLFF BOILER

21-21. A low-water relay automatically shuts down the boiler when water drops to an unsafe level. It works in conjunction with the low-water electrode which is mounted in the top of the boiler and which makes contact with the water. The programmer is a device which energizes the ignition transformer and oil solenoid when starting the unit. It de-energizes the ignition transformer and programmer when firing is accomplished. An ignition transformer produces a 10,000-volt spark (arc) across the ignition electrodes to ignite the fuel oil. The oil solenoid valve is a device electrically operated by the control system. It admits oil to the furnace and prevents any leakage of oil through the nozzle when the cyclotherm is not operating. The protector relay is an electronic flame safeguard used with a photocell to prove an oil flame. The absence of flame or an internal or external circuit failure that simulates absence of flame prevents opening of the oil solenoid valve or closes it if open.

Note: There are few moving parts in the protector relay. Therefore, it should not deteriorate in use. A periodic inspection of the photocell, an annual replacement of the vacuum tube, and a check of burner adjustments may forestall unnecessary shutdowns.

WAY-WOLFF BOILER STARTING ADJUSTMENTS AND TESTS

21-22. Before starting, make sure of the following:

- Boiler is filled to the proper water level.
- Heating system is vented of any trapped air.
- Circulating pumps are operating correctly.

It is good practice to test the operation of the safety relief valve before starting. This is done by increasing water pressure to 30 pounds, at which point the valve should open.

SETTINGS AND ADJUSTMENTS

Aquastat

21-23. Move temperature setting lever left or right until the indicator points to the desired control point. The range is 80° F to 220° F. To increase or decrease differential, move the adjusting lever until the lever is adjacent to the desired point on the differential scale. Differential adjustment is 12° F to 40° F.

Low-water Relay

21-24. There is no adjustment or setting, but it must be manually reset if the unit is on low water or power failure cutout.

Protector Relay

21-25. There is no adjustment or setting, but it must be manually reset if the unit is on safety switch lockout.

Fuel Pump

21-26. Fuel pressure is adjusted by turning the regulating screw clockwise to increase pressure. Fuel pressure is 100 psi.

NEW BOILER REQUIREMENTS

21-27. When starting up a new boiler, check the burner and controls wiring. If color-coded wiring has been used, check to be sure that connections are made in conformity with the wiring diagram in the applicable TM. This is done by removing the covers on the controls and burner motor junction box and observing the connections. On a new installation, there are apt to be metal chips from pipe threads or pipe compound which may clog the burner nozzle if the following special precautions are not taken in starting up:

- Remove the burner tailpiece assembly.
- Remove the nozzle tip and strainer assembly and reconnect the tailpiece assembly to the fuel pipe. Make sure the ignition transformer is disconnected and place a can or bucket under the outlet end of the fuel feed assembly.
- Start the burner and allow the pump to flush 1 to 2 gallons of oil through the system. This precaution saves time later, as nozzles clog most frequently when starting up a new system.

Normally, the stack switch will go to the lockout position before the full quantity of oil is flushed through the burner in the above described manner, since it is set to lock out in 90 seconds. However, this can be prevented by turning the helix spring clutch on the control counterclockwise and then releasing it after the burner has started. In this manner, the operating contacts of this control go to the "ON" position and prevent the flame control from locking out and shutting down the burner during the flushing period. The contacts to the starting or cold position must be reset before firing the burner. After the flushing operation is completed, the nozzle tip and strainer assembly are replaced on the fuel tube assembly and reinstalled on the burner. Be sure to check the electrode setting. The burner can now be put into operation and fired. If ignition does not immediately take place after the oil pressure has been established on the pressure gauge mounted on the fuel unit, immediately shut the burner down. This is done by opening the main starter switch. Check to be sure the electrodes (Figure 21-8) are properly set and there is spark for ignition. A hissing sound is heard when the ignition is on. The electrode setting is 1/8 to 3/16 inches apart. The air inlet shutter on the burner is set at approximately half-open on starting. If during starting any oil has been allowed to accumulate in the burner due to faulty ignition, caused by a partially plugged nozzle or incorrect electrode setting, the burner is apt to start with a rumbling noise. If this occurs, the air shutter should be opened to the full-open position and then gradually closed down. The burner air inlet damper is adjusted so there is absolutely no trace of smoke or even a haze when starting or running. If the burner is allowed to operate with a small amount of smoke or haze when the boiler is starting, it will usually clear up after the burner has been in operation a short time. However, this condition should not be tolerated as the continual smoking during starting will eventually cause the formation of soot in the boiler. The flame should have a smooth, steady sound without pulsations or rumbling. The fire can be observed through the observation port at the bottom of the boiler. The flame, as viewed through this port, should just about come down to the wet bottom plate and be a light orange color. A white flame indicates a partially clogged nozzle. A red or dark red flame is dangerous. It indicates insufficient air or a partially clogged nozzle. Shut the unit down and clean the burner nozzle. While starting the burner, be sure the water circulating pump is operating and actually circulating water through the boiler. Very often on new installations, air pockets in the system prevent the circulation of water even though the pump is operating. If the pump is not circulating water through the boiler, it can be observed by a rapid increase in the boiler temperature while the water outlet, a short distance from the boiler, remains cold. If the pump is operating satisfactorily, the burner should then be allowed to continue to operate until the boiler water reaches the temperature of the setting of the temperature limit control. This is normally set at 180^{0} F. The burner should then be allowed to cycle on and off to be sure the entire system is operating satisfactorily. After the burner has been in operation 2 or 3 hours, remove the fuel feed assembly again and check the nozzle tip itself to be sure that no dirt particles are lodged in it. Clogged nozzles usually occur in the first few days of operation due to dirt in the oil lines and to improper flushing. Therefore, inspect the boiler frequently during this period to be sure that it is not smoking, which usually is the first sign of a clogged nozzle. The oil burner nozzle usually does not clog solid but becomes partially clogged. This condition distorts the spray and causes poor combustion. It will eventually cause a smoky fire that causes soot to form in the tubes in the boiler.

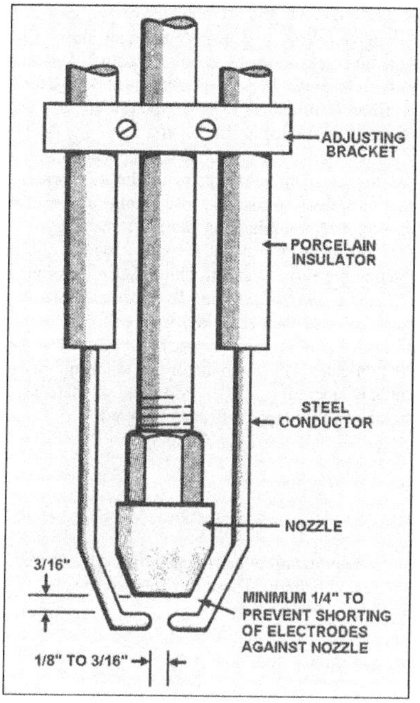

Figure 21-8. Electrode

BOILER MAINTENANCE INSTRUCTIONS

21-28. After the boiler has been installed and put into operation, several maintenance procedures are necessary for efficient operation. Maintenance instructions for the circulating pump are discussed.

BOILER

21-29. If the oil burner is properly cared for and not allowed to operate for extended periods with a smoky fire, inspect and clean the fire tubes and main combustion chamber every 6 months. Clean the boiler by removing the burner, top mounting cover, and insulation and punch the tubes with the flue-cleaning brush. Clean the main combustion chamber in the same manner by scraping down any accumulation of soot or carbon that has formed. At this time, inspect the heat-resisting combustion cone and clean it thoroughly if any carbon has accumulated due to a defective or partially clogged nozzle. Any accumulation of soot or carbon punched through the fire tubes into the main firing chamber is cleaned out at the bottom through the cleaned port. The entire cleaning operation is best done with an industrial vacuum cleaner. This does a more thorough job of picking up the loose soot and carbon accumulation than can be done with a hand brush.

OIL BURNER UNIT

21-30. During the periodic cleaning operation, remove and clean the burner fan. This is done by disconnecting the electrical connection to the motor and loosening the two bolts holding the motor to the burner housing. The motor and fan are then removed as an assembly. Rinse the fan in a cleaning solution or solvent to remove any grime or dirt that has accumulated on it. Inspect it for damage which might cause it to be out of balance. Remove, clean, and inspect the fuel unit. Thoroughly clean and inspect the electrodes to be sure there are no cracks in the porcelain insulator. Also inspect the electrode cables at this time and reassemble them after cleaning. Inspect the filter installed in the suction line to the fuel pump at least once a month. The filter is provided with a replacement cartridge which can be cleaned out. However, it should be replaced at least once a year depending on the grade and cleanliness of the oil used. At the same time this filter is inspected and cleaned, also inspect and clean the strainer in the fuel unit. The oil burner nozzle is one of the most important parts of the oil burner. Normally it does not require thorough cleaning more than once every 6 months. The slightest sign of the boiler smoking or operating erratically (such as a pulsating fire, rumbling on starting, shutting down, or a change in the normal combustion sound of the unit) usually indicates a partially clogged nozzle. The motors on both the oil burner and the circulating pump are furnished with either sleeve bearings or presealed ball bearings with grease to last for at least 2 years in normal service. It is recommended that ball bearings be flushed and thoroughly cleaned with kerosene and then repacked with ball bearing grease nearly one-third full after approximately 2 years of service. It is only necessary to oil sleeve bearings occasionally with light lubricating oil. Do not over lubricate. The commutator and slip ring brushes on DC motors are selected to give a life of many thousands of operating hours. However, standard vessel maintenance should call for inspection about every 3 months and for replacement as required.

NOZZLES

ATOMIZING

21-31. For nozzles, the term "atomizing" means separating oil into very small droplets and delivering them in the desired spray angle and pattern. These droplets are so small that in many cases the spray looks like a vapor. The nozzle has four essential parts:

- Fine mesh screen.
- Screw pin distributor having tangential slots.
- Conical swirl chamber.
- Orifice.

The strainer serves to protect the small passages in the distributor slots and the orifice from clogging dirt and solid particles in the oil. Oil under the usual pressure setting of 100 psi passes through the strainer and enters the swirl chamber. It rotates rapidly due to high velocity through the small tangential slots in the distributor. The oil is then ejected from the orifice in the form of a rotating tube at a high rotational velocity. As the oil emerges from the orifice, it takes the form of a film or sheet of oil in the shape of a cone. As this film travels away from the orifice, it is stretched into a thinner and thinner film. The rupturing of this film at a short distance in front of the orifice forms the minute droplets of the oil spray. The included angle of the spray as it is projected from the nozzle is determined by the design of the nozzle orifice, swirl chamber, and distributor slots. These factors also control the spray pattern whether the spray is in the form of a hollow cone or a full cone. A hollow cone pattern has relatively few droplets in the center of the spray. A full cone pattern has droplets across the whole cross section at a short distance in front of the nozzle. Figure 21-9 emphasizes the difference between these spray patterns. The spray angle and pattern of the nozzle are important. They must match the pattern of the burner combustion air as it is blown through the diffuser at the end of the blower tube. This air pattern is basically a function of burner design. However, it varies with the rate at which the burner is fired, the diffuser used, and the setting of the diffuser with respect to the nozzle. For this reason, it is important that a nozzle with the recommended spray angle and pattern be used. If the oil spray angle and pattern do not match the air pattern, some or all of the following conditions are noticed in the operation of the unit.

- Smoky fire even with excess air indicated by low CO_2.
- Excessive rumbling or pulsating fire.
- Rapid carbon formation in the combustion chamber and fire tubes.
- Erratic or unstable ignition. This can also be caused by improper electrode setting.

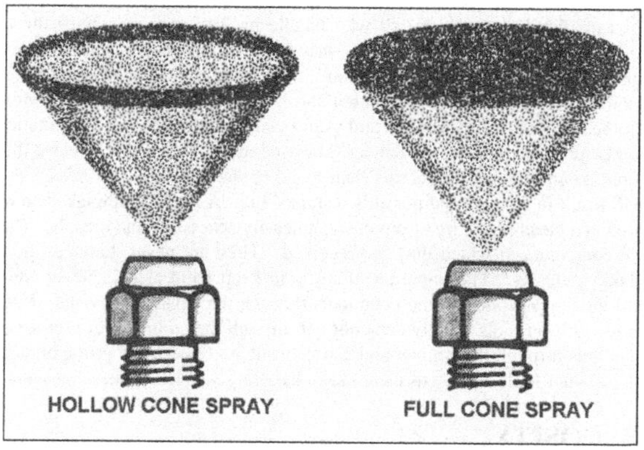

Figure 21-9. Spray Patterns

METERING

21-32. In addition to breaking up oil into small droplets, the nozzle is the means for metering the amount of oil supplied to the combustion chamber. Oil burner nozzles are calibrated in gallons per hour (gph) at a pressure of 100 psi. In the range of 2.00 to 3.00 gph, the difference is .25 gph, while in the range of 3.00 to 10.00 gph, the difference is .50 gph. The flow rates as marked on the nozzles are maintained within plus or minus 5 percent. Spray angles are maintained within minus 2° and plus 5° of the marking on the nozzle. The discharge rate from a nozzle is determined by the size of the slots and the orifice. For each flow rate, a definite relationship exists between the slot area and the orifice area. This must be maintained within very close limits to produce satisfactory sprays. It is not possible to change the flow rates of nozzles in the field by interchanging the parts. The parts in each nozzle are matched during manufacture and they must remain that way.

DIMENSIONS

21-33. Nozzles are made with the same degree of precision and care as a fine watch. By ordinary standards, the dimensions in a nozzle are extremely small, especially so in the smaller sizes. For example, orifices under .010 inches (ten thousandths of an inch) are common. As a comparison, a small sewing needle is .020 to .030 inches in diameter. The slots in these same nozzles are even smaller than the orifices. For example, in a .75 gph nozzle with two slots, they may be only .005 inches deep. Compare that with a good grade of writing paper .004 inches thick or human hair from .0015 to .002 inches thick. All of these dimensions must be maintained within very close limits. If the dimensions of the slots or of the orifice vary as much as .0005 inches (one-half thousandth of an inch) the nozzle may be the next higher or lower size. These small dimensions are required to maintain the correct flow rates and the proper spray quality for good performance. Not only must the dimensions be held but also the finishes on these surfaces must be extremely good. The smaller the nozzle, the more critical the finish.

CARE, CLEANING, AND HANDLING

21-34. These facts on nozzle dimensions are mentioned here simply to impress on you the importance of intelligent care of oil burner nozzles. Dirt particles do not have to be very large to cause trouble. In the factory, the nozzle parts are carefully handled to avoid contamination with dust or lint from clothing and wiping cloths. Test oil is filtered constantly and with much finer than average filters to prevent contamination during spray testing. Compressed air used for blowing parts is filtered to prevent moisture and rust from getting into the nozzles. The nozzles are handled in a way to avoid damage to any machined surfaces. Equal care should be used by all who handle nozzles or who service the oil burner. Loose material in a nozzle interferes with its operation. It is removed by using clean, filtered, compressed air or by rinsing in a clean solvent. Deposits of "carbon" in the slots, swirl chamber, and orifice are not easily removed. These deposits are not really carbon but varnish and gum formed by the heating of the oil under operating conditions. Some of this varnish and gum becomes very hard and is practically impossible to remove. In those cases, many nozzles are damaged beyond use in the attempt to clean them. In any type of cleaning operation, do not use any tool harder than wood on the slots and the orifice. The slightest scratch on the orifice will result in a streak and possibly a smoky fire. A round toothpick is an ideal cleaning tool. Nozzles are stored in a clean place, free from dust, and easily accessible and orderly. They are always kept in the individual containers in which they are received. They are never dumped into a toolbox, bin, or drawer or handled carelessly. The important thing is to keep them clean. Never carry nozzles in your pocket. Lint and dust may work into the containers through the smallest crevices. Use clean tools when installing a nozzle. Grit on tools usually does not get through the strainer, but grease on the tools, loaded with dust and lint, gets through the strainer and causes trouble. Grease and grime on your fingers also can cause trouble. Keep your tools and hands clean when handling nozzles in their original containers.

FUEL OIL VISCOSITY

21-35. Viscosity is by far the most important specification applied to fuel oil from the standpoint of nozzle performance. Viscosity is the measure of its resistance to flow. Oil with a high viscosity is more difficult to pump and to atomize. Oil with a low viscosity is much easier to handle in all respects and is easier to atomize. The common designation of viscosity in the fuel oil industry is in terms of seconds Saybolt universal (SSU). The higher the SSU number the greater the viscosity. The present Commercial Standards Specifications for number 2 fuel oil set a maximum permissible viscosity of 40 SSU at 100^{0} F. A fuel oil within these specifications is satisfactory for operation in nozzles down to about 1.00 gph at a pressure of 100 psi. When nozzles smaller than 1.00 gph are used, for best performance use lower viscosity range of number 2 oil. Because higher viscosity means higher resistance to flow, high-viscosity oil slows down the velocity in the swirl chamber and orifice of the nozzle. This results in a larger particle size in the spray from the nozzle. At the same time that the particle size increases, the spray angle decreases under these conditions. The result is a smoky, unstable, and noisy fire if the viscosity is too high. These effects can be minimized to some extent by using higher pressure on the nozzle. For example, if the pump pressure is increased to 110 psi, or 125 psi in many cases, the effects of high viscosity can be overcome. Lower viscosity gives finer atomization and generally improves burning.

SECTION III. CRANE BOILER

The Crane 50-30-9 boiler is a sectional, cast iron, steam boiler containing a tankless hot water heater. It is fired by a Conservoil Model 2A12C oil burner. The unit is automatic and rated at 450,000 BTU/hour. The burner is oil fired, uses the same fuel as the diesel engine, and is located at the bottom of the boiler unit. The burner unit is forced-draft fed and fires into a fire brick combustion chamber. The heating unit or water side of the boiler is of hollow cast iron sections.

BOILER CONSTRUCTION

21-36. Generally, the heating unit consists of six sections: a front, a rear, and four intermediates. The front section accommodates the pressure and vacuum gauge, water gauge, burner unit, sight glass, and door. The outlet connection is mounted on the rear section. All sections have machine-faced surfaces. To assemble this unit, first coat each machined surface with a suitable sealing compound. Insert an asbestos gasket between the sections. Secure the six sections into a single watertight unit by staybolts. After assembling the combustion chamber and heating unit, apply an insulating cement to the outside surface. The insulating cement is a noncombustible material used to retain heat and protect the outer shell of the boiler unit.

AUXILIARY EQUIPMENT

21-37. As a complete assembly, the boiler is fully automatic, depending on its auxiliary switches and gauges for starting and stopping. The oil burner unit (Figure 21-10) is an integral unit. It consists of a special motor flange-mounted on the burner case. It drives the burner blower, mounted on the motor shaft, and also the fuel pump by means of a flexible coupling. The special burner motor is an 1,800-rpm, 115-volt, DC-driven motor. It has, in addition to the DC commutator, a set of slip rings on the armature to supply AC to the primary side of the ignition transformer. The burner motor is furnished with either sleeve bearings or presealed ball bearings with enough grease to last 2 years. The ball bearing is flushed and cleaned with kerosene and repacked one-third full with ball bearing grease after approximately 2 years of service. The sleeve bearing requires occasional oiling with a light oil. The motor and combustion fan is removed for periodic maintenance by disconnecting the motor leads at the ignition transformer. Remove the mounting bolts holding the motor to the burner housing, then lift out the motor and fan. Clean the combustion fan by brushing off the accumulation of lint and dirt and washing it in solvent. Remove the brush, retaining the plugs and extract the brush from the brush holder. Inspect the brushes for arcing damage or abnormal wear. Replace any badly worn brushes. Reinstall the brushes in the motor and then mount the motor and fan in the burner housing. Connect the electrical leads to the transformer. The fuel pump is a positive-displacement pump which operates at 100 psi. It is mounted on the right of the burner housing. The fuel oil pump takes oil from the vessel's fuel oil day tanks through the fuel oil filter located in the suction line close to the fuel oil pump. The filter element is a replaceable cartridge and should be replaced after 30 days of operation. The filter housing is equipped with a drain cock located at the bottom of the housing. It should be drained weekly. Loss of fuel pressure can usually be traced to a leaking suction line, plugged line, or dirty filter. No maintenance is recommended on the fuel pump other than tightening leaky gaskets. The ignition transformer is mounted on the left side of the lower housing and held in place by two cap screws. The AC supply is fed into the primary coils of the ignition transformer through the stack switch. The secondary or high-tension side of the ignition transformer supplies 10,000 volts to the ignition electrode through the ignition cables. The transformer is a sealed unit and is replaced in case of failure. The ignition electrodes are mounted inside the lower housing, secured to the inner sleeve, and extended into the manifold. The electrode is a porcelain-covered wire connected by ignition cables to the ignition transformer. During the startup or ignition period, an electric spark between the ignition electrodes ignites the oil atomized by the nozzle tip. The ignition points are adjustable and must be set in relation with the burner nozzle and the combustion sleeve. The tip of the ignition points are set at 1/8 inches apart and 3/16 inches ahead of the end of the burner nozzle tip. A distance of 9/16 inches is set between the bottom of the electrode and the centerline of the burner nozzle. The nozzle is recessed 3/8 inches inside the combustion sleeve. Remove the ignition electrodes for inspection by disconnecting the ignition cables and fuel line from the fuel tube and electrode assembly. Withdraw the ignition and sleeve assembly from the lower housing. With the ignition sleeve removed from the lower housing, the electrodes and burner nozzle are accessible for cleaning, adjustment, or replacement. Check the ignition electrodes for cracked porcelain or badly burned tips. Cracked porcelain causes arcing, therefore no spark at ignition tips. Badly burned tips causes a weak spark or no spark at all. Replace any faulty parts. The burner nozzle, the same as that used on the Way-Wolff boiler, is one of the most important components of the oil burner. It provides for efficient and trouble-free operation of the boiler. This little metal unit, no more than 1 1/2 inches long, controls the atomizing and metering of the fuel oil. Maintenance of the burner tip is of utmost importance for it is the heart of the boiler and heating system. Almost any malfunction of the system, other than feed water trouble, can be detected by viewing combustion conditions in the combustion chamber. The combustion chamber and fire can be seen through the lower observation door. Here, the color of the fire indicates proper or improper combustion. The conditions to look for are as follows:

- **Orange flame.** Indicates proper combustion or perfect mixing of fuel oil and air. It gives the flame a smooth steady sound.
- **White flame.** Usually indicates a partially clogged nozzle. The most frequent cause is an improper setting of the air inlet shutter.
- **Dark red flame.** This flame is by far the worst and most dangerous and cannot be tolerated. Not even one ignition cycle should be allowed under this condition. The dark red flame is an

indication of a partially clogged nozzle or insufficient air for the amount of fuel being injected into the combustion chamber.

- **Pulsating or rumbling noise in the boiler.** A partially clogged nozzle disrupts the spray pattern, thereby causing the droplets of oil not to break up at the cone end. The heavier oil tends to spin but and burn against the firebox wall or floor. This causes the red flame and rumbling noise. When this condition exists, the burner unit must be removed and the burner nozzle cleaned.

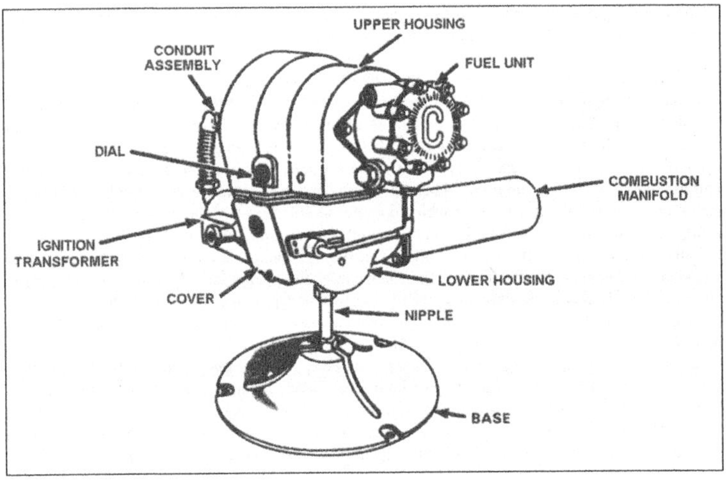

Figure 21-10. Burner Unit for Crane Boiler

TANKLESS HOT WATER HEATER

21-38. The tankless hot water heater is best described as a heat exchanger. Cold water from the vessel's potable water tank passes through a series of finned coils at approximately 210° F. Then it passes to a mixing valve where cold water is fed into the hot water to attain the desired temperature at the domestic faucets. The mixing valve is constructed and equipped with a dial setting so that a wide range of temperatures are available. Figure 21-11 shows a schematic of the tankless hot water heater.

BOILER UNIT COMPONENTS

21-39. Incorporated into the boiler unit are various switches, gauges, and controls that are called components. They are vital in the cycle of events that makes this a fully automatic heating unit. The components and their function are described below.

- **Aquastat.** The aquastat is mounted on the boiler with a capillary tube and bulb protruding into and immersed in the boiler water. The bulb contains a vapor or gas that reacts to heat by expanding and contracting. It operates a bellows connected to the movable side of a set of contact points. The temperature range is preset by the manufacturer at 180° F with a -40° differential. The temperature range setting may be adjusted by the engineer aboard the vessel to meet climatic conditions. However, the -40° differential is built in and should not be tampered with. On demand, the aquastat closes the circuit and feeds power to the stack switch.
- **Stack Switch.** This stack switch is the same as that used on the Way-Wolff boiler. It is located in the elbow at the back of the boiler.

- **Oil Burner Cutout Switch.** This switch is a technical term for the master or disconnect switch. Located in the DC power supply between the fuse box and the boiler, it is usually mounted close to the burner. The purpose of this switch is to give independent control of electric power to the burner unit. More specifically, this switch allows shutdown of the boiler and all components for maintenance, repairs, and tests. It is also used as an emergency cutoff switch.
- **Pressure Control.** This control is located at the top outboard end of the heating boiler. This safety control is pressure-operated and secures the boiler heating unit if pressure exceeds 10 psi. The electrical side of the unit is connected in series on one side of the main 115-volt feed line. When activated, it opens the circuit and shuts down the boiler.
- **Low-water Cutoff Switch.** This switch is located on the forward side of the boiler. Its purpose is to secure the boiler heating unit if the water level drops below a safe operating level. The electrical side of this unit is also connected in series between the disconnect and stack switch.
- **Safety Valve.** This valve is located at the top of the boiler. It is not an electrical unit and has no control over the operation of the boiler. It has a spring-loaded seat that remains closed and is set to open at 30 psi. Any pressure above 30 psi overcomes the spring tension and the valve relieves it into the atmosphere. When this happens, the boiler should be secured and all systems tested. There is also the possibility over a long period that the spring may lose some of its tension and have to be reset. If no test facilities are available, a new valve must be installed.

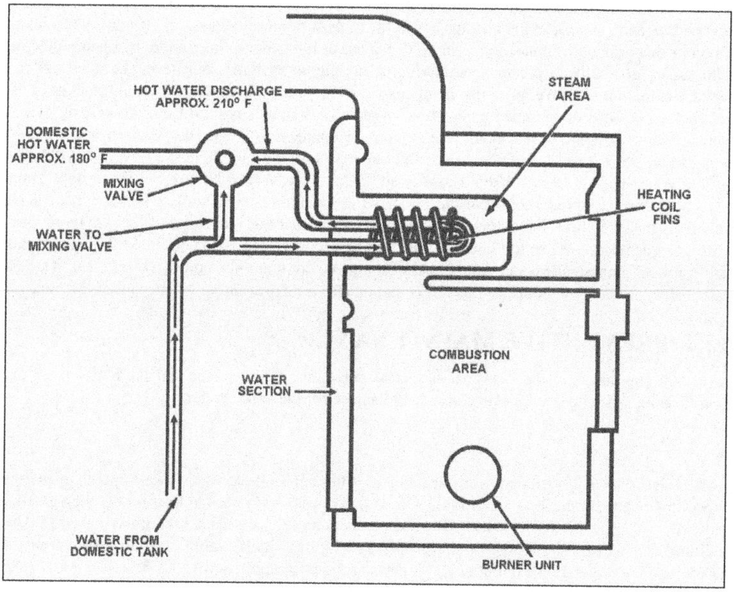

Figure 21-11. Schematic of Tankless Hot Water Heater

BOILER VISUAL INSPECTIONS

21-40. It is possible that the first vessel you are assigned to may have a boiler that has been idle for some time. If a boiler is to be idle for a long period, it should be cleaned, dried, and coated with a suitable preservative. The heating system throughout the vessel should be drained. The crew that secures a boiler must provide a checklist and detailed instructions giving step-by-step instructions for depreservation of the boiler. Each connection or part which needs attention must also be tagged. Several copies of the checklist should be located in the wheelhouse attached to the wheel. Using this checklist and the tags as guides, inspect the boiler and depreserve it. Reconnect all the apparatus to restore the plant to proper working order. Regardless of the type of boiler, it must have the correct amount of water in it to operate properly. Therefore, you must admit the prescribed amount of makeup water which registers by assuming the correct level in the sight glass or water column, or register the prescribed pressure on a gauge. Check the fuel system to make sure the day tank is properly filled with water-free fuel. Be sure that all strainers, filters, valves, couplings, and pumps are properly aligned for normal operation. Check the stack to make sure the weather boot or any other capping device is removed. This is to ensure the proper drawing needed for draft control. Check the safety valves or relief valves and remove any tape that may have been applied during the idle period. Inspect the safety valve to make sure the lead seal is not broken and that the valve's hand-relieving gear works. If the fire and water sides are preserved with P_1 or P_3 preservative, remove it with suitable solvents. Also, boil out the water side with boiler compound (usually phosphate or caustic soda). When boiling out is required, the cleaning solution has to be drained out through the bottom blow valves or water side cleanout connection. Observe the usual safety precautions of securing fires and relieving boiler pressure to safeguard personnel and the boiler. During this preliminary inspection, also inspect the boiler and allied fittings, mountings, and attachments for any leakage, rust, or other signs of abnormal condition. If any are found, secure the boiler and make appropriate repairs. When the inspection reveals that everything is working normally, you may light off and bring the boiler up to its maximum safe allowable working pressure. Before cutting the boiler in on the line leading to the many heaters throughout the vessel, open a petcock in the heating line at the highest point in the system. If none is provided, loosen a union or other suitable connection. Admit the steam or hot water to the system by opening the main stop on the boiler gradually to prevent hammering of the piping system. Whenever the main outlet valve is opened, the main inlet valve must also be opened. At this time, any entrapped air in the lines is forced out through the petcocks or loosened connection. When all the air is purged from the system, evidenced by steam or water escaping at that point, the petcock or connection is closed. Allow a few minutes for the heating to take effect and then feel all inlet and outlet pipes to be sure there is a flow through the heater. If the flow is not normal, the probable cause is an air lock. Relieve it by purging the system again. Once these checks are completed, observe the boiler while operating to make sure that steam or hot water is flowing normally. Check the automatic devices to ensure that air, ignition, and fuel are being provided in the proper quantities. When the boiler is functioning properly, the flame observed through the inspection port is orange, without streaks or smoke. Also, the gases coming out of the stack are clear. Once all these requirements are met, the boiler is ready to be put on a routine preventive maintenance schedule.

BOILER PREVENTIVE MAINTENANCE

21-41. All equipment in the Army is subject to preventive maintenance and the boiler is no exception. The following is a schedule for preventive maintenance for the crane boiler.

DAILY

21-42. Keep day tank supplied with clean fuel. Maintain correct water level, operating pressures and temperatures, and combustion conditions. On some vessels, it is possible to close off the vessel until it is practically airtight. If this is done with the engine running, a partial vacuum is created. Under this condition, fire will be drawn out of the burner unit into the boiler room creating a hazardous situation. Therefore, do not operate the boiler unless it has an adequate supply of air.

WEEKLY

21-43. When a water column is attached, use the blowdown connections weekly to prevent a buildup of scale and to make sure the column is in good order.

MONTHLY

21-44. Lubricate oil cups on motors and pumps. Check manually operated valves for correct operation. Repack stems, renew gaskets, or resurface seats as necessary. Secure the boiler and drain out all water through the bottom blowout valve. This is done to remove sediment and to protect the heating surfaces from internal damage.

QUARTERLY

21-45. Inspect commutators, slip rings, and carbon brushes on motors or converters. Clean this apparatus with low pressure (5 psi) dry compressed air, fine sandpaper, or commutator stone.

SEMIANNUALLY

21-46. Secure the heating system and dismantle the boiler to the extent necessary to gain access to the internal parts. Clean the combustion chamber and fire tubes with a wire brush and vacuum cleaner. Clean the helix of the stack switch. Renew any defective electrodes or brittle ignition wires. Clean the fuel nozzle and related parts. Inspect and repair furnace brick work and any other accessible boiler parts.

ANNUALLY

21-47. Repeat all semiannual maintenance procedures. Perform a hydrostatic test as directed by CFR 46. Where provided, renew fusible plugs.

BIANNUALLY

21-48. Repeat the annual inspection procedures. Inspect ball bearings on motors and pumps. If the bearings are grease-packed, it may be necessary to clean out the old grease and repack the bearings with fresh grease.

SECTION IV. HEATING AND COOLING SYSTEMS

Because of the necessity of keeping weight and size to a minimum, forced-circulation water heating and cooling systems are used on most Army vessels. The three types of circulation systems (Figure 21-12) are the one-pipe; the two-pipe, reverse-return; and the two-pipe, direct-return systems.

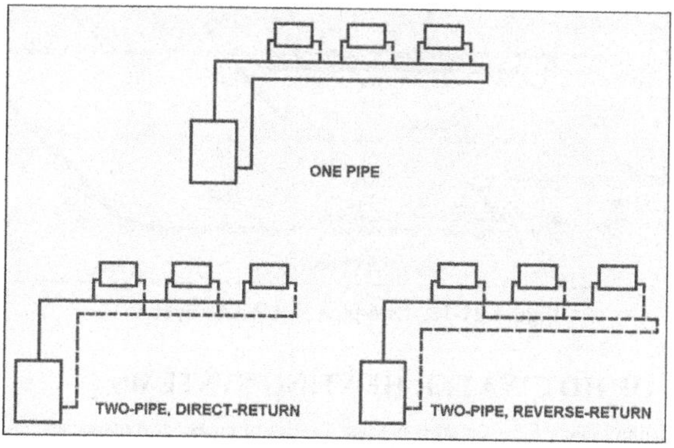

Figure 21-12. Heating Systems

HOT WATER HEATING SYSTEMS

21-49. The one-pipe system (Figure 21-13) is the most practical of the three systems. It provides good heating with the least amount of weight and space. However, the system is not desirable for combined heating and cooling because of the high friction losses in cooling applications. The two-pipe, reverse-return system is heavier than the one-pipe system. It also does not give appreciably better heating than the one-pipe system. It is used only for the combined heating and cooling application for which the one-pipe system is not suited. The two-pipe, direct-return system is the least desirable of the three systems. A direct-return installation may require less pipe. However, this slight advantage is offset by the poor operation of the system and the difficulty in balancing. In the reverse-return system, the circuit to each heating unit is approximately the same length. In the direct-return system, the unit nearest the boiler has the shortest loop. Consequently, it is a difficult system to balance so that each unit is properly supplied with water. Orifices or plug cocks (balancing fittings) are used extensively to obtain proper balance.

ONE-PIPE HOT WATER SYSTEM

21-50. Figure 21-13 illustrates the methods generally used for distributing hot water in a one-pipe system. For a small vessel, a single circuit is sufficient; for larger vessels, two circuits are used. In some cases where three or four decks are being heated, multiple circuits are required to maintain good distribution. By dividing the system into several loops, a reduction in the pumping head pressure and piping weight is obtained.

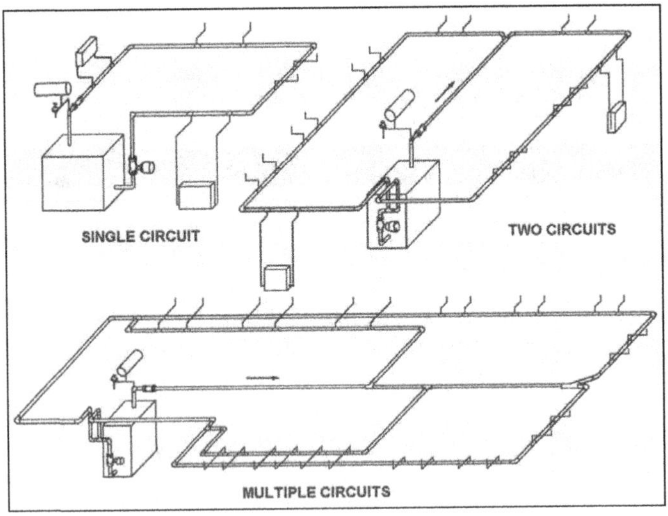

Figure 21-13. One-pipe Hot Water System

ELEMENTS OF HOT WATER HEATING SYSTEMS

21-51. The heating system for a vessel is not just a series of pipes. It contains several different types of components. Each of these components are described below.

REDUCING VALVES

21-52. Reducing valves are required on all hot water systems to protect the boiler from the full pressure of the freshwater system. These valves are set at a standard gauge pressure by the manufacturer and must be adjusted for higher pressure. The pressure the valve is set for is just enough to keep the highest radiator or heater filled with water. As the water is heated in high-temperature systems, its expansion usually provides extra pressure. This eliminates the necessity for resetting the reducing valve.

RELIEF VALVES

21-53. A relief valve is provided to keep the boiler from exceeding its designated pressure. Generally, the relief valve can be set at a specified pressure and needs no further adjustments. When the system is heated for the winter, pressure may increase to the point where the valve will relieve a few times. But, unless the water is cooled to the point where more water is admitted to the boiler or is heated to an unusually high temperature, no further relieving will occur. However, if the relief valve is located on the discharge side of the circulating pump and the pump head pressure is unusually high, the relief valve should be checked. This is so that it can be set to limit the pressure to required conditions without relieving too often. Drains for relief valves are piped to deck drains or to the bilge through traps and check valves.

FLOW CHECK VALVES

21-54. Flow check (swing check) valves are installed in the circulating supply mains of all heating systems used for heating quarters and potable water. These valves are of the vertical, angular, or horizontal types. When used with a gauge valve on the return end of the heating loop, a flow check valve permits the boiler to be drained for inspection or repair without draining the entire system.

CIRCULATING PUMP

21-55. The circulating pump is usually the centrifugal type. On small craft where heating is controlled by a room thermostat, the pump is operated automatically. In some cases, it is more practical to start and stop the pump manually. On large vessels, the pump runs continuously during cold weather and is stopped and started manually or automatically by an outside thermostat. In some installations, a standby pump is provided so that the regular pump can be inspected and repaired without having to shut down the heating system. Cutout valves are provided so that the pump can be repaired without draining the system.

FLOW FITTINGS

21-56. Flow fittings are of two general types (scoop and venturi) (see Figure 21-14). The scoop fitting is used in the supply branch and the venturi fitting in the return branch of the heating system.

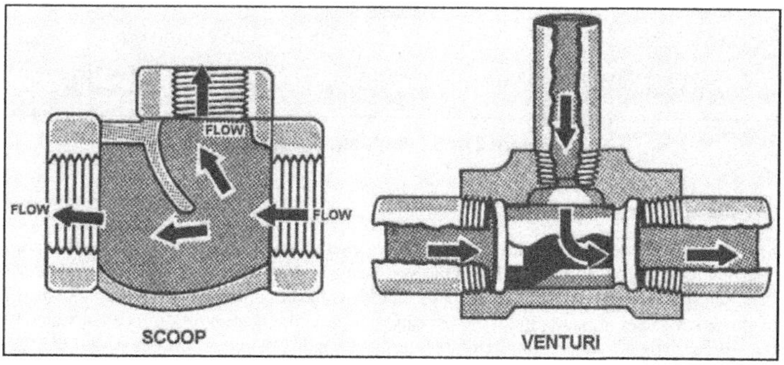

Figure 21-14. Flow Fittings

HEATING EQUIPMENT

DUCT HEATERS

21-57. Duct heaters are installed in duct systems and are used wherever feasible because of savings in weight, space, and piping. These heaters are built to withstand considerable shock and have standard connections to simplify piping. Steam flows through copper tubes which are arranged in a single row. Figure 21-15 shows ventilation heaters with two types of tube or coil arrangements. Two different types of coil arrangements are used. Figure 21-15, view B shows the "S" arrangement with the tubing serpentined. The S-type is used in small heaters, but it is not efficient for large heaters. Figure 21-15, view C shows a "T" arrangement which has a copper tube within a tube (a 3/8-inch distributing tube inside a 5/8-inch outer tube). Steam pressures up to 150 psi may be used in these heaters. Large ventilation supply systems, except those serving machinery spaces, have a pre-heater at or near the weather intake. By locating the heater near the intake, the duct temperature is kept high enough to avoid condensation during cold weather operation. In the majority of the systems that use a pre-heater, there is also a reheater. In circulating cooling systems, reheaters maintain specified space temperatures during cold water operation. Reheaters supply either single spaces or zones. Zones are made up of spaces that are expected to have similar heat loads. Reheaters are controlled by room thermostats. In recirculating cooling systems in newer ships, reheaters use a constant supply of steam to reheat air to any zone which would be overcooled. In small ventilation systems with short duct runs where one heater will produce the required temperature rise, a combination heater is used alone. This heater has two separate heating coils in one housing. Each coil has its own steam supply and is controlled by an individual thermostat. Combination heaters may also be used with reheaters. This is to supply more than one space where the combination heater is sufficient to maintain the required temperature rise in one of the spaces.

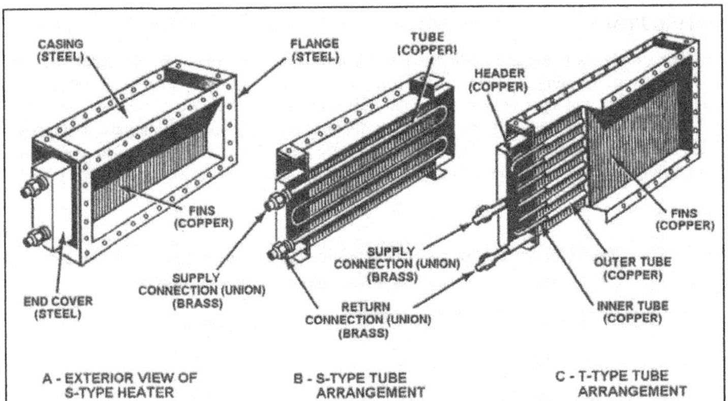

Figure 21-15. Ventilation Heaters

CONVECTOR HEATERS

21-58. Convector heaters (see Figure 21-16) are installed in small spaces or in spaces that are not fitted with mechanical supply ventilation. These heaters have a high heating capacity for their size and weight. They are considerably smaller than radiators or pipe coils of the same capacity. They will also withstand severe shock. A steam pressure up to 150 psi can be used in the heaters or a forced hot water system can be used. When they are used with steam pressure between 25 psi and 50 psi, temperature differentials at different levels in the room are reduced. Heating is regulated by the air bypass damper in the front part of the heater. The cabinets are generally of steel, though they may be of nonmagnetic stainless steel, copper, or aluminum.

UNIT HEATERS

21-59. Unit heaters (Figure 21-17) are self-contained heating units comprised of a fan, fan motor, heating coil, and adjustable louvers. They are used in special cases. For example, they are used when the amount of supply ventilation is too small to provide enough heat through ventilation heaters. Also, where there is no mechanical ventilation supply and heat requirements exceed the capacity of convector heaters. They can be used with either steam pressure up to 150 psi or with forced hot-water systems. Composite parts of the heaters include the heat transfer surface (fins and tubes), fan thermostatic control valve, strainer, trap, and directional louvers.

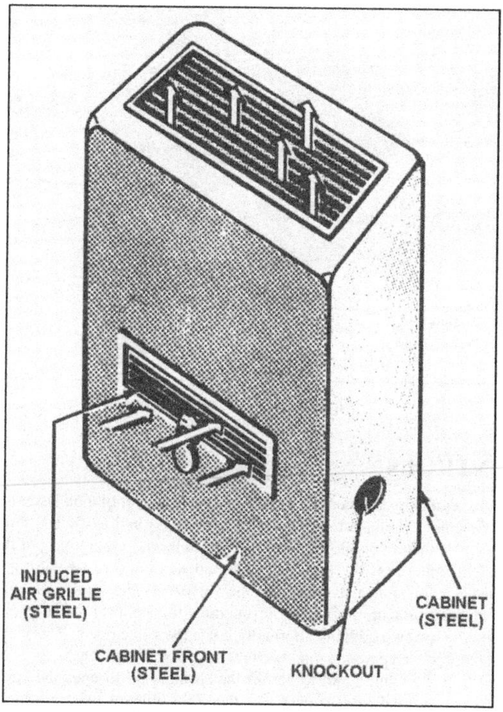

Figure 21-16. Convector Heater

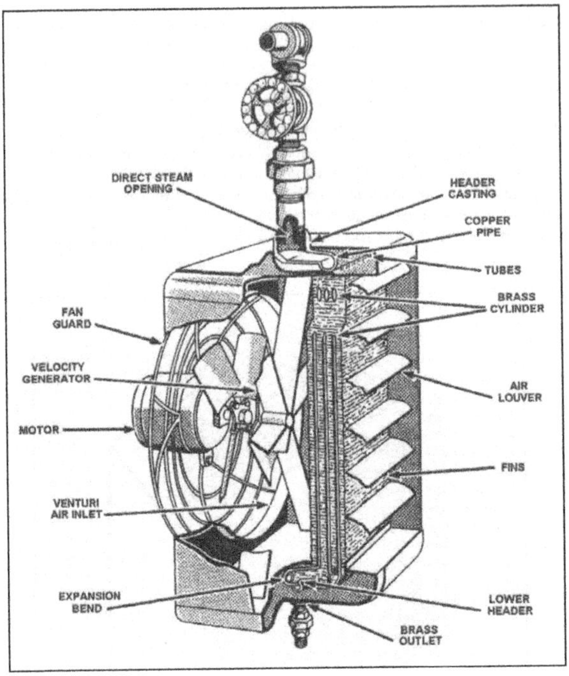

Figure 21-17. Unit Heater

HEATING CONTROLS

21-60. Temperature regulators that control the flow of steam to ventilation heaters and unit heaters consist of a thermostatic assembly (bulb, tubing, and motor bellows) and a valve assembly. The thermostatic assembly contains just enough volatile liquid to fill the bulb and tubing but not the motor bellows in the valve. A rise in temperature on the bulb causes liquid to flow through the tubing and into the motor bellows located in the hot-chamber assembly of the valve. As the liquid enters the motor bellows, it is vaporized by the temperature of the steam surrounding the bellows. The pressure (developed in the bellows) overcomes the spring load which normally holds the valve open. This causes a gradual closing of the valve. A decrease in temperature on the bulb reverses the process. It allows vapor to leave the bellows, to condense in the tubing, and to permit the spring load to open the valve. In this manner, the valve will open and close gradually to pass just the right amount of steam to react to the changes in temperature around the thermostatic bulb.

THERMOSTAT

21-61. The thermostatic bulb, tubing, and motor bellows are integral with the thermostat and they cannot be separated. When any part of the assembly is damaged, the entire assembly must be replaced. The tubing connects the bulb to the motor bellows and is armored for protection. The types of thermostats designed for use with ventilation and unit heater regulators are the R, L, and W thermostats. Each is thermostat is described below.

Type R Thermostat

21-62. The type R thermostat is shown in Figure 21-18. It is designed for mounting on a bulkhead or on a stanchion within the space served by a combination heater, a reheater, or a unit heater. The temperature regulator of a type R thermostat is adjusted by rotating the adjusting knob which extends or contracts the adjusting bellows. Clockwise rotation of the knob extends the adjusting bellows. This decreases the liquid capacity of the bulb. It thereby forces liquid into the motor bellows and closes the valve. The valve will not reopen until a lower temperature at the thermostat acts on the bulb to provide space for the previously ejected liquid. Reverse rotation of the adjusting knob raises the temperature setting by increasing the liquid capacity of the bulb. The type R thermostat is used with valves governing the flow of steam to combination heaters, reheaters, and unit heaters.

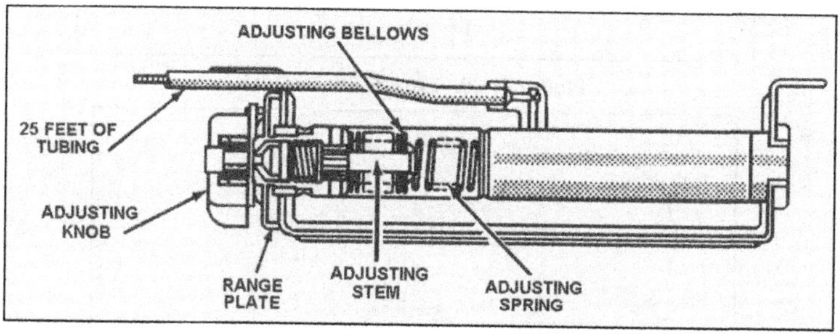

Figure 21-18. Type R Thermostat

Type L Thermostat

21-63. The type L thermostat is shown in Figure 21-19. It is a duct-mounted, adjustable thermostat used with valves governing the flow of steam to pre-heaters. This thermostat is flange-mounted in the duct, 4 to 6 feet beyond the pre-heater. The main control element of the bulb extends into the air stream inside the duct. The type L thermostat operates in the same manner as the type R thermostat.

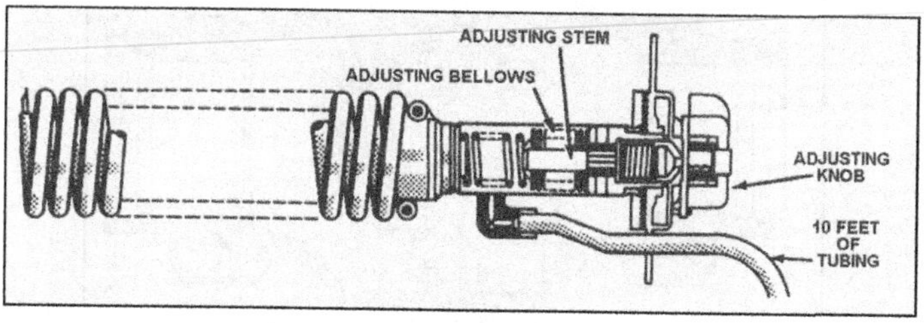

Figure 21-19. Type L Thermostat

Type W Thermostat

21-64. The type W thermostat shown in Figure 21-20 is nonadjustable. It is preset at the factory to be fully open when the air temperature at the thermostat drops to 35° F. The thermostat may be located in the airstream ahead of the pre-heater or in any other location where its operation will be governed by the weather. When the thermostat is located on the weather decks, it must be shielded from the effects of the sun (or other heat sources) by a bulkhead or deck. The type W thermostat is used with the Model D valve shown in Figure 21-21. It functions to prevent freezing of the condensate in the pre-heater tubes. This is done by allowing the valve to admit steam to the pre-heater when the temperature of the incoming air drops below 35° F or 1.7° C.

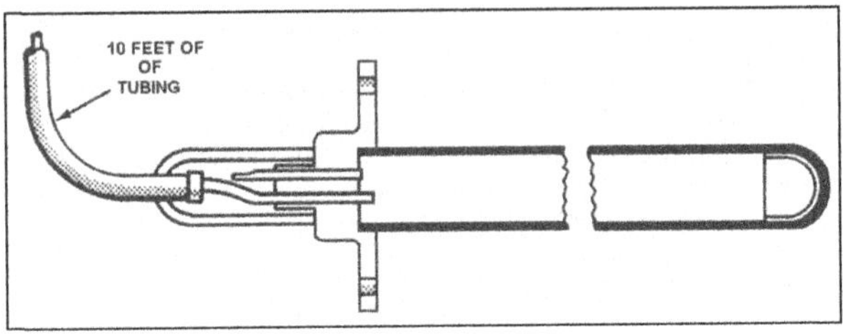

Figure 21-20. Type W Thermostat

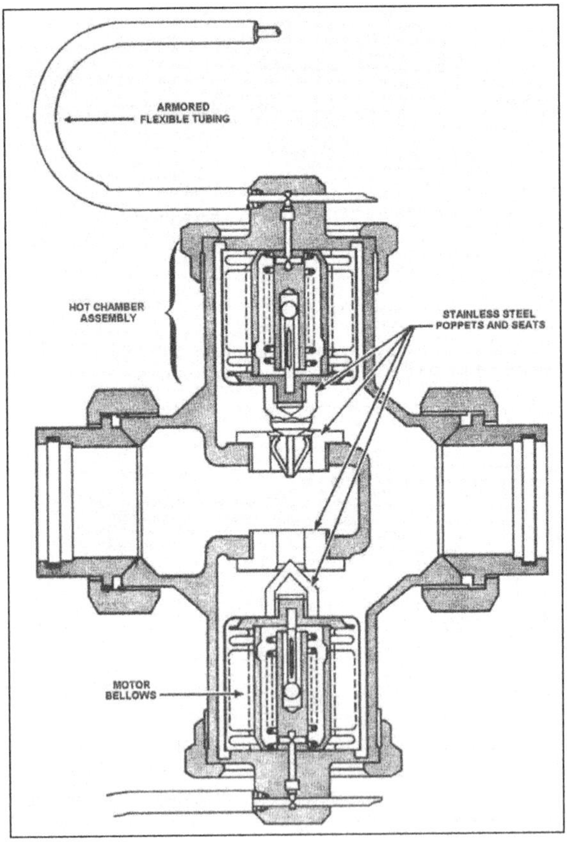

Figure 21-21. Model D Valve

VALVES

21-65. Three types of valves designed for use with ventilation and unit heater regulators are the E, G, and D models. Use of the model E, G, and D valves is determined by the designated capacity of the heaters. Each is valve is described below.

Model E Valve

21-66. The model E valve (Figure 21-22) is available in sizes to govern steam flow in the range of 5 to 350 pounds of steam per hour. The model E valve has a needle type poppet and seat. It can be used with a type W, L, or R thermostat to regulate steam flow to a ventilation or unit heater.

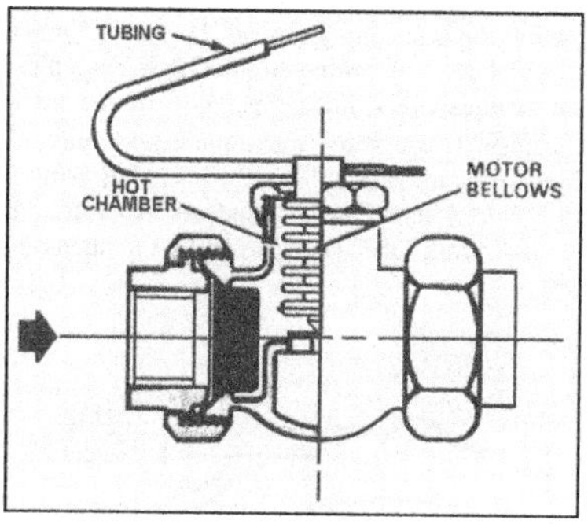

Figure 21-22. Model E Valve

Model G Valve

21-67. The model G valve (Figure 21-23) is similar to the model E valve. The exception is that it has an additional bellows (balancing bellows) located under the valve poppet and connected to the valve stem. This valve is available in various sizes to handle steam capacities in the range of 450 to 900 pounds of steam per hour. It is designed to be used with a type R or type L thermostat to regulate steam flow in a ventilation heater.

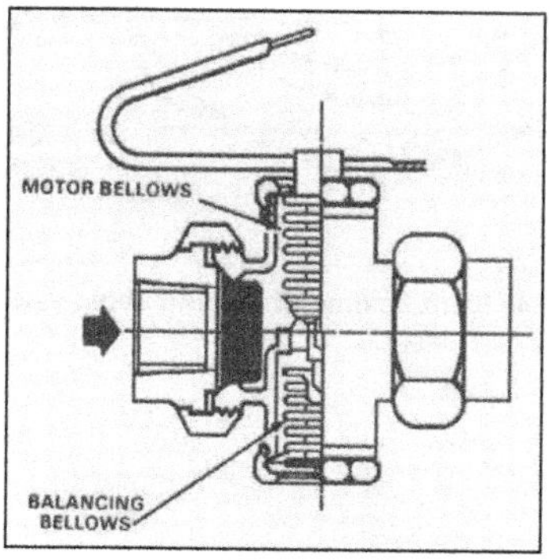

Figure 21-23. Model G Valve

Model D Valve

21-68. The model D valve (Figure 21-21) is actually two model E valves in a common valve body. It is intended for use only on pre-heaters or combination heaters. One of the valves is sized to pass about 75 percent of the load rating of the model D valve. It is actuated by a type L thermostat when used with a pre-heater or by a type R thermostat when used to regulate steam flow in a combination heater. The other valve is sized to pass about 25 percent of the load rating of the model D valve. It is actuated by a type W thermostat, when the temperature of the incoming air drops below 35° F or 1.7° C. This is to make sure there is enough steam in the pre-heater or combination heater to prevent freezing of the condensate in the heater tubes. The capacity of the model D valve is the combined capacity of two E-type valves. For example, a model D valve intended for use with a heater capacity of 40 pounds of steam per hour would use two E-type valves. The one for the W thermostat side would be rated at 10 pounds of steam per hour. The one for the L or R thermostat side would be rated at 30 pounds of steam per hour. The temperature regulator is adjusted by rotating the adjusting knob on the type L or type R thermostat to the desired temperature shown on the range plate. It is calibrated in degrees Fahrenheit. After 5 minutes of operation of the heater at the desired setting, check the temperature of the air in the space with a calibrated thermometer. If further adjustment is necessary, move the setting only 1° F or 2° F (.5° C to 1° C) and allow another period of 5 minutes to elapse before rechecking with the thermometer. Information concerning valve and thermostatic assemblies is found on the valve bonnet. This is stenciled, (see Figure 21-24) with numerals indicating tube length, operating temperature of the motor bellows, poppet number, and steam pressure.

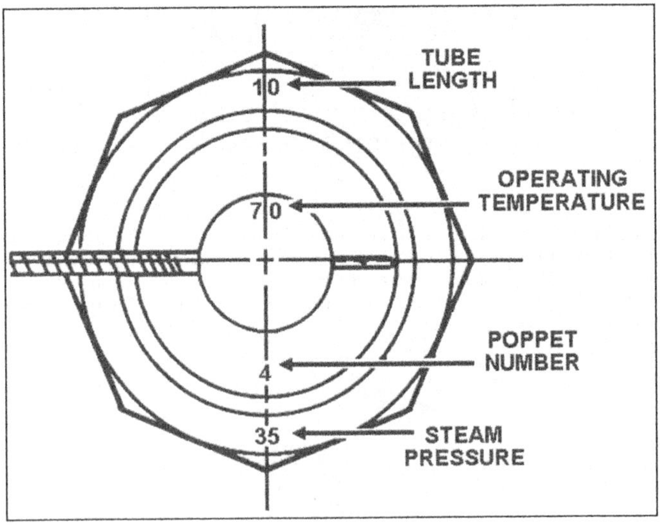

Figure 21-24. Identification on Thermostatic Valve Bonnets

MAINTENANCE OF VENTILATION EQUIPMENT

21-69. Shipboard ventilation must serve not only to supply, circulate, and distribute fresh air but be also to remove the used, contaminated, and overheated air from various spaces. If ventilation equipment fails to perform its functions properly, conditions may be created which will jeopardize the health or life of crew members. Therefore, individuals responsible for inspection and maintenance must be thoroughly familiar with the ventilation equipment. A shipboard ventilation system and its constituent parts cannot be isolated and separated from other component systems in a complete air conditioning system. For example, the air duct distribution system of a ship may be used for other systems in cooling, heating, and dehumidifying the ship's atmospheric air. In addition to ducts, a ventilation system may include the following:

- Weather openings.
- Screens.
- Filters.
- Fans.
- Gratings.
- Closures.
- Heaters.
- Cooling coils.
- Venturi tube.
- Dampers.
- Terminals.

If a ventilation system is to function effectively, all of its various units must be kept clean and in satisfactory operating condition. To maintain a ventilating system in the best condition applicable precautionary measures and prescribed maintenance procedures must be adhered to.

GUARDING AGAINST OBSTRUCTIONS TO VENTILATION

21-70. Such items as swabs, deck gear, and trash stowed in fan rooms or ventilation trunks not only restrict airflow but also increase dirt and odors taken inboard. Ventilation terminals must never be used for stowage. Wet clothing secured to ventilation terminals increases moisture content of the compartment air and restricts airflow. Stowage arrangements should be such that ventilation weather openings are never restricted.

KEEPING THE SYSTEM CLEAN

21-71. Dirt accumulation in a ventilation system not only restricts the flow of air but also creates a serious fire hazard. In a clean duct the cooling effect of the metal tends to act as a flame arrester. However, an accumulation of foreign matter within a duct becomes a potential source of combustion. One method of reducing the amount of dirt and combustible matter which may be carried into a ventilation system is to wet down the areas near the air intakes before sweeping. Since a great volume of air passes through or over the elements of a ventilation system, dirt will collect in the various units in spite of precautionary measures. The greatest accumulation of dirt will be within trunks and ducts where it is not readily noticeable. Therefore, periodic inspections and a definite service procedure are necessary to keep the system clean.

STORAGE WATER HEATERS SYSTEMS

21-72. There are two types of storage water heaters systems (direct and indirect). Each system is described below.

DIRECT SYSTEM

21-73. Heaters for direct heating of potable water are generally furnished as a unit with the tank. The tanks are of galvanized or black steel of welded construction with copper heating coils.

INDIRECT SYSTEM

21-74. In the indirect system, the heater is made with cast iron shells and copper coils. Heating water from the boiler is circulated through the shell. Water to be heated is circulated through the tubes or coils. Since freshwater collects sediment, the coils can be washed out and kept in proper condition by opening the drain valve periodically.

STEAM TRAPS

21-75. Steam traps are installed on drains of all heating units. Steam traps are automatic valves that release condensed steam (condensate) from a steam space while preventing the loss of live steam. They also remove non-condensable gases from the steam space. Steam traps are designed to maintain steam energy efficiency for performing specific tasks such as heating a building or maintaining heat for process use. Once steam has transferred heat through a process and becomes hot water, it is removed by the trap from the steam side as condensate and either returned to the boiler via condensate return lines or discharge to the atmosphere.

TYPES

21-76. The steam traps used are of five general types (thermostatic, float, float-thermostatic, impulse, and inverted bucket). Each type is described below.

Thermostatic

21-77. A thermostatic trap usually operates through a bellows filled with a volatile fluid. These traps are made so that under any pressure, the valve will remain closed and will not open until the condensate is cooled to at least 15^{0} to 20^{0} below the temperature of the steam. This type of trap is more efficient for radiators when they are directly connected.

Float

21-78. The snap-action, ball-float trap is used where sediment prevents a direct-action trap from seating tightly. They are also used for light loads where tight closing is required. All parts are made of heat-treated, precision-machined, stainless steel. This type trap is designed to vent air from hot water heating systems, water storage tanks, centrifugal pumps, and fuel lines. Operation of the trap is completely automatic.

Float-Thermostatic

21-79. The float-thermostatic regulates the flow of condensate by the action of the float. As long as there is sufficient condensate in the trap to hold up the float, the valve is kept open and condensate can flow out. The thermostatic element of this trap operates the same as the element of the ball-float trap. It will operate only when the temperature is well below the steam temperature. The purpose of this type trap is to permit any air that may accumulate in the system to pass on into the drain. The float-thermostatic trap is suitable for pre-heaters, hot-water heaters, and other equipment having high steam capacities. It also provides an almost continuous flow of condensate.

Impulse

21-80. The impulse trap is suitable for high capacities. It works on variations in vapor pressure in the control chamber of the valve. Therefore, the condensate must be cooled considerably before the valve will open. It does not operate satisfactorily in low-pressure systems.

Inverted bucket

21-81. With the inverted bucket trap, the condensate enters the bottom of the trap, passes through the strainer, and gradually fills the bottom of the trap. Air in the bucket is driven out through a small hole, and the valve remains wide open allowing air and water to be discharged through the outlet. As soon as steam enters the trap, it causes the bucket to become buoyant, and the bucket rises and closes the valve. The valve will remain closed until enough condensate has again accumulated to cause the bucket to sink and open the valve. The strainer is easily removed for cleaning by unscrewing the plug. These traps are suitable for large capacities and are used on hot-water tanks, steam tables, and pre-heaters.

Steam Trap Maintenance

21-82. Steam trap maintenance may be performed as follows:

- **Purpose.** For maximum trap life and steam economy, a regular schedule should be established for preventive maintenance and testing. Trap size and operating pressure are important to determine how frequently traps should be checked and tested.
- **Testing.** Low-pressure traps (up to 60 pounds) should be tested monthly.
  - *Pyrometer Testing.* File a clean spot on both the inlet and discharge lines. If the temperature of the trap is at or near true steam temperature and the temperature of the trap discharge is at or near a temperature corresponding with steam pressure in the common return line, the trap is functioning satisfactorily. Intermittent discharges can be detected by a rise and fall of the needle on the pyrometer when the thermocouple is contacting the discharge line. If the discharge line temperature is nearly as high as the trap inlet temperature, the trap is blowing through.
  - *Listening Device.* Hold one end of a steel rod against the trap cap and the other end against your ear. Intermittent discharges can be clearly heard. Excessive rattle indicates loss of prime. Considerable experience is required for this method of testing, since other noises are telegraphed along the pipelines.
  - *Feel Test.* Wearing canvas gloves, place one hand on the inlet line and the other hand on the discharge line. A great variation in temperature will be quickly apparent, indicating proper trap operation. Equalized temperature suggests prime is lost.
- **Maintenance.** When the trap fails to operate and the reason is not readily apparent, the discharge from the trap should be observed. If the trap is not installed with a test outlet, break the discharge connections. See Table 21-1 for a list of troubles, probable causes, and remedies of steam traps.

Table 21-1. Troubleshoot Steam Traps

TROUBLE	PROBABLE CAUSE	REMEDY
Cold trap, no discharge	Pressure too high.	Replace trap or replace pressure charge assembly trap.
	Defective trap.	Replace trap.
	Clogged piping to trap.	Repair or replace piping.
	Trap body filled with dirt.	Clean trap.
Hot trap, no discharge	Leaky bypass valve, broken siphon line, or water heater coils.	Repair or replace defective component.
Steam loss	Defective trap.	Replace trap.
	Incorrect installation.	Check installation.
Continuous flow	Trap too small.	Install larger trap or an additional trap in parallel.
	Boiler foaming.	Repair as necessary.
Sluggish heating	Trap too small.	Install larger trap.
	Incorrect trap installed.	Install correct trap.
	Insufficient traps installed.	Install additional traps.

Chapter 22

AUXILIARY EQUIPMENT

INTRODUCTION

22-1. In addition to all the machinery described in previous chapters of this manual, a number of other units of machinery are essential to the operation of a ship. They directly or indirectly concern you as an engineman. Such auxiliary machinery includes steering gears, anchor windlasses, deck winches, and capstans. Some of this machinery may be located within the engineering spaces of the ship, but many of the units are located outside the engineering spaces. This chapter provides information on auxiliary machinery with which you, the engineman, will be primarily concerned (steering systems). See FM 55-501 for more information on deck machinery (such as windlasses, winches, and capstans).

Note: FM 55-501 is currently under revision and is projected to be replaced with TC 55-501.

STEERING DEVICES

22-2. The direction of movement of a ship is controlled partially by steering devices which receive their power from steering engines. It is also controlled partially by the arrangement, speed, and direction of rotation of the ship's propellers. The steering device is called a rudder. The rudder is a more or less rectangular metal blade (usually hollow on large ships) which is supported by a rudder stock. The rudder stock enters the ship through a rudder post and a watertight fitting, as shown in Figure 22-1. A yoke or quadrant, secured to the head of the rudder stock, transmits the motion imparted by the steering mechanism. A ship's rudder is basically used to attain and maintain a desired heading. The force necessary to do this is developed by dynamic pressure against the flat surface of the rudder. The magnitude of this force and the direction and degree to which it is applied produces the rudder effect which controls stern movement. It therefore controls the ship's heading. To function most effectively, a rudder should be located aft of and quite close to the propeller. Many modern ships have twin rudders, each set directly behind a propeller to receive the full thrust of water. This arrangement tends to make a ship highly maneuverable. Three types of rudders in general use are the balanced, the semi-balanced, and the unbalanced. These three types are shown in Figure 22-2.

STEERING SYSTEMS

22-3. The engineering section is responsible for daily maintenance of the steering system. Yet, due to the many different designs and types of systems, this chapter gives only a brief description of how the overall system works. You must refer to applicable technical manuals (TMs) and lubrication orders (LOs) for the vessel that you are working in before attempting to check, maintain, or repair the steering system. The steering systems installed aboard Army vessels vary depending on the age, class, and type of vessel. The steering system may be manual, hydraulic, telemotor, or electrohydraulic. Some of the older small craft may still use the manual steering system. Landing craft use hydraulic steering. The newer and larger Army watercraft uses either the electrohydraulic or the hydraulic telemotor steering system.

MANUAL STEERING SYSTEM

22-4. The simplest type of manual steering (Figure 22-3) consists of a steering wheel with a drum on which a tiller cable winds and unwinds as the wheel is tuned by the helmsman. This causes a corresponding movement of the tiller quadrant, which is rigidly fastened to the rudder stock. The tiller cables are arranged on the drum so that turning the wheel in one direction, toward starboard for instance, causes a shortening of the starboard side. Turning the wheel in the opposite direction swings the bow of the vessel toward the portside. The tiller lines are supported on sheaves, which also serve to reduce cable friction. To maintain adequate tension on the cables, turnbuckles are sometimes used to take up any slack likely to occur. This is to ensure movement of the quadrant as at the steering wheel. In maintaining the system, the engineman is responsible for inspecting and ensuring the following:

- Steering cable is lubricated and not frayed.
- Sheaves in the fairleads are aligned and that they rotate freely.
- Sheaves, rudder post, and quadrant are lubricated.
- Turnbuckles are tight and that there is no slack in the cable.

In the event the sheaves in the fairleads are "frozen" or the steering cable is frayed, worn, or damaged, it should be reported immediately to the coxswain and a note made of it in the logbook.

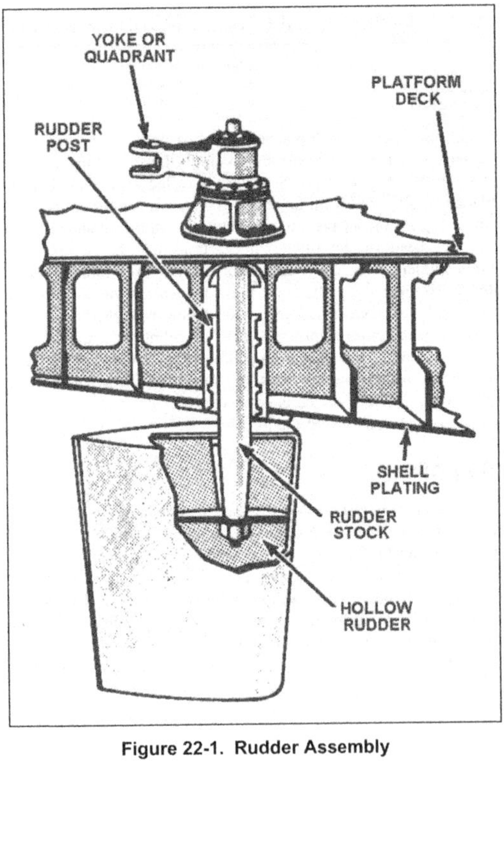

Figure 22-1. Rudder Assembly

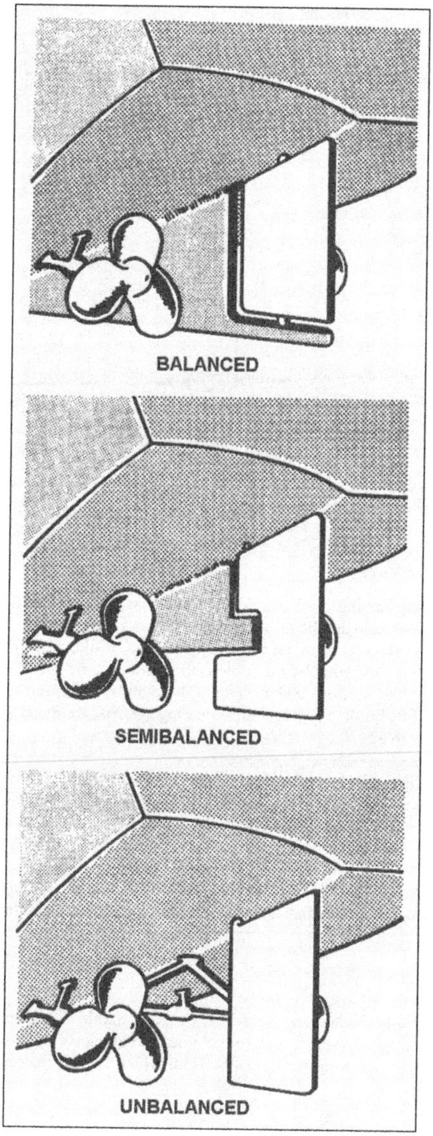

Figure 22-2. Types of Rudders

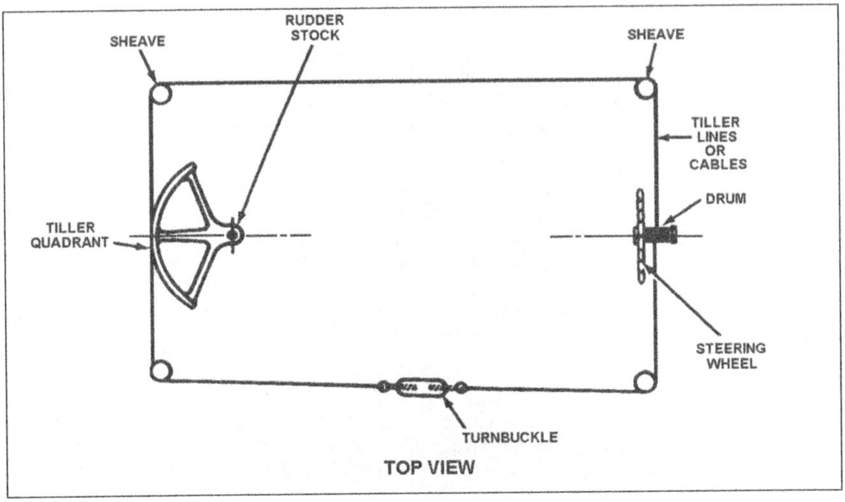

Figure 22-3. Manual Steering System

HYDRAULIC STEERING SYSTEM

22-5. As stated earlier, landing craft use the hydraulic steering system (Figure 22-4). This steering system uses medium pressure hydraulic oil to activate cylinders which position the rudders. Oil is supplied by the hydraulic pumps to the helm unit. The helm unit is the principle metering and directional controlling device. When the hydraulic oil is directed to one side or the other of the cylinders, they will extend or retract giving the desired rudder angle. The pumps, helm unit, and other valves control the direction and volume of hydraulic oil. Hydraulic steering systems are made up of devices such as pumps, reservoirs, and various valves.

Note: The steering system for LCM hull numbers 8500 through 8519 is designed to be supplied by one pump. Using both pumps will only cause a double bypass flow resulting in excessive heating of oil. The steering system pump discharge valves should be set with one valve open and one valve closed. Steering systems for hull numbers 8520 through 8560 and 8580 through 8618 is designed to use the flow from both pumps. However, the system will function with only one pump operating. This series of landing craft use a closed system. Return lines from the actuating cylinders and counterbalance valves return hydraulic fluid to the steering reservoir.

Pumps

22-6. The flow of fluid to the system is provided by two variable displacement axial piston pumps, one mounted on each engine. Each pump regulates volume delivery according to system flow demands. The flow from the pump is ported to a manifold line from which branch lines lead off to various actuating systems.

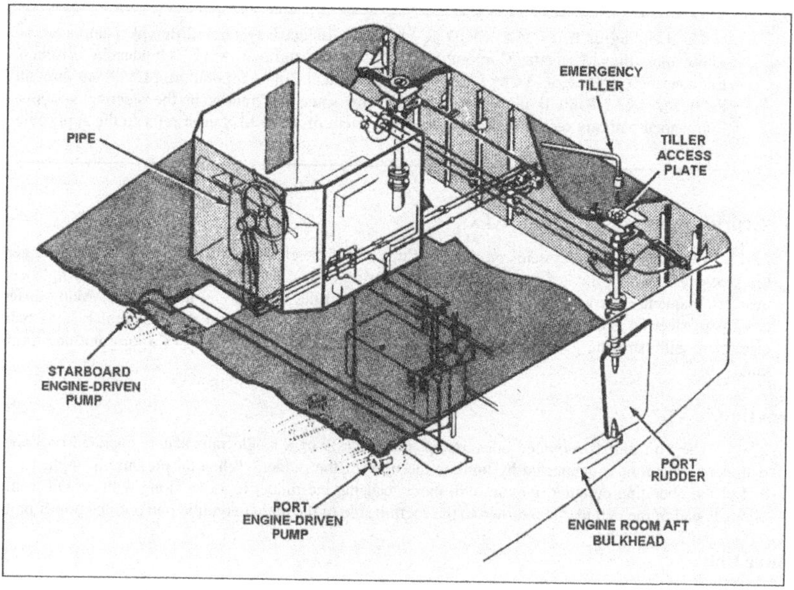

Figure 22-4. Hydraulic Steering System Functional Diagram

Reservoir

22-7. The reservoir is the source from which the hydraulic pumps receive their supply of fluid. It is also a storage container for the return fluid displaced by the actuating components. The reservoir is fluid-pressurized. This ensures a supply of fluid to the pumps at all times.

Check Valves

22-8. These are installed in the system return lines to direct return flow to the reservoir. Also, they prevent pressure from acting against the return ports of other system components.

Relief Valve

22-9. This valve is to be set for a maximum system pressure of 1,050 psi.

Flow Divider Valve

22-10. This valve is a preset priority type supplying a constant 2 gpm to the helm unit. Excess oil is bled back to the reservoir.

Counterbalance Valves

22-11. The counterbalance valves, of which there are two, induce an artificial pressure on the low pressure side of the steering cylinders. This is to prevent the rudder from running ahead of the pumps. These valves are set at 1,000 psi.

Note: Landing craft, LCM-8, MOD 1, has been produced by three different manufacturers. These are the Marinette Corporation, LCM 8500 through 8519; Gunderson Brothers Engineering Corporation, LCM 8520 through 8539; and Rohr Corporation, LCM 8540 through 8560 and LCM 8580 through 8618. There are some differences in the steering systems. Enginemen working on the hydraulic steering system of an LCM-8 must refer to the applicable TM.

ELECTROHYDRAULIC STEERING GEAR

22-12. Most steering gear systems on modern ships are of the electrohydraulic type. The steering gear is the mechanism which transmits power from the steering engine to the rudder stock. The term "steering gear" is frequently used to include the driving engine and the transmitting mechanism. Many different designs of steering gear are in use, but the principle of operation for all of them is similar. One type of electrohydraulic steering gear is shown in Figure 22-5. It consists essentially of a ram unit and a power unit.

Ram Unit

22-13. The ram unit is mounted athwartship and consists of a single ram that is operated by opposed cylinders. The ram is connected by links to the tiller of the rudder. When oil pressure is applied to one end of the operating cylinder, the ram will move, causing the rudder to move along with it. Oil from the opposite end of the cylinder is returned to the suction side of the main hydraulic pump in the power unit.

Power Unit

22-14. The power unit consists of two independent pumping systems. Two systems are used for reliability. One pump can be operating while the other is on standby. Each pumping system consists of a variable-delivery, axial-piston main pump and a vane auxiliary pump. Both are driven by a single electric motor through a flexible coupling. Also included in each system is the following:

- Transfer valve with operating gear.
- Relief valves.
- Differential control box.
- Trick wheel.

The whole unit is mounted on a bedplate which serves as the top of an oil reservoir. As stated previously, steering power is taken from either pumping system acting alone. The pumps of the power unit are connected to the ram cylinders by high-pressure piping. The two transfer valves are placed in the piping system to allow for the lineup of one pump to the ram cylinders with the other pump isolated. A hand lever and mechanical linkage (not shown) are connected to the two transfer valves so that both valves are operated together. This allows for rapid shifting from the on-service pumping unit to the standby unit. It also prevents lining up both pumps to the ram at the same time. The hand lever, usually located between the trick wheels, has three positions marked P, N, and S.

- P denotes the port pump connected to the ram.
- N denotes neutral (neither pump connected to the ram).
- S denotes the starboard pump connected to the ram.

The hand lever is also usually connected to motor switches. This permits the operator to connect the selected pump to the ram and start the pump drive motor in one operation.

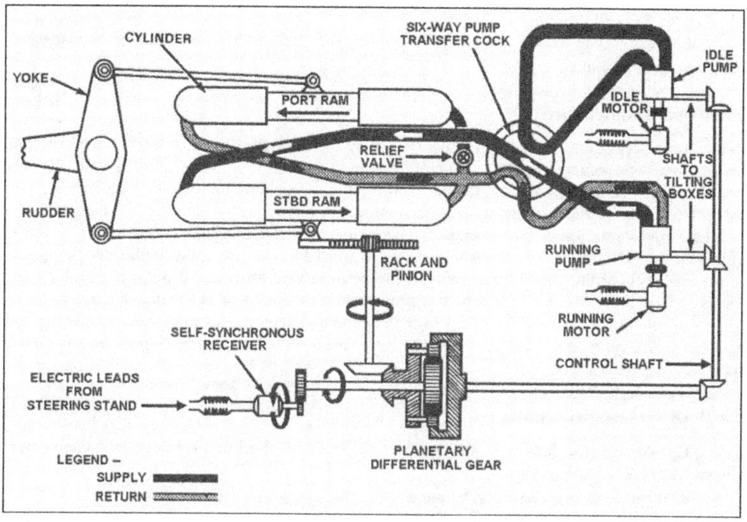

Figure 22-5. Electrohydraulic Steering Gear

Principles of Operation

22-15. The on-service hydraulic pump is running at all times and is a constant-speed pump. Unless steering is actually taking place, the tilt box of the main hydraulic pump is at zero stroke and no oil is being moved within the main system. The auxiliary pump provides control oil and supercharge flows for the system. When a steering order signal comes into the differential control box (from either the remote steering system in the ship's wheelhouse or the trick wheel), the control box mechanically positions the tilt box of the main hydraulic pump to the required angle and position. Remember that fluid and flow may be in either direction in a hydraulic speed gear depending on which way the tilt box is angled. For this reason, the constant speed, unidirectional motor can be used to drive the main hydraulic pump. The pump will also still have the capability to drive the ram in either direction. With the main hydraulic pump now pumping fluid into one of the ram cylinders, the ram will move, moving the rudders. A rack and gear are attached to the rudder yoke between the rudder links. As the ram and rudder move, the rack gear moves, driving the followup pinion gear. The pinions drive followup shafts which feed into the differential box. This feedback or servo system tells the differential control box when the steering operation has been completed. As the ordered rudder angle is approached, the differential control box will begin realigning the tilt box of the main hydraulic pump so that by the time the desired rudder angle is reached, the tilt box is at zero stroke. This means that the ordered signal (from the steering stand in the ship's pilothouse or trick wheel) and the actual signal (from the followup shafts) are the same. If either of these changes, the differential control box will react accordingly to cause the main hydraulic unit to pump oil to one end or the other of the ram. The trick wheels are provided for local-hydraulic control of the steering system in case of failure of the remote steering system. A hand pump and associated service lines are also provided for local-manual operation of the ram in case of failure of both hydraulic pump units. Advantages of the electrohydraulic steering gear are as follows:

- Little friction and inertia of moving parts (such as in heavy differential screws and gears).
- Lower power consumption peaks.
- Sensitive response, with little lag, to movement of the steering wheel.
- Small deck space and headroom required.

- Savings in weight.
- Flexibility in the arrangement of hydraulic cylinders, pumps, and control mechanisms.
- Dependability.

Operation and Maintenance

22-16. Actual operation of the steering equipment by the engineman watchstander is normally limited to abnormal and emergency situations. Therefore, you should be thoroughly familiar with all emergency procedures such as local-hydraulic steering with the trick wheel and local-manual steering with the hand pump. Operating instructions and system diagrams are normally posted near the steering gear. Diagrams describe various procedures and lineups for operation of the steering gear. Be sure that the standby equipment is ready for instant use. Maintenance required for most electrohydraulic steering gears is that the exposed parts of the steering gear rams be protected against water and damage from rolling or falling objects. Cover them with a thin film of rust-preventive compound or a heavy oil. To protect the exposed parts from rolling objects, place a guard over the parts and keep the steering gear compartment clear of loose gear. Always refer to the applicable TM for your vessel before starting to perform any maintenance on the steering engine.

CONTROL OF STEERING GEARS

22-17. The steering gear may be controlled from the steering wheel on the bridge by either of two of the following remote control systems:

- It can be done electrically, by means of an AC synchronous transmission system.
- It can be done hydraulically, by means of a telemotor system.

Synchronous Transmission

22-18. The AC synchronous transmission remote control consists of receiving and transmitting units which are similar to small motors. These units are connected to the same AC supply. When the transmitter motor is turned, the receiver motor turns at the same speed and in the same direction. The illustrated steering gears are equipped with this type of remote control. Only the receiver unit is shown in Figure 22-6. The transmitters are located in steering stands at remote control stations. They are mechanically connected through gearing to the steering wheels. A transmitter at each of the remote stations is electrically connected to a receiver in the steering room. The receiver is connected to the control shaft of the variable-displacement hydraulic pump through a differential. Where there is more than one remote steering station, a switch is provided for selecting the desired control station. Indicating lights are provided on the steering stands and at the selector switch. They indicate the selected circuit and also that power is available.

Electrohydraulic Telemotor Control

22-19. The electrohydraulic telemotor remote control shown in Figure 22-6 is found on some Army ships that are equipped with electrohydraulic steering engines. The system consists of the following:

- Steering console (hydraulic transmitter) in the pilothouse.
- Hydraulic receiver in the steering gear compartment.
- Connecting hydraulic tubing.

There is also an electric cable which connects the helm angle transmitter on the receiver housing with the helm angle indicator on the steering console. A hydraulic transmitter is located inside the steering console and under the steering wheel. The hydraulic transmitter components consist of the following:

- Pump.
- Hydraulic tubing.
- Two relief valves.
- Two check valves.

- Replenishing tank.
- Bypass valve.

The remotely located receiver is a hydraulic ram unit with two cylinders (one on each end of the receiver housing) in axial alignment. A double-acting plunger (ram) operates in the cylinders. On the middle portion of this plunger, a crosshead is connected for mechanical linkage to the steering gear control mechanism. The direction of the hydraulic fluid movement depends on the direction of rotation of the steering wheel. Rotating the steering wheel actuates bevel and spur gears. These gears in turn, operate a fixed displacement axial piston hydraulic pump. In this pump, the tilt or angle is set permanently at a fixed angle so that the pistons are always on stroke. When the pump shaft and cylinder barrel are rotated by the steering wheel, the pistons draw fluid in from one fluid line and discharges it to the other fluid line. Reversing the rotation of the steering wheel reverses the direction of fluid flow through the pump. The pump has external check valves and piping for replenishing the hydraulic system from the reservoir. Relief valves and a bypass valve are also included, as well as vents for purging air from the system. When the hydraulic pump shaft is rotated in one direction, the fluid output is discharged from one side of the pump to one of the receiver cylinders. In the other cylinder, hydraulic fluid is displaced to the hydraulic pump. Therefore, the hydraulic fluid under pressure moves the receiver plunger and produces a linear movement of the crosshead. This motion in turn is transmitted to the connected steering gear control mechanism. Travel of the crosshead and plunger is limited by adjustable stops on the receiver housing. Air valves and filling or charging connections are provided on the receiver cylinders for venting, filling, or purging the hydraulic system. Specific filling and purging instructions for telemotor systems should be obtained from applicable steering gear or telemotor manuals. These instructions vary, depending on the type of unit and the specific installations. Filling the telemotor system from the charging tank is done by opening the air cocks at the forward telemotor and starting the charging pump. After the oil has appeared at the air cocks, pumping is continued until the oil is free of bubbles. The air cocks are then closed and the oil is allowed to fill the replenishing tank. While the pump is still operating, the valve leading to the replenishing tank is closed and the system is subjected to a slight pressure. Effective operation of the telemotor system depends on the purging of all entrapped air and the elimination of leaks. Purging the system of air, under normal operating conditions, is done by opening the valve leading from the replenishing tank and opening the air cocks at the forward telemotor. Since the replenishing tank is located above the highest point of the telemotor system, air is forced out of the air cocks by gravity flow of the oil. The cocks are then closed when the oil flows smoothly without bubbles. Frequently inspect valves and joints to detect and eliminate leaks in the system. To correct a leaky piston in the internally packed telemotor, ensure that the leathers are in good condition. Also, make sure that the springs (if used) keep the leather in contact with the inside wall of the telemotor cylinders. To stop leaks in the externally packed telemotor, tighten the glands just enough to cause the packing to be compressed about the rams until the leak is stopped. Hydraulic fluid characteristics are especially important when the telemotor system is exposed to low ambient temperatures. Under such conditions, a high grade mineral oil having a cold pour point of -24° F to -40° F should be used. Where low temperatures are not involved or where special oil with the designated cold pour point cannot be readily obtained, the symbol 2075H oil may be used satisfactorily. The oil should have a low rate of expansion and be sufficiently viscous at 150° F to remain a good hydraulic fluid. When filling the charging tank with hydraulic fluid, strain oil into the tank through a funnel which has in it a fine mesh screen. This prevents the entry of foreign matter and at the same time prevents the entry of air bubbles into the hydraulic system. The oil should never be allowed to become contaminated with water.

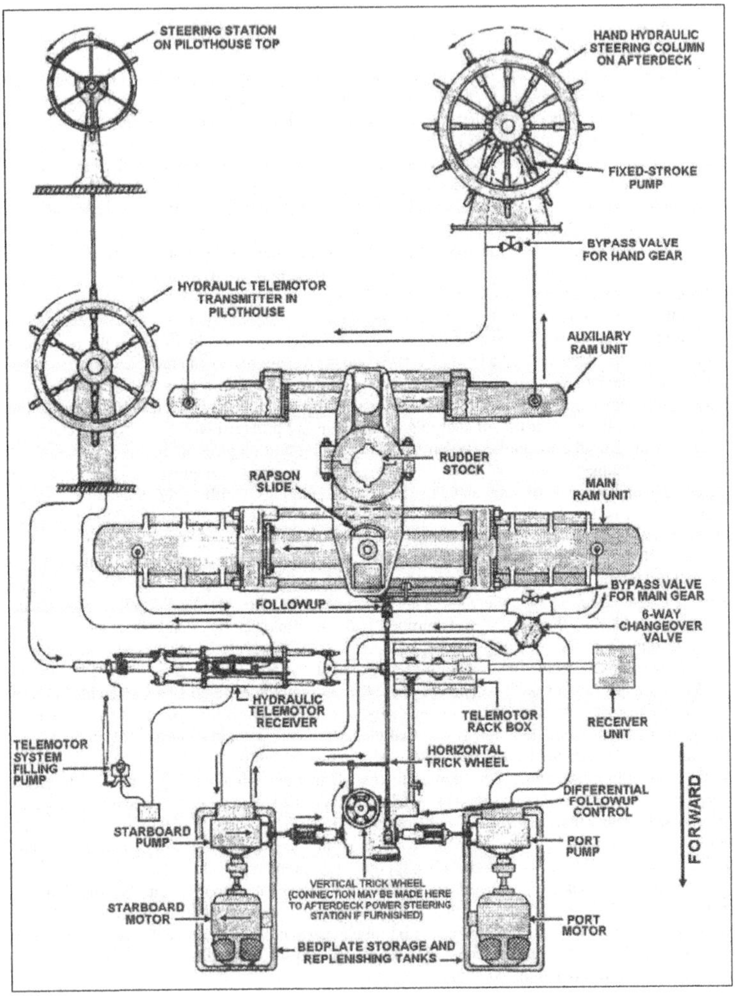

Figure 22-6. Electrohydraulic Steering Gear With Telemotor Control

Chapter 23

ENGINEER ADMINISTRATION, SUPERVISION, AND TRAINING

INTRODUCTION

23-1. Although it is possible to consider administration, supervision, and training as three separate areas of responsibility, remember that the three things cannot be totally separated. Much of your work requires you to administer, supervise, and train, all at the same time. This chapter discusses briefly some of your administrative and supervisory responsibilities. It also deals with certain aspects of your responsibility for training others.

ENGINEER ADMINISTRATIVE AND SUPERVISORY RESPONSIBILITIES

23-2. As a marine engineer, you will have administrative and supervisory responsibilities in connection with engine room operations and engine room maintenance and repair. For example, consider a pump overhaul job when performing the following:

- As an administrator, you will schedule the job, check on the history of the pump, and see that the required entries are made in appropriate logs and records.
- As a supervisor, you will actually oversee the work and make sure it is done correctly.
- As a trainer, you will provide information and instruction on repair parts, repair procedures and policies, safety precautions, and other matters.

These administrative, supervisory, and training tasks have a direct relationship to the job at hand (namely, overhaul of the pump). However, the pump overhaul job cannot get started unless a variety of administrative, supervisory, and training functions are performed on a continuing basis. Materials, repair parts, and tools must be available when they are needed. Jobs must also be scheduled in regard to the urgency of other work. Records must be kept and reports submitted. Personnel must also be in a continuous state of training so that they can assume increasingly important duties and responsibilities. To keep things running smoothly you must take your administrative, supervisory, and training responsibilities seriously. The engineering department administrative organization is set up to provide for proper assignment of duties and supervision of personnel. However, no organization can run itself. All personnel are needed to see that all pertinent instructions are carried out. Personnel will also ensure that all machinery, equipment, and piping systems are operated in accordance with good engineering practice. You must see that operating instructions and safety precautions are posted by machinery and obeyed by all engine room personnel. You must make sure that all watch-standers are properly supervised and that records and reports are filled in correctly and submitted as required. You must ensure that the entire engineering plant is operated with maximum reliability, efficiency, and safety. The experienced engineer will soon learn that administration and supervision are closely related. The administrator's and supervisor's role requires leadership and common sense to earn the respect of others. Many decisions are difficult to make. To be effective in decision-making, the supervisor assembles pertinent facts, policies, and procedures. He/She develops a course of action and evaluates possible alternatives. He/She then selects the best course of action and acts as required.

Records Administration

23-3. The basic source for information pertaining to forms and records that must be maintained on all equipment aboard ship is DA Pamphlet 750-8. These records are the equipment operational, maintenance, and historical records. This pamphlet gives instructions for use, preparation, maintenance, and disposition of records and reports for Army watercraft and amphibians.

Engine Logbook

23-4. It is mandatory that an engine logbook be maintained daily. It should give an account of all events occurring in the engine room each day. The log will also list the following:

- All repairs.
- All maintenance performed aboard ship.
- All repairs and maintenance to be performed by shore personnel.

The engine logbook will be maintained in accordance with AR 56-9 and DA Pamphlet 750-8.

Trash Logbook

23-5. All trash departing from the vessel is logged in a trash logbook. Log in what kind of trash, how bagged, and where disposed of in accordance with AR 56-9.

Oil Record Book

23-6. All class H-2and BGs will use the blank form located in CG-4602A. The information will include the following:

- Transfer of fuel oil (time, date, and amounts).
- Ballasting or cleaning of bunker fuel tanks.
- Disposal of oily residues.
- Other exceptional discharges of oil.

This information will be recorded in logbooks for class A-1, B-1 vessels, and other class C-1 vessels.

Night Order Book

23-7. This book is made up on a daily basis by the chief engineer. In it are listed things that are to be done during the night by the engine room watch. It will also list precautionary notes and specific problem areas that must be checked, maintained, or repaired. It is mandatory that when an engineer comes on watch, he/she read, initial, and comply with instructions that are entered for that day.

Maintenance and Repair

23-8. To fulfill your administrative and supervisory responsibilities in connection with maintenance and repair, you must have the ability to plan ahead. During an average workday, occasions will arise when personnel have to leave their working spaces for one reason or another, thereby delaying the completion of the scheduled work. Some delays can be anticipated and some cannot. Among those which can be anticipated are:

- Training schedules.
- Examinations.
- Meals.
- Watches.
- Other military duties.

Before making personnel work assignments, the engineer should determine what delays can be anticipated. It may be possible to arrange assignments so that work interruption is held to a minimum. When estimating the completion time of a maintenance task, the engineer should allow for these predictable delays. Some engine room maintenance and repair work just will not fit into a schedule but must be done whenever the opportunity arises. So, in addition to having the ability to plan, you must have a certain amount of flexibility so that you can alter your plans to fit existing circumstances. A few administrative and supervisory considerations that apply particularly to maintenance and repair are as follows.

Materials and Repair Parts

23-9. Responsibility for maintaining adequate stocks of repair parts and repair materials belongs at least as much to you as it does to marine maintenance. The duties of the marine maintenance officer are to procure, receive, stow, issue, and account for most types of stores required for support of the ship. However, marine supply is not the prime user of repair parts and repair materials. The initiative for maintaining adequate stocks of repair materials, parts, and equipment must come from personnel who are going to use them. However, a few points are noted here concerning identification of repair parts and materials. Identification is not usually any great problem when you are dealing with familiar equipment on your own ship. However, it may present problems when you are doing repair work for other ships, as you would be if assigned to the machine shop on a floating machine shop. When materials and repair parts are not specified in the instructions accompanying a job, this does not always mean that you are free to use your own judgment. Instead, it may merely mean that you are expected to know where to look for information on the types of material or repair parts to be used. Shipboard sources of information will be most helpful to you in identifying or selecting materials and repair parts. These include the following:

- **Nameplates on the equipment.** Nameplates on equipment supply information regarding the characteristics of the equipment. They are therefore a useful source of information concerning the equipment itself. Nameplate data seldom, if ever, includes the exact materials required for repairs. However, information given on characteristics of the equipment and on pressure and temperature limitations may provide useful clues for selection of materials.
- **Manufacturer's technical manuals (TMs) and catalogs.** Manufacturer's TMs are furnished with all machinery and equipment aboard ship. Materials and repair parts are sometimes described in the text of these TMs. However, more commonly, details of materials and parts are given on the drawings. Manufacturer's catalogs of repair parts are also furnished with some shipboard equipment. When available, these catalogs are a valuable source of information on repair parts and materials.
- **Stock cards maintained by the supply officer.** The set of stock cards that is maintained by the supply officer is often a useful source of information on repair materials and repair parts. One stock card is maintained for each type of machinery repair part carried on board.
- **Ship's plans, blueprints, and other drawings.** Ship's plans, blueprints, and other drawings are available on board ship. They are excellent sources of information on materials and parts to be used in making various kinds of repairs. Many of these plans and blueprints are furnished in the regular large sizes, but microfilm is being used increasingly for these drawings. Information obtained from plans, blueprints, and other drawings should always be checked for any changes made since the original installation.
- **Applicable Army TMs.** Army TMs are furnished with all machinery and equipment aboard ship. Materials and repair parts are sometimes described in the text of these TMs. However, more commonly, details of materials and parts are given on the drawings.

Requesting Parts

23-10. Whenever you find it necessary to request materials or repair parts, remember the following two things:

- If at all possible, find the correct national stock number (NSN) for each item requested. All materials now in the supply system have been assigned NSNs, and you should be able to locate them.
- Work informally with marine supply personnel who are actually responsible for identifying and requisitioning material. You have the technical knowledge and you know what you need. However, if you cannot find the correct stock number, give enough standard identification information so that supply personnel onboard ship or ashore can identify the item you want.

Experienced supply personnel are familiar with identification publications and can help you to locate the correct stock numbers and other important identifying information.

Scheduling Work

23-11. Careful planning is required to keep up with all engine room maintenance and repair work. Some of the factors that may be helpful in scheduling this work are as follows:

- Size up each job before you let anyone start working on it. Check the applicable maintenance requirement cards so that you will know exactly what needs to be done. Also, check all applicable drawings.
- Check on materials before you start. Be sure that all required materials are available before your personnel start working on any job. Do not overlook small items (such as nuts, bolts, washers, packing and gasket materials, tools, and measuring devices).

An inoperable piece of machinery may be useless as well as a nuisance and a safety hazard. It should not be spread around the engine room in bits and pieces while you wait for the arrival of repair parts or materials. Check on the priority of the job and of all other work that needs to be done before scheduling any job. When assigning work, carefully consider the capabilities and experience of your personnel. However, when possible, the less experienced should be given difficult work to do under supervision so that they may acquire skill in such jobs. When assigning work, be sure that the person who is going to do a job is given the necessary information. An experienced person may need only a drawing and a general statement concerning the nature of the job. A less experienced person is likely to require additional instructions and closer supervision. Keep track of the work as it is being done. In particular, check to be sure that proper materials and parts are being used and that the job is properly laid out or set up. Check to see that tools and equipment are being used correctly and that safety precautions are being observed. After a job has been completed, carefully inspect to be sure that everything has been done correctly and that all final details have been taken care of. Check to be sure that any necessary records or reports have been prepared. Remember that job inspections can serve at least two of the following very important purposes:

- The first is to make sure that the work has been properly performed.
- The second is to increase the skills and knowledge of the person who has done the work.

Do not overlook the training aspects of a job inspection. When your inspection of a completed job reveals any defects or flaws, be sure the person understands what is wrong, why it is wrong, and how to avoid similar mistakes in the future.

ENGINEER TRAINING RESPONSIBILITIES

23-12. As a senior engineman or engineer, you are responsible for training other personnel in engine room operations. The following are some of the skills and knowledges you must possess in order to train others:

- Construction details and operating principles and characteristics of all types of propulsion plants and associated engine room auxiliary machinery.
- Propulsion plant layout and piping system arrangement.
- Principles of steam engineering (including theory of combustion, theory of energy transformations, and factors governing engine room efficiency).
- Nature and theory of engine room operations.
- Operational troubleshooting.
- Engineering casualty control.
- Engine room maintenance and repair.
- Characteristics of metals and alloys.
- Tests and inspections of main engines.
- Characteristics and tests of lube oil.
- Records, reports, and other administrative requirements.

Even a very wide range of abilities and an extensive theoretical knowledge will not, in themselves, guarantee your success as an instructor. You must be technically competent to be able to teach others.

However, your technical competence must be supplemented by the ability to organize information and present it effectively. You will instruct lower-ranking personnel in the correct procedures to be used for casualties involving the engine room. It is your responsibility to see that personnel, under your supervision, learn the capabilities and limitations of the equipment with which they work and the procedures to follow should casualties occur. Safety is a responsibility of all personnel. As an engineer or senior engineman, instruct your personnel in all shipboard safety precautions. Most importantly, watch for careless methods of work (a frequent source of accidents). Set a good example for following safety practices because the example you set will have a great influence on other personnel. Always report any unsafe condition.

Glossary

A	area
AC	alternating current
AR	Army regulation
ATTN	attention
BDC	bottom dead center
bhp	brake horsepower
bmep	brake mean effective pressure
Btu	British thermal unit
C	Celsius
CFR	Code of Federal Regulation
CG	Coast Guard
CPR	cardiopulmonary resuscitation
cu.	cubic
CYL	cylinder
d	distance
DA	Department of the Army
D.C.	District of Columbia
F	Fahrenheit; force
FM	field manual; Fairbanks-Morse
ft.	feet/foot
gal.	gallon(s)
GM	General Motors
gpm	gallons per minute
hp	horsepower
HQ	Headquarters
HQDA	Headquarters, Department of the Army
hr.	hour(s)
ihp	indicated horsepower
imep	indicated mean effective pressure
in.	inch(es)
IPS	inches per second
ite	indicated thermal efficiency
KE	kinetic energy
L	length
lb.	pound(s)
LCM	landing craft, mechanized
LO	lubrication order

mep	mean effective pressure
min.	minute(s)
MOS	military occupational specialty
mph	miles per hour
MQB	Marine Qualification Board
N	number
NAVSEA	Naval Sea Systems Command
No.	number
NSN	national stock number
p	pressure
P	power
PAM	pamph let
POI	program of instruction
psf	pounds per square foot
psfa	pounds per square foot absolute
psi	pounds per square inch
psia	pounds per square foot absolute
psig	pounds per square inch gauge
RC	reserve component
RPM	revolutions per minute
sec.	second(s)
sq.	square
STBD	starboard
t	time
T	torque
TC	training circular
TDC	top dead center
TM	technical manual
US	United States
USAML	United States Army Marine License
USATRADOC	United States Army Training and Doctrine Command
USCG	United States Coast Guard
VA	Virginia

References

Army Regulations

AR 25-400-2	The Army Records Information Management System (ARIMS) 2 October 2007
AR 56-9	Watercraft, 7 February 2002
AR 750-59	Army Corrosion Prevention and Control Program 9 December 2005

Department of the Army Forms

DA forms are available from the APD website at ________________

DA Form 2028	Recommended Changes to Publications and Blank Forms
DA Form 4640	Harbor Boat Deck Department Log for Class A&B Vessels
DA Form 5273	Harbor Boat Deck and Engine Log for Class B Vessels

Department of the Army Pamphlets

DA PAM 611-21	Military Occupational Classification and Structure, 22 January 2007
DA PAM 750-8	The Army Maintenance Management System (TAMMS) Users Manual, 22 August 2005

Field Manuals

FM 4-25.11	First Aid, 23 December 2002
FM 55-501	Marine Crewman's Handbook, 1 December 1999

Miscellaneous Publications

CFR 46	Shipping, 1 October 2005
CG 4602A	Oil Record Book for Ships, January 2007

This page intentionally left blank.

Index

Made in the USA
Monee, IL
05 November 2026

67810973R00252